Lecture Notes in Civil Engineering

Volume 282

Lecture Notes in Civil Engineering (LNCE) publishes the latest developments in Civil Engineering - quickly, informally and in top quality. Though original research reported in proceedings and post-proceedings represents the core of LNCE, edited volumes of exceptionally high quality and interest may also be considered for publication. Volumes published in LNCE embrace all aspects and subfields of, as well as new challenges in, Civil Engineering. Topics in the series include:

- Construction and Structural Mechanics
- Building Materials
- Concrete, Steel and Timber Structures
- Geotechnical Engineering
- Earthquake Engineering
- Coastal Engineering
- Ocean and Offshore Engineering; Ships and Floating Structures
- Hydraulics, Hydrology and Water Resources Engineering
- Environmental Engineering and Sustainability
- Structural Health and Monitoring
- Surveying and Geographical Information Systems
- Indoor Environments
- Transportation and Traffic
- Risk Analysis
- Safety and Security

To submit a proposal or request further information, please contact the appropriate Springer Editor:

- Pierpaolo Riva at pierpaolo.riva@springer.com (Europe and Americas);
- Swati Meherishi at swati.meherishi@springer.com (Asia - except China, and Australia, New Zealand);
- Wayne Hu at wayne.hu@springer.com (China).

All books in the series now indexed by Scopus and EI Compendex database!

More information about this series at https://link.springer.com/bookseries/15087

Pavel Akimov · Nikolai Vatin · Aleksandr Tusnin · Anna Doroshenko
Editors

Proceedings of FORM 2022

Construction The Formation of Living Environment

Springer

Editors
Pavel Akimov
Moscow State University
of Civil Engineering
Moscow, Russia

Nikolai Vatin
Peter the Great St. Petersburg
Polytechnic University
Saint Petersburg, Russia

Aleksandr Tusnin
Moscow State University
of Civil Engineering
Moscow, Russia

Anna Doroshenko
Moscow State University
of Civil Engineering
Moscow, Russia

ISSN 2366-2557 ISSN 2366-2565 (electronic)
Lecture Notes in Civil Engineering
ISBN 978-3-031-10852-5 ISBN 978-3-031-10853-2 (eBook)
https://doi.org/10.1007/978-3-031-10853-2

This Springer imprint is published by the registered company Springer Nature Switzerland AG
The registered company address is: Gewerbestrasse 11, 6330 Cham, Switzerland

Contents

Improvement of Strength Properties of Arbolite Concrete Composites by Impregnation with Sulfur—By-Products of Oil and Gas Industry

Yulia Sokolova, Marina Akulova, Baizak Isakulov, Alla Sokolova, Zhumabay Tukashev, and Zhanna Konysbaeva

Abstract One of the most perspective methods of enhancing durability of building materials and articles exposed to various aggressive media is application of sulfur for their manufacturing. For the research purposes, there has been applied the method of deep impregnation of arbolite articles with sulfur—a by-product of Zhanazhol oil production located in Aktobe region of Kazakhstan. During impregnation process of the solid grain surface, the molecules of impregnation liquid affected by physical and chemical phenomena penetrate into voids and remain there in their initial state or when exposed to elevated temperatures, catalysts and radiation transform into irreversible state. The results of the experiments have proved that in all the impregnated samples the strength had increased by a factor of $2.0 \div 4.8$. The data have also demonstrated that depth of sulfur penetration into capillary-porous article had augmented twofold and more. Impregnation of arbolite concrete samples by sulfur gives the evidence of high strength features of arbolites. As seen from these data,

---

Y. Sokolova · A. Sokolova (✉)
Moscow State University of Civil Engineering, Moscow 129337, Russian Federation
e-mail: as.falconi@yandex.ru

Y. Sokolova
Scientific Research Institute of Building Physics NIISF RAASN, 21 Lokomotivny proezd, Moscow 127238, Russian Federation

M. Akulova · Z. Konysbaeva
Ivanovo State Polytechnic University, 21 Sheremetyesky prospect, Ivanovo 153000, Russian Federation

M. Akulova
Ivanovo Fire Rescue Academy of State Firefighting Service of Ministry of Russian Federation for Civil Defense, Emergencies and Elimination of Consequences of Natural Disasters, 33 Prospekt Stroiteley, Ivanovo 153011, Russian Federation

B. Isakulov
Kazakh-Russian International University, Aktobe Aiteke bi, 52, Aktobe, Kazakhstan

Z. Tukashev
Baishev University, Bratiev Zhubanovikh, 302a, Aktobe, Kazakhstan

P. Akimov et al. (eds.), *Proceedings of FORM 2022*, Lecture Notes in Civil Engineering 282, https://doi.org/10.1007/978-3-031-10853-2_1

strength and resistance to aggressive media of arbolites impregnated with sulfur by-products is rather high. This proves that arbolite blocks impregnated with sulfur by-products could be recommended for masonry of inspection manholes, underground engineering structures.

1 Introduction

Application of sulfur is considered one of the most perspective methods for enhancing durability of building materials and articles exposed to various aggressive media. Lately in Kazakhstan and overseas, investigation of sulfur with the purpose of its application in various spheres of construction industry is gaining momentum. In Kazakhstan, the principal deposits of oil and gas are located in the western regions of the Republic. The specific feature of Kazakhstan oil is the significant content of sulfur in its composition.

Recently, among researchers in CIS-countries and abroad there has been significant amount of interest in the method of concrete consolidation by means of its impregnation with monomers or oligomers followed by their polymerization [1–25].

When monomers or oligomers were used for impregnation purpose, polymer-impregnated concretes with improved strength properties and augmented resistance to some aggressive media were obtained. However, high cost of monomers and oligomers, their scarcity and complicated technology of polymer-impregnated concretes manufacturing are the constrains limiting their application. Sulfur is an accessible and economic product; its cost is significantly lower if compared with traditional monomers and oligomers used for impregnation. At the same time, concretes impregnated with sulfur fall short of their physical and mechanical properties insignificantly with respect to polymer-impregnated concretes. Moreover, the process of sulfur crystallization is noticeably simpler and more accessible than polymerization of monomers in concrete. Properties of building materials depend on porosity at a greater extent. Thus, if pores are filled with melted sulfur that hardens at cooling down, performance characteristics of materials could be refined.

2 Materials

For the research purposes, the authors used sulfur—by-products of Atyraus and Zhanazhol gas and oil productions of Kazakhstan as an impregnating material. Sulfur is a solid crystalline substance with melting temperature slightly higher than 119 °C, and boiling temperature of 445 °C. The chemical composition of sulfur is given in the Table 1.

Table 1 Chemical composition of sulfur

Grade	GOST (All-Union Standard)	Sulfur, % wt.	Ash, % wt.	Organic substances, % wt.	Water, % wt.
9998	127.1-93	99.060	0.400%;	0.053%;	0.010%

Table 2 Chemical composition of cotton plant

Content of components, %				
Crushed cotton plant stalks	Cellulose	Lignin	Pentosan	Polysaccharides
	39.1	19.2	17.7	24

Table 3 Chemical composition of crushed reed fibers, %

Cellulose $C_6H_{10}O_5$	Lignin $C_{41}H_{40}O_{15}$	Pentosan $C_5H_8O_4$	Resins and soluble components
46.17	29.76	22.00	2.07

Table 4 Chemical composition of fly-ash

Loss at burning, % wt.	Oxides' content, % wt.						
	SiO_2	$Al_2O_3 + TiO_2$	Fe_2O_3	CaO	MgO	NaO_2	SO_2
7.33	48.3	23.92	5.94	9	1.9	0.18	0.52

In order to solve the set tasks, crushed stalks of cotton plant and reed fibers were used as an organic filler. Chemical composition of cotton plant was assessed empirically in compliance with GOST (All-Union Standard) 19,222-84, 25,820-2000, and based on the reference and literature data [1–6] (Table 2).

For comparison of the results, crushed reed fibers have been used in the research. Their chemical composition is given in the Table 3.

For production of slag-lime arbolite samples, there were used electric thermophosphoric slag of Chimkent industrial association and electric steel-making slag of Aktobe integrated iron-and-steel works. Slags were used in the form of powders milled until they reached specific surface area under PSK-2 (320–330 m^2/kg) in compliance with the All-Union Standard GOST 10,180-90. As alkali components, there were applied soluble glass and soda-sulfate mixture, being by-products of caprolactam production. As an active mineral additive, there was used fly-ash of Actobe power plant complying with the requirements of GOST 10,181-2000. Chemical composition of fly-ash is laid out in the Table 4. Portland cement, grade 400, produced by Navoinsky cement plant was used for manufacturing cement-ash-slag arbolite samples. In the paper, there was also used Krasnooktyabr'sky bauxite slurry of Kustanayskoye deposit. Slurry was used in the form of powders milled until they reached specific surface area under PSK-2 (320–330 m^2/kg) in conformance to GOST 13,015.0, GOST 3476, GOST 10,180, GOST 7076, and OST 67-11. Bauxite slurries are co-products of aluminum oxide production. Visually, bauxite slurry is medium-grained sand of beige-fallow color, with impurities of shelly clumps of various sizes. Moisture of bauxite slurry samples should not exceed 20–30%, density—2.6–2.86 g/cm^3, tamped density in loose state—from 1.1 up to 1.3 g/cm^3.

In competitive experiments, there was also used Portland cement, grades 400 and 500 (GOST 10,178-85) and synthesized clinker minerals.

For research purposes, there were prepared four series of samples form slag-lime arbolites with binding composition, % wt. (cement: ash: electric thermophosphoric slag: soluble glass: soda sulfate mixture), whereas the fifth series was made for comparison purposes—from cement-ash-slag arbolite with binding compositions, % wt. (cement: ash: slag). Each series consisted of six samples with the dimensions of 100 × 100 × 100 mm, three of which consisted of crushed cotton plant only, whereas fibers of crushed reed with diameter around 18–20 mm were inserted into the core of three remaining samples. Two compositions of arbolite mixture were used for impregnation of arbolite-concrete samples with sulfur by-products.

3 Methods

Characteristics of initial and activated binder were determined in compliance with the GOST 30,515-97, GOST 31,108-2003 and 7473-2010. Ultimate tensile and bending strength was tested on test beams with the dimensions 40 × 40 × 160 mm on the apparatus IP 2710. Ultimate strength of arbolite samples impregnated with sulfur by-products was determined under the standard methodics. Phase composition of impregnated arobolite samples was defined by X-ray diffraction analysis. X-ray photography was done on the diffraction meter DRON-3. Differential thermal analysis of hydrated sulfur-containing cement powders was implemented on the photo recording derivatograph under the standard method. Impregnation of arbolite samples by sulfur by-product was done in the following sequence:

- to carry out preliminary heating and drying (dewatering) of the structure of capillary-porous samples at the temperature of 120–145 °C during 6–10 h;
- to carry out impregnation of arbolite samples dried until the permanent weight with melted sulfur by-product at the temperature of 120–185 °C during 8–16 h;
- to conduct even cooling of impregnated samples to a required depth of building articles (for 2–4 h) until the ambient temperature is reached.
- Thus, the duration of complete technological cycle of arbolite samples impregnation with sulfur by-product is fluctuating from 16 up to 32 h.

4 Results

Our investigation has shown the following results:

Grade strength of reference samples impregnated with melted sulfur for the compositions ***I*** and ***II*** accounted for correspondingly:

Table 5 Change of properties of slag-lime arbolite after impregnation with sulfur

Binder's composition, % wt. (cement: ash: electro thermophosphoric slag: soluble glass: soda-sulphate mixture)	Duration of impregnation, hours	Sulfur gain increment, %	Strength of reference samples, MPa	Strength of impregnated samples, MPa	Strain hardening **g coefficient,** Strength of impregnated samples/Strength of reference samples
14: - : 43: 30: 13	2	3.1	1.7	3.5	2.1
	5	3.8		4.6	2.7
	10	5.9		7.9	3.7
	24	6.9		8.2	4.8
18: 39: - : 31: 12	2	2.9	1.8	3.7	2.2
	5	3.8		4.7	2.6
	10	6.0		8.3	4.6
	24	6.8		8.5	4.7
14: 39: 17: 30: -	2	3.2	1.83	3.6	1.97
	5	3.9		4.7	2.6
	10	6.0		7.9	4.3
	24	6.95		8.6	4.7

- for slag-lime arbolite 1.7; 1.8; 1.83 MPa;
- for cement-ash-slag arbolite 3.1; 3.1; 3.25 MPa.

It was stated from the experimental data that with the increase in the content of ash, slag and bauxite slurry, sulfur weight increment in the samples had been distributed in the following manner:

- after 24-h' impregnation at standard atmosphere pressure samples for the composition ***I***—by 6.9, 6.8 and 9.5% (Table 5);
- after 24-h' impregnation for the composition ***II***—by 7.1, 7.2 and 7.2% (Table 6).

After 5-h' vacuum impregnation, these values have grown up to 10.2% for the composition ***I*** and up to 8.1% for the composition ***II***. After 5-h' impregnation with preliminary vacuuming of samples, these values have reached 10.5 and 9.7%, correspondingly.

Weight increment of samples as a function of the content of ash, slags and slurries in arbolite could be explained by the change of structural porosity of impregnated arbolite concrete. Impregnation method and duration have a significant impact on this value. It is worth noting that the presence of sulfur in all case leads to enhancement of mechanical properties of samples.

Testing of arbolite samples impregnated with sulfur on compression has shown that all the compositions had improved their mechanical strength by a factor of 2.0 ÷ 4.8 without any exemption.

As seen from the data enclosed in the Tables 5 and 6, the depth of sulfur penetration into capillary-porous article would increase twofold and more if an impregnated

Table 6 Change of properties of cement-ash-slag arbolite after impregnation with sulfur

Binder's composition, wt. % (cement, slag)	Duration of impregnation, hours	Sulfur gain increment, %	Strength of reference samples, MPa	Strength of impregnated samples, MPa	Strain-hardening coefficient, Strength of impregnated samples/ Strength of reference samples
Cement:ash:slag (60:40:0)	2 5 10 24	2.8 3.9 6.75 7.1	3.1	6,3 7.2 8.9 10.2	2.03 2.3 2.88 3.3
Cement:ash:slag (60:35:5)	2 5 10 24	2.8 3.85 6.7 7.2	3.1	6.3 7.3 9.1 10.5	2.03 2.4 2.9 3.4
Cement:ash:slag (60:30:10)	2 5 10 24	2.7 3.8 6.9 7.2	3.25	6.5 7.5 9.4 11.2	2.0 2.3 2.9 3.5

sample was dried until constant weight. Depending on the amount of residual moisture of samples prepared for impregnation, mechanical properties of impregnated arbolite would change.

From the given tables, it is obvious that the longer impregnation lasts (up to 24 h), the more intensive the growth of mechanical strength of sulfur impregnated arbolite is. Further sample curing in melted sulfur does not affect concrete strength properties significantly. Kinetics of ultimate strength growth for arbolite in the process of its impregnation with sulfur has demonstrated that maximum relative gain in ultimate compressive strength had been observed for samples with the composition cement-ash-slag content (cement (60): ash (30): slag (10)).

However, this difference is not that prominent and after 2–3 h' impregnation it is practically smoothed. Along with that, the value of strain hardening coefficient for the composition I is equal to 4.8, and for the composition II—3.5, respectively.

Impregnation of cement paste with sulfur proves enhanced strength properties of arbolites.

5 Discussions

The implemented research gave the grounds for validating the hypothesis for strength formation of slag-lime and cement-ash-slag arbolite impregnated with sulfur. According to the theories of the authors [7, 22–25], in the case of sulfur use, from our point of view, the most significant features were the presence of three-dimensional frame in the pores of arbolite matrix, the increase in density of contact zone of fillers with cement paste due to joint adhesive impact of cement gel and sulfur as well as dimensional filling of pores, cracks and other defects with low-viscosity monomers facilitating strengthening of cement paste contact zone.

6 Conclusions

It was stated that technical sulfur is regarded as a perspective impregnating material for enhancing durability of building materials and articles exposed to diverse aggressive media. Impregnation of arbolite concrete samples with sulfur gives evidence of the improved strength properties of the researched sample. It had also been proven that strength and resistance to aggressive media of arbolite concretes impregnated with sulfur was rather high that enabled to draw a conclusion that arbolite concrete blocks impregnated with sulfur by-product could be recommended for masonry of inspection manholes, and underground engineering structures.

References

1. Assakunova BT, Jussupova MA, Baimenova GR, Kulshikova ST (2019) News of the national academy of sciences of the Republic of Kazakhstan series of geology and technical sciences 3(435):67–72 (2019)
2. Akulova MV, Isakulov BR, Dzhumabaev MD (2014) Scientific and technical herald of the Volga Region 1:49–52
3. Akchabaev AA, Bisenov KA, Uderbaev SS (1999) Reports of the Ministry of Science and higher education, vol 4. NAS RK, Almaty, pp 57–60
4. Akulova MV, Isakulov BR, Fedosov SV, Shchepochkina YUA (2014) Wood concrete mix contains portland cement, rush cane stems, technical sulphur, chrome-containing sludge, pyrite stubs and water. Patent RU2535578-C1, 20 December 2014, C04B-028/04, Russia
5. Akulova MV, Isakulov BR, Fedosov SV, Shchepochkina YuA (2013) Method to produce wood concrete products with making base for plastering on their surface. Patent RU2517308-C1, 08 July 2013, Russia
6. Beysenbayev OK, Umirzakov SI, Tleuov AS, Smaylov BM, Issa AB, Dzhamantikov Kh, Zakirov BS (2019) News of the national academy of sciences of the Republic of Kazakhstan series of geology and technical sciences 1(433):80–89
7. Bazhenov YuM (1976) Perspectives of application of polymer-impregnated concretes and polymer concretes in construction. In: Proceedings. Stroyizdat, Moscow, pp 33–38
8. Borokhovsky VA, Salyk AI et al New type of chemical products – polymer sulfur. Publishing House NIITEKhIM, Moscow 198, 40 p
9. Vashchuk VYa, Dolyuk VP (1976) The impact of porosity on the strength of polymer concretes. Stroyizdat, Moscow, 208 p
10. Sokolova Yu, Akulova M, Isakulov BR, Sokolova AG, Isakulov AB (2020) IOP Conf Ser Mater Sci Eng 869(3):032005
11. Sadieva KhR, Massalimova BK, Abisheva RD, Tsoy IG, Nurlybayeva AN, Darmenbayeva AS, Ybraimzhanova LK, Bakibaev AA, Sapi AK (2019) News of the national academy of sciences of the Republic of Kazakhstan series of geology and technical sciences 4(436):158–166

12. Sokolova YuA, Akulova MV, Isakulov BR, Sokolova AG, Kul'sharov BB, Isakulov AB (2020) News of the National Academy of sciences of the Republic of Kazakhstan-series chemistry and technology 6: 65–72(2020)
13. Zhiv AS, Galibuy S, Isakulov BR (2013) Mechanization of construction 3(825):14–17
14. Dzhumabaev MD (2016) Lightweight arbolitic concrete based on compositcementitious slurry binding and solid organic waste (for example by-products of agriculture of the Republic of Kazakhstan): dissertation on competition of the scientific degree of the candidate tehn. Sciences, Ivanovo, 59 p
15. Tulaganow AA (1998) BRD Heft 1/2, 44, Jahragang 222–225
16. Suleimenov ST (1996) Physico-chemical processes of structure formation in building materials from mineral waste industry (Moscow, Manuscript), 138 p
17. Zhiv AS, Isakulov BR (2014) Scientific herald of the Voronezh State University of Architecture and Civil Engineering. Constr Archit 23:61–74
18. Issakulov BR, Zhiv AS, Zhiv YuA, Strelnikova AS (2010) Light concrete on the base of industrial and agricultural waste. In: Proceedings of the 2nd international conference on sustainable construction materials and technologies
19. Isakulov BR, Akulova MV, Kulsharov BB, Sartova AM, Isakulov AB (2020) News of the National Academy of Sciences of the Republic of Kazakhstan. Ser Geol Tech Sci 4(442):28–34
20. Isakulov BR, Jumabayev MD, Abdullaev HT, Akishev UK, Aymaganbetov MN (2019) Periodico Tche Quimica 16(32):375–387
21. Isakulov BR, Dzhumabaev MD, Abdullaev KhT, Konysbaeva ZhO, Shalabaeva SI (2020) Int J Eng Res Technol 13(12):4880–4884
22. Orlovsky YuI, Semchenkov AS, Khorzhevsky VI (1995) Concrete and reinforced concrete 3:21–24
23. Orlovsky YuI (1990) Application of sulfur and sulfur containing by-products in construction industry, 3–5
24. Pautroyev VV, Volgushev AN, Orlovsky YuI (1985) Concrete and reinforced concrete 5:16–17
25. Parphenyuk SA (1987) Automobile roads 2:16

Trends in BIM Tools Adoption in Construction Project Implementation: A Case Study in Vietnam

Nguyen Quoc Toan, Pham Xuan Anh, and Nguyen Van Tam

Abstract This study aims to investigate the emerging trends in the adoption of BIM tools in construction project implementation and the effectiveness of BIM implementation in the Vietnamese construction industry. A total of 159 valid samples were collected from the construction workforce based on their previous participation in or directly implementation construction projects adopting BIM tools. Data were calculated and assessed on the basis of their simple percentage and descriptive statistics (i.e., mean and standard deviation) approaches. The findings indicated that three BIM tools consist of Revit, Autodesk BIM 360, and Sketchup are the most frequently and commonly software in Vietnam construction project implementation. Also, adopting BIM tools can up to save 10% estimating cost suffice the value of the project and a savings of from 30 min up to 2 h in time using BIM tools per day from BIM users' perspective. In addition, BIM tools have been most popularly adopted in three responsibilities including the design phase, quantity take-off, and project management in the implementation of construction projects in Vietnam.

Keywords BIM tools · Adoption · Users · Construction projects

1 Introduction

For the construction industry, the adoption and implementation of Building Information Modelling (BIM) are steadily rising in many nations [1]. The paradigm shift in the construction process is experiencing an increased transfer of technological advancement from developed countries to developing countries which makes

N. Q. Toan (✉) · P. X. Anh · N. V. Tam
Hanoi University of Civil Engineering, 55 Giai Phong Road, Hanoi, Vietnam
e-mail: toannq@huce.edu.vn

P. X. Anh
e-mail: anhpx@huce.edu.vn

N. V. Tam
e-mail: tamnt3@huce.edu.vn

P. Akimov et al. (eds.), *Proceedings of FORM 2022*, Lecture Notes in Civil Engineering 282, https://doi.org/10.1007/978-3-031-10853-2_2

a deep and fundamental change that is rapidly transforming the global construction sector [2, 3]. BIM becomes an inevitable development trend to get along with the development of the fourth industrial revolution, or 'Industry 4.0', contributing to improving the quality, efficiency, and competitiveness of each enterprise. With its outstanding features compared to traditional methods, BIM has fostered increase construction labor productivity, increase work efficiency in terms of quality, progress, and cost, and reduce waste on construction projects [4–7]. As a result, BIM has adopted several different levels in many countries such as UK, USA, Germany, Japan, Australia, Brazil, Canada, France, New Zealand, Korea, Singapore, Hong Kong, Russia, China. The BIM adoption and implementation are effective at various levels depending on the scope of BIM application for stakeholders, including investors, secondary investors in purchasing and leasing apartments, consultants, contractors, equipment installation contractors, management and operation contractors as well as local authorities and state management agencies on construction. BIM technologies bring advantages to construction project quality and delivery improvement [8]. In Vietnam, the construction industry has made significant progress towards modernization and globalization in recent decades, which saw the strongly developing of many fields such as construction technologies, construction project management, construction materials, architectures and construction planning, urban and housing development. Although construction productivity has recently been improved, it is still lower than in other countries in Southeast Asia [9, 10]. Hence, the adoption of BIM in construction project implementation is an effective measure to improve construction productivity and enhance construction project performance [11, 12]. The goal of this study identified and assessed emerging trends in the adoption of BIM tools in the construction project implementation through data collected in an investigation in Vietnam.

2 Methodology

A review of the literature was conducted for the aim of articulating issues regarding the concept of BIM tools in the construction sector with particular emphasis on the Vietnam construction industry. The review also aimed at identifying the BIM tools adoption in the construction project implementation, a questionnaire survey was formulated and sent to users working on construction projects. Data were collected from respondents who completed a questionnaire survey according to their previous participation in or directly implementation construction projects adopting BIM tools. A total of 250 questionnaires were distributed both by way of an interview (180) and by means of an online survey platform (70). The authors conducted the interview BIM users working on small, medium, and large construction enterprises in Vietnam. As for the online method, the questionnaires were sent directedly to preselected people who had been ensured to at least have first-hand knowledge of BIM. Any questionnaires that included incomplete data or missing values were removed. Finally, 159 valid questionnaires were collected, in which, valid questionnaires received from the

Table 1 Demographics of the respondents (N = 159)

Items	Categories	Frequency	Percentage (%)
Gender	Male	125	78.62
	Female	34	78.62
Education levels	Bachelor's degree	151	94.97
	Master's degree	7	4.40
	Doctor's degree	1	0.63
Work experience	1–5 years	93	58.49
	6–10 years	35	22.01
	11–15 years	16	10.06
	16–20 years	4	2.52
	Above 20 years	11	6.92
Role in construction project	Project manager	29	18.24
	Site manager	5	3.14
	Site supervisor	11	6.92
	Site engineer	13	8.18
	Designer	39	24.53
	Architect	25	15.72
	Estimator	35	22.01
	Company manager	2	1.26

former method were 107 (67%) and from the online survey were 52 (33%). The valid response rate for face-to-face interviews was 61% while 100% of distributed online forms were completed. Table 1 presents the demographics of the respondents under the investigation.

3 Results and Discussions

In this study, two software applications were applied to examine the findings, which are MS Excel 365 and SPSS 20. The analysis results have been calculated and assessed based on their simple percentage and descriptive statistics (i.e., mean and standard deviation). This section consists of four main items which include the demographics of respondents, emerging trends in the adoption of BIM tools, the effectiveness of BIM tools adoption in construction project implementation, and the impact of stakeholders on BIM tools adoption.

Emerging Trends in Adoption of BIM Tools in Construction Project Implementation

Figure 1 illustrates the proportions of participants choosing one or more programs among BIM software as their daily tools at work. Revit is a multi-discipline BIM

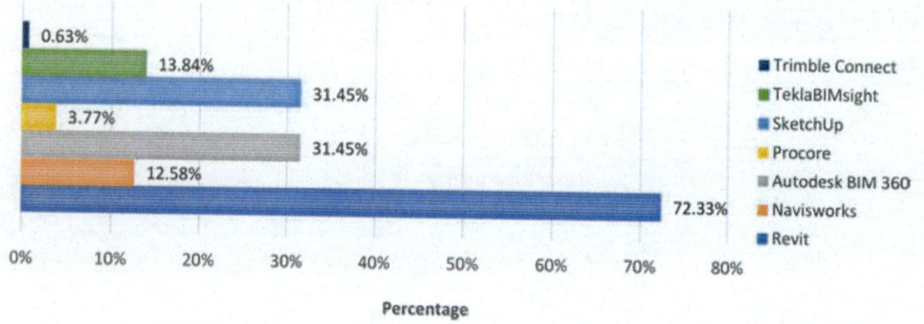

Fig. 1 List of used BIM tools

Table 2 The extent of using BIM tools

BIM tools	Minimum	Maximum	Mean	Std. deviation	Rank
Revit	1	5	2.99	1.410	1
Autodesk BIM 360	1	5	2.23	1.442	2
Sketchup	1	5	2.09	1.260	3
Navis	1	5	1.86	1.287	4
Tekla	1	5	1.75	1.153	5
Procore	1	5	1.62	1.095	6
Trimble	1	5	1.48	0.920	7

(Note: 1-Never, 2-Seldom, 3-Frequently, 4-Often, 5-Usually)

Table 3 The extent of popular BIM tools

BIM tools	Minimum	Maximum	Mean	Std. deviation	Rank
Revit	1	5	3.59	1.249	1
Sketchup	1	5	3.06	1.271	2
Autodesk BIM 360	1	5	2.84	1.247	3
Tekla	1	5	2.62	1.113	4
Navis	1	5	2.61	1.108	5
Trimble	1	5	2.21	0.970	6
Procore	1	5	2.15	0.929	7

(Note: 1-Not at all, 2-Slightly, 3-Neutral, 4-Very, 5-Extremely)

software with toolsets for architectural design, MEP and structural engineering, as well as construction. As demonstrated in Fig. 1, Revit has the largest percentage of respondents using it, at 72.33%. It is also the most frequently used program, which is proved in Table 2 with Mean = 2.99. Therefore, this software is being considered as the most widely used, as shown in Table 3, with Mean = 3.59 which proves that Revit is the most popular software in implementing BIM technology in

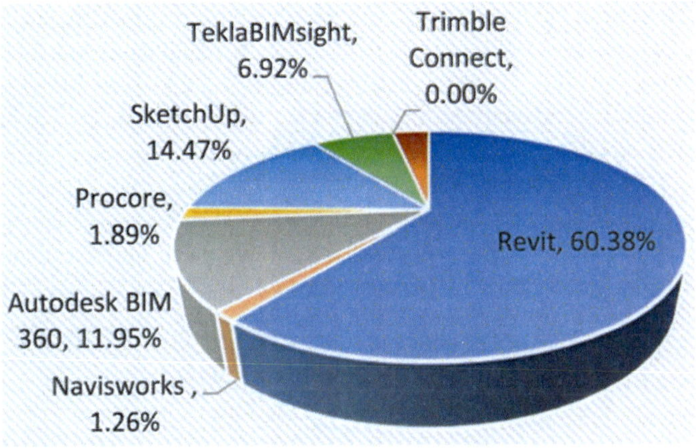

Fig. 2 The best BIM tool recommended by users

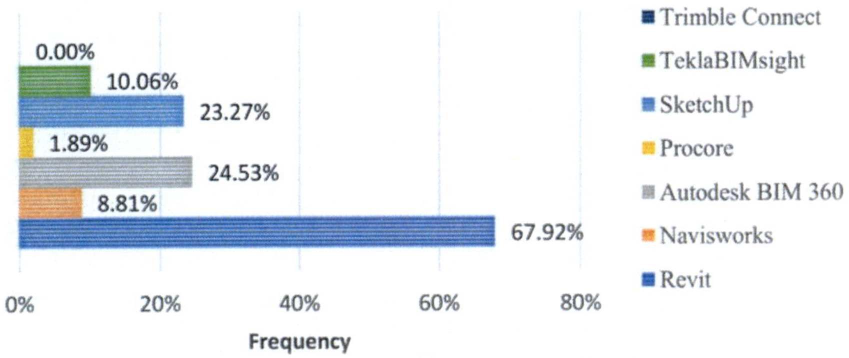

Fig. 3 Training course for users regarding BIM tools

Vietnam construction projects. Revit can import, export, and link user data in the most commonly used file formats in the industry, such as IFC, DWG, and DGN. Other than that, the solution enables multiple project coordinators to manage tasks and centrally shared models to facilitate streamlined collaboration. In terms of the best BIM tool, Revit is still in the first place with 60.38% of the respondents agreed that it is the most effective program for their regular tasks and 67.92% of the respondents have been taken Revit courses (Figs. 2 and 3). However, [13] indicated that 54% of the Nigerian respondents use 2D CAD only, 12% uses 3D CAD only, and 20% and 14% for 2D and 3D CAD and No CAD respectively, whereas, 57% of the users applied 3D and 8.6% don't use CAD in India [14].

Autodesk BIM 360 is a comprehensive project management platform designed for the construction industry. The software is designed to help project, field, and BIM managers to speed up the delivery of their projects while staying within their project budget and adhering to industry standards, safety rules, and project specifications.

The analysis results show that Autodesk BIM 360 and SketchUp have an equal percentage of users, at 31.45%. They are respectively being the second and the third frequently programs that have been used, with Mean = 2.23 for Autodesk BIM 360 and Mean = 2.09 for SketchUp. Although Autodesk BIM 360 has a greater mean value of frequency than SketchUp, its figure for the popular aspects is slightly lower, at Mean = 2.84 compared to Mean = 3.06 of SketchUp from Vietnamese users' perspective. It indicates that SketchUp has influenced more users, which has been strengthened by the percentage shown in Fig. 1. There is 14.47% of users agree that SketchUp is the best BIM tool, compared to 11.95% for Autodesk BIM 360. These figures are less than half of the figure for Revit but still make SketchUp and Autodesk BIM 360 be the second and the third popular BIM tools. Besides, 24.53% of the respondents taken Autodesk BIM 360 course while there is 23.27% chosen SketchUp course for training.

Followed by TeklaBIMsight and Navisworks software, with 13.84% and 12.58% of the participants using these tools, respectively. Tekla BIMsight streamlines the whole construction cycle by allowing project managers to resolve conflicts using the automated clash detection tool. The program also provides the entire construction industry a no-cost solution for modern, model-based coordination. Also, Navisworks is a project review and management software for AEC professionals and teams. It offers a holistic review of integrated models and data that involves stakeholders during the preconstruction phase for better control of project conclusions. As provided in Table 2, TeklaBIMsight has more users, its frequency of use is lower than Navisworks. The mean value of frequency for TeklaBIMsight is 1.75, compared to 1.86 of Navisworks. These two software have similar mean value in terms of popular BIM tools, at 2.62 and 2.61, respectively (Table 3). However, there is a difference when it comes to the question of "Which one is the best BIM tool?", 6.92% of users believed Tekla is the best BIM tool, while only 1.26% of the respondents had the same thought to Navisworks (Fig. 2). Surprisingly, the similarity in the proportion of respondents choosing TeklaBIMsight and Navisworks courses to train, at 10.06% and 8.81%, respectively (Fig. 3).

Procore is an easy-to-use construction project management software that helps users manage multiple projects in one place, whereas, Trimble Connect is a powerful collaboration tool for AEC professionals including architects, engineers, sub-contractors, general contracts, and project owners. The surveyed respondents explain that Procore and Trimble Connect are the least popular tools in Vietnam when there is under 4% of participants using them (3.77% for Procore and only 0.63% for Trimble Connect). They are also not being used quite often, with the frequency mean value at 1.62 and 1.48, respectively. Although there are more Procore users than that Trimble Connect, more participants believe that Trimble Connect is better well-known than Procore, as illustrated in Table 3. Procore has Mean = 2.15, slightly lower than the figure of 2.21 for Trimble Connect. However, no one thinks Trimble Connect is the best BIM tool while 1.89% of respondents chose Procore. Therefore, none of the participants took Trimble Connect course, while this figure for Procore is 1.89%.

Effectiveness of BIM Tools Adoption in Construction Project Implementation
Figure 4 indicates the practical use of BIM software among the respondents. Overall, most of the users spend less than 4 h using BIM tools, in which, 27.67% working less than 1 h using BIM software, 19.50% using from 1–2 h, and 22.01% spending 2–4 h on BIM programs. The rest spend more than 4 h a day on BIM software. The proportion for respondents working their whole day using BIM software is 14.47% with more than 8 h of time consumption in BIM. The figures for "4–6 h" and "6–8 h" categories are 11.32% and 5.03% respectively.

The effectiveness of using BIM software is illustrated in Fig. 5. Mainly, BIM tools help users save under 2 h per day, while 25.79% of the respondents reported that less than 30 min were saved by using BIM software. The amount of time-saving by using BIM tools per day for 30–60 min and 1–2 h are 23.09% and 23.27%, respectively.

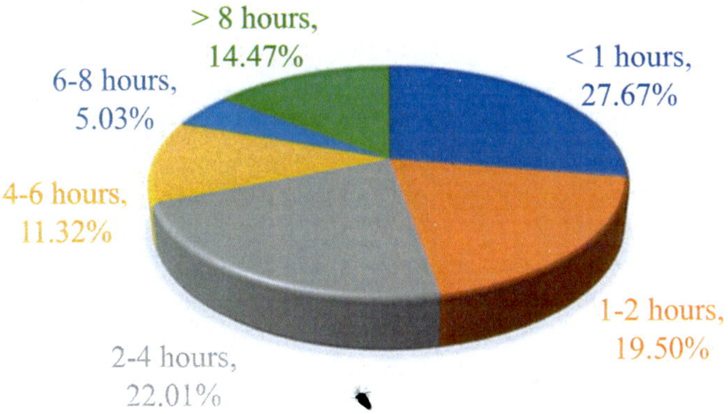

Fig. 4 The number of hours spending on using BIM tools per day

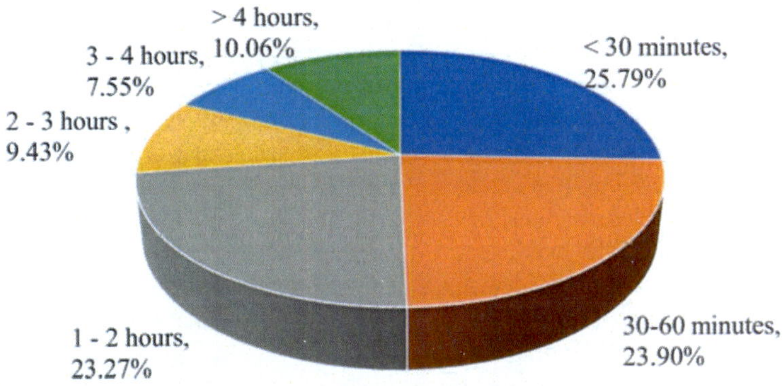

Fig. 5 The amount of time savings by using BIM tools per day

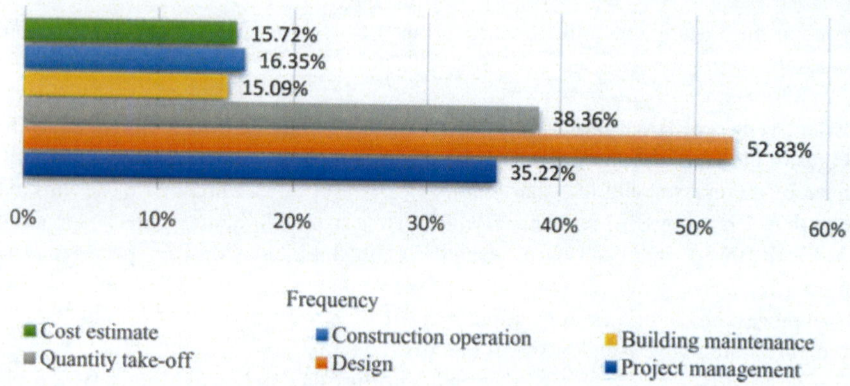

Fig. 6 Scope of work/responsibilities do BIM tools

However, 10.06% of users can save more than 4 h per day at their work, whereas, the proportions for "2–3 h" and "3–4 h" are slightly lower, at 9.43% and 7.55%, respectively.

BIM is an intelligent digital representation or a model-based process for planning, designing, building, managing logistics, cost estimation, and carrying out the process of operations and maintenance [15–17]. In Vietnam, six main work which has been done by BIM tools identified in Fig. 6 (i.e., project management; design; quantity take-off; building maintenance; construction operation, and cost estimate). More than half of the respondents (52.83%) use BIM tools in the design stage. BIM is possible to work and create multiple design options for the buildings that represent different layouts, color combinations, landscapes, facades. It becomes very easy to explore multiple options and comparing each option digitally. It also becomes possible to perform energy analysis and research the behavior of the building during a specific part of the year, month, and day [18]. Based on these analyses, designers can change the designs or the direction of the overall building. After finalizing a conceptual design there comes a need to get a structural design engineer and MEP design team to finalize the remaining disciplines. These MEP consultants and structural consultants can design based on the standards and codes applicable so that they can send the design intent to the BIM team for further modeling. Later all three architectural, structural, and MEP design models can be integrated for detection design issues, clashes between multiple disciplines, and coordination purposes. This is where the collaborative side of BIM comes into existence. Clashes and interferences can be checked and resolved right in the design stage leaving very fewer chances of design changes during the construction process [6, 19]. This saves money, time, and increases construction productivity overall.

Quantity take-off and project management are the second and the third main work that participants prefer using BIM software, at 38.36% and 35.22% of the users, respectively. Quantity takeoff based on BIM is a faster and more reliable method than the traditional 2D-based quantity takeoff approach. However, the quality of BIM models affects the accuracy of the extracted quantities. Incomplete details and inappropriate modeling methods cause deviations in extracted quantities [20, 21]. BIM can be utilized by project managers as a good tool to simulate project conditions to avoid redundant works and waste of time and cost [22]. Although Vietnamese project managers generally have little knowledge concerning BIM, which makes it hard for them to see their applications, BIM can help construction project managers in the task of delivering successful projects and provide a better basis for decisions is provided in comparison with traditional projects.

Other responsibilities, which are cost estimating, construction operation, and building maintenance, has similar proportions of BIM users that are 15.72%; 16.35%%, and 15.09%, respectively. Due to the fragmentation of the construction industry and the linearity of the design process, cost estimating is typically performed at a time when the conceptual design is quite advanced or even completed, which is much too late in the design process to help the different stakeholders make informed decisions. Performing value engineering and cost estimation based on BIM from the beginning of the design process would potentially enable a faster and more cost-effective project delivery process, higher-quality buildings, and increased control and predictability for the owner [23]. BIM provides a process where information stays consistent throughout the project and changes can be readily accommodated. BIM supports the full project lifecycle and offers the capability to integrate costing efforts throughout all project phases [24]. Besides, BIM uses precise geometry and relevant data to can improve the working conditions at construction sites and during building maintenance service of facilities. Using a BIM-based facility maintenance management system for maintenance staff in the operation and maintenance phase can be an effective solution for building maintenance management [25–27].

4 Conclusions

BIM is a set of digital tools that can manage construction projects' effectiveness. It provides significant benefits for the stakeholders in construction projects but seems that it is not fully implemented yet in the Vietnamese construction industry. This study aimed to identify and assess emerging trends in the adoption of BIM tools in the construction project implementation through data collected in an investigation in Vietnam. The data was collected by 159 valid surveyed questionnaires with participants of the construction practitioners (i.e., project manager, site manager, site supervisor, site engineer, designer, architect, estimator, company manager). Based on simple percentages and descriptive statistics (i.e., mean and standard deviation) approaches, the authors evaluated the emerging trends in the adoption of BIM tools in construction project implementation and the effectiveness of BIM implementation in the Vietnamese construction industry. Among numerous BIM tools that are adopting in Vietnamese construction project, the analysis results indicated that three BIM tools consist of Revit, Autodesk BIM 360, and Sketchup are the most frequently and commonly software in construction project implementation. Also, adopting BIM tools can up to save 10% estimating cost suffice the value of the project and a savings of from 30 min up to 2 h in time using BIM tools per day from BIM users' perspective. BIM tools have been most popularly adopted in three responsibilities including the design phase, quantity take-off, and project management in the implementation of construction projects.

References

1. Chan DW, Olawumi TO, Ho AM (2019) Perceived benefits of and barriers to Building Information Modelling (BIM) implementation in construction: the case of Hong Kong. J Build Eng 25:100764. https://doi.org/10.1016/j.jobe.2019.100764
2. Enegbuma WI, Ali KN (2011) A preliminary critical success factor (CSFs) analysis of building information modelling (BIM) implementation in Malaysia. In: Proceedings of the Asian Conference on real estate sustainable growth, management challenges
3. Abubakar M et al (2014) Contractors' perception of the factors affecting Building Information Modelling (BIM) adoption in the Nigerian Construction Industry. In: Computing in civil and building engineering, pp 167–178. https://doi.org/10.1061/9780784413616.022
4. Gamil Y, Rahman IAR (2019) Awareness and challenges of building information modelling (BIM) implementation in the Yemen construction industry. J Eng Des Technol. https://doi.org/10.1108/JEDT-03-2019-0063
5. Succar B (2009) Building information modelling framework: a research and delivery foundation for industry stakeholders. Autom Constr 18(3):357–375. https://doi.org/10.1016/j.autcon.2008.10.003
6. Abd Hamid AB et al (2018) Building information modelling: challenges and barriers in implement of BIM for interior design industry in Malaysia. In: IOP conference series: earth and environmental science. IOP Publishing
7. Chan AP et al (2018) Critical review of studies on building information modeling (BIM) in project management. Front Eng Manage. https://doi.org/10.15302/J-FEM-2018203
8. Diaz P (2016) Analysis of benefits, advantages and challenges of building information modelling in construction industry. J Adv Civil Eng 2(2):1–11. https://doi.org/10.18831/djcivil.org/2016021001
9. Van Tam N, Huong NL, Ngoc NB (2018) Factors affecting labour productivity of construction worker on construction site: a case of Hanoi. J Sci Technol Civil Eng (STCE)-NUCE 12(5):127–138. https://doi.org/10.31814/stce.nuce2018-12(5)-13
10. Van TN, Quoc TN, Le Dinh L (2021) An analysis of value chain in the Vietnam construction industry. Int J Sustain Constr Eng Technol 12(3):12–23. https://doi.org/10.30880/ijscet.2021.12.03.002
11. Van Tam N et al (2021) Factors affecting adoption of building information modeling in construction projects: a case of Vietnam. Cogent Bus Manage 8(1):1918848. https://doi.org/10.1080/23311975.2021.1918848
12. Van Tam N et al (2021) Impact of BIM-related factors affecting construction project performance. Int J Build Pathol Adapt. ahead-of-print. https://doi.org/10.1108/IJBPA-05-2021-0068
13. Ezeokoli F, Okoye P, Nkeleme E (2016) Factors affecting the adaptability of building information modelling (BIM) for construction projects in Anambra State Nigeria. J Sci Res Rep 11(5):1. https://doi.org/10.9734/JSRR/2016/26563
14. Sawhney A et al (2014) State of BIM adoption and outlook in India. RICS School of Built Environment, pp 1–30
15. Kumar JV, Mukherjee M (2009) Scope of building information modeling (BIM) in India. J Eng Sci Technol Rev 2(1):165–169. https://doi.org/10.25103/jestr.021.30

16. Fatima A, Saleem M, Alamgir S (2015) Adoption and scope of building information modelling (BIM) in construction industry of Pakistan. In: Proceedings of 6th international conference on structural engineering and construction management
17. Masood R, Kharal M, Nasir A (2014) Is BIM adoption advantageous for construction industry of Pakistan? Proc Eng 77:229–238. https://doi.org/10.1016/j.proeng.2014.07.021
18. Wang Y et al (2013) Engagement of facilities management in design stage through BIM: framework and a case study. Adv Civil Eng. https://doi.org/10.1155/2013/189105
19. Jin Z et al (2019) Using 4D BIM to assess construction risks during the design phase. Eng Constr Archit Manage. https://doi.org/10.1108/ECAM-09-2018-0379
20. Khosakitchalert C, Yabuki N, Fukuda T (2019) Improving the accuracy of BIM-based quantity takeoff for compound elements. Autom Constr 106:102891. https://doi.org/10.1016/j.autcon.2019.102891
21. Khosakitchalert C, Yabuki N, Fukuda T (2020) Automated modification of compound elements for accurate BIM-based quantity takeoff. Autom Constr 113:103142. https://doi.org/10.1016/j.autcon.2020.103142
22. Fazli A et al (2014) Appraising effectiveness of Building Information Management (BIM) in project management. Procedia Technol 16:1116–1125. https://doi.org/10.1016/j.protcy.2014.10.126
23. Forgues D et al (2012) Rethinking the cost estimating process through 5D BIM: a case study. In: Construction research congress 2012: construction challenges in a flat world. https://doi.org/10.1061/9780784412329.079
24. Sabol L (2008) Challenges in cost estimating with Building Information Modeling. IFMA world workplace, pp 1–16
25. Lin Y-C, Su Y-C (2013) Developing mobile-and BIM-based integrated visual facility maintenance management system. Sci World J. https://doi.org/10.1155/2013/124249
26. Motawa I, Almarshad A (2013) A knowledge-based BIM system for building maintenance. Autom Constr 29:173–182. https://doi.org/10.1016/j.autcon.2012.09.008
27. Cortés-Pérez JP, Cortés-Pérez A, Prieto-Muriel P (2020) BIM-integrated management of occupational hazards in building construction and maintenance. Autom Constr 113:103115. https://doi.org/10.1016/j.autcon.2020.103115

Modelling of Filtration with Two Capture Mechanisms

Liudmila Kuzmina, Yuri Osipov, and Marina Sosedka

Abstract Filtering fluids with tiny particles in porous media occurs in many technological processes of environmental and civil engineering. To strengthen the foundation, a slurry is pumped into loose rock and, when hardened, strengthens it. Deep bed filtration is accompanied by the retention of particles. Various forces act on the particles and cause them to settle on the framework of the porous medium. The most common are the capture of particles in the inlets of narrow pores and attachment to the walls of wide pores. A model with two simultaneous particle capture mechanisms is considered. The problem is reduced to a standard model with an implicit aggregated filtration function. To simplify the calculations, it was previously proposed to use an approximate model with an explicit linear-constant filtration function. The linear-constant function has a break and does not optimally approximate the aggregated filtration function. The paper introduces a linear-fractional (hyperbolic) filtration function with a free parameter. The parameter is selected from the condition of the best approximation to the aggregated function. The system of equations in partial derivatives is integrated and reduced to a system of transcendental equations. It is shown that the solution to the model with a hyperbolic filtration function is closer to the true solution than the approximate solution with a linear-constant function.

Keywords Filtration · Porous medium · Particle capture · Mathematical model · Analytical solution

L. Kuzmina
HSE University, Moscow, Russian Federation
e-mail: lkuzmina@hse.ru

Y. Osipov (✉)
Moscow State University of Civil Engineering, Moscow 129337, Russian Federation
e-mail: yuri-osipov@mail.ru

M. Sosedka
Plekhanov Russian University of Economics, Moscow, Russian Federation

P. Akimov et al. (eds.), *Proceedings of FORM 2022*, Lecture Notes in Civil Engineering 282, https://doi.org/10.1007/978-3-031-10853-2_3

1 Introduction

Filtering the smallest particles in a porous medium is often found in nature and in industrial technologies. Filtration describes the spread of bacteria and viruses in groundwater, the purification of industrial and domestic liquid waste, the strengthening of the foundation in loose soil, the reduction in oil production when the rock is clogged near the well, and much more [1–5]. Filtration modeling, based on laboratory experiments and mathematical calculations, makes it possible to predict natural processes and improve industrial technologies.

A porous medium is a single solid or free-flowing solid (such as rock or sand) that contains pores—empty channels of complex shape that allow fluid to seep through the porous medium. The smallest suspended particles carried by a liquid through a porous medium can linger in the pores and form a precipitate. Settled particles are immobile and cannot be lifted into suspension by a liquid flow or other particles [6].

Depending on the physicochemical properties of suspended particles, a porous medium, and a carrier liquid, there are various mechanisms for particle capture associated with electrostatic and hydraulic forces, diffusion into blind pores, arch bridging, straining, etc. If the particle sizes and cross sections of pores are close, the size capture mechanism becomes decisive: the particles penetrate deep through large pores, but get stuck at the entrance of small-diameter pores. Another common trapping mechanism is the adhesion of small particles to the walls of wide pores [7] (see Fig. 1).

As a rule, when constructing a filtration model, a single main trapping mechanism is chosen; other possibilities for blocking particles are neglected [8–10]. The mathematical model of suspension or colloid filtration in a porous medium includes the equations of mass balance and kinetic growth of sediment [11]

$$\frac{\partial C}{\partial t} + \frac{\partial C}{\partial x} + \frac{\partial S}{\partial t} = 0, \qquad \frac{\partial S}{\partial t} = \Lambda(S)C \tag{1}$$

with boundary and initial conditions

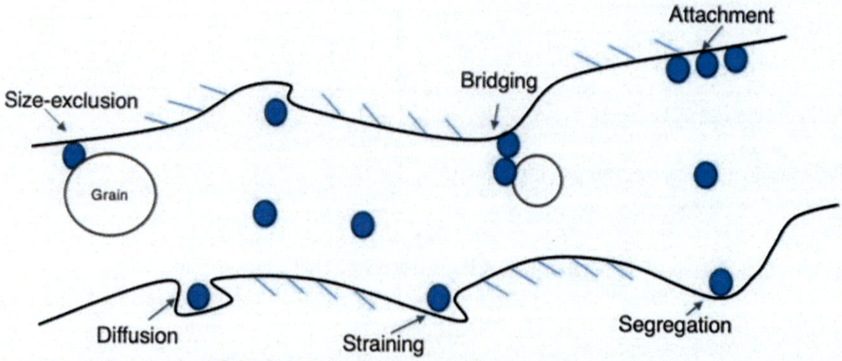

Fig. 1 Mechanisms for capturing particles in a porous medium

$$C|_{x=0} = 1, \quad C|_{t=0} = 0, \quad S|_{t=0} = 0. \tag{2}$$

Here $C(x,t)$, $S(x,t)$ are the unknown volumetric concentrations of suspended and settled particles, $\Lambda(S)$ is the filtration function, the form of which depends on the particle capture mechanism.

According to conditions (2), a suspension with concentration 1 is injected into the porous medium filled with the same fluid without any particles. The front moves in a porous medium at a speed $v = 1$. The inconsistency of the conditions at the origin of coordinates generates a discontinuity of the solution $C(x,t)$ at the front, which in the plane of variables x, t is given by the formula $t = x$. The solution is zero ahead of the front, because there is no suspension and sediment in this area. There are suspended and settled particles behind the front, the decision is positive. The exact solution to system (1), (2) behind the concentration front $t \geq x$ is defined by implicit relations [12]

$$\int_0^{S_0(t)} \frac{ds}{\Lambda(s)} = t, \quad \int_{S_0(t-x)}^{S(x,t)} \frac{ds}{s\Lambda(s)} = -x, \quad C(x,t) = \frac{S(x,t)}{S_0(t-x)}. \tag{3}$$

However, experiments show that particles are often captured by several simultaneously acting capture mechanisms. Complex models with different particle capture mechanisms, as a rule, do not allow a compact exact solution and are studied by numerical methods [13–16]. In some cases, a model with multiple capture mechanisms can be reduced to an aggregated model with a single capture mechanism. This gives way to find the solution (3), but the fact that the aggregate filtration function is implicitly specified makes calculations more difficult. To simplify the calculations in Eq. (2), an explicit filtration function close to the aggregated one can be used. In [17], for an approximate calculation of a model with two capture mechanisms—a size-exclusion and attachment—a linear-constant filtration function was proposed. However, the break of the function contradicts the smooth character of the filtration and the smoothness of the exact solution.

The article considers a mathematical model which describes filtration of suspensions and colloids with two capture mechanisms. In the aggregated model, a fractional linear function (hyperbola) with a free parameter is chosen as a simplified filtration function. The parameter is selected from the condition of the best approximation of the aggregate filtration function. The resulting hyperbolic function is closer to the aggregated function than the linear-constant filtration function. It is shown that solution to problem (1), (2) with a hyperbolic filtration function approximates the aggregated model better than the solution with a linear-constant filtration function.

2 Mathematical Model of Filtration with Two Capture Mechanisms

Consider the action of two mechanisms of particle capture—the size-exclusion and sticking to the walls. Particles are blocked in the necks of small pores and adhere to the walls of wide pores. With a small retained concentration, both capture mechanisms act with the same intensity. As the deposit grows, the number of small vacant pores that can capture particles decreases, and the number of new trapped particles blocked in small pores decreases. The mechanism of adhesion to the walls of wide pores becomes predominant.

Denote the concentrations of deposited particles retained by size-exclusion and sticking mechanisms by S_1 and S_2. The linear blocking filtration function of the size-exclusion capture mechanism is the Langmuir coefficient [18]

$$\Lambda^1(S_1) = \Lambda_0(1 - S_1/S_m), \tag{4}$$

the attachment mechanism corresponds to the constant

$$\Lambda^2(S_2) = \Lambda_1, \tag{5}$$

The mathematical model of filtration with two mechanisms for capturing particles in the domain $\{(x, t) : 0 \le x \le 1,\ t \ge 0\}$ is given by the equations

$$\frac{\partial C}{\partial t} + \frac{\partial C}{\partial x} + \frac{\partial S}{\partial t} = 0; \tag{6}$$

$$\frac{\partial S_i}{\partial t} = \Lambda^i(S_i)C, \quad i = 1, 2 \tag{7}$$

with boundary and initial conditions

$$C|_{x=0} = 1, \quad C|_{t=0} = 0, \quad S_i|_{t=0} = 0, \quad i = 1, 2. \tag{8}$$

Problem (6)–(8) can be reduced to the standard model (1), (2) with an aggregated filtration function. We express the solution C from each Eq. (7) and equate the resulting expressions

$$\frac{\partial S_2/\partial t}{\Lambda^2(S_2)} = \frac{\partial S_1/\partial t}{\Lambda^1(S_1)}.$$

Integrate the resulting relation with respect to the variable t using formulae (4) and (5)

$$\frac{S_2}{\Lambda_1} = -\frac{S_m}{\Lambda_0}\ln\left(1 - \frac{S_1}{S_m}\right) + K(x).$$

The integration constant is obtained from initial conditions (8): $K(x) = 0$. Express the solution S_2 in terms of S_1:

$$S_2 = -S_m \frac{\Lambda_1}{\Lambda_0} \ln\left(1 - \frac{S_1}{S_m}\right). \tag{9}$$

Denote the concentration of the total deposit

$$S = S_1 + S_2.$$

According to formula (9), the total deposit S is a function of the partial deposit S_1

$$S(S_1) = S_1 - S_m \frac{\Lambda_1}{\Lambda_0} \ln\left(1 - \frac{S_1}{S_m}\right). \tag{10}$$

From conditions (8) and formula (10)

$$S(0) = 0, \quad S(S_m) = \infty.$$

Differentiate function (10):

$$S'(S_1) = 1 + \frac{\Lambda_1}{\Lambda_0} \cdot \frac{1}{1 - S_1/S_m} > 0,$$

Since the derivative $S'(S_1)$ is positive, then function (10) has an inverse for $0 \leq S < \infty$. Adding Eqs. (7) for $i = 1$ and $i = 2$, we obtain system (1) with conditions (2) for the aggregated filtration function

$$\Lambda_{agg}(S) = \Lambda_0(1 - S_1(S)/S_m) + \Lambda_1, \tag{11}$$

The implicit form of the aggregated filtration function (11) complicates the calculations. To simplify calculations, an explicit filtration function close to the implicit function (11) is used. To model the retention of particles by two capture mechanisms, the article [17] introduces a linear-constant filtration function

$$\Lambda_{l-c}(S) = \begin{cases} \Lambda_0(1 - S/S_m) + \Lambda_1, & 0 \leq S \leq S_m; \\ \Lambda_1, & S \geq S_m. \end{cases} \tag{12}$$

The linear part of function (12) corresponds to the simultaneous action of two capture mechanisms, and the constant part corresponds to only one mechanism (attachment). At the point $S = S_m$, function (12) has a break. When the retained particles concentration reaches S_m, all small pores are closed by particles and the action of the dimensional mechanism stops. The deposit continues to grow only due to adhesion to the walls of wide pores. The proposed model has an exact solution.

For each point of the porous medium, the moment of transition time from a linear formula to a constant one. The model as a whole agrees with experiments.

In our opinion, the assumption of a sudden termination of the size-exclusion mechanism seems to be unjustified. In models with a single size-exclusion capture mechanism, it operates at any time, weakening with increasing sediment. The simultaneous action of size-exclusion, which weakens with sediment growth, and the constant attachment mechanism can be described by a linear-fractional filtration function. Hyperbola

$$\Lambda_{hyp}(S) = \frac{\Lambda_1 S + (\Lambda_0 + \Lambda_1)a}{S + a}, \tag{13}$$

where $a,\ \Lambda_0,\ \Lambda_1$ are positive constants, has no breaks and singularities at $S \geq 0$ and describes a smooth transition from the size-exclusion capture mechanism prevailing at the initial moment to the constant attachment mechanism. Filtration functions (11), (12) and (13) coincide for $S = 0$ and $S \to \infty$.

The parameter a is chosen so that the hyperbolic function (13) best approximates the aggregated filtration function (11). Since the maximum distance between the aggregated and linear-constant functions (11) and (12) is reached at the break point, the aggregated and hyperbolic filtration functions (11) and (13) are required to coincide at this point. For this purpose, functions (11), (13) are expressed in terms of the variable S_1 using formula (10).

$$\Lambda_{agg}(S_1) = \Lambda_0\left(1 - \frac{S_1}{S_m}\right) + \Lambda_1, \quad \Lambda_{hyp}(S_1) = \frac{\Lambda_1\left(S_1 - S_m \frac{\Lambda_1}{\Lambda_0}\ln(1 - S_1/S_m)\right) + (\Lambda_0 + \Lambda_1)a}{S_1 - S_m \frac{\Lambda_1}{\Lambda_0}\ln(1 - S_1/S_m) + a}. \tag{14}$$

The break point of the function $\Lambda_{l-c}(S_1)$ is defined by the equation

$$S_1 - S_m \frac{\Lambda_1}{\Lambda_0}\ln\left(1 - \frac{S_1}{S_m}\right) = S_m. \tag{15}$$

After changes

$$\lambda = \Lambda_1/\Lambda_0, \quad u = 1 - S_1/S_m$$

Equation (15) takes the form

$$u + \lambda \ln u = 0. \tag{16}$$

Since the left side of Eq. (16) changes sign on the interval (0, 1) and the derivative $g'(u) = 1 + \lambda/u$ is positive, the transcendental Eq. (16) has a unique solution on the interval $0 < u < 1$.

Consider the difference of functions (14) at the break point

$$\delta = \left(\Lambda_{agg}(S_1) - \Lambda_{hyp}(S_1)\right)\Big|_{S_1=S_m(1-u_0)} = \Lambda_0 u_0 - \frac{\Lambda_0 a}{S_m + a}.$$

The parameter a can be found from the condition $\delta = 0$:

$$a = S_m \frac{u_0}{1-u_0}. \tag{17}$$

Substituting function (13) into formulae (3) and performing integration, we obtain a system of transcendental equations for solutions $S_0(t-x)$ and $S(x,t)$:

$$\frac{S_0}{\Lambda_1} - \frac{\Lambda_0 a}{(\Lambda_1)^2} \ln \frac{\Lambda_1 S_0 + (\Lambda_0 + \Lambda_1)a}{(\Lambda_0 + \Lambda_1)a} = t - x, \quad \ln \frac{S}{S_0} + \frac{\Lambda_0}{\Lambda_1} \ln \frac{\Lambda_1 S + (\Lambda_0 + \Lambda_1)a}{\Lambda_1 S_0 + (\Lambda_0 + \Lambda_1)a} = -(\Lambda_0 + \Lambda_1)x. \tag{18}$$

With known functions $S_0(t-x)$ and $S(x,t)$ the solution $C(x,t)$ is determined from the last formula (3).

3 Numerical Solution

The numerical calculation was performed for the values obtained experimentally for suspension particles with a radius of 0.225 μm transported through sand with pores of an average radius of 11.5 μm [17]. The calculation was made by the finite difference method and an explicit upwind scheme [19–21]. Figure 2 shows graphs of three filtration functions (11)–(13).

Figures 3 and 4 show graphs of solutions for three models with filtration functions Λ_{agg}, Λ_{l-c}, Λ_{hyp} at the moment $t = 1$ of a suspension breakthrough through a porous medium and at the outlet $x = 1$.

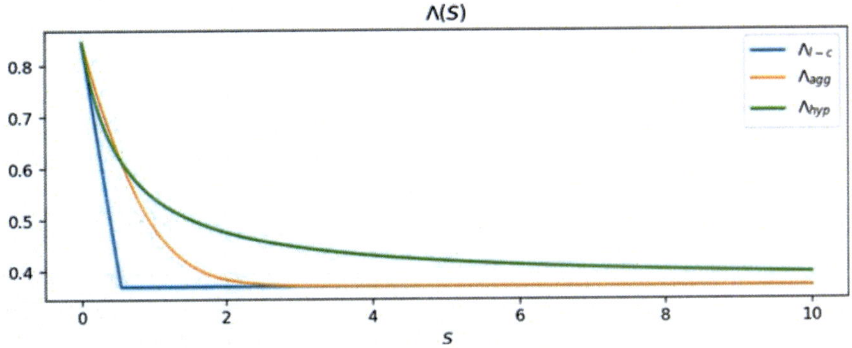

Fig. 2 Filtration functions Λ_{agg}, Λ_{l-c}, Λ_{hyp}

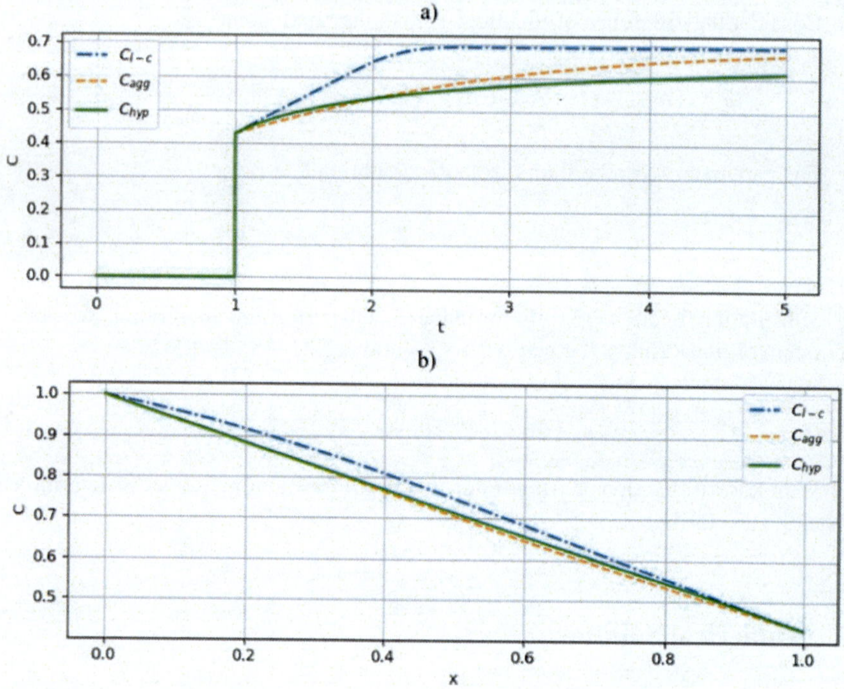

Fig. 3 Concentrations of suspended particles. **a** dynamics at $x = 1$. **b** profiles at $t = 1$

It follows from Figs. 3 and 4, that the graphs of solutions to problem (1), (2) with the aggregated and hyperbolic filtration functions practically coincide, and the graph of the solution of the model with a linear-constant function deviates from them. Table 1 shows the maximum relative deviations of solutions with hyperbolic and linear-constant functions from a solution with an aggregated filtration function in the domain $\{(x, t) : 0 \leq x \leq 1,\ 0 \leq t \leq 5\}$, Table 2 presents maximum absolute deviations for $5 \leq t \leq 10$.

Tables 1 and 2 show that the solution with hyperbolic filtration function approximates the exact solution with aggregated function better than the solution with linear-constant function. Note that for the aggregated filtration function, solution C is closer to the exact solution than S. This is due to the fact that $C(x, t)$ is limited and tends to 1, while $S(x, t)$ is not limited and tends to infinity at a large time.

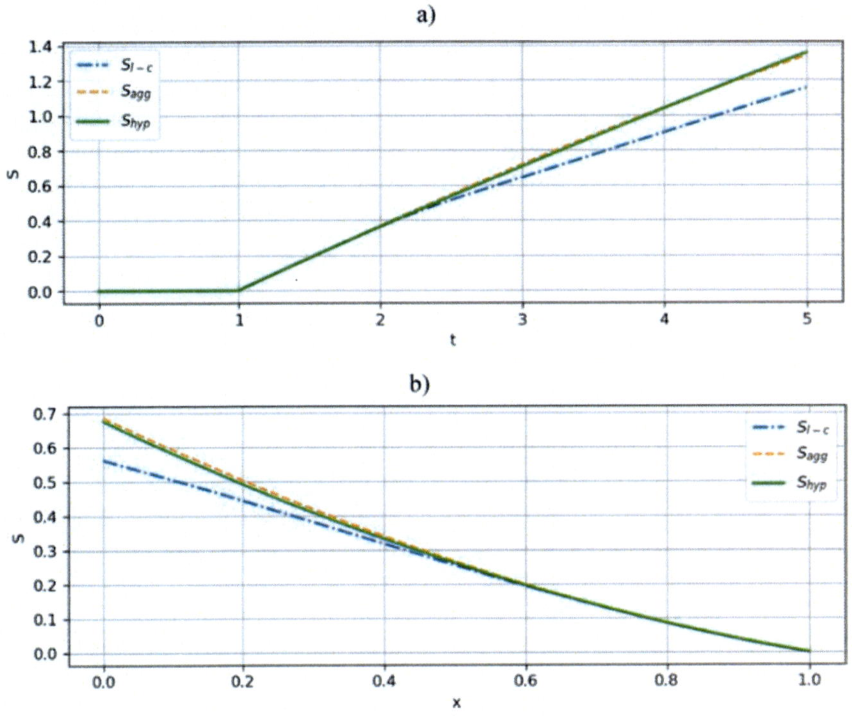

Fig. 4 Concentrations of retained particles. **a** dynamics at $x = 1$. **b** profiles at $t = 1$

Table 1 Maximum relative deviations from the aggregated solution for $0 \le t \le 5$

Model with filtration function	$C(x,t)$ (%)	$S(x,t)$ (%)
Λ_{hyp}	8,18	12,00
Λ_{l-c}	21,74	21,60

Table 2 Maximum absolute deviations from the aggregated solution for $5 \le t \le 10$

Model with filtration function	$C(x,t)$	$S(x,t)$
Λ_{hyp}	0.055	0.286
Λ_{l-c}	0.121	0.339

4 Discussions

Accounting for several particle capture mechanisms acting simultaneously complicates the filtration model. The mathematical model, which takes into account the size-exclusion capture mechanism and adhesion of particles to pore walls, consists of 4 equations with 4 unknowns (6), (7). The problem is reduced to a standard system

(1) with an implicitly defined aggregated filtration function. The replacement by a linear-constant function makes it possible to obtain the solution in explicit form, but the relative error of the approximate solution exceeds 20%. The hyperbolic filtration function proposed in the article approximates the solution better, but to find an approximate solution, it is required to solve a system of transcendental equations. Is it possible to choose a filtration function for which there is an explicit solution with a good approximation to the exact solution? This is the subject of the next study.

5 Conclusions

The study of filtration models with two mechanisms of particle capture in a porous medium allows the following conclusions.

- Accounting for several simultaneously acting particle capture mechanisms makes it possible to bring the model closer to real filtration processes in nature and technology.
- The model with two capture mechanisms is reduced to a standard model with an implicitly defined aggregated filter function.
- To simplify calculations, an approximate explicit filtration function is used.
- The approximate solution obtained using the hyperbolic filtration function approximates the exact solution better than the solution of the model with a linear-constant function.

References

1. Zhou Z, Zang H, Wang S, Du X, Ma D, Zhang J (2018) Filtration behaviour of cement-based grout in porous media. Transp Porous Media 125:435–463
2. Zhu G, Zhang Q, Liu R, Bai J, Li W, Feng X (2021) Experimental and numerical study on the permeation grouting diffusion mechanism considering filtration effects. Geofluids 6613990 (2021)
3. Tsuji M, Kobayashi S, Mikake S, Sato T, Matsui H (2017) Post-grouting experiences for reducing groundwater inflow at 500 m depth of the Mizunami Underground Research Laboratory, Japan. Proc Eng 191:543–550
4. Chrysikopoulos CV, Sotirelis NP, Kallithrakas-Kontos NG (2017) Cotransport of graphene oxide nanoparticles and kaolinite colloids in porous media. Transp Porous Media 119:181–204
5. Johnson WP, Rasmuson A, Pazmiño E, Hilpert M (2018) Why variant colloid transport behaviors emerge among identical individuals in porous media when colloid-surface repulsion exists. Environ Sci Technol 52:7230–7239
6. Bedrikovetsky P (2013) Mathematical theory of oil and gas recovery with applications to ex-USSR oil and gas fields. Springer, Dordrecht
7. Bedrikovetsky P (2008) Upscaling of stochastic micro model for suspension transport in porous media. Transp Porous Media 75:335–369
8. Santos A, Bedrikovetsky P, Fontoura S (2008) Analytical micro model for size exclusion: pore blocking and permeability reduction. J Membr Sci 308:115–127

9. Galaguz YuP, Kuzmina LI, Osipov YuV (2019) Problem of deep bed filtration in a porous medium with the initial deposit. Fluid Dyn 54(1):85–97
10. Kuzmina LI, Nazaikinskii VE, Osipov YV (2019) On a deep bed filtration problem with finite blocking time. Russ J Math Phys 26:130–134
11. Herzig JP, Leclerc DM, le Goff P (1970) Flow of suspensions through porous media—application to deep filtration. Ind Eng Chem Res 62(8):8–35
12. Vyazmina EA, Bedrikovetskii PG, Polyanin AD (2007) New classes of exact solutions to nonlinear sets of equations in the theory of filtration and convective mass transfer. Theor Found Chem Eng 41(5):556–564
13. Kuzmina L, Osipov Yu (2019) Particle capture in porous medium. IOP Conf Ser Mater Sci Eng 661:012122
14. Bradford SA, Torkzaban S, Shapiro A (2013) A theoretical analysis of colloid attachment and straining in chemically heterogeneous porous media. Langmuir 29:6944–6952
15. Yuan H, Shapiro AA (2011) A mathematical model for non-monotonic deposition profiles in deep bed filtration systems. Chem Eng J 166:105–115
16. Nurgul S, Beket K, Aigul M, Galymzhan K, Raigul T (2021) Numerical methods for solving improper problems of filtration theory. J Appl Eng Sci 19(1):98–108
17. Zhang H, Malgaresi GVC, Bedrikovetsky P (2018) Exact solutions for suspension-colloidal transport with multiple capture mechanisms. Int J Non-Linear Mech 105:27–42
18. Altoe JE, Bedrikovetsky PG, Siqueira AG, de Souza AL, Shecaira F (2006) Correction of basic equations for deep bed filtration with dispersion. J Petrol Sci Eng 51:68–84
19. Osipov Yu, Safina G, Galaguz Yu (2018) Calculation of the filtration problem by finite differences methods. MATEC Web Conf 251:04021
20. Osipov YuV, Safina GL, Galaguz YuP (2020) Filtration model with multiple particle capture. IOP Conf Ser J Phys Conf Ser 1425:012110
21. Safina G (2021) Calculation of retention profiles in porous medium. Lecture Notes Civil Eng 170:21–28

Dynamics of Reinforced Concrete Slab-Frame Structures with Punching Damages

Anatoly Alekseytsev, Michael Antonov, and Natalya Kurchenko

Abstract An approach to assessing the dynamic effects for Reinforced Concrete (RC) frames of structures with flat plate in extraordinary event load combination is considered. It is possible that initial actions can exceed the service design load that can lead to the punching damage of flat column connection before secondary component action. Secondary component action considers the removal of any of the columns from the design model over a finite period of time. The analysis of dynamic effects is carried out on the basis of two approaches: the first way is based on the energy method of G.A. Geniev and assumes a quasi-static assessment of the stress–strain state of the damaged system; the second way is the analysis of the transient dynamic process taking into account the physical and geometric nonlinearity. An approximate approach to modeling the damage of a reinforced concrete slab with punching damage is proposed and further prospects for its improvement are considered. The degree of a danger from accidental actions for structures pre-damaged by punching and influence of these damages on the survivability has been established. An example of assessing survivability of underground parking frame in the presence of punching damage for one of the joints between the slab and the column is considered.

Keywords Slab-frame structures · Punching damage · Progressive collapse · Nonlinear dynamic

1 Introduction

Modern requirements for ensuring the safety of RC frame buildings and structures imply ensuring their survivability in case of accidental actions. Regulations [1] and in scientific research [2–4] have several requirements to provide the survivability of structures. At the same time, special attention is paid to the behaviour of the flat slab-column connection. The various approaches is taking into account to consider the opportunity of punching failure of this connection. [5–7]. The most of the research

A. Alekseytsev (✉) · M. Antonov · N. Kurchenko
Moscow State University of Civil Engineering, Moscow 129337, Russian Federation
e-mail: aalexw@mail.ru

P. Akimov et al. (eds.), *Proceedings of FORM 2022*, Lecture Notes in Civil Engineering 282, https://doi.org/10.1007/978-3-031-10853-2_4

about punching is devoted to prevention this failure during usual design life. Nevertheless punching damage do not lead to progressive collapse according to experience of structural survey. However, the influence of primary damages on the survivability of structures in case of occurrence accidental actions remains practically unexplored.

The process of local damages associated with column removal [8–11] has the character of a transient dynamic process, which can be modeled both using a quasi-static energy approach or step-by-step methods of direct integration of the equations of motion of a damaged structure [12, 13]. At the same time, for the design, it seems necessary to identify the level of dynamic effects associated with local damage to columns. Coefficient of dynamic that determine the level of dynamic loading of structures can be used to optimize load-bearing structures of this type using modern heuristic algorithms [14–19]. As well as to analyze possible destruction mechanisms, including the effects of concrete destruction with the formation of catenary action, as well as possible buckling of elements during their reloading or changing the stiffness of the support nodes [20–22]. It should be noted that the punching effect can also occur as a result of local fire impacts, which lead to a significant decrease in the modulus of elasticity of concrete, and, consequently, its strength [23–25].

The review of studies allows us to conclude that the research of survivability of flat slab structures in the presence of preliminary punching damage of slab column connection is a relevant issue to make the possible methodological approaches to its solution.

2 Methods

At the beginning, it was described the description of modeling the reinforced concrete structure of the flat slab structures. Modern software systems make it possible to simulate the stress–strain state of reinforced concrete structures in a three-dimensional model with deviding the volumes of concrete and reinforcement into hexahedral or tetrahedral finite elements. However, in general, even with the use of modern 12–16 core processors such an approach for calculating frames, is constrained by the highest computational capacity and requirements for RAM, which must accommodate a stiffness matrix with a size of 32 GB or more. Therefore, we will form a analytical model from the condition of the possibility of its reproduction in real design conditions. To assess the dynamic effects, we will use a plate-bar model, that have a reduced to concrete stiffness of the elements:

$$E_{b,red} = \frac{E_b A_b + E_s A_s}{A_b + A_s}, \; G_{b,red} = \frac{0,4(1,5h_0 E_b + E_s A_{sw})}{1,5h_0 + A_{sw}} \tag{1}$$

where E_b, E_s—are the initial modulus of elasticity of concrete and reinforcement, A_b, A_s—are the cross-sectional areas of concrete and longitudinal work rebars, A_{sw}—is the cross-sectional area of stirrups in the projection of an inclined section with a length of 1.5 and a width of 1 m.

The nonlinear concrete strength model provides for different tensile and compressive resistance of concrete, so the Drucker-Prager model modified for concrete is used:

$$\sqrt{\frac{1}{6}[(\sigma_1-\sigma_2)^2+(\sigma_2-\sigma_3)^2+(\sigma_1-\sigma_3)^2}=A+B(\sigma_1+\sigma_2+\sigma_3); \quad (2)$$

$$A=\frac{2}{\sqrt{3}}\left(\frac{R_bR_{bt}}{R_b+R_{bt}}\right),\; B=\frac{2}{\sqrt{3}}\left(\frac{R_{bt}-R_b}{R_b+R_{bt}}\right), \quad (3)$$

where $\sigma_1, \sigma_2, \sigma_3$—are the principle stresses in concrete, R_b, R_{bt}—specified compressive and tensile strength of concrete. The principle stresses are determined on the yield surface specified by two parameters: C—stresses of cohesion (adhesion) in concrete and the angle of internal friction φ. These parameters are determined on the basis of physical or numerical experiments, where $C = f(R_{bt})$, $\varphi = 30\ldots35^{\circ}$. This model can also have an opportunity to take into account the expansion of concrete. The nonlinear behavior of reinforcement is modeled within the Prandtl model (bilinear diagram according to SP 63.13330.2018). Physically and geometrically, a nonlinear problem is solved using the Newton–Raphson method.

The quasi-static approach for taking into account dynamic effects will be carried out on the basis of the energy theory of G.A. Geniev in accordance with the dependencies that determine the degree of danger of the stress state for normal and inclined sections, as well as dynamic loading by the longitudinal force:

$$M_{ult,D}=2M_{ult,SDam}-M_{ult,S0};\; Q_{ult,D}=2Q_{ult,SDam}-Q_{ult,S0};\; N_{ult,D}=2N_{ult,SDam}-N_{ult,S0} \quad (4)$$

where M, Q, N—internal forces (bending moment, longitudinal and transverse forces; $|\bullet|_{ult,D}$—dynamic factor; $|\bullet|_{ult,SDam}$—static factor in the system damaged with accidental action; $|\bullet|_{ult,S0}$—Is a static factor in the original non-damaged system.

Consideration of the transient dynamic process associated with local damage to the bearing system is implemented within the step by step method of direct integration in accordance with the equation with a simplified consideration of Rayleigh structural damping:

$$[M]\ddot{y}(t)+[K](\beta\dot{y}(t)+y(t))=F(t), \quad (5)$$

where $[M]$, $[K]$ these are the general mass and stiffness matrices of the finite element model, respectively; $\ddot{y}(t)$, $\dot{y}(t)$, $y(t)$ these are vectors of accelerations, velocities and nodal displacements, respectively; $F(t)$ is the vector of the external load reduced to the nodes, β—is the structural damping coefficient, $\beta = 0,05$ in the absence of damage and nonlinear work in the service range of deformations. If there is punching damage and significant geometric nonlinearity, then according to experimental data, for example [25], the structural damping coefficient is taken $\beta = 0,1 \div 0,2$. So, to take into account the presence of punching damage the structural damping coefficient of $\beta = 0,15$ is considered. According to the preliminary

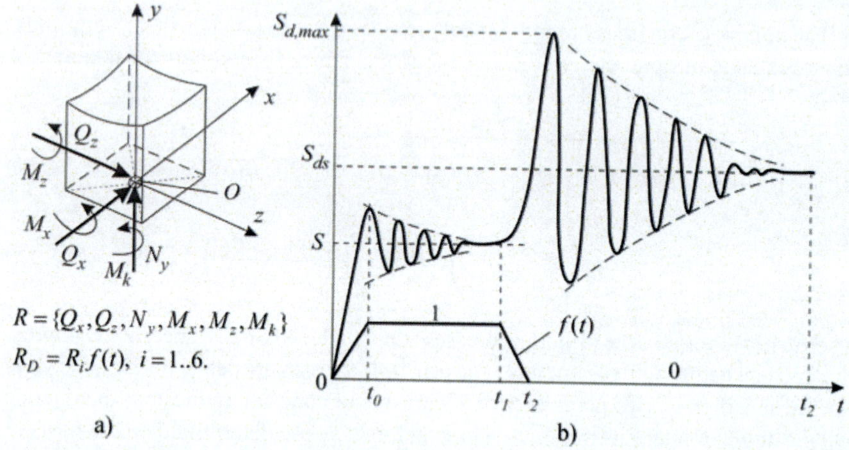

Fig. 1 Modelling of local damage: system of equivalent static force reaction (**a**); plot and graph f (t) of changing dynamic reaction in dangerous section (**b**)

model analyses, it was obtained the interval of natural frequencies when the mode of shape corresponds to the accidental design scenario. Damping is taken into account precisely in this interval.

Modeling of damage in the form of removal of support O in a finite time is modeled with a dynamic equilibrium taking into account the time of dynamic relaxation, Fig. 1. In this case, the system of support links is replaced by an equivalent system of support reactions R, which change over time.

These reactions are applied to the system over time t_0. The time interval $[0;\ t_0]$ is appointed with attention to the stability of the numerical integration process. During the time interval $(t_0;\ t_1]$ the system oscillates and the process of dynamic relaxation is realized, which ends with the onset of a state equivalent to static loading in the presence of support links. After that the simulation of removing the support in $(t_1;\ t_2]$ period. In this time interval reaction values decrease to zero. In the next time interval $(t_2;\ t_3]$ the reactions do not change; the system makes oscillations caused by accidental action. If the oscillations stabilize and damp, and the deflection corresponds to the limitations adopted in regular (SP 385.1325800.2018) $[S_{ds}]$, then we consider that the system has survivability to this accidental action.

Modeling of punching shear damage is implemented as follows. The first method is the physical erosion deleting of plate elements from the abutment zone of the column and the slab. The identification of these elements should be carried out in the course of a preliminary nonlinear calculation with an assessment of the level of principal deformations Fig. 2, a.

In this case, in the area of the damaged node, the mesh should be thickened. In the place of removal support reinforcement falling on the removed element should be installed. It also need to take into account the eccentricity of the location of the reinforcement, rods with the actual area (Fig. 2, b). The second method involves the correction of the reduced modulus of the longitudinal and transverse deformation

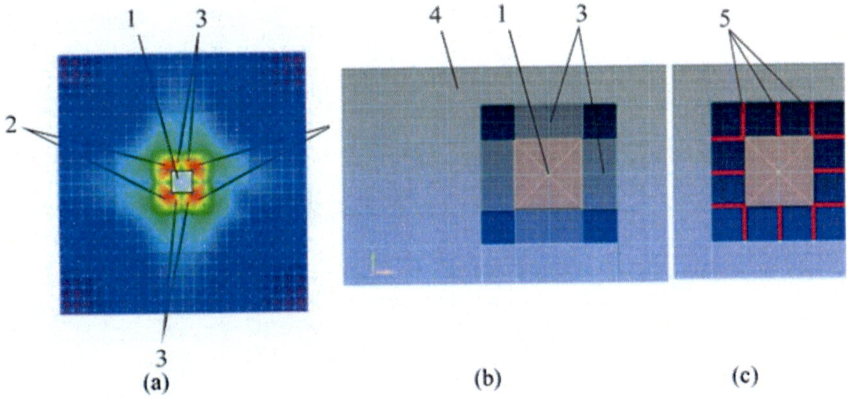

Fig. 2 Modeling punching shear damage: identification of the area of possible damage (**a**); modeling of damage by the introduction of elements with membrane deformations (**b**); modeling the shutdown of concrete (**c**); 1-zone of the slab over the column; 2—elements to be removed. 3—elements experiencing predominantly membrane deformation without shear; 4—reinforced concrete elements without damage; 5—bar elements of reinforcement

of concrete by introducing a softening function into the Drucker-Prager model, or by introducing membrane plate elements with characteristics corresponding to the removal of concrete from the reinforced section diagram (Fig. 2, b).

3 Results

Verification of the Drucker-Prager model. The parameters of the conditional adhesion and the angle of internal friction were selected for a concrete prism of class B25 of standard dimensions 150 × 150 × 600 according to the criterion of equality of the ultimate load to the value $N_{ult} = R_b A_b = 326, 25$ kN. The results of this calculation when the level of breaking load reaches 325 kN are shown in Fig. 3. After reaching the stress level of 14.5 MPa, quasi-plastic deformation begins and leads to the failure. Wherein C = 4, 25 MPa, $\varphi = 30^{\circ}$. These values can be corrected, if there is reinforcement in concrete, modeled by specifying bar elements that have common nodes with concrete. The steel of the reinforcement was modelled with the Prandtl diagram without hardening relationship.

Consider the results of calculating a fragment of the bearing system of a one-storey underground parking with a cell size of 6 × 6 m, a height of 3 m, loaded with a load evenly distributed over the area, including the weight of the slab and the weight of the coating. As the reinforced structure has a flat slab foundation as the base nodes of a column were restrained against translational and rotational movement. The slab has a solid section with a thickness of 200 mm, with 40 × 40 columns in plan. The reinforcement of the slab corresponds to the usual design life conditions with a load of 16.8 kN/m^2.

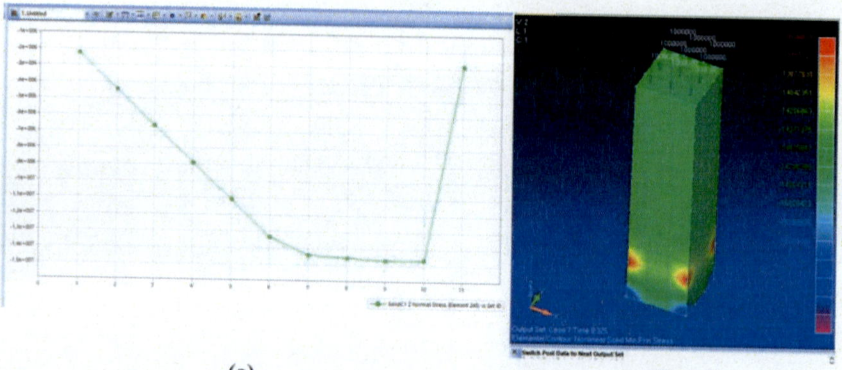

(a) (b)

Fig. 3 Numerical modelling of compression of a standard prism based on the use of the Drucker-Prager model: dependence of compressive stresses on dimensionless parametric load (**a**), prism model with visualization of the minimum principle stress (**b**)

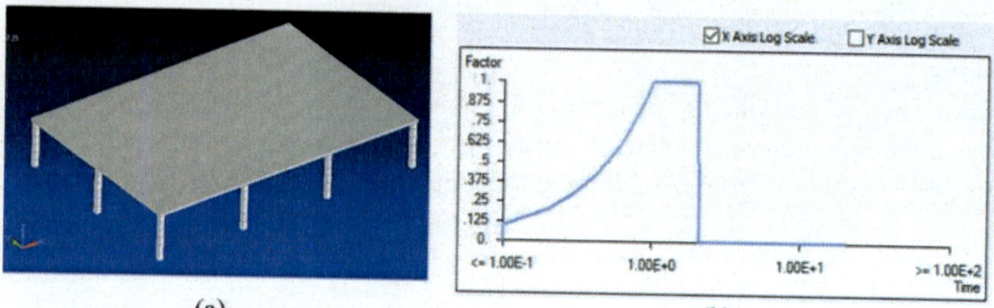

(a) (b)

Fig. 4 Investigated structure (**a**), function of changing in time the removal reaction (**b**)

The following calculations were considered:

- fast local damage to the middle column of the middle row without taking into account the existing damage from the punching of the slab by the column (A);
- the same, taking into account the modeling of damage from punching the slab by the column with the safety of the longitudinal working reinforcement (B).
- The beam-shell FE model of the object is shown in Fig. 4a. The numerical integration process lasted 12 s. Single links in the damaged column were replaced by reactions, the static equivalents that were multiplied at each moment of time by the values of the function shown in Fig. 4b, c.

The dynamic effect was estimated using formula (4) and based on the estimation by the method of direct integration (DI) of the equilibrium equations of the damaged system. The analysis results are shown in Fig. 5, where the cross indicates the column for which the support is removed. Using the DI method allowed to take into account, physical and geometric nonlinearity. While the level of deformations in the slab when the support link of the column was removed allowed us to judge its plastic behaviour

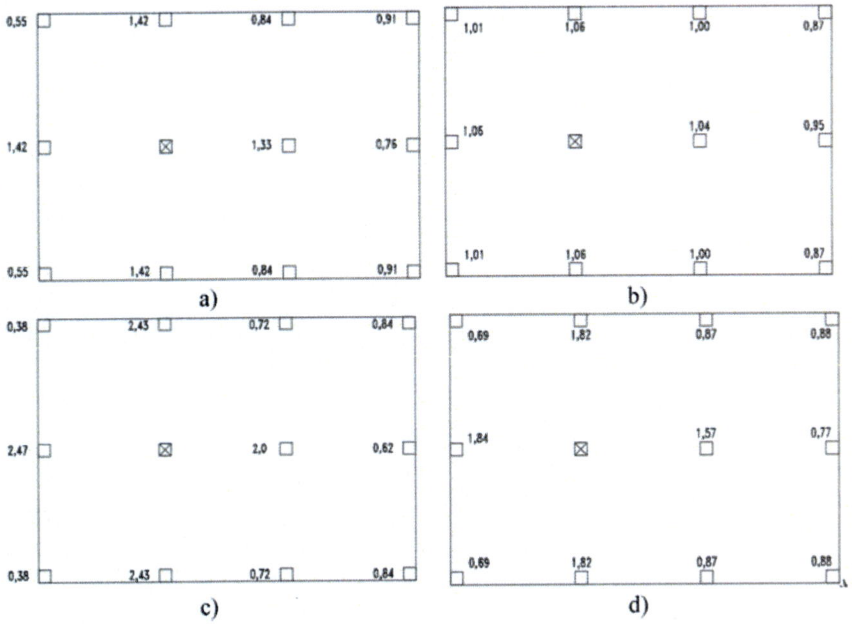

Fig. 5 Dynamic effects for longitudinal forces in columns, case (A): dynamic coefficients obtained on the basis of efforts calculated by the formula (3) (**a**); dynamism coefficients calculated by (DI method) (**b**); dynamic coefficient obtained on the basis of efforts according to (3) (**c**); and by (DI method) (**d**)

of failure. As can be seen in Fig. 5, the method of G.A. Geniev gives a certain degree of load caring capacity that allows us to use it in the design of such structures without using the expanded dynamics apparatus. At the same time, it is interesting to investigate the level of dynamic additional loading of columns. In Fig. 6, it can be seen that the DI method reduces the level of additional loading by 25% in comparison with the assessment according to (4).

There is also a decrease in the coefficient of dynamic reloading of bending moments in a slab to the level of the ultimate load bearing capacity of the section. In this case, the maximum coefficient of additional loading, determined by formula (4) was 3.83. The use of the direct dynamic method has shown that this coefficient is 2.55 to ensure the prevention of local destruction.

Now the case B is considered. The punching shear damage was specified for the right middle support relative to the removed one. Damage modeling was carried out in accordance with the scheme in Fig. 2b. In this case, the mesh of elements in the region of the reference section was condensed. The membrane stiffness of the plate elements was set to the equivalent tensile stiffness of the reinforcement; Only reinforcement was taken into account in the damaged zone. The calculation to remove the support at a load level of 16.8 kN/m^2 led to the total failure of the structure. The survivability of the structure in the presence of punching damage was ensured at a

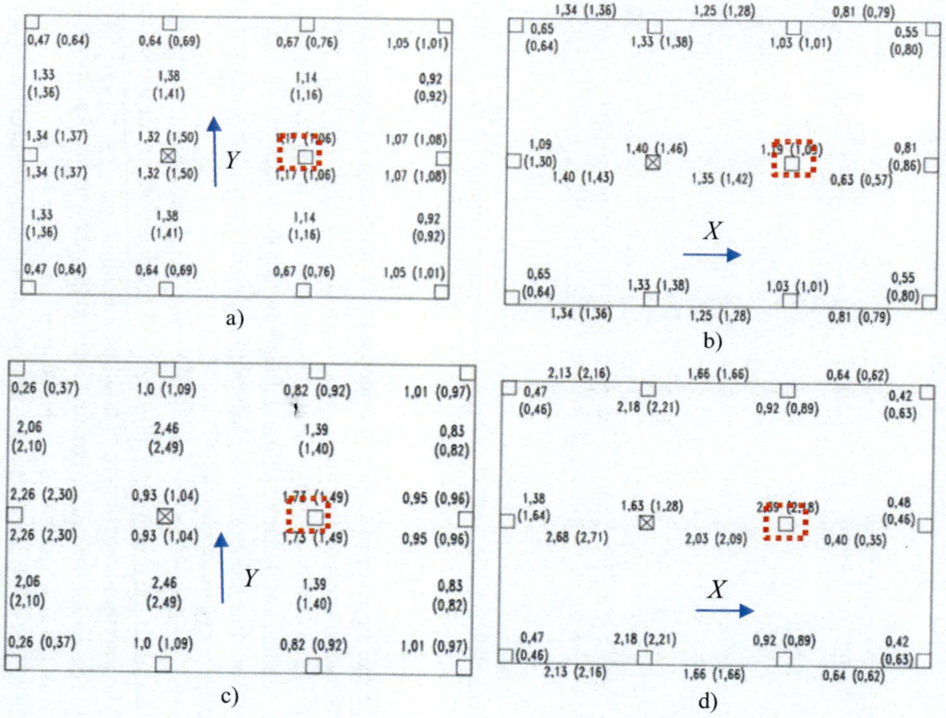

Fig. 6 Dynamic effects for bending moments in a slab, case (B): Dynamic coefficients obtained on the basis of the moment My by (DI) (**a**); the same for the moment Mx (**b**); dynamic coefficients obtained by (PI) based on the moments My (**c**); the same for the moment Mx (**d**)

load of 10 kN/m^2. When the structural damping coefficient was changed from 0.15 to 0.1, the value of this load decreased to 9 kN/m^2, which indicates the most important role of damping and the need for experimental research in this direction.

The calculation results showed that there are practically no changes in the dynamic effect for longitudinal forces in columns in the presence of punching damage in comparison with a solid slab. The differences do not exceed 5%. Some results of the dynamic effects for bending moments with a reduced loading level and punching shear damage is presented in Fig. 6. Here, the area of puncture damage is indicated by a dashed line. The values of the coefficients obtained without taking into account the damage from punching are given in parentheses.

As can be seen in Fig. 6, the presence of punching damage leads to an insignificant decrease in the dynamic factor for slab fragments associated with the damaged area and support areas per column for all frame cells, including the damage site. In the spans of the slab, the dynamic coefficient change insignificantly. This feature may indicate that at extraordinary extreme event in which the design scheme of the structure does not change, for example, exceeding the service load, the presence of these damages will not significantly affect the resistance of the structure to progressive destruction.

The analysis of the dynamic loading of the structure when changing the design scheme showed that the presence of damage from punching for the scheme we are considering leads to an increase in additional loading in moments in the places of local elimination of the support and damage to the slab from punching by no more than 10%.

4 Discussion

The process of punching slabs by columns is actively studied by a number of scientists [5–7, 25, 26], and it seems complicated and poorly studied both from the standpoint of theoretical justification and from the standpoint of modeling. Modern approaches to punching shear calculations described in the regular need to be improved. The study under consideration poses a number of tasks necessary for a comprehensive approach both to modeling flat slabs damaged by punching and to taking into account their danger. In particular, it seems relevant to identify the most dangerous combination of loads for slabs damaged by punching when there is a moving load on the surface, for example, from cars. In addition, the influence of the number and location of punching damage on the resistance of buildings and structures to progressive destruction should be investigated. It also need to take into account three-dimensional behaviour Also, the issues of assessing the dynamic effects in the longitudinal armature and the degree of punching damage, at which the survivability of the structure at a given level of loading is still ensured, remain topical.

5 Conclusions

The presented method of deformation modeling makes it possible with satisfactory accuracy to analyze the dynamics of reinforced concrete frames of buildings with flat slab structures and to simulate in a simplified way the damage caused by punching slabs by columns. While for the reduced section the flexural stiffness of the plate elements decreases, and the membrane stiffness is set equal to the stiffness tensile reinforcement in the damaged zone.

It has been established that the presence of damage from punching in the junction of the plate and the column when considering the three dimensional behaviour of the structure significantly affects its survivability if an accidental dynamic loading leads to a change (structural restructuring) of the design scheme and transfers the power flow through this node.

References

1. Shapiro GI, Eysman YA, Zalesov AS (2005) Recommendation for protection reinforced residential structures from progressive collapse. Moscow Committee of Architecture Publishing, Moscow, pp 5–10
2. Kolchunov VI, Klyuyeva NV, Androsova NB, Bukhtiyarova AS (2014) Survivability of buildings and structures under beyond-design actions. ASV Publishing, Moscow, pp 10–15
3. Tamrazyan AG (2012) Reducing risks in construction in natural and technological accident. ASV Publishing, Moscow, pp 5–12
4. Travush VI, Fedorova NV (2017) Russ J Build Constr Archit 1:6–14
5. Tamrazyan AG (2020) Conference series materials safety of structure. Probl Solutions 1:101–109
6. Karpenko NI, Karpenko SN (2012) Beton i zhelezobeton 5:10–16
7. Trekin NN, Krylov VV (2018) Sci Aspect 4:771–778
8. Savin SY, Kolchunov VI, Emelianov SG (2018) IOP Conf Ser Mater Sci Eng 456:012089
9. Tamrazyan AG, Alekseytsev AV (2019) E3S Web Conf 97:04005
10. Fedorova NV, Il'yushchenko TA, Medyankin MD, Insafutdinov AR (2019) Build Reconstr 2(82):72–80
11. Fedorov VS, Mednov YA (2010) Build Reconstr 6(32):48–53
12. Fialko SY, Kabantsev OV, Perelmuter AV (2021) Mag Civil Eng 2(102):10214
13. Kabantsev OV, Mitrovitch B (2018) IOP Conf Ser Mater Sci Eng 456:012002
14. Bondarenko VM, Klyuyeva NV, Degtyar' AN, Androsova NB (2007) Eng Transp 4–16:5–10 (2007)
15. Tamrazyan AG (2020) Conference series materials safety of structure. Probl Solutions 1:115–122
16. Alekseytsev AV, Al Ali M (2018) Mag Civil Eng 7(83):175–185
17. Serpik IN, Alekseytsev AV (2018) IOP Conf Ser Mater Sci Eng 365:052003
18. Alekseytsev AV, Botagovsky M, Kurchenko NS (2019) E3S Web Conf 97(61):03002
19. Alekseytsev AV, Gaile L, Drukis P (2019) Mag Civil Eng 91(7):3–15
20. Savin SY, Kolchunov VI, Kovalev VV (2020) Build Reconstr 87(1):71–80
21. Kolchunov VI, Prasolov NO, Morgunov MV (2007) Stroitel'naya mekhanika inzhenernykh konstruktsiy i sooruzheniy 4:40–44
22. Morgunov MV, Grishin RA (2019) Conf Ser Mater MGSU Inst Eng Archit 2:75–76
23. Tamrazyan AG (2016) Proc Eng 153:715–720

24. Fedorov VS, Levitskiy VY, Solov'yev IA (2015) Build Reconstr 5(61):47–55
25. Russell M, Owen JS (2015) Eng Struct 99:28–41
26. Marques MG, Liberati EA, Gomes RB (2021) J Struct Eng 7(147):04021090

High-Strength Modified Concrete for Monolithic Construction

Madina Salamanova, Djokhar Medjidov, and Aset Uspanova

Abstract Intensive high-rise construction requires the improvement of existing technologies for the construction of buildings and structures, the modeling of higher quality concrete, which guarantees the safety and operational reliability of facilities. It is monolithic construction using high-strength concrete composites of a new generation based on high-quality materials, mineral powders and effective superplasticizers that will raise the level of concrete science. This work presents the possibility of obtaining high-strength concrete for monolithic construction using the complex modification of the concrete matrix with an organic and mineral component, high strength and durability results can be achieved. The effect of additives based on polyaryl and polycarboxylites on the properties of the concrete mix and concrete has been studied, and the positive effect of the mineral powder of limestone flour on the formation of the structure and properties of the composite and their interaction with microfillers of various types has been established. It has also been found that additives based on polycarboxylates have higher strength indicators. The studied compositions of concrete compositions "portland cement—filler—superplasticizer—sand—crushed stone" provide the strength characteristics of concrete of classes B 40–50.

Keywords High-strength concrete · Monolithic construction · Modifier · Chemical additives · Microsilica · Superplasticizer · Broken brick · Limestone

M. Salamanova (✉) · D. Medjidov · A. Uspanova
Department of Construction Production Technology, Grozny State Oil Technical University named after Academician M.D. Millionshtchikov, 100, KH. Isaev Avenue, Grozny 364051, Russia
e-mail: madina_salamanova@mail.ru

M. Salamanova
Division for Materials Science, Kh. Ibragimov Complex Institute of the Russian Academy of Sciences, 21, Staropromyslovskoe Highway, Grozny 364051, Russia

P. Akimov et al. (eds.), *Proceedings of FORM 2022*, Lecture Notes in Civil Engineering 282, https://doi.org/10.1007/978-3-031-10853-2_5

1 Introduction

Modern construction is intensively developing and special attention is paid to the construction of high-rise buildings and structures, since their presence determines the prestige and attractiveness of the city, region, and country. Modeling and creation of high-rise objects requires the improvement of technological processes and building products that guarantee the safety and operational reliability of buildings. Monolithic construction, using high-strength concrete composites of a new generation, based on high-quality materials, mineral powders and effective superplasticizers, contributes to the development and achievement of a high level of high-rise technology [1–4, 11, 13].

Such well-known materials scientists as Yu.M. Bazhenov, V.S. Demyanova, V.I. Kalashnikov were engaged in the development of formulations for high-strength concretes of class B40 and higher [1, 2, 5]. The industrial experience of domestic high-rise construction is associated with the construction of such unique facilities as the ultra-modern Moscow City business center (Moscow), the Lakhta Center public and business complex (St. Petersburg), the Grozny City multifunctional complex (Grozny). For several years, leading research and development organizations have been working on the development of recipes for high-quality and high-strength concretes with a strength class of B40–B150 for the construction of the Akhmat Tower skyscraper (Grozny) [6–9].

Employees of the Grozny State Oil Technical University named after academician M.D. Millionshchikov contribute by designing high-strength concrete composites with physical, mechanical and technical and economic indicators, using local natural and technogenic raw materials in combination with modern chemical modifiers. It is the complex use of organic and mineral compounds as additives in concrete that will increase by several orders of magnitude indicators of strength, resistance to various aggressive environments, density, frost resistance, etc.

Mineral additives play a certain role in the composition of the concrete composition, and, depending on the degree of dispersion and reactivity, they participate in the processes of structure formation, increasing the strength characteristics and properties of the cement stone as a whole [10–12]. To obtain high strength concrete composites, it is necessary to increase the consumption of Portland cement, but subsequently problems such as alkaline corrosion, high exotherm, shrinkage, and creep are observed. To eliminate these shortcomings and obtain a dense and impermeable structure of cement stone, it is desirable to use expanding additives of aluminosilicate nature, which help to neutralize the effect of high cement consumption and create prestress in the hardening structural element due to the formation of ettringite needle crystals and low-basic hydrosilicates and aluminates calcium, thereby regulating volumetric and linear changes in the process of formation of a single conglomerate.

In addition to active mineral additives, it is recommended to use finely dispersed filler powders that do not interact with clinker miners, but fill intergranular voids and cavities in the concrete structure, creating a more dense and impermeable composite. Microfillers are made by fine grinding of rocks of sedimentary origin and technogenic

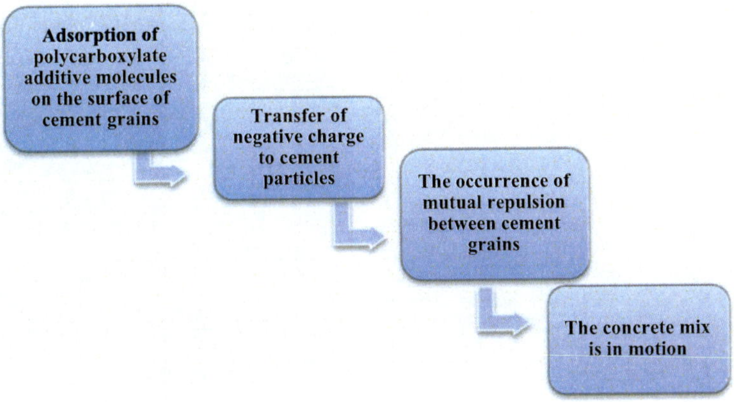

Fig. 1 The principle of operation of polycarboxylate superplasticizer

raw materials, which include limestone, dolomite, quartz sand, concrete scrap, broken brick, ash and slag waste, etc.

And of course, effective superplasticizers based on polyaryls, polycarboxylates or acrylates play a special role in obtaining high-strength concretes, which significantly reduce the water demand of the concrete mixture and the consumption of Portland cement. The mechanism of the steric action of modifiers based on polycarboxylate esters can be represented as the following scheme (Fig. 1).

Polyaryls have a deeper and more complex effect on the rheological characteristics of the concrete mixture, such as viscosity, fluidity, ultimate shear stress, etc., creating favorable conditions for the dissolution of clinker minerals and mineral powders in an aqueous medium. The chemical additive MasterPolyheed is synthesized on the basis of polyaryl ether, reducing the water-cement ratio without compromising the viscosity of the concrete mixture, leads to a significant reduction in energy and labor costs, which is associated with a reduction in the cost of the composition and simplification of the process of its laying. It should also be noted the stability of MasterPolyheed regardless of changes in the mineral composition and properties of Portland cement and coarse and fine aggregates. The listed qualities of the polyaryl additive guarantee the stability of the production process, the ability to use inert materials of various qualities, and the quality of the surface of concrete structures is significantly improved.

2 Materials and Methods

To study the complex effect of the organic and mineral components on the properties of concrete, compositions with chemical modifiers and finely dispersed mineral powders of various nature were modeled. MasterGlenium 115 based on polycarboxylate esters (pH 9–11) and Master-PolyHeed 3043 based on polyaryl esters (5–8)

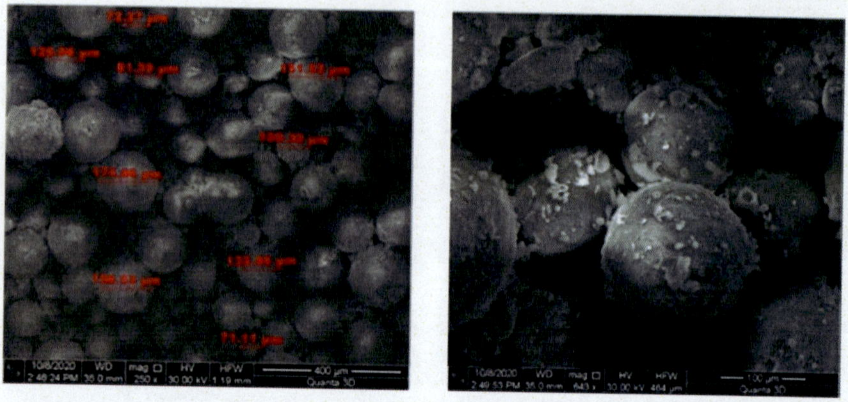

Fig. 2 Micrographs of microsilica particles

were used as chemical additives. Condensed microsilica MK-85 was used as an active mineral additive, the oxide composition is mainly represented by SiO_2 up to 90–92%, Fig. 2 shows microphotographs of micro-silica particles characterizing the spherical, rounded shape of grains in the form of microspheres. As mineral filler powders, finely divided additives from crushed bricks and limestone with a specific surface area of 360 and 245 m^2/kg, respectively, were used. Micrographs of lime and brick dust particles indicate closed porosity and angularity of grains (Fig. 3). Chemical composition of mineral fillers:

- lime powder in %: $MgO = 0.72$; $Al_2O_3 = 1.55$; $SiO_2 = 5.05$; $K_2O = 0.60$; $CaO = 90.14$; $Fe_2O_3 = 1.40$; $SO_3 = 0.49$; loss on ignition = 0.05.
- brick powder in %: $MgO = 2.32$; $Al_2O_3 = 15.61$; $SiO_2 = 58.88$; $K_2O = 2.74$; $CaO = 12.2$; $Fe_2O_3 = 6.44$; $SO_3 = 0.48$; $Na_2O = 1.34$.

As a binder, Portland cement of the M500 strength grade was used without additional manufacturer joint-stock company "Chechencement", the beginning and end of setting were 2 h 15 min and 3 h 40 min, respectively; the normal density of the cement paste is 25%. Mineralogical composition, %: C_3S—59; C_2S—16; C_3A—8; C_4AF—13. Aggregate was used imported from the Alagirskoe deposit: granite-diabase crushed stone of a fraction of 5–20 mm with a crushing grade in the range of 1200–1400; quartz sand with fineness modulus 2.8.

Concrete formulations were selected by calculation and experiment, the consumption of molding sand components is shown in Table 1, the dosage of chemical additives varied within 1.2–1.8% of the mass of Portland cement.

Sample cubes 10 cm in size were made from the designed compositions (9 samples of each composition), which hardened under normal humidity conditions and were tested on a P-100 hydraulic press at certain intervals according to the methods of GOST 10,180-2012 Concrete. Methods for determining the strength of control samples; GOST 12,730.1-78 Concrete. Density determination methods and the test results are presented in Table 1 and Fig. 4.

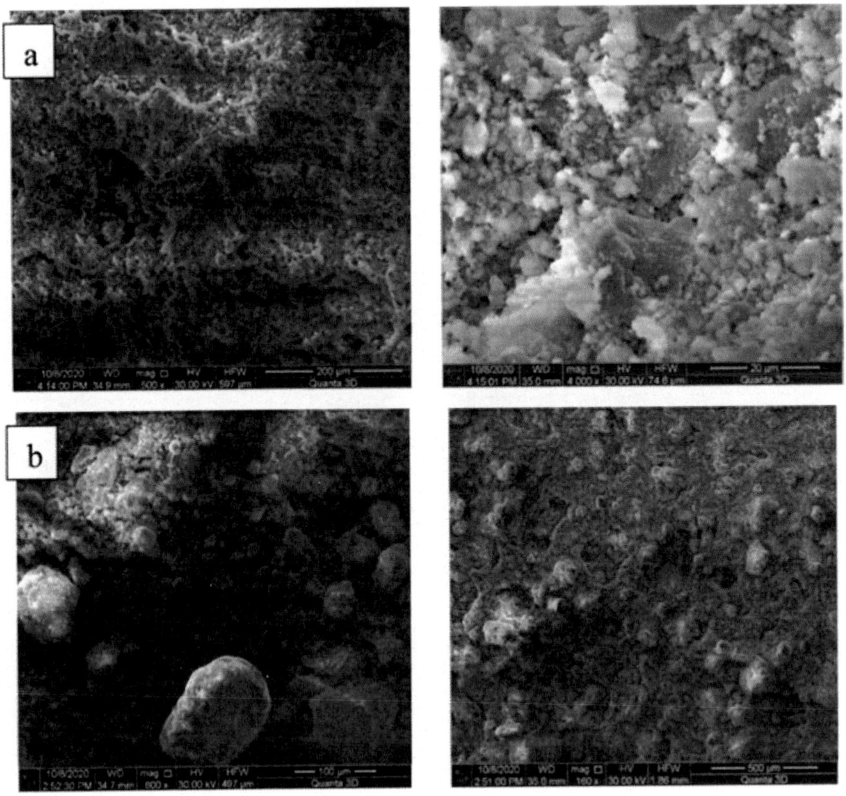

Fig. 3 Micrographs of limestone particles **a** and broken bricks **b**

Table 1 Recipes and physical and mechanical properties of modified concrete

Composition	Consumption, kg/m³									Water-cement ratio	Density of the concrete mixture, kg/m³	Tensile strength MPa, aged day		
	Cement	Sand	rubble	Water	Lime powder	Microsilica	Brick powder	MasterPolyHeed 3043	MasterGlenium 115			3	7	28
1	460	870	850	193	80	25	-	5.5	-	0.42	2389	23.4	44.3	53.8
2	450	870	860	176	80	25	-	8.1	-	0.39	2426	25.9	49.5	60.1
3	460	870	850	197	80	25	-	-	5.5	0.43	2395	24.3	44.7	58.4
4	450	870	860	180	80	25	-	-	8.1	0.40	2440	29.2	57.5	67.3
5	465	870	870	195	-.	25	80	5.6	-	0.42	2360	24.4	44.8	53.1
6	465	870	870	200	-	25	80	-	5.6	0.43	2372	25.1	48.4	53.3

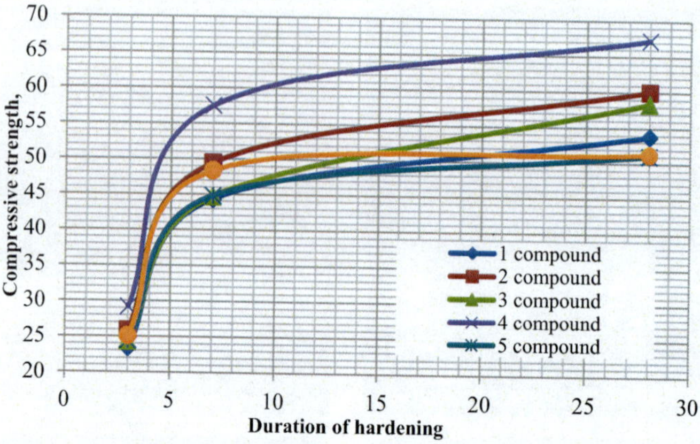

Fig. 4 Strengthening dynamics

3 Results and Discussions

The studied compositions of high-strength self-compacting concretes are characterized by high rates of strength development; in the first 7 days, the samples gained more than 70% of the design strength. The complex use of limestone flour, microsilica and MasterGlenium 115 (1.8%) at the age of 28 days contributed to the achievement of a maximum strength of 67.3 MPa (Fig. 4).

It should be noted that the developed concrete compositions correspond to concrete strength classes B40–50, the mobility of concrete mixtures was obtained P5 (cone flow 57–62 cm). The results of the research made it possible to establish the dependence of the strength of concrete on the types of chemical modifiers and mineral filler used (Fig. 5), the complex use of lime powder with a specific surface area of 245 m^2/kg and the MasterGlenium 115 polycarboxylate additive increases the strength by up to 10% in comparison with MasterPolyHeed 3043 polyaryl, and the composition with brick powder showed similar results.

The studied compositions of high-strength self-compacting concretes are distinguished by high rates of strength development; in the first 7 days, the samples gained more than 70% of the design strength. The complex use of lime powder, microsilica and MasterGlenium 115 (1.8%) at the age of 28 days contributed to the achievement of a maximum strength of 67.3 MPa (Fig. 4).

It should be noted that the developed concrete compositions correspond to concrete strength classes B40–50, the mobility of concrete mixtures was obtained P5 (cone flow 57–62 cm). The results of the research made it possible to establish the dependence of the strength of concrete on the types of chemical modifiers and mineral filler used (Fig. 5), the complex use of limestone flour with a specific surface area of 245 m^2/kg and the MasterGlenium 115 polycarboxylate additive increases the strength by up to 10% in comparison with MasterPolyHeed polyaryl 3043, and the composition with brick powder showed similar results.

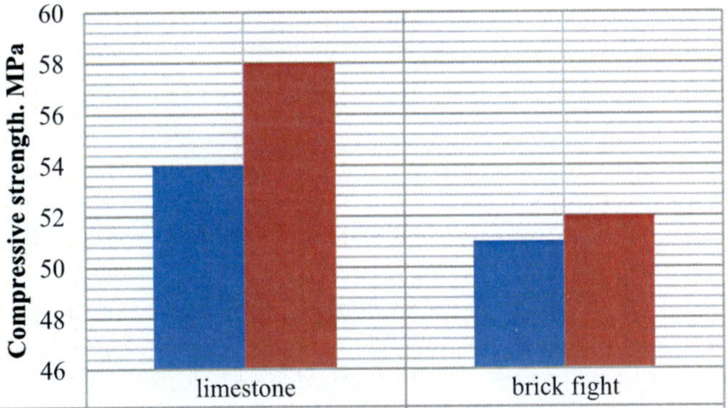

	limestone	brick fight
■ MasterPolyHeed	54	51
■ MasterGlenium	58	52

Fig. 5 Dependence of the influence of types of superplasticizer and microfiller on the strength of concrete

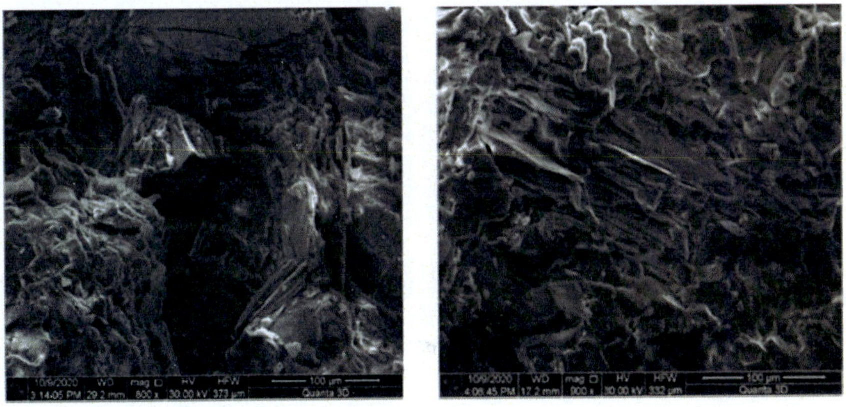

Fig. 6 Micrographs of modified concrete (composition 4)

Concrete samples with a composition of 4 28 days old were examined using a Quanta 3D 200 i scanning electron microscope (Fig. 5), microphotographs at a magnification of 800–900 times indicate the formation of a dense fine-grained structure, lamellar aggregates indicate the presence of low-basic calcium hydrosilicates, needle-like crystals confirm the presence of calcium hydroaluminates. Neoplasms of cement stone provide guaranteed strength and predict the durability of concrete composites (Fig. 6).

The conducted studies have shown that the complex use of the organic and mineral components in the composition contributed to the production of high-strength concretes, the fine powder of lime powder and the MasterGlenium 115 polycarboxylate additive create an impermeable and dense structure, calcium carbonate in a finely divided state reinforces the composition, monolithizing it and thereby increasing the strength of concrete. The chemical modifier MasterPolyHeed 3043 is slightly inferior in properties to MasterGlenium 115, but its cost is higher by several orders of magnitude, and its use in concrete is not economically viable.

4 Conclusions

The conducted studies have shown that the complex use of the organic and mineral components in the composition contributed to the production of high-strength concretes, the fine powder of lime powder and the MasterGlenium 115 polycarboxylate additive create an impermeable and dense structure, calcium carbonate in a finely divided state reinforces the composition, monolithizing it and thereby increasing the strength of concrete. The chemical modifier MasterPolyHeed 3043 is slightly inferior in properties to MasterGlenium 115, but its cost is higher by several orders of magnitude, and its use in concrete is not economically viable.

References

1. Bazhenov YuM, Demyanova BC, Kalashnikov VI (2006) Modified high quality concretes. ACB, Moscow, p 289
2. Bazhenov YuM (2009) Concrete: technologies of the future. Construction: new technologies - new equipment. ID “Panorama”, No 8, Moscow, pp 29–32
3. Hillemeier B, Buchenau G, Herr R, Huttl R, Klubendorf St, Schubert K Spezialbetone (Betonkalender Ernst &Sohn, No 1), pp 534–549
4. Murtazaev SAYu, Salamanova MSh, Bisultanov RG (2015) Influence of finely dispersed micro-fillers from volcanic ash on the properties of concrete. Collection of articles of the international scientific-practical conference dedicated to the 95th anniversary of GSTOU named after acad. M.D. Millionshchikov, 24–26 March, Grozny, vol 1, pp 171–176
5. Kalashnikov VI (2008) Industry of non-metallic building materials and the future of concrete. Constr Mater (3):20–24
6. Garkavi MS, Yakubov VI (2006) Crushing screenings - an effective way to improve the quality of concrete. Constr Mater (11):13–17
7. Murtazaev SAYu, Salamanova MSh, Saidumov MS, Gishlakaeva MI (2011) The use of rock processing waste in fine-grained concrete. In: Proceedings of the All-Russian scientific and practical conference “science and education in the Chechen Republic: status and prospects”, dedicated to the 10th anniversary of the formation of the KNII RAS, Grozny, pp 181–184
8. Shlain IB (1985) Development of deposits of non-metallic raw materials. Nedra, Moscow, p 344
9. Nakhaev MR, Salamanova MSh, Ismailova ZKh (2020) Regularities of the processes of formation of the structure and strength of a clinker-free binder of alkaline activation. Constr Mater Prod (3):21–29
10. Kaprielov SS (2008) Modified high-strength concretes of classes B80 and B90 in monolithic structures. Constr Mater (3):9–13
11. Salamanova MSh, Saidumov MS, Murtazaeva TSA, Khubaev MSM (2015) High-quality modified concretes based on mineral additives and superplasticizers of various nature. Innov Invest (8):159–163

12. Salamanova MSh, Alashanov AKh, Gabashev AA, Movsulov MM (2021) Development of concretes modified on polycarboxylate superplasticizer. Sci Tech J Bull GSTOU Tech Sci XVII(1(23)):76–82
13. Salamanova MSh, Murtazaev SAYu, Alashanov AKh, Murtazaeva TSA (2021) Prospects for the use of cement industry waste to obtain modern concrete composites. Constr Mater (5):54–63

Epoxy Materials Filled with Buckwheat Husk Ash

Elena Gotlib, Rutthe Schelton Ntsoumou, Alina Valeeva, Engel Galimov, and Alla Sokolova

Abstract The present paper is dedicated to utilizing of agricultural by-products as the fillers for polymer materials, namely buckwheat husk ash. This material is an environmentally-green large tonnage by-product, though the scientific developments of its application are rare and not practically entrenched. Application of buckwheat husk ash in the composition of epoxy materials could be seen as rather upcoming trend as it enables to obtain the coating with enhanced wear resistance and improved tribological properties. It was proven by the authors that applying buckwheat husk ash as a filler of epoxy compositions is an upcoming trend as it improves physical and mechanical properties of compositions such as wear resistance by 37%, hardness by 3%. At the same time, the coefficient of static friction is reduced by 54%. It was concluded that the optimal temperature of obtaining buckwheat husk ash ensuring its maximum modifying effect is 800 °C.

Keywords Epoxy composites · Buckwheat husk ash · Modification · Filler

1 Introduction

Protection of natural environment and effective use of secondary resources are gaining momentum and becoming ever-more important for building a circulation economy.

Along with that, agricultural by-products are considered environmentally-green materials, whereas the scale of their manufacturing is augmenting from one year to

E. Gotlib · R. S. Ntsoumou
Kazan National Research Technological University, 68 Karl Marx Street, Kazan 420015, Russian Federation

A. Valeeva · E. Galimov
Kazan National Research Technical University named after A.N. Tupolev – KAI, 10 Karl Marx Street, Kazan 420111, Russian Federation

A. Sokolova (✉)
Moscow State University of Civil Engineering, Moscow 129337, Russian Federation
e-mail: as.falconi@yandex.ru

P. Akimov et al. (eds.), *Proceedings of FORM 2022*, Lecture Notes in Civil Engineering 282, https://doi.org/10.1007/978-3-031-10853-2_6

another. Thus, in Russia the by-products of buckwheat grinding are accumulated in the amount of 122 000 tons annually [1]. It is generally recognized that buckwheat is one of the most important national grocery products in Russia, being the essential component of childhood and dietary nutrition [2].

At the same time, the scientific developments in relation to utilizing of the by-products of buckwheat production are rather scarce and, for the most part, are not practically entrenched [3].

Due to this, proficient utilizing of buckwheat husk as a raw material for manufacturing fillers for polymer materials could be seen as a relevant objective of "green chemistry".

2 Materials and Methods

In order to obtain epoxy polymers, there was used the diane resin of the grade ED-20 (the All-Union Standard GOST 10,587-84) hardened by aminoalkylphenol (APh-2) (Technical Conditions 2494-052-00,205,423-2004) during 7 days at room temperature. The content of the hardener was determined by the equimolar ratio [epoxygroups]:[amine].

As a filler, there was applied: buckwheat husk ash (BHA) obtained from the buckwheat husk, the grinding was preliminarily done on the rotor cutting mill RM120, then buckwheat husk was burnt in the blind roaster SNOL 3/11 at the temperatures: 350, 500 and 800 °C and screened through the sieve with the mesh size of 64 mcm.

The content of buckwheat husk ash in the composition accounted for 10% wt. per 100% wt. of epoxy polymer.

The content of mineral components in the composition of buckwheat husk ash was investigated on the X-ray fluorescence spectrometer Rigaku ZSX Primus II.

pH of the fillers' surface was determined by the pH—meter produced by SevenMulti.

Friction coefficient was estimated by the automated friction apparatus «Tribometer, CSM Instruments» (Switzerland) (ASTM G99–959, DIN50324 and ISO 20808). Linear velocity at testing was accounted for 8.94 cm/sec, sampling frequency—10 Hz, temperature—50 °C, and humidity—20%.

Testing on abrasive resistance of epoxy materials was done on the vertical optimeter ISV-1 at the following mode: specific pressure of the counter body on the tested sample surface P = 1 MPa, slip velocity V_{sl} = 1 m/sec, no lubrication.

Hardness was evaluated on the portable Shore hardness tester according to the D-scale (GOST 24,621-91).

Electronic microscopic analysis was carried out on the focused-beam microscope Jeol JSM7001F by using the energy- dispersive detector Oxford INCA X-max 80 for evaluation of the elemental composition.

Sol–gel analysis was implemented in the Soxhlet-apparatus in the boiling acetone during 6 consecutive hours.

The area of pores' specific surface was determined under the method BET (Brunauer, Emmet and Teller method) ISP 9277, on the analyzer of surface area and pores size "Nova 1200e" ISO 15901-2. Sample's degasation was done during 3 h at the temperature 150 °C in the vacuum 2 Pa.

X-ray diffraction quantitative phase analysis (Q-PXRD) was done by means of multifunctional X-ray diffraction meter Rigaku Smart Lab, at the following parameters: angular spacing from 30 to 650, with scanning step 0.02, exposal 1 s in a point, the scanning rate 10 °C/min. Handling and deciphering of X-ray diffraction patterns was implemented by the software package PDXL-2 and the database ICDD PDF-2.

3 Results and Discussion

The data obtained by X-ray fluorescence analysis have shown (Table 1) that elemental composition of the buckwheat husk ash's mineral part does not essentially depend on the temperature of its manufacturing.

In fact, when the temperature of husk burning grows, some decrease in the chlorine's content takes place, whereas the concentration of other elements varies within the measurements accuracy for the given method (+_5%).

BHA contains noticeable quantities of magnesium, potassium, and most of all, calcium. At the same time, the presence of silicon is not large (Table 1).

Elemental composition of buckwheat husk should, naturally, have a significant impact on the modifying effect in epoxy composites. This is related to the fact that in the process of epoxy oligomers hardening at room temperature in the presence of ash, no change of elemental composition of the given filler can occur [4].

Buckwheat husk particles have relatively small thickness 0.13–0.18 mm and rather large particle size, it has high specific surface that preconditions good affinity to epoxy binder. The analysis of BHA-structure by the method of scanning electronic microscopy has shown (Fig. 1) that BHA has imperfect surface with raised lumps

Table 1 The content of mineral components in the composition of buckwheat husk ash obtained at various temperatures

Filler	Na	Mg	Al	Si	P	S	Cl	K	Ca	Mn	Fe	Cu	Zn
BHA 350 °C	–	9.59	0.75	2.10	3.21	1.87	1.96	61.17	20.18	1.13	1.50	0.30	0.99
BHA 500 °C	0.25	10.65	0.32	1.93	3.51	1.78	1.34	65.63	18.46	0.56	1.54	0.5	0.06
BHA 800 °C	0.33	11.51	0.55	1.81	3.2	1.77	1.26	67.04	18.57	0.7	1.42	0.8	0.02

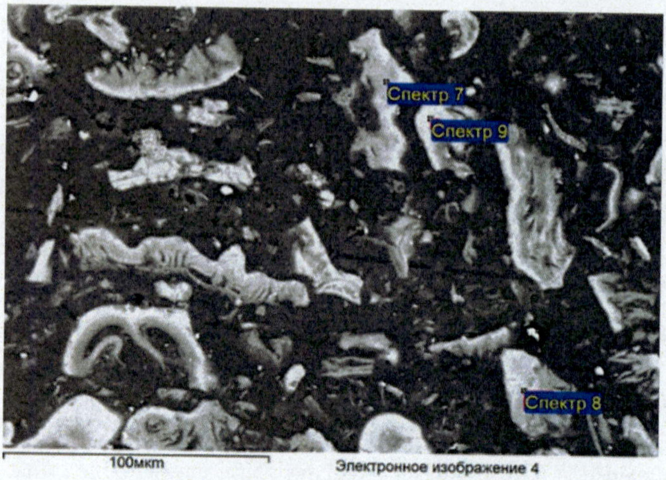

Fig. 1 Electronic microphotograph of buckwheat husk ash (image magnification 10^6)

formed by the fibrils oriented longitudinally or transversely. As a whole, volumetric microporous fibrous structures is formed [5].

Buckwheat husk ask has elongate, petaled inclusions (Fig. 1) that can give micro reinforcing effect when applied as a filler for epoxy materials. Diffraction patterns of buckwheat husk samples are presented by wide, not evident peaks, i.e. it contains mostly so-called x-ray amorphous phase (Fig. 2).

As the nature of this amorphous phase is unknown [6], it is impossible to estimate its content in terms of quantity. From the Q-PXRD data it follows that with the growth of burning temperature of buckwheat husk, the degree of crystallinity of its ash is increasing.

It is not realistic to single out crystalline phases in the composition of buckwheat husk, and the composition of crystalline part of its ash is presented in the Table 2. Phase composition of buckwheat husk ash has been selected according to the database ICDD PDF-2 and the results of X-ray fluorescence analysis.

Buckwheat husk ash has an alkaline nature of its surface (Table 3). Along with that, BHA obtained at the temperature 800 °C has higher values of aqueous dispersion pH. This is probably related to slightly higher content of K and Na in its composition (Table 1) that form alkali interacting with water.

The study of physical and mechanical properties of epoxy coatings not filled and filled with buckwheat husk ash (BHA) (Table 4) has shown that filling significantly lowers their tear.

Wear resistance of epoxy coatings filled with BHA does not noticeably depend on the burning temperature of buckwheat husk. However, with the growth of temperature of its obtaining, wear resistance of epoxy coating slightly decreases. Maximum lowering of wear resistance of epoxy coatings takes place when applying ash obtained at the temperature 800 °C.

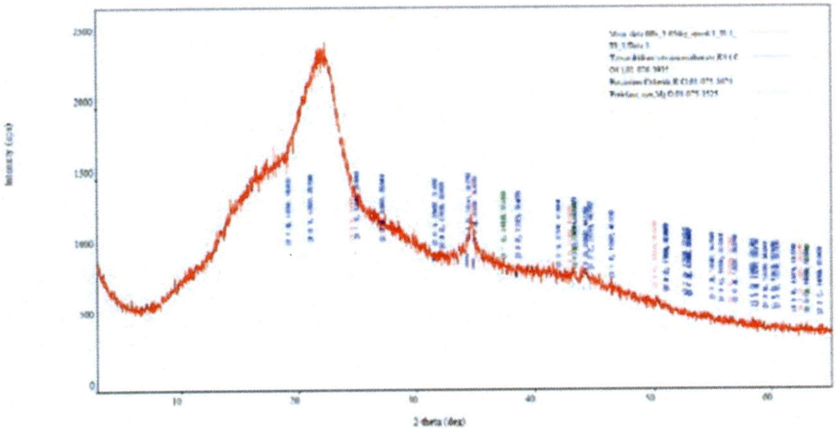

a) buckwheat husk

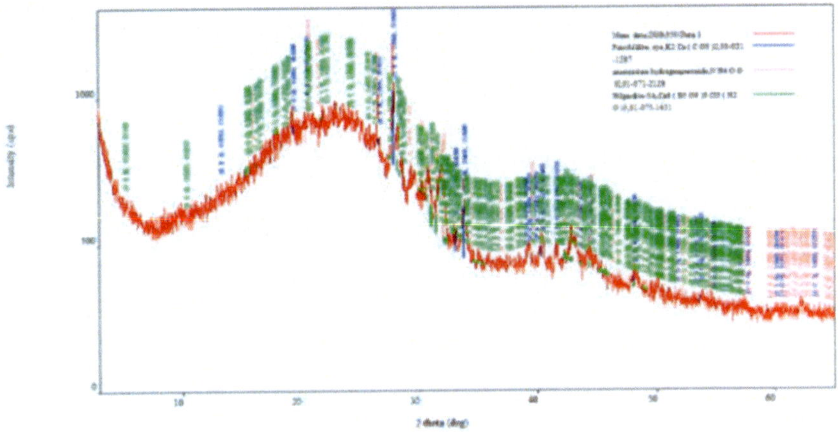

b) RHA_1

Fig. 2 X-ray diffraction patterns for the samples: **a** buckwheat husk, **b** RHA_1, **c** RHA_2, **d** RHA_3

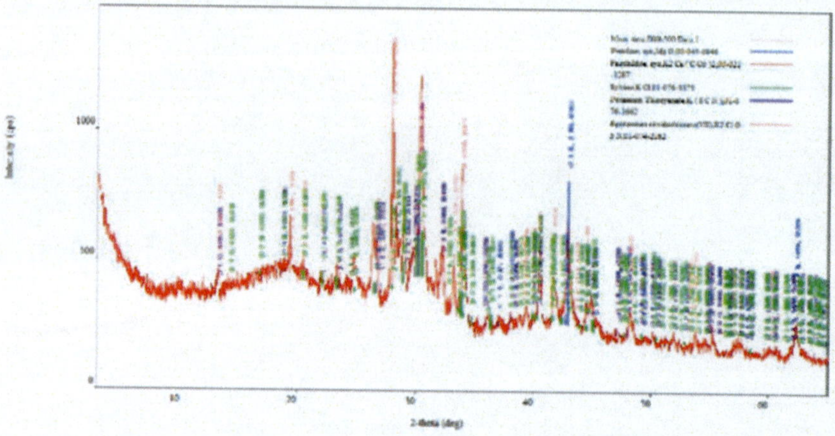

c) RHA_2

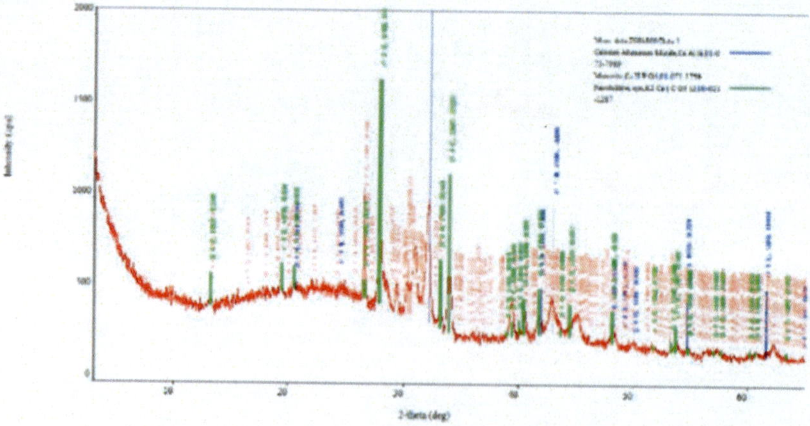

d) RHA_3

Fig. 2 (continued)

The effect of wear reduction is explained by the fact that the particles of metals oxides, being the part of BHA composition, prevent the development of strains in a polymer matrix, performing as "spikes" bonding structural elements of a cross-linked polymer and impede sliding along the plane of sliding [7, 8].

At the same time, buckwheat husk ahs obtained at 350 and 500 °C lowers the hardness of epoxy materials filled with it. In case of applying BHA obtained at 800 °C, there takes place insignificant growth of hardness of epoxy materials (Table 4).

The researched fillers obtained at the base of buckwheat husk ash at temperatures 500 and 800 °C enhance antifriction properties, i.e. significantly lower the coefficient of static friction of epoxy coatings (Table 4).

Table 2 Phase composition of buckwheat husk ash

Filler	Phase composition	Content, % wt
BHA_1	$K_2Ca(CO_3)_2$	51
	NH_4OOH	29
	$Ca_6(B_5O_9)_3Cl(H_2O)_3$	20
BHA_2	MgO	18
	$K_2Ca(CO_3)_2$	20
	KCl	1
	K(SCN)	4
	K_2ClO_3N	58
BHA_3	$K_2Ca(CO_3)_2$	2
	$CaHPO_4$	96
	$MgCO_3$	2

Table 3. pH of aqueous dispersions of buckwheat husk and their oil consumption

Type of a filler	Oil consumption, g/g	pH
BHA_1	53	9.91
BHA_2	61	9.98
BHA_3	68	11.41

Table 4 Performance characteristics of epoxy compositions filled with BHA

NN	Filler	Wear, $\times 10^{-6}$ m	Hardness, HSD	Coefficient of static friction	Bending strength (σ_f), MPa
1	Reference sample	19	46.01	0.39	64.49
2	BHA_1	13.5	38.86	0.38	16.60
3	BHA_2	12.5	43.22	0.19	−21.7
4	BHA_3	11.9	47.37	0.18	24.70

The maximum effect of this indicator's lowering is achieved at application of BHA is twofold. The BHA obtained at 350 °C practically does not affect the friction coefficient of epoxy materials (Table 4).

The growth of wear resistance of filled epoxy coatings is caused by the ash particles that perform as the local inhibitor of bonding and impede the cracks formation in a polymer matrix [9].

Reduction of friction coefficient and the rate of composites wear, according to the authors' opinion, depends on the effectiveness of interphase interactions of filler's particles with the matrix, damping capacity and resistance of filled materials to temperature strains [10].

Table 5 Porosity of buckwheat husk and its ash

Filler	Area of specific surface of pores, m^2/g
Buckwheat husk	0.146
BHA_1	0.719
BHA_2	1.045
BHA_3	3.564

These conditions ensure formation of homogeneous structure of filled epoxy composites with smaller internal stresses [11].

Thus, BHA obtained at the temperature 800 °C could be used as the most effective filler of epoxy coatings.

Bending strength of epoxy coatings is lowered at filling with BHA, at a lesser extent at its obtaining at more elevated temperature (Table 4).

As the presence of a developed porous structure is one of the key factors defining effectiveness of fillers at manufacturing of polymer materials, the greater modifying effect of BHA_3 is preconditioned by a larger specific surface of its pores (Table 5).

Actually, when the temperature of ash obtaining grows, its porosity consistently rises.

During the thermal action in cellulose-containing materials, such as buckwheat husk ash, two types of competing reactions are observed: condensation, with formation of carbocyclic structures and water, and decomposition leading to formation of low-molecular volatile compounds. This preconditions pores formation [9].

The impact of this parameter is explained by the fact that interaction of a filler and a binder occurs mainly due to sorption processes ensuring enhancement of performance properties of filled compositions.

In order to determine the completeness of epoxy material hardening, sol–gel analysis was applied. The investigation has shown that the content pf gel-fraction in epoxy compositions filled with buckwheat husk ash is decreased by 15% approximately (Fig. 3).

Thus, at the presence of BHA, less dense cross-linked structure is formed. Along with that, the elevation of burning temperature from 350 up to 800 °C increases the content of gel-fraction in epoxy compositions filled with it, i.e. facilitates complete hardening of an epoxy material.

This could be related with both reduction of the organic phase in the composition of buckwheat husk ash and activation of the process of epoxy ring opening due to the presence of active hydroxyl containing groups with the catalytic effect on the surface of buckwheat husk ash [12].

Thus, application of BHA in the composition of epoxy materials could be seen as rather upcoming trend as it enables to obtain the coating with enhanced wear resistance and improved tribological properties.

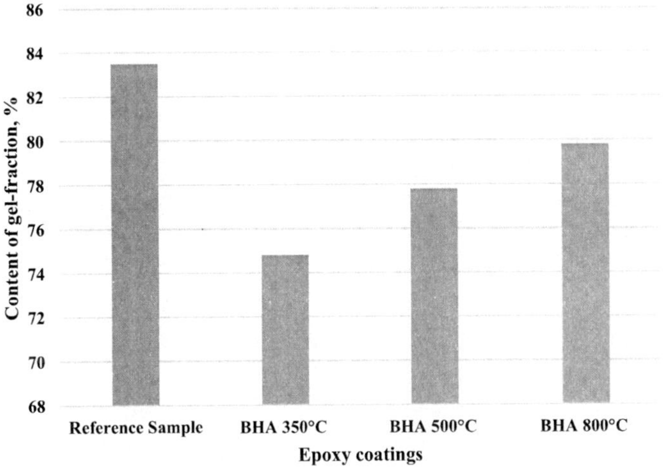

Fig. 3 The content of gel-fraction in epoxy compositions filled with buckwheat husk ash

4 Conclusions

Application of buckwheat husk ash as fillers of epoxy compositions improves their physical and mechanical properties, namely wear resistance by 37%, hardness by 3%. At the same time, the coefficient of static friction is reduced by 54%.

It was proved that the optimal temperature of obtaining buckwheat husk ash ensuring its maximum modifying effect is 800 °C.

References

1. Zabolotnaya AM et al (2015) Some aspects of complex technology of buckwheat husk processing. In: Proceedings of th III Int. Sc.-pract. Conference, Khanty-Mansyisk, part 1, pp 99–101
2. Sveshnikova ES, Chelysheva IA, Panova LG (2008) The use of by-products of agricultural industry for polymer filling. Plast Massy 1:29–21
3. Zemnukhova LA, Shkorina ED, Fedorishcheva GA (2005) Composition of inorganic components of buckwheat husk and straw. Russian J Appl Chem 78(2):324–328 (2005)
4. Pankeev VV, Nikiforov AV, Sveshnikova ES, Panova LG (2012) New fillers for epoxy compounds on the base of modified cellulose-containing by-products. Plast Massy 5:50–52
5. Gotlib E, Nya HPT, Nguyen TLA, Sokolova A, Yamaleeva E, Musin I (2019) Agricultural by-products as advanced raw materials for obtaining modifiers and fillers for epoxy materials. Key Eng Mater 822:343–349
6. Onishchenko DV, Chakov VV (2011) Renewable vegetable stock as the base for obtaining function nanocomposite materials with universal function. J Appl Chem 84(9):1562–1566
7. Nakamura Y et al (2019) Characterization of cellulose microfibrils, cellulose molecules, and hemicelluloses in buckwheat and rice husks. Cellulose 26:6529–6541

8. Yamaguchi T et al (2006) Friction and wear properties of new hard porous carbon materials made from Rice Chaff. In: Proceedings of the 3rd Asia international conference on tribology, Kanazawa, pp 379–380
9. Sveshnikova ES et al (2009) Use of agricultural waste materials for the filling of polymers. Int Polymer Sci Technol 36(5):15–19
10. Włoch M, Landowska P (2022) Preparation and properties of thermoplastic polyurethane composites filled with powdered buckwheat husks. Materials 15:356
11. Majewski L, Gaspar A (2018) Evaluation of suitability of wheat bran as a natural filler in polymer processing. BioRes 13(3):7037–7052
12. Valeeva AR, Gimranova AR, Gotlib EM, Galimov ER (2020) Anti-friction epoxy coatings modified with rice husk. IOP Conf Ser Mater Sci Eng 3:1–8

The Building Materials Production Process and Uneven Heating

Evgeny Radkevich, Olga Vasil'eva, and Georgiy Filippov

Abstract We consider the simplified building material production process based on autocatalytic reaction of two reagents in a finite cylinder. We investigate the effect of uneven heating of the mixture on the duration of the production process. To describe the process, we used a new recently proposed model taking into account the terms of the second order. The mathematical model under consideration is an initial boundary value problem for a nonlinear system of second-order partial differential equations with variable coefficients. We have obtained analytical stationary solutions to the problem, one of which is the equilibrium position, the achievement of which means the end of the process The influence of the uneven distribution of the mixture temperature is studied by numerical investigation of the specified mathematical model. For the numerical solution of the initial boundary value problem for a system of second-order equations with variable coefficients, a second-order finite-difference method is used. To study the obtained solution, quadrature formulas of the second order of accuracy and spline approximation methods are used. The results of the study of the effect of uneven heating of the mixture on the duration of the production process are presented. The obtained results are compared with the results obtained using other mathematical models of autocatalytic reactions. Based on the comparison of the results, it is concluded that taking into account the uneven heating of the mixture makes it possible to more accurately estimate the time of the production process for different criteria of its completion.

Keywords The building materials · Mathematical model · Production process

E. Radkevich
Faculty of Mechanics and Mathematics, Lomonosov Moscow State University, Vorobiovy gory , Moscow 119991, Russian Federation

O. Vasil'eva (✉) · G. Filippov
Moscow State University of Civil Engineering, Moscow 129337, Russian Federation
e-mail: vasiljeva.ovas@yandex.ru; vasilievaoa@mgsu.ru

O. Vasil'eva
Higher Chemical College, D. Mendeleev University of Chemical Technology of Russia, Miusskay Sq. 9, Moscow 125047, Russian Federation

P. Akimov et al. (eds.), *Proceedings of FORM 2022*, Lecture Notes in Civil Engineering 282, https://doi.org/10.1007/978-3-031-10853-2_7

1 Introduction

We consider the simplified building material production process based on autocatalytic reaction of two reagents in a finite cylinder. We investigate the effect of uneven heating of the mixture on the duration of the production process. The kinetic system of Carleman equations can be used for the mathematical description of this process [1–5]. To describe the process, we used a new model proposed and verified in the works [6, 7]. This model, unlike the kinetic Carleman system, takes into account the terms of the second order. Taking into account the terms of the second order makes it possible to correctly formulate the initial boundary value problem for a system of partial differential equations from a mathematical point of view and allows to describe the process more accurately [6, 7]. The mathematical model under consideration is an initial boundary value problem for a nonlinear system of second-order partial differential equations with variable coefficients. Therefore, obtaining an analytical solution of the system in the general case is a difficult task. We have obtained analytical stationary solutions to the problem, one of which is the equilibrium position, the achievement of which means the end of the process. To study the mathematical model in general, numerical solution methods based on the construction of a finite-difference scheme of second orders of accuracy are used. The numerical results obtained are compared with the results obtained using other mathematical models.

2 Problem Formulation

We consider the simplified building material production process based on autocatalytic reaction of two reagents in an finite cylinder [8–13]. Taking into account the final dimensions of the cylinder is not a fundamental difficulty, as the mathematical model under consideration allows us to correctly set the boundary conditions on the left and right sides of the cylinder [6, 7]. To simplify the analysis and comparison of the results obtained using different mathematical models [14–20], we formulate a mathematical model in the reduced variables. Deviations of the reagent densities from the equilibrium position are denoted by $u(t, x)$ and $w(t, x)$. The length of the cylinder in which the reaction takes place is set to 1. The particle velocities of the first and second components of the mixture due to the temperature of the mixture at a given point x are denoted by $c(x) > c_0 > 0$ and $-c(x)$. In these notations, the mathematical model has the form [6, 7]

$$\begin{aligned} \frac{\partial u}{\partial t} + c(x)\frac{\partial u}{\partial x} &= -\frac{1}{\varepsilon}(u^2 - w^2) + \alpha(\frac{\partial^2 u}{\partial x^2} - \frac{\partial^2 w}{\partial x^2}) \\ \frac{\partial w}{\partial t} - c(x)\frac{\partial w}{\partial x} &= \frac{1}{\varepsilon}(u^2 - w^2) - \alpha(\frac{\partial^2 u}{\partial x^2} - \frac{\partial^2 w}{\partial x^2}) \end{aligned} \quad (1)$$

the initial conditions

$$u(0,x) = u^0(x),$$
$$w(0,x) = w^0(x), \tag{2}$$

the boundary conditions

$$u(t,0) = 0,\ u(t,1) = 0,$$
$$w(t,0) = 0,\ w(t,1) = 0. \tag{3}$$

Here α constant small parameters, initial and boundary conditions satisfy the following relations

$$u(0,0) = u^0(0),$$
$$w(0,0) = w^0(0), \tag{4}$$

$$\int_0^1 u^0(x)dx = 0,\ \int_0^1 w^0(x)dx = 0. \tag{5}$$

We obtain stationary solutions of the initial boundary value problem. Adding up the equations of the system, we get

$$(u - w)_x = 0,\ u = w + c_1 \tag{6}$$

and finally we have two types of stationary solutions to the problem

$$u(t,x) = \frac{1}{2c_1} e^{\frac{2c_1}{\varepsilon}\int\frac{dx}{c(x)}} + \frac{c_1}{2}, \tag{7}$$

$$w(t,x) = \frac{1}{2c_1} e^{\frac{2c_1}{\varepsilon}\int\frac{dx}{c(x)}} - \frac{c_1}{2}, \tag{8}$$

And

$$u(t,x) = 0,$$
$$w(t,x) = 0. \tag{9}$$

The first solution does not satisfy the boundary conditions of the problem. Thus, we have the only trivial stationary solution, which is the equilibrium position of interest to us. We introduce the notation of the deviation of the solution from the equilibrium position

$$Du(t) = \max_{0\le x\le 1}|u(t,x) - 0|,\ Dw(t) = \max_{0\le x\le 1}|w(t,x) - 0|, \tag{10}$$

And

$$e_1(t) = \Big(\int_0^1 (u(t,x) - 0)^2 dx\Big)^{\frac{1}{2}},\ e_2(t) = \Big(\int_0^1 (w(t,x) - 0)^2 dx\Big)^{\frac{1}{2}},\ E(t) = (e_2^2(t) + e_2^2(t))^{\frac{1}{2}}. \quad (11)$$

We consider two criteria for terminating the process. The first one, the reaction is considered complete if the maximum of deviations $Du(t)$, $Dw(t)$ does not exceed the specified value γ. The second one, $E(t)$ does not exceed the specified value γ. Thus, the time of the process in accordance with the selected criterion is the minimum time $T_i(\gamma)$, after which the selected criterion is executed [15, 16].

3 Numerical Results

We consider the case of a linear dependence of the velocity of reagent particles on a spatial variable $c(x) = c_0 + \beta x$, and the initial conditions for the case of finding the density of particles in the opposite phase

$$\begin{aligned} u(0, x) &= -0.2 \sin(2\pi x), \\ w(0, x) &= 0.3 \sin(2\pi x). \end{aligned} \quad (12)$$

We consider the mathematical model for values of the parameter $\varepsilon = 0.27$, $\alpha = 0.27$, $c_0 = 0.5$ and various parameter β values.

In Fig. 1, we can see the change in time of the maxima of the modules of the deviation of the solution components from the equilibrium position. The curves correspond

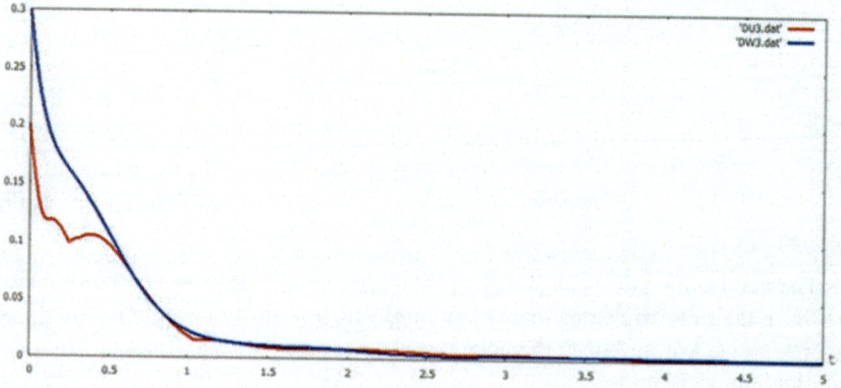

Fig. 1 Changing the maxima of modules of the solution deviation from the equilibrium position $Du(t)$, $Dw(t)$

to the value of the parameter $\beta = 0.5$. Here we see a practically monotonous tendency to zero of the maxima of the modules of both components of the solution.

Figure 2 shows the change in the deviation "energy" of the solution components $e_1(t)$ and $e_2(t)$ for $\beta = 0.5$ In Fig. 3 we can see the trajectories in the phase plane $e_1(t)$, $e_2(t)$ for previous case.

Figure 4 shows change in the total "energy" of deviation of the solution of the problem from the equilibrium position $E(t)$ for this case.

Figure 5 shows the change in the total "energy" of the deviation of the solution of the problem from the equilibrium position for different values of p. The curves $E1$, $E2$, $E3$ and E4 correspond to the following values of the parameter β: 0, 0.25, 0, 0.5, 1. In the figure you can see that the rate of reduction of the total "energy" at the beginning of the reaction ($0 < t < 0.1$) does not depend on the parameter β. Although at $0 < t < 0.1$, the rate of reduction of the total "energy" reaches a maximum.

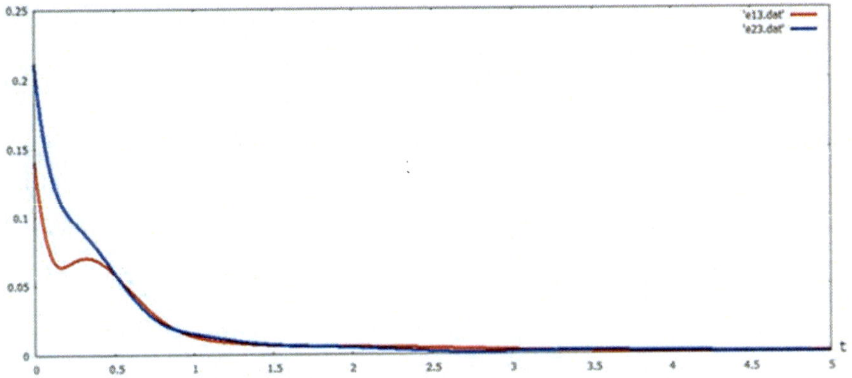

Fig. 2 Changing the solution deviation from the equilibrium position $e_1(t)$, $e_2(t)$

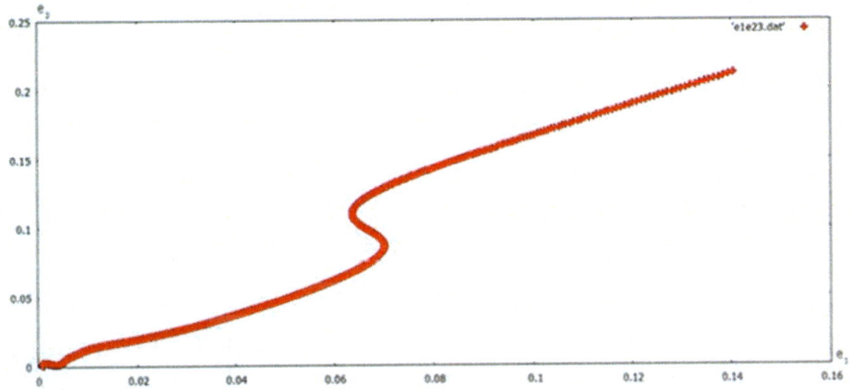

Fig. 3 Trajectories in the phase plane $e_1(t)$, $e_2(t)$

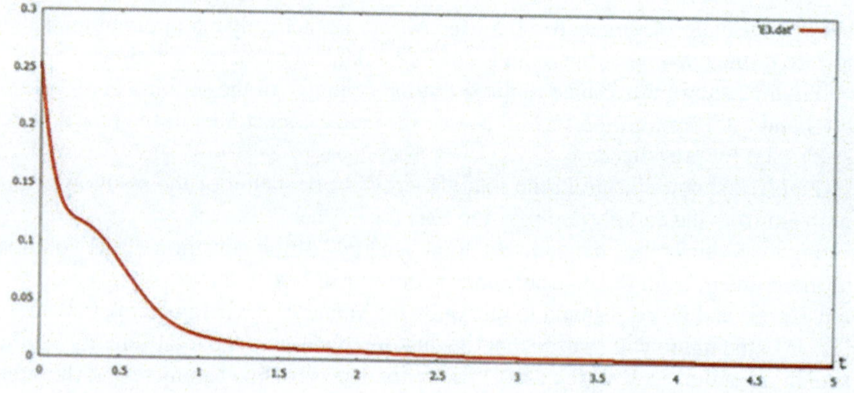

Fig. 4 Total "energy" of deviation $E(t)$

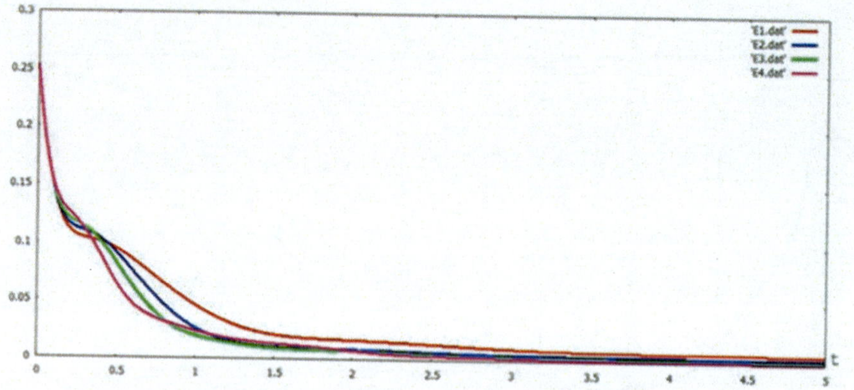

Fig. 5 Total "energy" of deviation $E1(t)$, $E2(t)$, $E3(t)$, $E4(t)$ for $\beta_1 = 0$, $\beta_2 = 0.25$, $\beta_3 = 0.5$, $\beta_4 = 1$

Figure 6 shows the dependences of the reaction time on β, obtained using the first criterion for $\gamma = 0.015$ and $\gamma = 0.075$. Figure 7 shows the dependences of the reaction time on β, obtained using the second criterion for $\gamma = 0.015$ and $\gamma = 0.075$.

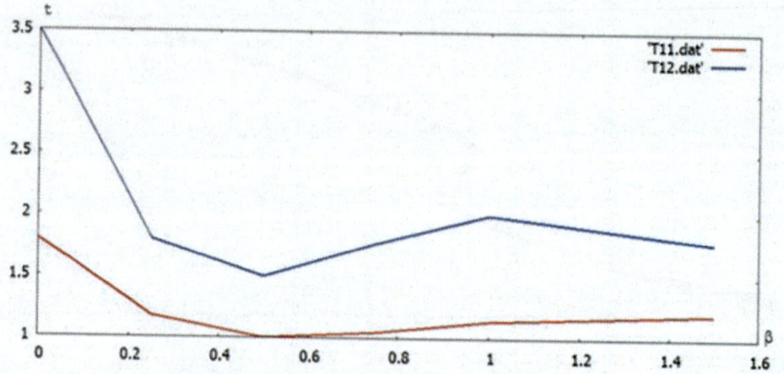

Fig. 6 Dependence reaction completion time on β, according to the first criterion for $\gamma = 0.015$ and $\gamma = 0.0075$

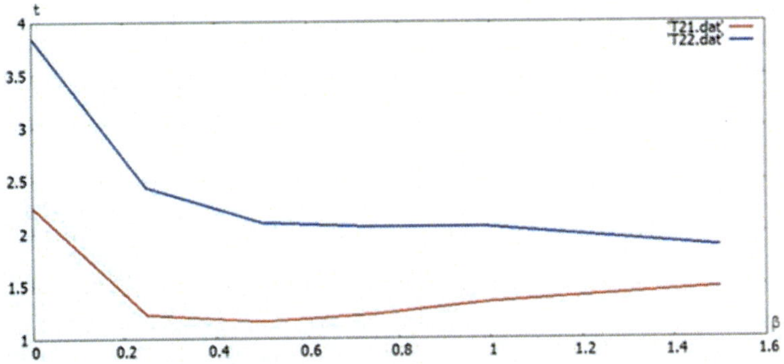

Fig. 7 Dependence reaction completion time on β, according to the second criterion for $\gamma = 0.015$ and $\gamma = 0.0075$

4 Conclusion

We consider the simplified building material production process based on autocatalytic reaction of two reagents in an finite cylinder. The construction of a mathematical model describing this process is carried out taking into account the dependence of the temperature of the mixture on the spatial coordinate. The mathematical model is an initial boundary value problem for a system of two nonlinear partial differential equations of the second order. The existence of a unique stationary solution of the problem satisfying the given boundary conditions and being the equilibrium position for the process under consideration is proved. A numerical study of the mathematical model is carried out. Examples of the numerical results obtained are given and discussed.

References

1. Godunov SK, Sultangazin UM (1971) Russ Math Surv 26:1–56
2. Carleman T (1957) Publications Scientifiques de l'Institut Mittag–Leffler. Almqvist & Wiksells, Uppsala, p 112
3. Aristov VV, Ilyin OV (2010) AIP Conf Proc 649–54
4. Ilyin OV (2007) Comput Math Math Phys 47:1990–2001
5. Illner R, Reed MC (1084) SIAM J Appl Math 44:1067–1075
6. Vasil'eva OA (2020) J Phys Conf Ser 012157
7. Vasil'eva OA (2021) IOP Conf Ser Mater Sci Eng 1030:0120009
8. Radkevich EV, Vasil'eva OA, Dukhnovskii SA (2015) J Math Sci 207:296–323
9. Vasil'eva OA, Dukhnovskii SA, Radkevich EV (2018) J Math Sci 235:392–454
10. Radkevich EV, Vasil'eva OA (2017) J Math Sci 224:763–795
11. Vasil'eva OA (2015) Proc Moscow State Univ Civ Eng 7–15. (in Russian)
12. Euler N, Steeb W-H (1989) Aust J Phys 42:1–10
13. Vasil'eva OA (2017) Matec VebConf 117:00174
14. Dukhnovskii SA (2017) Construction. Sci Educ 7:10–18 (in Russian)

15. Dukhnovskii SA (2021) J Math Sci 259:349–375
16. Dukhnovskii SA (2016) Proc Moscow State Univ Civil Eng 11:7–14. (in Russian)
17. Dukhnovskii SA (2017) J Samara State Tech Univ Ser Phys Math Sci 21:7–41
18. Aristov VV (2001) Kluwer Academic Publishing, p 312
19. Vasil'eva OA (2019) E3S Web Conf 97:02011
20. Dukhnovskii SA (2019) Moscow Univ Math Bull 74:246–248
21. Samarskii AA (1977) The theory of difference schemes, Moscow, p 666. (in Russian)

Behavior of Crane Beams in Bending and Restraint Torsion

Olga Tusnina

Abstract A common solution in foreign worldwide practice is the use of crane beams without brake bracing structures or only with braces to the upper flange of the beam in the support zones. In this case, when lateral forces from the crane act on the thin-walled crane beam of an open profile (as usual it is I-beam), it experiences restrained torsion, and requires a special approach for the analysis of its stress–strain state. The article considers the calculation of a crane I-beam without a brake structure and with braces to the upper chord in the support zones. It is shown that the account of restraint torsion for a free I-section crane beam is mandatory. For this case of combined bending and restraint torsion of the beam, finite element analysis with the use for modeling of the beam either simple beam finite elements with 7 degrees of freedom or plate elements shows the most accurate results. In addition, for engineering analysis, it is possible to use a modified beam model using finite elements with 6 degrees of freedom, the calculation results of which turned out to be quite close to the theoretical ones. It is shown that the braces in the supporting zones fix the beam from torsion quite well, allow reducing stresses and beam deformability, and can be used for beams for light-duty cranes with small load capacity. For medium and heavy duty cranes, such solutions are not recommended due to the possibility of fatigue failure of the beam, and full-length brake structures should be provided along the upper flange of the beam.

Keywords Crane beam · Brake structure · Restraint torsion · Warping · Bimoment

1 Introduction

I-section crane beams are widely used as supporting structures for overhead cranes of various load capacities and operating modes. In foreign design practice (USA, Canada) the design solutions of crane beams are often used without brake structures, or only with the braces to the upper flange in the support zone of the beam. This

O. Tusnina (✉)
Moscow State University of Civil Engineering, Moscow 129337, Russian Federation
e-mail: tusninaoa@mgsu.ru

P. Akimov et al. (eds.), *Proceedings of FORM 2022*, Lecture Notes in Civil Engineering 282, https://doi.org/10.1007/978-3-031-10853-2_8

simplifies the mounting of the structures, but it is not always an effective solution in terms of metal consumption, crane ease of use and ensuring the fatigue resistance of beams.

In Russian standard design solutions (for example, typical series 1.426.2-7, issue 3 "Crane steel beams for overhead support cranes. Split beams with a span of 6 and 12 m for general purpose cranes with a lifting capacity of up to 50 tons") can be made without a brake structure, but with a enlarged upper belt [1]. Such a solution is acceptable only if there is no need to pass along the crane tracks, i.e. for light duty cranes (A1–A3).

Transverse horizontal forces (lateral forces) are transmitted from the crane wheels to the crane beam. The lateral forces can be caused by different reasons—these can be both forces from the braking of the trolley on the crane bridge, and forces from skew of the crane bridge.

According to the current Russian standard SP 20.13330 "Loads and actions", the consideration of lateral forces caused by the skew of the crane bridge is required only for heavy-duty cranes (A7 and A8). However, in practice, the skew of cranes also occurs for cranes of light and medium operating modes. At the same time, the lack of passageways along the crane tracks for light-duty cranes does not allow for regular inspections and repairs of the crane tracks, and, as a result, the operating conditions of the crane become more complicated.

It should be noted that the magnitude of the lateral forces caused by the skew of the crane bridge, determined according to the current Russian standards, is greater than from the braking of the trolley. In addition, the lateral forces from crane skews calculated using different theoretical formulas are different by several times [2]. This indicates a fairly high uncertainty in choosing the correct loading of crane beams with crane loads, which depend on many factors [3] and, in addition to the crane beam itself, also have a significant impact on the operation of the entire frame [2, 4] and need to be studied.

Lateral forces are applied at the points of contact of the crane wheels with the rail, i.e. with an eccentricity relative to the bending center of the I-beam, which causes it to twist.

The brake lattice or solid structures at the level of the upper flange of the beam prevents its torsion. However, in the absence of such, the thin-walled I-beam experiences restraint torsion, which causes additional sectorial stresses to appear in its flanges.

Thus, a free I-beam crane beam without brake structures works in combined bending and restraint torsion [5, 6].

Restraint torsion can be taken into account directly by applying the theory of thin-walled members developed by V.Z. Vlasov [7], or using approximate methods [8, 9], which is a common approach in European and American standards.

For example, according to European standards (EN 1993-6: Eurocode 3: Design of steel structures—Part 6: Crane supporting structures), a simplified model is used—the bimoment acting in the section is taken into account by considering the bending of the flanges in the horizontal plane from the fictitious additional horizontal load acting to the flanges.

The use of such a simplified approach for free beams can lead to an overestimation of stresses up to 63% [10] compared to the calculation performed using the Vlasov's theory. This result is very conservative and significantly increases the metal consumption of the beam.

According to Russian standard SP 16.13330 "Steel structures", when a beam works in combined bending and restraint torsion, the strength proof is performed directly taking into account the sectorial stresses caused by bimoment acting in the section. The bimoment can be determined with the use of theoretical tabular solutions [11] but not all cases of beam loading are given in the literature, and it is difficult to use tabular values directly. The most effective and correct is the use of modern software systems that implement the finite element method [12], performing the calculation either using plate finite elements or beam thin-walled finite elements with 7 degrees of freedom [13].

It should be noted that in addition to performing the strength analysis, it also becomes necessary to check the lateral-torsional buckling of the free crane beam [14–16], which is obviously provided for beams with brake structures, reliably attached to the top flange of the beam.

2 Materials and Methods

To analyze the possibility of using various options for finite element analysis for free crane beams, we consider a hinged I-beam with a span of 9 m, loaded along the length with a uniformly distributed moment of 1 kN/m. Linear displacements of one of the ends of the beam in the direction of its longitudinal axis are fixed, while the other is free. The warping of both end sections of the beam is limited. The considered beam is a symmetrical welded I-beam with the dimensions of the wall 25 × 828 mm and the flanges 36 × 350 mm made of steel S255.

Consider the following finite element models:

Model 1. A simple beam model using beam finite elements with 6 degrees of freedom in the node (Fig. 1).

Model 2. Modified beam model using beam finite elements with 6 degrees of freedom in the node (Fig. 2).

The modified beam model [17] consists of three beams (two beams at the center of the flanges and one beam at the center of the web) connected by beam rigid elements with increased longitudinal and flexural stiffness and zero torsional stiffness.

The beams modeling the flanges work only in tension/compression and bending only in the horizontal plane, having a stiffness corresponding to the section of the flange. The beam modeling the wall is specified with the geometric characteristics of the wall and with the total torsional stiffness of the entire section in pure torsion (see Table 1).

Linear displacements of the supporting nodes of the shelves are fixed from linear displacements from the plane of the beam.

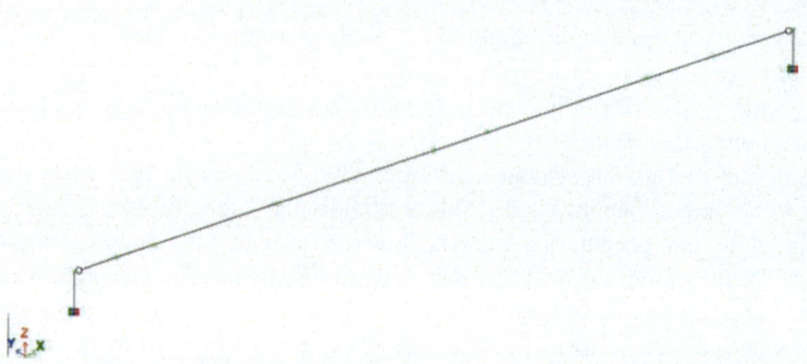

Fig. 1 Simple beam models: model 1 (Lira-SAPR) with the use of beam finite elements with 6 degrees of freedom and model 3 (STK) with the use of thin-walled beam finite elements with 7 degrees of freedom

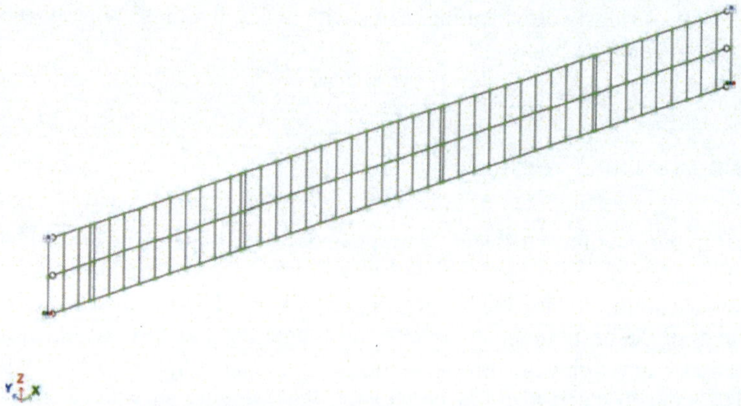

Fig. 2 Modified beam model using beam finite elements with 6 degrees of freedom in the node (model 2, Lira-SAPR)

Model 3. A simple beam model using thin-walled beam finite elements with 7 degrees of freedom in the node [18, 19] (see Fig. 1).

The calculation using thin-walled bar finite elements was performed using the STK software package [20].

Model 4. Plate model (Fig. 3).

In the model using plate finite elements to fix the warping of the support sections, bar finite elements are specified along the median axis of the I-beam section, which have high torsional and bending stiffness in the horizontal plane. The stiffness value was selected on the basis of test calculations and comparison of numerical calculations result with the theoretical one.

If necessary, using these beam elements, it is possible to simulate not only an absolutely rigid limitation of the deformation of the beam support sections, but also

Table 1 Stiffness characteristics of the beam finite elements used for calculations

Parameter		Value
Profile (symmetrical I-beam)		−25 × 828 + 2–36 × 350
Flange width b_f, sm		35
Flange thickness t_f, sm		3.6
Web height h_w, sm		82.8
Web thickness t_w, sm		2.5
A, sm^2		459
I_y, sm^4		588,830.1
I_z, sm^4		25,832.8
I_t, sm^4		1418.8
I_ω, sm^6		44,091,621.0
Steel		S255
E, kN/sm^2		206,000,000
Poisson ratio ν		0.3
G, kN/sm^2		79,230,769.23
Simple beam model with 6 degrees of freedom in the node (model 1)		
Simple beam finite element with 6 degrees of freedom in node	EA, kN	9,455,400
	EI_y, kN·m^2	1,212,990
	EI_z, kN·m^2	53,215.6
	GI_t, kN·m^2	1124.09
Modified beam model with 6 degrees of freedom in the node (model 2)		
Elements used for modeling flanges	EA, kN	2,595,600
	EI_y, kN·m^2	–
	EI_z, kN·m^2	26,496.75
	GI_t, kN·m^2	–
Elements used for modeling web	EA, kN	4,264,200
	EI_y, kN·m^2	243,626.84
	EI_z, kN·m^2	222.1
	GI_t, kN·m^2	1124.09
Simple beam model with 7 degrees of freedom in the node (model 3)		
Thin-walled beam finite element	EA, kN	9,455,400
	EI_y, kN·m^2	1,212,990
	EI_z, kN·m^2	53,215.6
	GI_t, kN·m^2	1124.09
	EI_ω, kN·m^4	9082.87
Plate model (model 4)		
Elements simulating warping constraint of support sections of the beam	EA, kN	100,000,000
	EI_y, kN·m^2	10,000
	EI_z, kN·m^2	10,000
	GI_t, kN·m^2	100,000

by varying their rigidity, it is possible to take into account the pliable fastening, depending on the actual design solution of the beam support.

Calculations of all models (model 1, model 2 and model 4), except for the model using thin-walled beam finite elements, were performed using the Lira-SAPR software package (Fig. 3).

Table 1 shows the stiffness characteristics of the beam finite elements used for calculation.

In addition, a theoretical calculation was performed using the theory of V.Z. Vlasov [7]. According to theoretical data [11], the bimoment in the cross section x along the length of the beam is determined by the formula:

$$B = \frac{m}{k^2}\left(1 - \frac{kl \cdot chk\left(\frac{l}{2} - x\right)}{2sh\frac{kl}{2}}\right) \tag{1}$$

The beam twist angle is determined by the formula:

$$\theta = \frac{m}{k^4 E I_\omega}\left(\frac{k^2 z(l - x)}{2} - \frac{kl \cdot sh\frac{kx}{2} sh\frac{k(l-x)}{2}}{sh\frac{kl}{2}}\right) \tag{2}$$

Normal stresses are defined as follows:

Fig. 3 Plate model (model 4, Lira-SAPR)

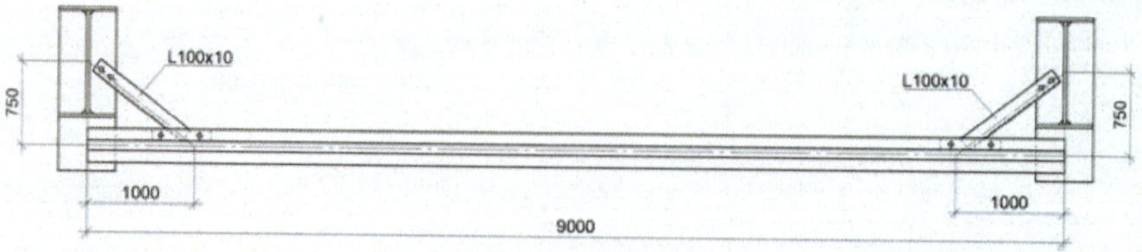

Fig. 4 Structural solution of a crane beam with braces to the upper flange in the supporting zones

$$\sigma_\omega = \frac{B}{I_\omega}\omega \tag{3}$$

Sectorial moment of inertia of an I-beam:

$$I_\omega = \frac{I_{fz}(h_w + t_f)^2}{2} \tag{4}$$

Sectorial coordinate:

$$\omega = \frac{bh}{4} \tag{5}$$

Formulas (1)–(2) contain elastic, bending-torsional characteristic, sm^{-1}:

$$k = \sqrt{\frac{GI_t}{EI_\omega}}, \tag{6}$$

where EI_ω—sectorial stiffness; GI_t—pure torsional stiffness.

Table 2 shows the comparison of the results of the calculation of the described finite element models and the theoretical values determined analytically by formulas (1)–(3).

As can be seen, the use of the modified beam model makes it possible to obtain results close to the theoretical values (the difference is no more than 5%).

The most accurate results were obtained when using plate finite elements and thin-walled beam elements—the difference from the theoretical solution was no more than 1 and 2%, respectively.

The use of a simple beam model with 6 degrees of freedom in a node does not allow taking into account the effect of restraint torsion—there are no bimoment and sectorial stresses in this calculation, but the beam rotation angle increases significantly (almost 5.5 times).

To analyze the influence of the design solution of the brake structure on the load-bearing capacity of the crane beam, we will calculate the split crane beam considered

Table 2 Comparison of results

Model	Rotation angel, rad*1000		Bimoment B, kN·sm^2		Normal stress σ_ω, kN/sm^2	
	Value	Error (%)	Value	Error (%)	Value	Error (%)
Model 1	8.17	441.06	–	–	–	–
Model 2	1.44	−4.64	2.652	2.16	0.445	2.06
Model 3	1.53	1.32	2.600	0.15	0.437	0.23
Model 4	1.52	0.66	2.597	0.05	0.437	0.23
Theory	1.51	–	2.596	–	0.436	–

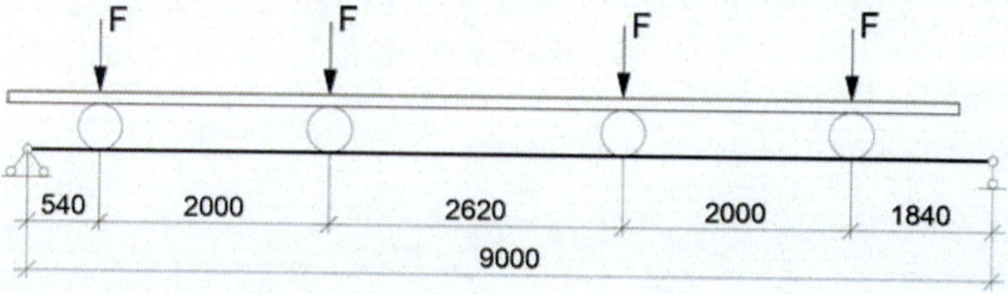

Fig. 5 Scheme of crane loads action

above, with an I-section with a span of 9 m, on the effect of crane loads. The following two options for constructive solution of the crane structure are considered:

- the free I-beam without any braces of the upper flange;
- the I-beam with braces to the upper flange in the support zones made of single angles L100 × 10 (Fig. 4).

There is 1 crane with a lifting capacity of 50 tons and operating mode A3 on the beam. According to the passport data of the crane, the maximum standard vertical pressure of the crane wheel is $F = 310$ kN, the maximum standard transverse load from the wheel is $T = 15$ kN. The scheme of the most unfavorable location of the crane is shown on Fig. 5.

3 Results

The results of the analysis for the transverse brake load for the cases of the free I-beam and the I-beam with the braces to the upper flange in the support zones are shown in Tables 3 and 4, respectively.

For a free I-beam, it is necessary to take into account the effect of restraint torsion and the additional sectorial normal stresses in the flanges, which are about 20% of the total stress value (see Table 3).

The braces in the support zones reduce the twisting of the beam, as well as the overall level of normal stresses in the section, including one from bending in the horizontal plane (see Table 4).

Table 4 shows a comparison of the total normal stresses in the beam section from the action of the vertical load of the crane wheel and the transverse load from the braking of the trolley.

The total normal stresses are determined at the most loaded point of the upper flange, by the following equation:

$$\sigma_{sum} = \sigma_N + \sigma_{My} + \sigma_{Mz} + \sigma_B = \frac{N}{A} + \frac{M_y z}{I_{yn}} + \frac{M_z y}{I_{zn}} + \frac{B\omega}{I_\omega} \quad (7)$$

Table 3 Results of the beam analysis on the transverse brake load

Parameter	Model 1	Model 2	Model 3	Model 4	Error		
					Model 3/Model 4 (%)	Model 2/Model 4 (%)	Model 1/Model 4 (%)
Free I-beam							
M_z, kN·m	86.75	86.73	91.28	86.23	5.85	0.58	0.61
B, kN·m ^{2}	–	8.63	9.20	9.02	2.07	−4.35	–
σ_{Mz}, kN/sm^2	5.88	5.87	6.18	5.84	5.85	0.57	0.60
σ_B, kN/sm^2	–	1.45	1.55	1.51	2.07	−4.32	
σ_{total}, kN/sm^2	−5.88	−7.32	−7.73	−7.36	5.07	−0.44	−20.11
Rotation angel, rad*1000	30.21	5.23	5.93	6.04	−1.83	−13.41	400.00
Displacement of the upper flange, mm	27.03	16.58	17.45	16.50	5.73	0.45	63.79
I-beam with braces							
N, kN	−58.20	−80.74	−79.49	−89.83	−11.51	−10.12	−35.20
M_y, kN·m	25.14	34.83	35.77	38.60	−7.33	−9.76	−34.87
M_z, kN·m	43.01	26.24	27.11	26.63	1.79	−1.48	61.49
B, kN·m^2	–	6.67	6.61	6.59	0.32	1.29	–
Axial force in brace, kN	−72.75	−100.81	−99.36	−99.89	−0.53	0.91	−27.17
σ_N, kN/sm^2	−0.13	−0.18	−0.17	−0.20	−11.74	−10.33	−35.21
σ_{My}, kN/sm^2	0.18	0.26	0.26	0.28	−7.47	−9.74	−35.06
σ_{Mz}, kN/sm^2	2.91	1.78	1.84	1.80	1.79	−1.48	61.51
σ_B, kN/sm^2	–	1.12	1.11	1.11	0.33	1.33	–
σ_{total}, kN/sm^2	−3.23	−3.33	−3.38	−3.39	−0.22	−1.77	−4.84
Rotation angel, rad*1000	29.65	3.09	2.89	3.17	−8.72	−2.33	373.84
Displacement of the upper flange, mm	26.73	4.06	4.11	4.15	−1.02	−2.22	543.49

The underestimation of normal stresses level in the flange of the free I-beam using of simple beam model with 6 degrees of freedom in node (model 1) is about 14%. In this case it turned out to be critical for the free I-beam—obtained value of maximum normal stress in the flange (21.66 kN/sm^2) is less than the yield strength of steel S255 (R_y = 22.5 kN/cm^2) and this show that strength of the beam is provided. But

Table 4 Comparison of the total normal stresses in the section

	Model 1	Model 2	Model 3	Model 4	Error		
					Model 3/Model 4 (%)	Model 2/Model 4 (%)	Model 1/Model 4 (%)
Free I-beam							
$\sigma_{total,\ max}$, kN/sm^2	−21.66	−23.11	−24.34	−23.23	4.93	−0.48	−14.20
With braces							
$\sigma_{total,\ max}$, kN/sm^2	−19.77	−21.32	−21.37	−20.45	0.54	−0.20	−5.90

in fact, the bearing capacity of the beam, taking into account the restraint torsion, is not provided—the stresses are bigger than yield stress.

It is common in Russian practice for the free beam to consider only top flange taking the bending in horizontal plane. Total normal stresses in the top flange calculated using this simplification are 27.5 kN/sm^2, that is 18.4% higher than actual stresses (see in Table 4 for plate model 23.23 kN/sm^2), so this method seems to be too conservative due to the increased material consumption.

The braces make it possible to ensure the bearing capacity of the beam, reducing the overall level of stresses, and the twisting of the beam. The use of a simple model with 6 degrees of freedom in the node gives a not so much underestimated result (about 6%) and, in general, can be allowed in this case for simplified practical analysis.

The stiffness of the beam with braces increased significantly—the horizontal displacements of the upper flange of the beam are 17 mm for the free I-beam and 4.2 mm for the beam with braces. That is 4 times less, while the rotation angle decreased by almost 2 times.

It should be noted that in this article a crane beam without transverse ribs is considered. A test analysis was carried out taking into account one-sided transverse ribs −100 × 8. According to the analysis results, the stress–strain state of the flanges, taking into account the ribs, is very close to the analysis without the ribs, which does not allow considering the ribs as structural elements that prevent restraint torsion and influenced on the beam behavior.

One more thing is that deformability of a beam with braces is higher than that of a beam with a full-length lattice or sheet brake structure. Due to the alternating twisting of the beam, with the regular operation of cranes, there is a risk of early fatigue failure of the welded beam in the zone of flange-to-web welds [21, 22].

So it is required to carry out experimental and theoretical studies of the endurance of crane beams with braces only in support zones. And in order to prevent the problem of fatigue failure, for cranes of medium and heavy operating modes, it seems appropriate to apply constructive measures that facilitate the operation of crane beams for high-cycle loads [23, 24], with providing traditional solutions, with fixing the upper flange of the beam by full-length brake structure design and prevent beam torsion before performing these researches.

4 Conclusions

Based on the analyses performed, the following conclusions can be made:

1. The calculation of I-section crane beams without brake structures or other fastenings of the upper flange must be carried out, taking into account restraint torsion and sectorial normal stresses, the contribution of which to the total stress is about 20% and cannot be ignored.
2. The calculation of thin-walled open profile crane beams without brake structures subjected to restraint torsion is most preferably performed using modern software systems that implement the finite element method using plate finite elements.

   When modeling it is necessary to provide a correct mesh of plate finite elements, boundary conditions corresponding to real ones, stiffness characteristics, load application schemes, and so on.
3. In order to simplify the analysis of the beam subjected to restraint torsion, it is possible to use beam finite elements:

   - the use of beam finite elements with 7 degrees of freedom in the node, taking into account the warping of the section [17, 18];
   - using a modified beam model, composed of beam finite elements with 6 degrees of freedom in the node, that allowing to take into account the occurrence of bimoments in the flanges of the I-beam [24].

   Both of these methods provide sufficient calculation accuracy for engineering purposes (error is less than 5%).

   The use of a simple beam model made of elements with 6 degrees of freedom in this case is not possible, due to a significant underestimation of the normal stresses level in the section.
4. Braces to the upper flange in the support zones of split crane beams of I-section instead of full-length typical lattice or shell brake structures are a very effective way to reduce torsion. Such design solutions can be successfully implemented for cranes of light duty.
5. For medium and heavy duty cranes, the use of typical Russian solutions with the brake trusses or decks along the upper flange of the beams is the most preferable—this allows to reduce the deformability of the beam and thereby increases its endurance.

References

1. "Steel structures of buildings", vol 2. In: Kuznetsov VV (ed) Metal constructions. Designer's guide. ASV, Moscow, 512 p
2. Zolina TV, Tusnin AR (2015) Ind Civil Eng 5:17–23
3. Yao J, Gu H, Chen L (2018) Math Probl Eng. Article ID 6527307
4. Wei-Jun Y, Zhi-Jian Y, Chuan-Zhi Z (1995) J Changsha Commun Univ 01
5. Tusnin AR, Prokić M (2014) J Appl Eng Sci 12(3):179–186
6. Trahair NS, Pi Y-L (1997) Eng Struct 19(5):372–377
7. Vlasov VZ (1959) Thin–walled elastic members. Stroyizdat - Fizmatgiz, Moscow, 568 p
8. Melcher J, Karmazinova M (2012) Procedia Eng 40:262–267
9. Hudz S, Gasii G, Pents V (2018) Int J Eng Tech 7:141–148
10. Gramza G, Zamorowski J (2015) J Civil Eng Env Arch 52(3/II):135–148
11. Bychkov DV (1962) Structural mechanics of bar thin-walled structures. Gosstroyizdat, Moscow, 476 p
12. Kindmann R, Kraus M (2011) FEM for nonlinear calculations of beam structures. In: Steel structures: design using FEM. Ernst & Sohn Verlag, Berlin, pp 168–216
13. Tusnin AR (2015) Sci Rev 11:79–82
14. Bijak R, Kolodziej G, Siedlecka M (2019) Lateral-torsional buckling resistance of crane runway girders. In: IOP conference series: materials science and engineering, vol 471, p 052076
15. Movaghati S (2019) Strengthening beam sections of industrial buildings against lateral torsional buckling. In: Proceedings of the annual stability conference, St. Louis, Missouri, 2–5 April 2019
16. Bijak R (2015) Arch Civ Eng 61(4):127–140
17. Demenev MG (1990) Development of coating structures using membrane panels. Ph.D. thesis, Moscow
18. Tusnin AR (2016) Procedia Eng 150:1673–1679
19. Tusnin AR (2004) Calculation and design of structures of thin-walled rods with open profile. Dr. Sc. thesis, Moscow
20. Tusnin AR, Tusnina OA (2012) Ind Civil Eng 8:62–64
21. Citarelli S, Feldmann M (2019) Fatigue Failure of runway beams due to wheel loads. In: Proceedings of Nordic Steel, vol 3, Issue 3–4. The 14th Nordic steel construction conference, Copenhagen, Denmark, 18–20 September 2019, pp 621–628
22. Rykaluk K, Marcinczak K, Rowiński S (2018) Arch Civ Mech Eng 18:69–82
23. Takki VF, Egorov IF, Tusnina OA (2020) Ind Civil Eng 12:61–67
24. Danilov A, Tusnina O (2021) E3S Web Conf 263:02022

Facade Insulation Systems Using Polyethylene Foam

E. A. Mednikova, K. A. Ter Zakaryan, A. K. Ter Zakaryan, B. A. Efimov, and A. A. Medvedev

Abstract Foamed polymers are one of the most effective and reliable types of thermal insulation. The area of use of materials of this group can become much broader if their combustibility decreases. This is achievable by chemically modifying the polymer matrix, introducing finely milled mineral flame retardants into the matrix, replacing the flammable foaming agent with inert ones and, in particular, with CO_2. It is possible to use products in which the core of combustible polyethylene foam is protected by a film of non-combustible modified polymer with flame retardants introduced.

The purpose of the research presented in the article was to study the possibility of implementing facade systems within the framework of existing design solutions using polyethylene foam. The study of the material properties was carried out according to standard methods. As a result, it was found that the tensile strength in the longitudinal direction of the polyethylene foam samples is 80–92 kPa, and the strength of the weld is 29–32 kPa. Polyethylene foam with an average density of 18–20 kg/m^3 has a thermal conductivity of 0.032–0.034 W/(m·K), diffusion moisture absorption without a metalized coating is 0.44 kg/m^2, and with a coating of 0.37 kg/m^2; water absorption at partial immersion in water for 24 h–0.013 kg/m^2; water absorption by volume when wholly immersed in water for 28 days 0.96%. The material practically does not change its properties under conditions of long-term alternating temperature changes from −60 to +70 °C.

In the case of new construction and reconstruction of facades, insulation is carried out along the outer contour, followed by cladding with facade products fixed on vertical rails. It is possible to implement facade heat-insulating composite systems with external plaster layers arranged along with the reinforcing mesh. With new

E. A. Mednikova (✉)
National Research University "Higher School of Economics", 20 Myasnitskaya Street, Moscow 101000, Russian Federation
e-mail: lisamednikova97@gmail.com

K. A. Ter Zakaryan · A. K. Ter Zakaryan
"TEPOFOL" Ltd., Shcherbakovskaya Street 3, Moscow 105318, Russian Federation

B. A. Efimov · A. A. Medvedev
Moscow State University of Civil Engineering, Moscow 129337, Russian Federation

P. Akimov et al. (eds.), *Proceedings of FORM 2022*, Lecture Notes in Civil Engineering 282, https://doi.org/10.1007/978-3-031-10853-2_9

construction, it is possible to use methods of layered combined masonry. Depending on the thickness of the thermal insulation, the heat transfer resistance of layered structures ranges from 3.2 $m^2{}^\circ C/W$ or more.

Keywords Slab thermal insulation · Insulating shell · Thermal resistance · Polyethylene foam · Lock connection · Flammability

1 Introduction

In facade thermal insulation systems with a ventilated gap, with surface plastering, or with external brickwork, stone wool slabs are the principal thermal insulation material. The strength properties, density, and thermal conductivity of these products are adapted to the conditions of use [1–3]. These products belong to the group of non-combustible, which determines their widespread use in construction. At the same time, the properties of stone wool-based boards depend on atmospheric humidity. Depending on the operating conditions, the thermal conductivity of mineral wool boards can increase by 50–60% compared to dry products [4–6].

Firstly, sorption moistening of the heat-insulating layer leads to a decrease in the thermal resistance of the structure by 30–40%, which causes a significant increase in heat loss through the insulating shell [7–9]. Secondly, moisture negatively affects both the adhesion of external plaster layers (in the facade heat-insulating, composite system) and the state of the supporting structure to which the insulation system is attached. In facade systems with a ventilated space, wet thermal insulation accelerates corrosion processes in the metal elements of the mounting and suspension system [10–12].

Foamed plastics used in construction, namely extruded polystyrene foam, polyurethane foam, polyisocyanurate foam, polyolefin foams have a thermal conductivity in the dry state, the lower thermal conductivity of mineral wool products, and very low water absorption [13–16], which determines the stability of their characteristics about changing atmospheric humidity. The combustibility of foamed plastics predetermines the features of their use in structures [16–18].

In facade systems with ventilated space, the use of combustible boards based on foamed plastics is not allowed; in systems with external plastering, such boards are used, with certain restrictions. Given the low water absorption and water permeability, as well as resistance to aggressive environments, foamed plastics are widely used in insulation systems for structures in contact with wet soil: floating floors on the ground, insulation of the basement of the building, and isolation of basements.

It should be noted that fundamental research is currently being carried out to reduce the flammability of foamed polymers and organic materials are gradually being created with a flammability group approaching G1-G2 [19, 20].

The use of any plate heat-insulating materials creates the problem of cold bridges at the joints between the plates and when they are adjacent to the supporting structures

Fig. 1 Thermal imaging of the insulating shell: "bridges" of increased heat transfer are visible

(Fig. 1). At the same time, there is a group of materials, the installation of which is carried out without joints. This is the so-called seamless insulation.

This group of materials includes products based on polyethylene foam. This material is elastic, that is, it can deform with the base to which it is mechanically attached; the material is waterproof and also has a low vapor and moisture permeability and thermal conductivity, which does not depend on the humidity of the atmospheric air. Products are produced in two types: with a tight fit of layers (Fig. 2a) and products of the AirLayer line (Fig. 3b) with spot welding of layers and air cavities between them (RF patent No. 199048) [21].

Hotel products (usually rolled polyethylene foam) are overlapped along with the side faces and welded with hot air (Fig. 3), thereby forming a seamless shell. Considering the properties of the material, the insulating shell has low thermal conductivity, water and gas permeability, that is, the interior becomes completely protected. This engineering solution allows for complete control of the microclimate parameters in the room, depending on its purpose: living space, sports facility, warehouse, storage, refrigerators (RF patent No. 2645190) [22].

(a) (b)

Fig. 2 Products made of polyethylene foam: **a** - layered products; **b** - products of the AirLayer line

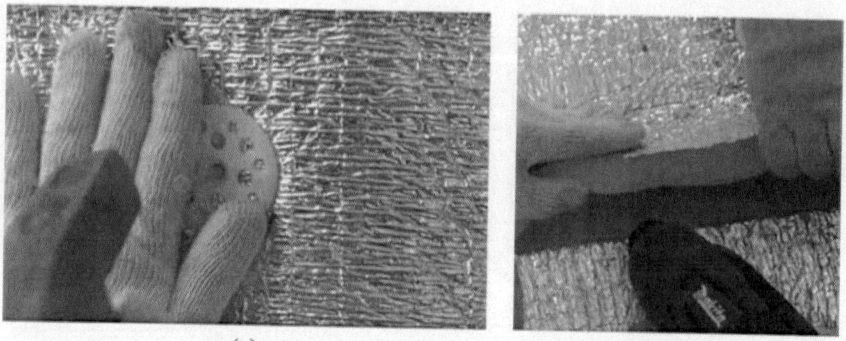

(a) (b)

Fig. 3 Installation of thermal insulation: **a** - mechanical fastening of the insulation sheet; **b** - welding of joints of rolled polyethylene foam

2 Methods

Like most polymeric building materials, traditional polyethylene foam is classified as combustible. Therefore, studies are currently being carried out aimed at reducing its flammability due to the chemical modification of the polymer matrix, the introduction of mineral finely ground fire retardants into the matrix, and the replacement of a fire-dangerous foaming agent with inert ones and, in particular, with CO_2. As an option, it is possible to use laminated products, in which the core of combustible polyethylene foam is protected from the outer and inner surfaces by a film of non-combustible polyethylene (a modified polymer with introduced flame retardants).

Studies of polyethylene foam and systems for its use were carried out at the Moscow University of Civil Engineering (NIU MGSU), the Institute of Building Physics (NIISF RAASN). Part of the research presented in this paper was carried out using the equipment of the Central Collective Use Center named after prof. Yu. M. Borisov VSTU, supported by the Ministry of Science and Higher Education of the Russian Federation, Agreement No. 075-15-2021-662.

The evaluation of tensile strength in the longitudinal direction was carried out according to the methods of GOST EN 1608-2011 (Fig. 4); diffusion moisture absorption of foamed polyethylene was carried out according to the methods of GOST EN 12,088-2011, and water absorption of samples with partial or complete immersion was carried out according to the methods of GOST EN 1609-2011). Operational resistance under conditions after climatic exposure was assessed by the compression deformation of the material and creep over time, carried out according to the method of GOST EN 1606-2011.

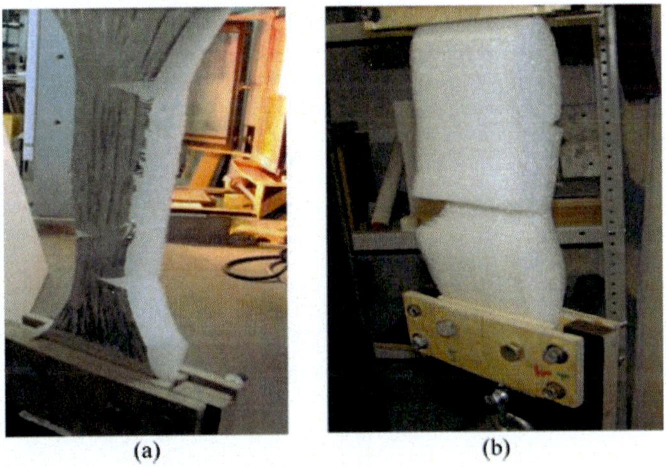

Fig. 4 Tests to determine the tensile strength of polyethylene foam samples in the longitudinal direction: **a** - for a solid product; **b** - along the weld

3 Experimental and Results

As a result of the research, it was found that the tensile strength in the longitudinal direction for samples with or without a reflective coating is 80–92 kPa; in this case, the strength of the welded joint (seam) is 29–32 kPa. Such indicators of tensile strength are sufficient to maintain good adhesion between the structure and the supporting base and maintain the integrity of the insulating shell due to base movements or its thermal deformation.

With an average density of polyethylene foam samples of 18–20 kg/m^3, its thermal conductivity does not exceed 0.034 W/(m·K), even in environments with high humidity. The thermal conductivity of layered products of the AirLayer brand is 10–20% lower. Diffusion moisture absorption of samples is 0.37–0.44 kg/m^2; water absorption at partial immersion in water for 24 h is 0.013 kg/m^2; water absorption by volume when wholly immersed in water for 28 days is 0.96%. The breaking stress on the polyethylene-metal contact surface was 12–17 kPa. The material is resistant to both temperature (in the range from minus 40 to 60 °C) and alternating temperature effects.

4 Discussions

In constructing the facades of residential buildings, systems based on foamed polyethylene can be used both in new construction and in the reconstruction of facades. In façade heat-insulating structural plaster coatings, the insulating material is protected both from atmospheric influences and fire's possible effects. The structure of the plaster coating is such that it is an analog of textile concrete, in which the role of the reinforcing component is played by a mesh of glass or polymer fibers (Fig. 5).

During new construction and reconstruction of buildings, the load-bearing walls of which are made of brickwork, the masonry of ceramic stones or cellular concrete blocks, insulation are usually carried out along the outer contour, followed by cladding with facade products fixed on vertical rails (Fig. 6).

With new construction, it is possible to use systems of layered combined masonry (Fig. 7). Depending on the thickness of the thermal insulation, the resistance to heat transfer of layered structures ranges from 3.2 $m^2{}^\circ C/W$ or more.

Systems of seamless insulation using plaster coatings, with insulation along the outer contour, as well as insulation in a layered combined masonry, make it possible to meet the regulatory requirements for thermal resistance; normative indicators

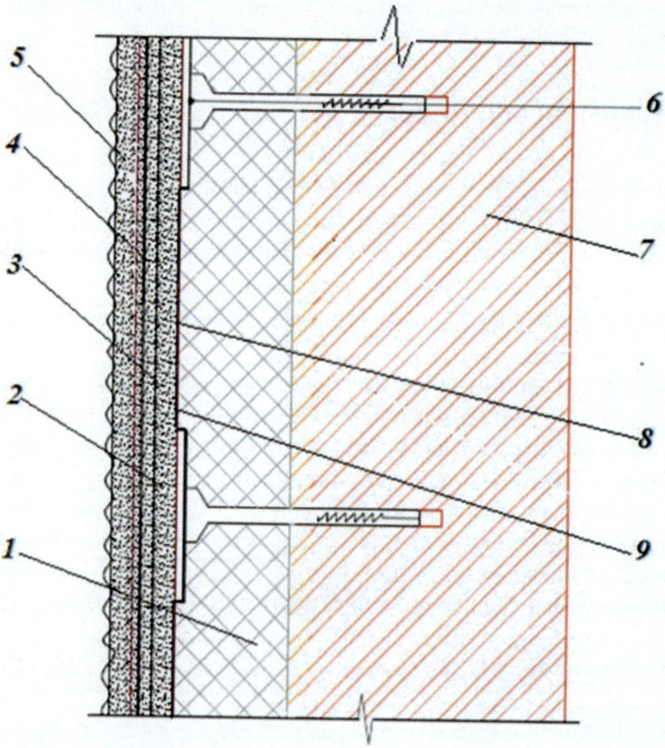

Fig. 5 Facade heat-insulating structural system using polyethylene foam (SFTK-NPE): 1 - thermal insulation (polyethylene foam rolls or mats); 2 - basic plaster composition; 3 - facade fiberglass alkali-resistant mesh; 4 - finishing primer; 5 - protective and decorative plaster; 6 - plate-shaped dowel with a spacer element; 7 - load-bearing wall; 8 - adhesive primer; 9 - fiberglass mounting mesh

(a) (b) (c)

Fig. 6 Insulation with rolled polyethylene: **a** - the facade of the building, **b** - the corner of the building; **c** - finished insulating shell

(a) (b) (c)

Fig. 7 Layered brickwork: a - layout of materials in layered masonry; b - view of the masonry from the side of the bearing wall; c - view from the side of the outer cladding

of temperature and humidity conditions inside the premises; protect non-existent structures from excessive humidity, negative and alternating temperatures.

All these results are an example of the implementation of an integrated approach to meeting energy efficiency requirements: keeping the heat in rooms, ensuring the durability and integrity of building structures, and using materials with low technological energy intensity.

5 Conclusions

The use of foamed polymers as one of the components of insulating shells makes it possible to solve the issues of energy saving and the formation of optimal operating conditions for structures. Systems for effective heat-insulating materials should take into account both the characteristics of the materials themselves and the practical orientation of the facilities.

Like most polymeric building materials, traditional polyethylene foam is classified as combustible. Therefore, studies are currently being carried out aimed at reducing its flammability due to the chemical modification of the polymer matrix, the introduction of mineral finely ground fire retardants into the matrix, and the replacement of a fire-dangerous foaming agent with inert ones and, in particular, with CO2. As an option, it is possible to use laminated products, in which the core of combustible polyethylene foam is protected from the outer and inner surfaces by a film of non-combustible polyethylene from a modified polymer with flame retardants introduced.

Acknowledgements The research presented in the article was carried out within the framework of the plan of the fundamental scientific study of the Ministry of Construction of Russia and the Russian Academy of Architecture and Civil Sciences for 2022–2023 on the topic 3.1.2. "Development of scientific foundations for creating fine-grained concrete reinforced with textile material (textile concrete) ". Part of the research was carried out using the equipment of the Center for Collective Use (CCU) named after Prof. Yu.M. Borisov VSTU, supported by the Ministry of Science and Higher Education of the Russian Federation, Agreement No. 075-15-2021-662.

References

1. Bobrova E, Pilipenko A, Zhukov A (2019) Insulating sheath system and energy efficiency of buildings 02019 (TPACEE 2018). https://doi.org/10.1051/e3sconf/20199102019
2. Umnyakova N, Chernysheva O (2016) Thermal features of three-layer brick walls. Proc Eng 153:805–809. XXV Polis-Russian-Slovak seminar "Theoretical Foundation of Civil Engineering"
3. Umnyakova NP, Tsygankov VM, Kuzmin VA (2018) Experimental heat engineering research for the rational design of wall structures with reflective heat insulation. Housing Constr 1–2:38–42
4. Zhukov A, Medvedev A, Poserenin A, Efimov B (2019) Ecological and energy efficiency of insulating systems135:03070 (2019). https://doi.org/10.1051/e3sconf/201913503070. (ITESE-2019)
5. Shitikova MV, Bobrova EYu, Popov II, Zhukov AD (2019) Energy efficiency technical thermal insulation. In: International multi-conference on industrial engineering and modern technologies, FarEastCon, 8934917 (2019). https://doi.org/10.1109/FarEastCon.2019.8934917. https://ieeexplore.ieee.org/stamp/stamp.jsp?arnumber=8934917
6. Zhukov AD, Bobrova EYu, Bessonov IV, Medvedev AA, Demissi BA (2020) Application of statistical methods for solving problems of building materials science. Nanotechnol Constr Sci Online J 12(6):313–319. https://doi.org/10.15828/2075-8545-2020-12-6-313-319
7. Ter-Zakaryan KA, Zhukov AD (2021) Short overview of practical application and further prospects of materials based on crosslinked polyethylene. In: Thomas J, Thomas S, Ahmad Z (eds) Crosslinkable polyethylene. Materials horizons: from nature to nanomaterials. Springer, Singapore (2021). https://doi.org/10.1007/978-981-16-0514-7_12
8. Pilipenko A, Bobrova E, Zhukov A (2019) Optimization of plastic foam composition for insulation systems, 02017 (TPACEE 2018). https://doi.org/10.1051/e3sconf/20199102017

9. Zhukov AD, Ter-Zakaryan KA, Bobrova EYu, Pilipenko AS (2019) Insulation sheath materials for cold preservation. Mater Technol Constr Archit II:452–457. https://doi.org/10.4028/www.scientific.net/MSF.974.452
10. Zhukov A, Dovydenko T, Kozlov S, Ter-Zakaryan K, Bobrova E (2019) Innovative technologies for low-rise construction 91:02032. E3S web of conferences (TPACEE 2018). https://doi.org/10.1051/e3sconf/20199102032
11. Ter-Zakaryan KA, Zhukov AD, Bobrova EYu, Bessonov IV, Mednikova EA (2021) Foam polymers in multifunctional insulating coatings. Polymers 13(21):3698. https://doi.org/10.3390/polym13213698
12. Aldawi F, Alam F, Date A, Kumar A, Rasul M (2012) Thermal performance modelling of residential house wall systems. Proc Eng 49:161–168
13. Gnip IJ, Keršulis VJ, Vaitkus SJ (2006) Predicting the deformability of expanded polystyrene in long-term compression. Mech Comp Mater 41(5):407–414
14. Gnip IJ, Keršulis VJ, Vaitkus SJ (2005) Analytical description of the creep of expanded polystyrene under compressive loading. Mech Comp Mater 41(4):357–364
15. Jelle BP, Gustavsen A, Baetens R (2010) The path to the high-performance thermal building insulation materials and solutions of tomorrow. J Build Phys 34(2):99–123. https://doi.org/10.1177/1744259110372782
16. Semenov VS, Bessonov IV, Ter-Zakaryan KA, Zhukov AD, Mednikova EA (2020) Energy-saving seamless insulation systems for frame buildings using foamed polyethylene. Regional energy problems (Problemele energetic regional), 4. https://doi.org/10.5281/zenodo.4018999. UDC: 691.175.2./6./8
17. Zhukov A, Semyonov V, Gnip I, Vaitkus S (2017) The investigation of expanded polystyrene creep behavior. In: MATEC Web of Conferences, Seminar 2017 Theoretical Foundation of Civil Engineering, vol 117
18. Stepina I, Kotlyarova I (2018) Modification of wood with monoethanolamine-borate by the data of x-ray photoelectron spectroscopy. MATEC Web Conf 04005 (2018). https://doi.org/10.1051/matecconf/201819604005
19. Stepina I (2018) Fire protection of timber building structural units by mono-and di-ethanolamine-(NB)-phenyl borates. MATEC Web Conf 03016. https://doi.org/10.1051/matecconf/201819303016
20. Ter-Zakaryan KA (2018) Patent of the Russian Federation 2645190
21. Ter-Zakaryan KA (2020) Patent of the Russian Federation for utility model 199048

Hydrological Studies of Mereb-Gash Basin in the Context of Water Management

Dmitry Kozlov and Anghesom Ghebrehiwot

Abstract The present work attempts to obtain unique hydrological information and effective tools for streamflow modelling from insufficiently gauged river basin. To this end, physically-based semi-distributed and conceptual models are investigated using climate reanalysis datasets and geomorphologic catchment characteristics. Prior to feeding model forcing variables, their statistical and spatial patterns, methods of potential evapotranspiration estimations, and basin drought conditions are studied. Two or more approaches have been employed for each of these tasks. The findings from statistical and spatial analyses indicate presence of predominantly significant monotonic trends: precipitation and humidity tend to decrease, whereas temperature and potential evapotranspiration tend to increase with considerable spatial variability in the basin. Similarly, the drought indices exhibit predominantly humid conditions from 1979 through the end of 1990s but persistent dry conditions over the period 2000 to 2013. Out of the three methods for the estimation of potential evapotranspiration, Hargreaves and Penman–Monteith produced consistent estimations, the latter being with the highest values. Concerning the rainfall-runoff models, the goodness-of-fit results for single-event-based hydrographs indicate that GIUH-Nash model has the potential to produce acceptable values in most cases irrespective of sources and resolutions of digital elevation models, but it is found to be sensitive to the type of algorithms selected for stream network generations. The findings of the continuous simulation models suggest considerable overestimation of most of the precipitation in the reanalysis datasets, which in turn has a significant effect on other variables, such as potential evapotranspiration and thereby leading to a substantial discrepancy between simulated and observed. While there is no substitute for restoring and enhancing ground observation networks, further research that aims at identifying effective mechanisms to properly utilize reanalysis datasets are suggested.

Keywords Drought · Mathematical modelling · NAM · Potential evapotranspiration · SWAT model · SWOT analysis · Mereb-Gash

D. Kozlov · A. Ghebrehiwot (✉)
Moscow State University of Civil Engineering, Moscow 129337, Russian Federation
e-mail: bahghi2012@gmail.com

P. Akimov et al. (eds.), *Proceedings of FORM 2022*, Lecture Notes in Civil Engineering 282, https://doi.org/10.1007/978-3-031-10853-2_10

1 Introduction

Climate fluctuations affect the availability, quantity and quality of water thereby affecting food security, energy, sustainability of ecosystems, and so on. Hydrological changes caused by climate change create additional challenges for water resources management. East Africa's water resources are subjected to strong hydrological and climatic variability, both in space and time, and are key factors constraining further economic development. As a result, these regions are ecologically, socially and economically vulnerable to natural hazards and climate change. Global and regional projections [1] indicate that these regions are expected to experience increasing water scarcity and land desertification. So, adaptation and mitigation of these challenges and their consequences through, inter alia, effective water resources management is important for achieving sustainable development goals. In this connection, the possibility of obtaining reliable and scientifically sound assessments of the current state of water resources and forecasts about future scenarios is of utmost priority. The establishment of new water infrastructural facilities and keeping the existing ones fully operational can only be done by taking in to account the spatio-temporal variability and accurate quantification of river flows. The development of effective regulatory mechanisms and storage facilities in regions with limited natural water resources can increase water availability, for example, through the effective use of seasonal streamflows. So, streamflow predictions from catchments with insufficient in-situ hydrometric and climatic observations, as is the case in many regions of Eritrea, have been the interest of researchers in the field of engineering hydrology.

More than 72% of the Eritrean population is engaged in subsistence agriculture. The growing population will irreversibly lead to an increase in the population's demand for food. To meet this demand, the potential arable lands of Eritrea (1.5 million ha) that are located mainly in arid and semi-desert agro-ecological zones need to be properly developed. However, these areas receive an annual rainfall ranging from 200 to 400 mm, which is low for agricultural production of the required level if not supplemented by irrigation. The widely practiced irrigation system under such situations is called runoff irrigation or spate irrigation. Spate irrigation is defined as "an ancient irrigation practice that involves the diversion of flashy spate floods running off from mountainous catchments where flood flows, usually flowing for only a few hours with appreciable discharges and with recession flows lasting for only one to a few days, are channeled through short steep canals to bunded basins, which are flooded to a certain depth".[1] Since recently, significant efforts have been made to capitalize the Western Lowlands and Eastern Lowlands of Eritrea by constructing improved runoff irrigation systems. For example, the number of runoff irrigation systems in the Gash-Barka region alone increased from 16 (in 2003) to 35 (in 2016). However, none of these projects, which are usually implemented on hydrometeorologically unguaged catchments, have fully achieved the intended goals as of 2016

[1] van Steenbergen F., Lawrence P., Haile A. M., Salman M., и Faures J., *Guidelines on spate irrigation. FAO irrigation and drainage paper 65*. Rome: Food and Agriculture Organization of the United Nations, 2010.

[1]. The absence or insufficiency of reliable hydrological data, among others (e.g., management, operation, technical, etc.), is seemingly the main reason for the failure of these systems. Considering the unavailability of sufficient data on one hand and the ongoing development activities, on the other hand, the utilization of technology-based data and effective hydrologic response prediction tools in the interest of addressing the complex water resources management becomes timely and relevant.

Hydrological models simulate natural hydrological and biogeochemical processes. They are not only useful tools to underpin our understanding of the dynamic interactions between climate and land surface hydrology, but also provide the missing information as a basis for decisions regarding the development and management of water and land resources. A broad spectrum of critical environmental and water resources problems are addressed using hydrological models [2–6]. Hydrological models primarily require streamflow data, climate data, and basin characteristics [4], which are inadequately available in most practical cases and aggravated by the increasing effects of human activities on the components of the hydrological cycle. Hydrometric observatory stations have been dwindling at global level due to probably the shift in using global climate reanalysis datasets as model forcing variables.

The objective of the study is, therefore, to obtain unique hydrological information and effective tool for runoff predictions in Mereb-Gash river basin in Eritrea. The specific objectives are: to analyze the current status of the water resources management system and propose a strategy for its sustainable development; to investigate the efficacy of climate reanalysis datasets as forcing variables in physically-based mathematical models; and to evaluate the effectiveness of the geomorphologic instantaneous unit hydrograph (GIUH) based Nash model for single-event-based simulations.

2 Methods

2.1 *Description of the Study Area*

The Mereb-Gash basin whose outlet is located near the town of Kassala (in Sudan) possesses a complex catchment landscape. The upper reaches of Mereb-Gash are dominated by mountainous terrains and high-gradient hills with limited plains. This landform type accounts for 42% of the total area with more than 15% slope. Out of the total area, the middle and lower portions of the basin (29%) are predominantly flat plains and medium-gradient hills with slopes between 0 to 5%. The remaining landforms are undulating (18%) and strongly sloping (11%) with 5–10 and 10–15% slopes, respectively. The main river system in the upper reaches originates near the capital city (Asmara, Ertirea) and flows southwards. Then, it turns west to form the border between Ethiopia and Eritrea in the central part of the country. Finally, it flows westwards to Sudan, forming an inland delta north of Kassala. The maximum and mean annual flows at Kassala are reported to be 1000 m^3/s and 680 million m^3, respectively [7]. The total basin area and main channel length at the same outlet are approximately 22,850 km^2 and 550 km, respectively, and an altitudinal range from 510 to 3164 m (in Ethiopia) above mean sea level.

River basins are variously classified [14], depending on size, main river length, flow conditions, etc. Accordingly, Mereb-Gash is large river basin when its size is considered otherwise it is medium river basin when its main channel length is considered. It is one of the two major Eritrean river basins that drain into the Nile basin, the world's longest river. Out of the total area, about 76% is located within the Eritrean territory. It is a long, narrow and elongated basin, having a basin shape factor and average basin slope of 11 and 7%, respectively [7]. The soil cover of Mereb-Gash is as diverse as the topography and climate of the country. The soil information within the Mereb-Gash (Fig. 1) can be better described on the basis of the information obtained from the digital soil map of the world. As such, the Western Lowland plains

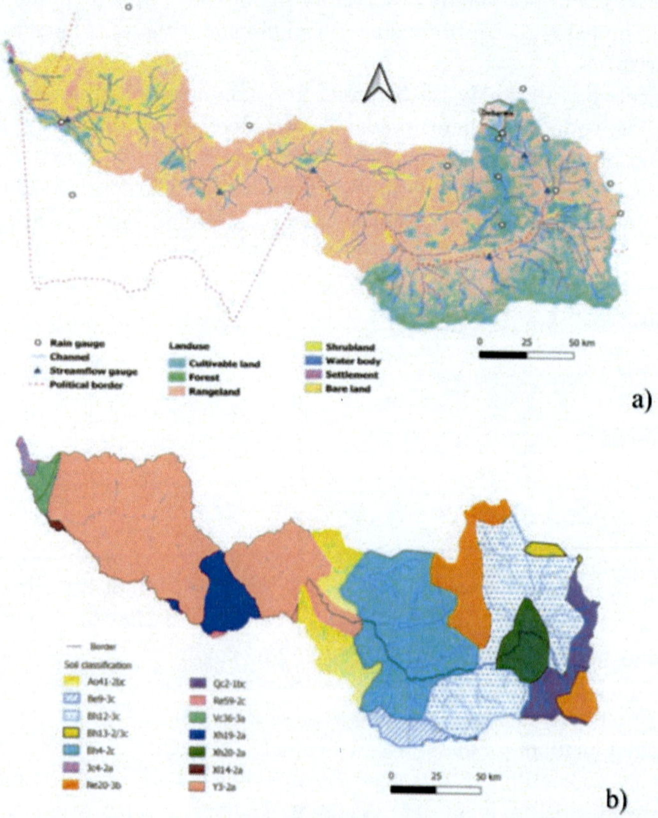

Fig. 1 Landuse along with rain gauges and streamflow gauges **a** and soil mapping units composition **b** of Mereb-Gash basin adapted from GlobeLand30 and FAO–UNESCO global soil map, respectively

and Central Highlands are dominated by Yermosols (32%) and Humic Cambisols (37%). Intensive soil erosion is visible and worrisome, evidently linked to the nature of the rainfall and inherited degraded natural environment.

The climate of the Mereb-Gash basin within the Eritrean territory can be fairly represented by Moist Highlands and Moist Lowlands agro-ecological zones. The Moist Highlands zone, which is located in the eastern and south-eastern part of the basin where the main river starts, has an altitude of over 1600 m above mean sea level with average annual rainfall between 500 and 700 mm. The Moist Lowlands zone, located in the central parts of the basin, has an altitudinal range from more than 500 to 1600 m above mean sea level, warm to hot semi-arid climate and average annual rainfall between 500 to 800 mm. The climate of the extreme lower part of the basin is more of Arid Lowlands zone, with a hot arid climate and an average annual rainfall of 200 to 500 mm. The climate setting of the remaining tributaries (24%), which are positioned outside of the Eritrean territory, can be assumed to be similar to their immediate proximities. Rainfall in the basin usually occurs as moderate to high-intensity events of short duration with limited areal extent. Evapotranspiration is also variable with more than 2000 mm/year in the lowlands and relatively around 1500 mm/year in the highlands.

To better understand the current status of water resources in the basin, SWOT analysis was carried out through selection of representative indicators, followed by their quantitative and qualitative descriptions.

2.2 *Model Setup*

To achieve the outlined objectives, different approaches to the modelling processes were adopted and are schematically represented by Fig. 2. Firstly, the GIUH-Nash model was selected to generate hydrographs of an individual storm event. This model mainly depends on the information derived from physiographic and geomorphologic catchment characteristics. Apparently, a digital elevation model (DEM) and geographic information system are powerful tools for obtaining these characteristics, even in the absence of topographical maps. To understand the effects of DEM and algorithm selection on GIUH-Nash-based streamflow predictions from poorly gauged Debarwa sub-basin within the upper reaches of the Mereb-Gash river basin, four frequently used open-source DEMs with different resolutions and two QGIS algorithms (GRASS and SAGA) were investigated. Secondly, two of the widely used physically-based mathematical models that possess varying structures and assumptions were applied: SWAT and NAM. Climate Forecast System Reanalysis (CFSR) from the National Center for Environmental Prediction (USA) was used as forcing model inputs so as to produce continuous streamflow simulations. CFSR data covered over 35 year period (from 1979 to 2013) on daily basis for 32 stations within and around the Mereb-Gash river basin. Prior to feeding the climate variables to the models, there was a need to evaluate these variables such as statistical analysis, computation of potential evapotranspiration, and drought indices. As such, three

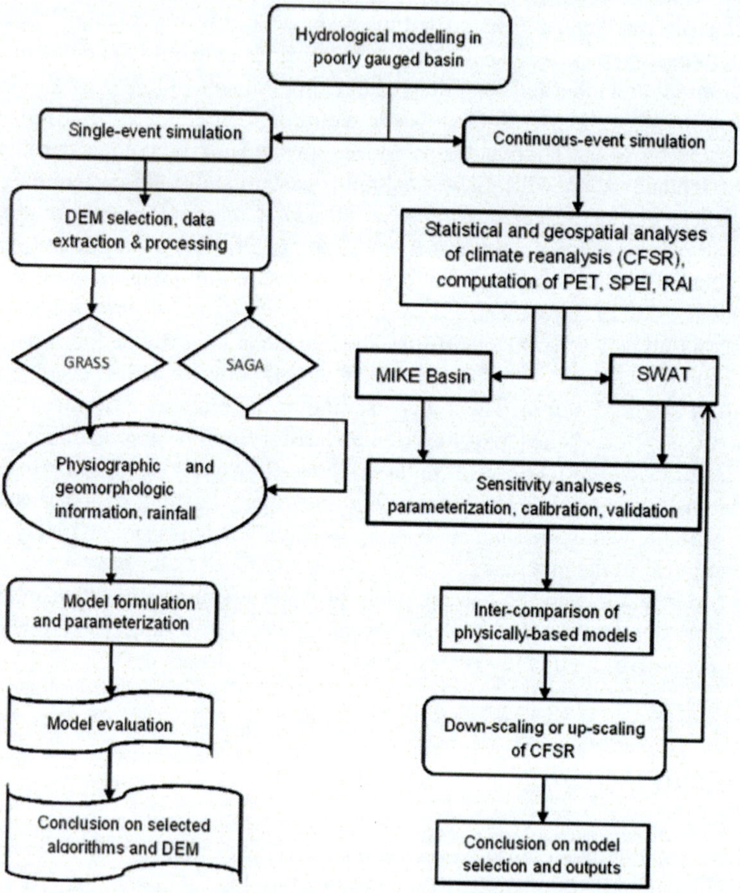

Fig. 2 Schematic representation of the hydrological modelling of Mereb-Gash

potential evapotranspiration estimation methods were employed, namely Thornthwaite, Hargreaves, and Penman–Monteith. Besides, two drought indices, namely standardized precipitation evapotranspiration index (SPEI) developed by Vicente-Serrano and others [8] and rainfall anomaly index (RAI) proposed by Rooy [9] were used in order to understand the long-term climate characteristics of the study area. Finally, statistical and visual inter-comparisons were done among the outputs of the selected models, out of which conclusions were drawn.

2.3 Model Calibration and Validation

Calibration, which is intimately linked to uncertainty analysis, is adjusting model inputs to achieve the best simulation match with observation. There are three important concepts that require proper understanding during calibration: parameterization, objective function definition, and non-uniqueness (uncertainty). In parameterization, the questions on which parameters that represent hydrological processes to use and how to regionalize them are answered. In this respect, the significance of one parameter or a combination of parameters with respect to the objective function or a model output needs to be carefully investigated. In uncertainty analysis, the errors in the model inputs through the calibration process are propagated and quantified. Finally, the calibrated parameters with an independent set of data without further changes to the parameters or model structure are tested in the validation process. In other words, validation is used to build confidence in the calibrated parameters. Two rules that should be followed during validation process are generally recommended [10]: (1) using the same range of parameters as the ones obtained in the calibration, and (2) the data used for validation should have almost the same statistics as the data used for calibration.

In this study, both manual and automatic calibration procedures were employed with consideration for different objective functions, depending on which model is applied. For example, an automatic optimization routine based on a multi-objective optimization strategy, in which four objectives can be simultaneously optimized, is part of the NAM model. These objectives relate to coherence between different simulated and observed characteristics of hydrographs, attributed to flow volume, peak, duration, and overall water balance. In manual calibration, a trial-and-error parameter is employed so that the goodness-of-fit of the calibrated model could be identified based on the visual examination by means of comparing predicted against observed flows. The initial selection of parameters depends on the behaviour of the initial model result before any calibration [11]. For example, parameterization, uncertainty analysis, etc. procedures were done in SWAT model using a standalone program SWAT-CUP that comprises various objective functions and hundreds of parameters. The outputs corresponding to each objective function are normally unique, leading to the conditionality of objective functions. The objective functions reflect the degree of matching between simulated and observed. Practically, perfect matching is unrealistic due to various sources of uncertainty in hydrologic modelling [21], including input data, model assumptions and simplifications, selection of optimal parameters, and many other activities that happen in the watershed.

3 Results

3.1 SWOT Analysis

Global threats, such as climate change, have already been realized in various sectors of Eritrea, affecting the economy, among which the water management system may become the most sensitive to such changes. Global climate projections also predict increased extreme phenomena (drought and flood), which in turn will intensify the vulnerability to climate change. A recent study [13] demonstrated that more than 88% of the farming households in Eritrea were found to be vulnerable or highly vulnerable to climate change as a result of the combined effect of their exposure to external factors, sensitivity to internal factors, and lower adaptive capacity. The Nile River, with an estimated length of over 6800 km and the world's longest river, is located in the Horn of African region. As noted earlier, two major Eritrean basins, Mereb-Gash and Setit, are tributaries of this river basin whose annual rainfall ranging from a minimum of 240 mm to a maximum of 665 mm. These two basins combined have an area of 24 921 km^2 (0.8% of the total basin area or 20.4% of the country area). With the increase in water demand, the Nile has become at the center of the complex dispute, involving several countries with Ethiopia, Egypt and Sudan being at the forefront of this dispute. So, these issues along with other external factors are potential regional threats to equitable utilization of shared waters. The findings from SWOT analysis are summarized and presented in Table 1.

The viability of the water resources system at either the national or the basin level is seriously threatened by inextricably linked factors. Generally speaking, the current actions to overcome the threats and weaknesses are seemingly incommensurate. Global and regional threats to the sustainability of the water system obviously require urgent and special attention. However, it is ostensibly clear that the rate of change (for example, climate change, land degradation, etc.) is more than the speed with which the system can adequately respond. As the rate of change overwhelms the ability to respond, the water resources system loses its viability and sustainability. In addition to global and regional threats, the social sustainability in Eritrea is now threatened by the dynamics of technology, economy, and population that may accelerate the environmental and social rates of change. On the other hand, growing structural resistance against reorganization—also known as "structural inertia" reduces the ability to respond in time. In view of the above circumstances, strategies, intended to improve the sustainability of the water resources system of Eritrea, are graphically represented in Fig. 3.

Table 1 Summary of the SWOT analysis for water resources system in Eritrea

		Strengths	Weaknesses
Internal conditions		– Political will and commitment – Provision of water facilities and infrastructure – Technology transfer and human development – Regulations on waste management – Watershed management	– Water scarcity and spatial variability – Low economic growth – Population pressure and user competition – Absence of enacted water laws – No scientific support for the activities of authorities – Insufficient coordination among concerned parties – Low efficiency of project implementation
		Opportunities	**Threats**
External factors	National	– Coordinated implementation of IWRM plan – Climate change adaptation and mitigation programmes – Increasing efficiency in use and productivity of water resources – Integration of elements of environmental management systems – Political stability	– Environmental, economic, military, and political insecurity – Water security and high dependence ratio – Pollutant pressure – Low level of compliance with regulation and legislation – Interdependence of water resources and environment systems – Weak public financing mechanisms
	International	– World experience in water resources and ecosystem management – Regional cooperation on shared water resources utilization	– Impacts of transboundary basins mismanagement on water resources system – Global threats (e.g., climate change, its consequences, etc.)

3.2 *Event-Based Streamflow Model*

The event-based streamflow simulation indicated that the selected algorithms resulted in drainage networks of similar stream orders but different geomorphologic characteristics (for example, stream ratios). Despite incongruences among physical and geographical parameters and elevation profiles derived from various DEMs, the results of the study (Debarwa sub-basin) indicated that their implications on runoff predictions based on GIUH-Nash model are negligible unlike other models, such as TOPMODEL [14] and SWAT [15]. On the other hand, algorithm selection affected the model performance in agreement with other works (e.g., [16]). That is, GRASS generated streamflows with acceptable accuracy both visually and statistically (Fig. 4a),

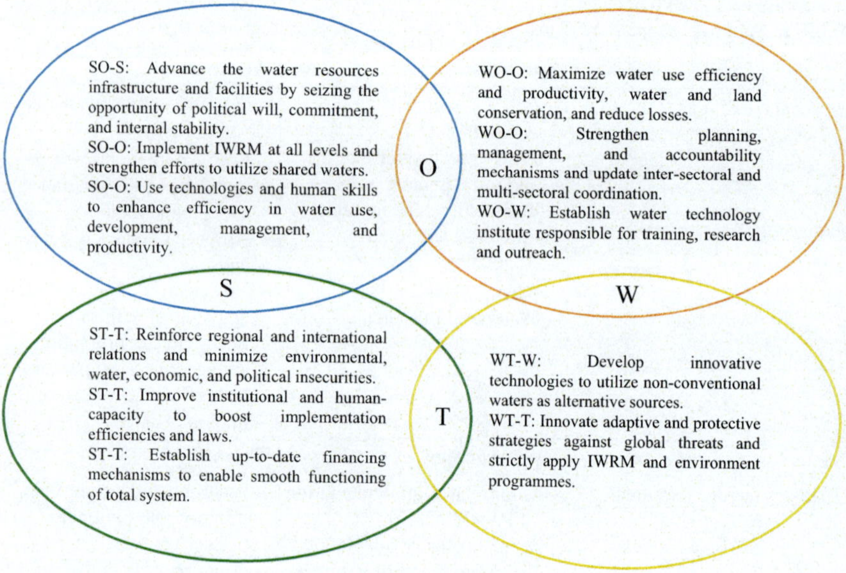

Fig. 3 Strategy for improving status of water resources system

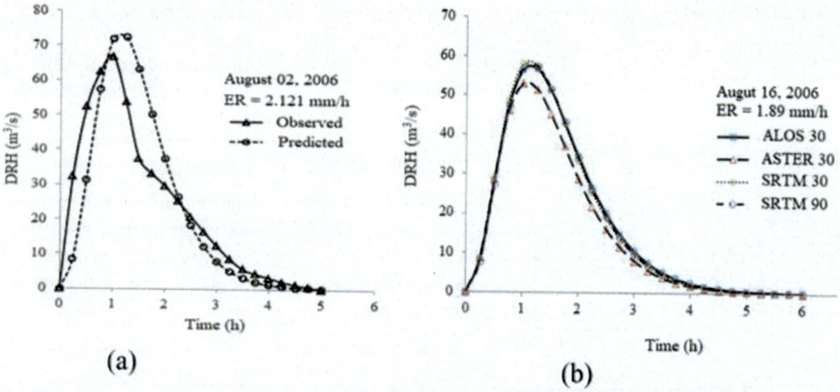

Fig. 4 Comparison of event-based streamflow predictions for given effective rainfalls (ER) from GRASS: **a** against observed data, and **b** between different DEM

whereas SAGA was found to be unsatisfactory. Besides, source and resolution of DEMs insignificantly affected GIUH-Nash model performance as compared to algorithm selection (Fig. 4b). As such, GRASS-based streamflow predictions are less influenced by source and resolution of DEM than SAGA-based predictions. Lower resolution yields relatively consistent and acceptable performance than higher resolution which complies with the conclusions drawn by others (e.g., [17]). Manning's coefficient is the most sensitive parameter greatly affecting the overall performance of the GIUH-Nash model; hence, reasonable and appropriate value corresponding to the catchment area under consideration, must be identified before applying the model for runoff predictions.

3.3 Statistical Analysis, Potential Evapotranspiration and Drought

The non-parametric statistical analysis of daily precipitation data is processed to understand the monthly, seasonal and annual variability. As such, precipitation was highly variable, ranging from over 1100 mm in the upper reaches to less than 700 mm, in the lower and middle reaches of Mereb-Gash basin. In most cases, quantitative overestimation of precipitation was easily noticed when statistical and visual comparisons with observed data in a few selected stations were analyzed. Generally, the relative humidity and precipitation analyses showed decreasing trend, whereas maximum temperature showed an increasing trend. Similar trends were observed in almost all global stations that were considered in the study.

Mann–Kendall and Modified Mann–Kendall monotonic trend tests were applied for various time series of climate data before and after modifications of the time series along with Sen's slope estimates. Comparing Mann–Kendall and Modified Mann–Kendall tests, significant modifications were obtained on the acceptance of the null hypothesis possibly due to the removal of serial correlation from the time series, leading to the agreement of these findings with other studies (e.g., [18]). Moreover, Sen's slope of precipitation from all the stations was less than 0 at 95% confidence level, indicating significant decreasing trend. But, the statistical analyses temperature and potential evapotranspiration implied that the trend patterns were found to be predominantly increasing at 95% confidence level. The climate classification of Mereb-Gash based on precipitation to potential evapotranspiration ratio was found to be largely (68.8%) experiencing semi-arid climate.

Calculation of potential evapotranspiration and analysis of drought conditions were inevitable prior to hydrologic modelling. The former being one of the important variables, the methods for its estimation may affect the performance of the model outputs. The latter uses the potential evapotranspiration and other climate variables to evaluate the drought conditions of the area of interest. Figure 5 intercompares potential evapotranspiration between Thornthwaite, Hargreaves, and Penman–Monteith methods at yearly and monthly time scales. While Thornthwaite and Penman–Monteith provide the lowest and highest values, Hargreaves-based produces values between the two methods. The Thornthwaite was found to be the most inconsistent method. The findings are actually in agreement with the conclusion from Shuttleworth [19], that is, apart from the Hargreaves and Blaney-Criddle methods, other temperature-based methods are less recommended. In almost all cases, Penman–Monteith produced the highest estimations. Numerous studies have demonstrated that Hargreaves has significantly overestimated than Penman–Monteith in humid

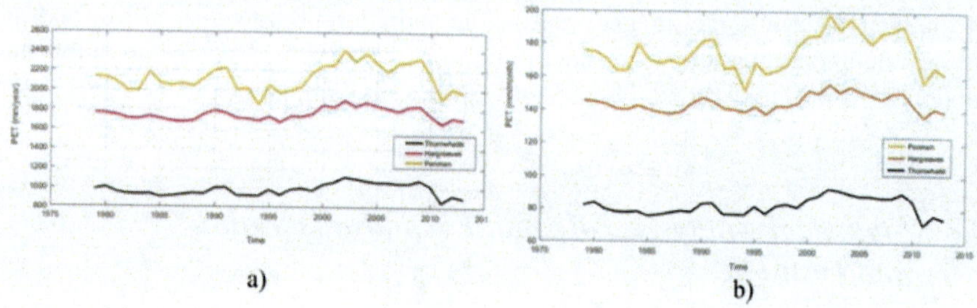

Fig. 5 Comparison of potential evapotranspiration estimation methods: **a** annual, and **b** monthly

environments [20] and vice versa in arid and semi-arid environments [21]. Normally, a polynomial line of 2nd degree was regressed when Hargreaves and Thornthwaite were plotted against Penman–Monteith. However, it doesn't mean that the same relationships can be established for areas with different climatic conditions. As expected, the spatial variability analyses of long-term annual potential evapotranspiration in the Mereb-Gash basin varied from 1500 mm upstream reaches to 2400 mm in downstream plains.

It would be unrealistic to provide a representative SPEI value for the whole Mereb-Gash basin because of the geographical and climate variability, among others. As such, analysis was done at the 32 representative stations. Figure 6 provides an illustration of drought conditions at monthly time scales at Debarwa station. In general, the period up to the end of 1990s was exposed to quasi-continuous humid conditions, with few intermittent drought periods. The persistent drought conditions during 2000 to 2010 are also clearly identified, independent of the time scale. The spatial variability of the annual SPEI for the driest year (2003) demonstrated that the whole study area had been exposed to severe and extreme drought conditions. The same procedure was employed to the wettest year (1998), providing positive SPEI values. As per the SPEI classification, the basin was found to be predominantly under normal and moderate wet conditions.

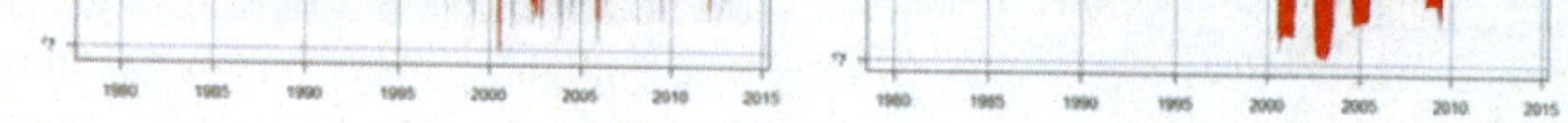

Fig. 6 Examples of SPEI representing the drought conditions of the Debarwa sub-basin at different time scales

The spatio-temporal distribution of the monthly, seasonal and annual RAI and its correlation with SPEI for all the stations were also investigated. The positive RAI values represent rainy or wet years and the negative values represent dry years, with different degrees of intensity. It was found that during the period 1979 up to the end of 1990s, RAI values were predominantly wet, varying from extremely humid to humid with few intermittent dry years. On the contrary, from the end of 1990s to 2013, RAI values are under predominantly dry conditions, varying between very dry and dry with few humid conditions. While the upper-third and lower-third of the Mereb-Gash basin experienced dry conditions, whereas the majority of the middle-third portion of the basin had been experiencing relatively humid conditions during the period of analysis. A regional study on drought in the Horn of Africa [22] and IPCC show similar findings. Thus, knowledge-based countermeasures that address desertification, land degradation and food security should be taken in the near-term, while supporting longer-term responses that enable adaptation and mitigation to climate change.

3.4 Continuous Streamflow Models

NAM comprises parameters, such as runoff coefficient, low and high peak flows, baseflow characteristics, surface and root zone storages. Soil type, vegetation cover, topography, geologic conditions, etc. are some among other factors that affect runoff coefficient. In the absence of detailed information, runoff parameters were assigned intuitively. As such, smaller runoff coefficient values are assigned for relatively flat areas with coarse, sandy soils and large unsaturated zones (mainly downstream areas), whereas large values for sub-basins with possibly low, permeable soils such as clay or bare rocks (upstream reaches). The peak runoff events are caused by large quantities of overland flow. The peak volume is adjusted by changing the overland flow runoff coefficient, whereas the shape of the peak is dependent on the time constant used in the runoff routing. Surface storage and root zone storage are observed to be the major parameters that affect the amount of evapotranspiration; the higher these values the higher rate of evapotranspiration and vice versa. Besides, the amount of base flow is affected by other runoff components; a decrease in overland flow or interflow is resulting in a higher baseflow, and vice versa. The shape of the baseflow recession is a function of the baseflow time constant. Though visual comparisons of the streamflows (Fig. 7) seem to capture seasonality in most cases, considerable water balance and peak flows (both low and high) mismatches can easily be noticed, especially when the accumulated flow volume is plotted.

Calibration and validation analysis indicate that timing, frequency of high and low peaks, volume, and shape of hydrographs varied overwhelmingly from that of the corresponding observed runoffs. The Nash-Sutcliff efficiency (NSE), coefficient of determination (R^2 or bR^2), PBIAS, and water balance and root mean square error (RMSE) calibration and validation statistical evaluation measures showed no difference from the conclusion drawn based on the visual graphical comparisons.

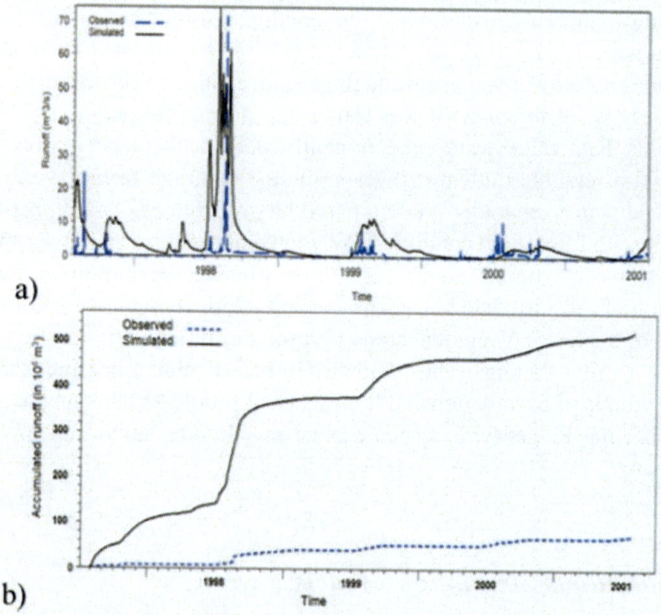

Fig. 7 Continuous streamflow (NAM) for Gala Adi-Nfas sub-basin: **a** flow rate, and **b** accumulated runoff

A typical example of these statistical indicators for Debarwa sub-basin for monthly flow simulations from both models during calibration is presented in (Table 2). Most of NSE for the various sub-basins in Mereb-Gash were less than zero meaning that the observed mean is a better predictor than the model. Moreover, the water balance also depicted a significant discrepancy between the simulated and observed runoffs. RMSE are greater than zero, representing imperfect fit among the observed and simulated runoffs. Overall statistical measures showed the poor performance of the NAM model in the area of interest. As noted earlier, findings indicated that visible overestimation of precipitation input data, which in turn could have influenced the other derived variables (potential evapotranspiration), thereby producing significant incongruity among the simulated and observed hydrographs. Therefore, though it is too premature to deduce a generalized comprehension based on the present work alone, the suitability of CFSR data as input variables in NAM is unsatisfactory.

Table 2 Objective functions for monthly flow simulations from SWAT and NAM during calibration

Type of model	Objective function				
	R^2	bR^2	*NS*	*RSR*	*PBIAS*
SWAT	0.80	0.77	0.73	0.52	−42
MIKE 11-NAM	0.20	0.19	−1.62	1.61	188

In SWAT model, however, there were significant improvements as compared to that of NAM. One more point that could be noticed from all the simulated hydrographs in NAM is the presence of baseflow throughout the analyzed period, even during the dry periods. However, this was not the case in almost all the outputs from SWAT model. Of course, overestimation and underestimation of peak flows are unequivocally distinguishable. But, the overall shape, seasonality and other flow characteristics from SWAT seem more realistic than that of NAM not only visually but also statistically. This argument is further justified by the daily and monthly streamflow outputs at the Debarwa sub-basin as shown in Fig. 8. For further details, the reader may refer to the works of the authors [23, 24].

However, we hypothetically assumed that the adjustment of reanalysis datasets with reference to the available field data can produce better results. Accordingly, adjustment was made only to precipitation because of the available data being only precipitation and the possibility of generating most of the other variables in SWAT model itself. Finally, the statistical evaluation of annual precipitation indicated that

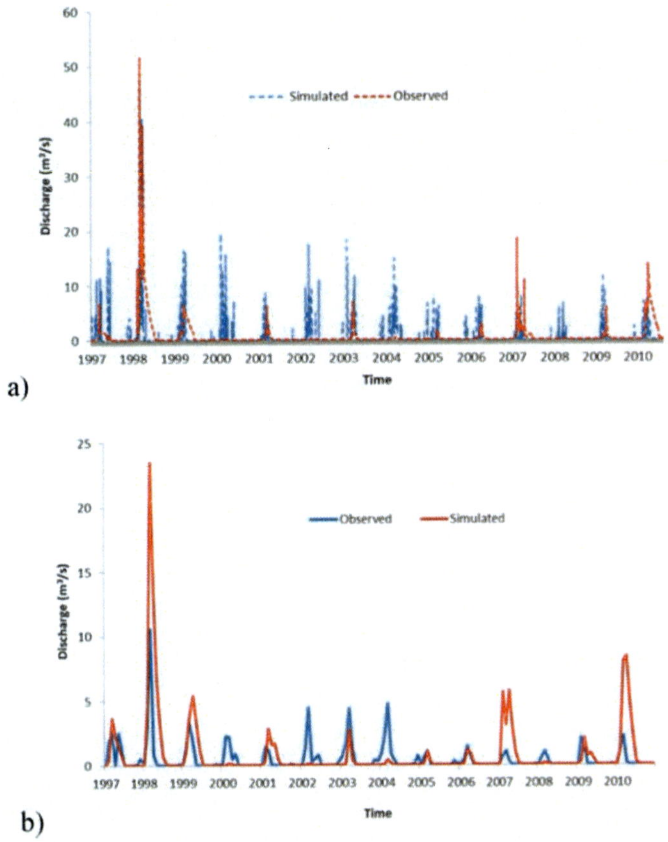

Fig. 8 Comparison of observed and simulated streamflows at Debarwa: **a** daily, and **b** monthly

most of the global stations overestimated precipitation than ground observations. Precipitation from 66% of the 32 global stations showed overestimation, 28% underestimation, and only 6% were the same; the overestimation in 31% of the stations were twice higher than the in-situ observations.

4 Discussion

The overall situation of water resources management system in Eritrea is currently compounded by weak socio-economic and environmental development, as well as, among other things, organizational, legal and political problems. Some of the challenges also stem from the very nature of water resources, characterized by scarcity, spatio-temporal variability, high dependence ratio, etc. High probability of devastating natural hydrological phenomena occurrence (e.g., floods and droughts) is directly connected to transboundary water bodies and development of extreme conditions, including the adverse effects of climate change. Statistical analysis of CFSR for the period from 1979 to 2013 showed that mainly monotonous trends of the main climatic characteristics were observed in the Mereb-Gash basin at almost all observation stations; precipitation tends to decrease, whereas air temperature and potential evapotranspiration tend to increase continuously. The SPEI and RAI drought indices at all observation stations showed predominantly persistent arid and severe arid conditions to some extent during the period 2000 through 2013, whereas there were predominantly wet conditions from 1979 to the end of the 1990s. The causes of persistent and significant trends in most climatic characteristics and severe prolonged droughts in recent years in the Mereb-Gash basin in particular and in the region as a whole are most likely related to human activities, conflicts and climate change. These trends agree with global and regional predictions that increased land surface air temperature and evapotranspiration and decreased precipitation amount, in interaction with climate variability and human activities, led to desertification in sub-Saharan Africa.

The political will and commitment of the government of Eritrea with regard to the establishment of water resources infrastructure and facilities shown in the last two decades have made a significant contribution to improvement of spatial extent in water-related ecosystems. This in turn created opportunities for the transfer of modern technologies and the development of human and institutional capacities. Therefore, there is a need to capitalize on this opportunity of political will and social mobilization of the population for additional expansion of the existing water infrastructure and water management systems. This strategy scenario is expected to enhance technology transfer, capacity-building, and IWRM implementation. In addition, concerted efforts are needed to reverse human and climate-driven environmental repercussions through the construction of check-dams and reservoirs on tributaries of river basins, terracing, afforestation, and closures.

The evaluation on the potential evapotranspiration estimation methods revealed that Thornthwaite method turned out to be incompatible as compared to the Hargreaves and Penman–Monteith. Even though the last two methods showed no significant differences, Penman–Monteith was used in the selected models evaluation. Accordingly, the applicability of the NAM model for modelling rainfall-runoff in Mereb-Gash river basin using CFSR turned out to be very uncertain. A comparison of physically-based models with distributed and semi-distributed parameters, in which CFSR were used as input parameters, showed that the SWAT model somewhat outperformed the NAM model by a big margin, especially it being effective at sub-basin level. Therefore, it is necessary to take into account the size of the basin when SWAT model is used to simulate river flow. In addition, although most of the objective functions confirmed the acceptable effectiveness of the SWAT model at sub-basin level, significant uncertainty was observed during calibration and validation periods. We strongly believe that in addition to the errors due to input data, the application of the so-called "principle of correct neglect" might have aggravated the errors. This means that reservoirs and ponds unaccounted for in the calculations, pumping of groundwater, irrigation activities and crop conditions most likely aggravated the insufficient efficiency of the models.

5 Conclusions

Currently, the water management systems are overwhelmingly dominated by weaknesses than strengths and threats are expected to outweigh opportunities unless the speed with which the system can adequately respond is more than the rate of change. The overall situation is compounded by weak social, economic, environmental developments as well as organizational, legal and policy gaps. From a practical point of view, it would be advantageous to seize the opportunity of political will and social mobilization to further expand the water resources facilities and infrastructures in the immediate future so as promote technology transfer and capacity-building. An improved ecosystem will definitely create a diverse environment, empower communities and institutions which will enable them selectively and appropriately respond to a multitude of environmental challenges, and strengthen adaptation capacities to cope up with changing environments.

Intercomparison of the selected physically-based models for streamflow simulations indicated that SWAT relatively outperformed NAM in terms of overall model performance. It is worth suggesting that when CFSR-based SWAT is to be applied in the region, it is better to pay attention to the size of the basin to be modelled because of two reasons. Firstly, there would be smaller spatio-temporal variations of climate variables and catchment characteristics, such as soil, geologic, topography, etc. Secondly, the errors generated from each sub-basin get accumulated when many smaller catchments are combined to get the outflow at the outlet of the whole basin. In addition, CFSR has a great advantage over in-situ observations in terms of its simplicity and accessibility. However, the results of this work have proven that in order to minimize model uncertainties and hence, improve forecasting capabilities, there is no alternative to restoring existing systems for collecting ground-based information and creating new hydrometeorological observations.

Considering the simplicity of the GIUH-Nash model development and application for practical purposes, ease of access to DEM and processing tools, incongruence of the theoretical and practical results from carefully selected algorithms and regularity in model outputs due to its negligible dependence on source and resolution of DEM, it has the potential to be a useful tool in streamflow prediction in ungauged regions. However, it is critical to consider the limitations of unit hydrograph applications and catchments that have an outlet located near an upstream confluence of two major rivers that may possibly give smaller values of higher order length which has major implications on the model outputs. In order to take a broad view of its applicability for effective event-based streamflow simulations in the region, inter-comparison with other methods and available reference information systems is suggested.

Finally, the design of water management systems should be carried out taking into account the limited water resources of rivers and lakes, the lack of comprehensive studies of water bodies as well as reliable reference data: hydrological, meteorological, geological, geodetic and environmental information. Under these conditions, the most effective tools are likely to be methods and technologies of engineering hydrology based on cautiously selected mathematical models.

References

1. Ghebrehiwot AA, Kozlov DV (2021) Vestn. MGSU 16:8
2. Singh VP, Frevert DK (2006) Watershed models. CRC Press, Boca Raton
3. Chowdhury KR, Eslamian S (2014) Climate change and hydrologic modeling. In: Eslamian S (ed.) Handbook of engineering hydrology – fundamentals and applications. CRS Press, London, pp 77–81
4. Ghebrehiwot AA, Kozlov DV (2019) Vestn. MGSU 14:8
5. Refsgaard JC (1996) Terminology, modelling protocol and classification of hydrological model codes. In: Refsgaard JC, Abbott MB (eds) Distributed hydrological modelling. Kluwer Academic Publishers, Dordrecht, pp 17–39
6. Singh VP, Woolhiser DA (2002) J Hydrol Eng 7:4
7. Ghebrehiwot AA, Kozlov DV (2020) Vestn. MGSU 15:7
8. Vicente-Serrano SM, Beguería S, López-Moreno JI (2010) J Clim 23:7
9. van Rooy MP (1965) Notos 14
10. Abbaspour KC, Rouholahnejad E, Vaghefi S, Srinivasan R, Yang H, Kløve B (2015) J Hydrol 524
11. Abbaspour KC, Vaghefi S, Yang H, Srinivasan R (2019) Sci Data 6:1
12. Arnold JG et al (2012) Trans ASABE 55:4
13. Debesai MG (2020) IOP Conf Ser Mater Sci Eng 1001:1
14. Wolock DM, Price CV (1994) Water Resour Res 30:11
15. Chaplot V (2005) J Hydrol 312:1–4
16. Wechsler SP (2007) Hydrol earth. Syst Sci 11:4
17. Reddy AS, Reddy MJ (2015) J Earth Syst Sci 124:7
18. Yue S, Wang C (2004) Water Resour Manag 18:3

19. Shuttleworth W (1993) Evaporation. In: Maidment DR (ed) Handbook of hydrology. McGraw Hill Inc., New York, pp 4.1–4.53
20. Pandey PK, Pandey V (2016) Earth Syst Environ 2:3
21. Jabloun M, Sahli A (2008) Agric Water Manag 95:6
22. Ghebrezgabher MG, Yang T, Yang X (2016) Adv Meteorol
23. Kozlov DV, Ghebrehiwot AA (2021) Water Sect Russ Probl Technol Manag 1
24. Ghebrehiwot AA, Kozlov DV (2021) E3S WoC 264

Comparison of Rolled and Leaned Concrete Gravity Dams

Victor Tolstikov and Sohaib Tareq

Abstract The article compares a roller compacted concrete gravity dam (RCC) and Trapezoid-shaped Cemented sand and gravel (CSG) dam. Calculations for the main and special combination of loads were carried out in the CRACK software package.

Keywords Gravity dam · Roller compacted concrete (RCC) · Cemented sand and gravel (CSG) · Symmetrical profile · Coefficients · Stability · Stress–strain state "CRACK"

1 Introduction

The article compares a traditional roller compacted concrete (RCC) gravity dam with a Trapezoid-shaped dam made of Cemented sand and gravel (CSG). The article notes the differences between the materials of the body of the dams, spelled out the main advantages and disadvantages of rolled concrete and CSG concrete. A comparative analysis of changes in the distribution of stresses in the body of the dam and in the base has been carried out. Calculations for the main and special combination of loads were carried out in the CRACK software package, which implements the finite element method in a strength setting and allows modeling various discontinuities in seams, cracks, and contacts. Based on a comparison of options, a choice was made in favor of a dam, the main volume of the body, which will be occupied by rolled concrete with reduced requirements for the selection of aggregates, preparation of construction joints, and minimum binder content. In the presence of a poor foundation or in conditions of high seismicity, the traditional triangular shape of the gravity dam is not optimal. An alternative design can be a trapezoidal dam profile (CSG), in which the binder content can be selected depending on the types of aggregate available. The concept of the trapezoidal dam is to design a concrete gravity dam built almost entirely of material (CSG) that meets the requirements for strength and stability.

V. Tolstikov · S. Tareq (✉)
Moscow State University of Civil Engineering, Moscow 129337, Russian Federation
e-mail: sohaib85sabah@gmail.com

P. Akimov et al. (eds.), *Proceedings of FORM 2022*, Lecture Notes in Civil Engineering 282, https://doi.org/10.1007/978-3-031-10853-2_11

Table 1 The top ten high (RCC) gravity dams in the world

			Volume ($м^3 \times 10^3$)		Content (Kg/$м^3$)	
Dam	Country	Height (m)	RCC	General	Cement	Pozzolan
Longtan	China	217	4952	7458	99	121
Guangzhou	China	201	820	2870	61	91
Mile 1	Colombia	188	1669	1730	85 ÷ 160	0
Urawama	Japan	156	1294	1860	91	39
Miyagazi	Japan	156	1537	2001	91	39
Jinangiao	China	156	2400	3920	72	108
Ralco	Chile	155	1596	1640	133	57
Takizawa	Japan	140	810	1800	84	36
Son La	Vietnam	139	2677	4800	60	160
Sayne	Turkey	137	1560	165	85	105

2 Roller Compacted Concrete Gravity Dams RCC

Roller Compacted Concrete (RCC) has gained worldwide acceptance as an alternative to conventional concrete in dams as well as other types of structures due to its economy, high performance and high speed of construction. Concrete gravity dams have a number of advantages that explain their wide application [6, 7]. Such dams are built in various climatic conditions and with high intensity. For example, the Beni Harun dam (Algeria, 2000), 95 m high and with a concrete volume of 1.9 million m^3, at temperatures up to +43° C. The dam built in Russia in a harsh climate, up to –36 °C in winter, the dam of the Bureyskaya hydroelectric dam, which has a height of 140 m.

In more detail, the situation with the construction of rolled concrete dams is presented in Table 1, which provides reference data on the use of rolled concrete in dam structures in different regions of the world [8, 10, 11, 22].

3 Advantages and Disadvantages of RCC Dams

- High reliability and safety.
- High initial strength.
- Reduction of construction time by 1.5 ÷ 2 times.
- Reducing the total cost of the hydroelectric complex [1].
- The construction of dams of this kind on weak semi-rock foundations is difficult due to the lack of shear resistance.
- There are tensile stresses under seismic action.
- Increased heterogeneity leads to increased filtration.
- High shears characteristics in horizontal joints.

Table 2 Based on the content of the cementing material, (RCC) can be classified as

Types of RCC	Binder content (cement + pozzolan) (kg /m^3)
Hard fil-Low dosage l	65–100
Medium dosage	100–150
Rich mix-High dosage	150–270
Japanese type	120–130

Table 3 Density and values (cohesion, angle of internal friction) for (RCC) of 4 dams built in different countries of the world [21]

Name	Content of cement (cement + ash)Kg/м^3	Filler type	Density of RCC (ton/m^3)	C (MPa), φ (degrees)for RCC
Dams	70 + 100 = 170	Sandstone	2,438–2,450	C = 3,17, φ = 42
Capanda	70 + 130 = 200	Granite	2,325–2,475	C = 2,2, φ = 58
Sierra Brava	60 + 0 = 60	Basalt	2,632–2,694	C = 2,46, φ = 48
Uruguay	91 + 39 = 130	–	–	C = 3, φ = 49

- The weakest point of rolled concrete dams is the numerous seams.
- Low strength, uniformity and high-water resistance.
- The impossibility of conducting work in the winter.
- High manual labor costs in preparing the base before laying.

4 Properties of Rolled Concrete

Depending on the content (cement + pozzolan) in the composition, there are four types of roller compacted concrete shown in Table 2.

According to previous studies of experimental samples of rolled concrete, different types of rolled concrete were established, from different dams the characteristics of which depend on their composition, the influence of different properties of materials and hardening conditions are shown in the Tables (3 and 4) [9, 10, 15, 17].

5 Trapezoid-Shaped CSG Dam

The idea was to design a dam with a profile somewhere between a gravity dam and an earth dam, using a material that has the characteristic properties of both concrete and rock fill [18]. The concept of symmetrical Trapezoid-shaped CSG dams is to design a concrete gravity dam constructed entirely from extra lean compacted concrete (CSG) to meet strength and stability requirements. With a more spread out profile of the dam, the requirements for the strength characteristics of the dam material

Table 4 Compressive strength of (RCC) in some massive structures

Dam	The country	Content of cement kg/m^3	Average compressive strength, MPa
Tarbela	Pakistan	111–142	17,5–18,7
Guri	Venezuela	100	14,2
Ogawa	Japan	120 (24 + 96)	14,0–18,0
Kurpsaiskaya	USSR/Russia	80 (56 + 24)	18,4
Zeyskaya	USSR/Russia	100	13,0
Tamagawa	Japan	130 (90 + 40)	15,0

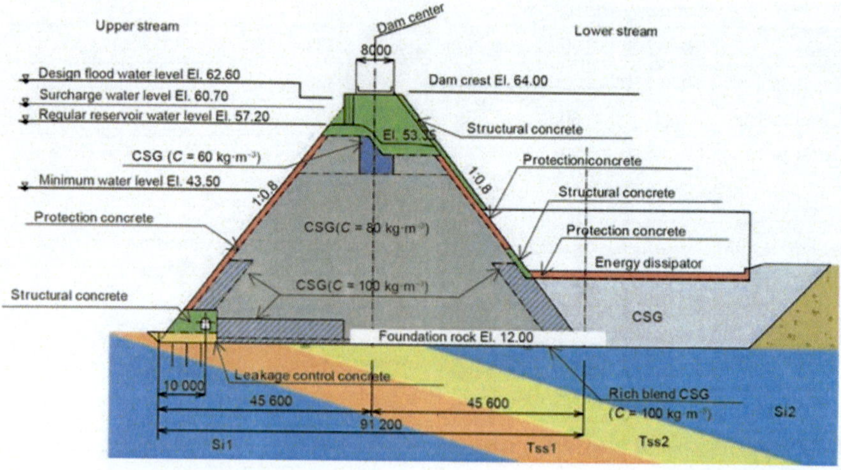

Fig. 1 Tobetsu dam in Japan (H = 52 m)

are much reduced, since the stress level in the body of the dam and in the base is noticeably lower and they can be built on weak semi-rock foundations [18]. The cost-effectiveness of a symmetrical Trapezoid-shaped dam from a "CSG" consists in the cheapness of the material, because the filler is soil taken near the construction site, which does not require processing, sorting, washing, fractionation, soil with the addition of a small amount of cement (50–80 kg/m^3).

For example, The coffer-dams of Nagashima, Tokuyama, and Takizawa, some slope treatment projects, and some sediment control dams were built using this method, as were Okukubi dam (H = 39 m) and Tobetsu dam H = 52 m, Fig. 1, which were completed in 2012 [19, 20]. And also the Safsaf dam built in Algeria in 2011. More details on the construction of trapezoidal symmetrical dams (CSG) are presented in Table 5 [16].

Table 5 Trapezoidal dam from CSG [16]

Dam	The country	Height(m)	Year of completion
Cindere	Turkey	107	2008
Gongguoqiao	China	56	2009
Tobetsu	Japan	52	2010
Hongkou cofferdam	China	35,5	2006
Nagashima	Japan	34	2000
Ano Mera	Greece	32	1997
Jiemian cofferdam	China	16,3	2004

6 Advantages and Disadvantages of Material Dams Cemented Sand and Gravel (CSG)

- High resistance to seismic conditions.
- Better stability against sliding.
- Can be built on weak rock and half weak rock foundations.
- Significantly less tensile stresses on the foot of the dam under seismic impact.
- Better shear strength at contact and construction joints.
- Flexible, better adaptation to foundation settlements.
- Lower and more uniform stress distribution.
- Lower strength requirements.
- No need to cool the concrete mix.
- Less cement content reduces cost.
- Flexibility (compliance) of the material reduces the uneven sediment.
- Shear stresses are almost half that of conventional gravity dams.
- Local materials can be used as a filler without significant processing.
- Reducing the amount of crushing and sorting work.
- High construction speed due to the absence of the need to process "cold joints".
- Waterproof is ensured through the creation of a top face of traditional concrete, reinforced concrete screen or polymer film coating.
- Minimal environmental impact.
- Low density.
- Low modulus of elasticity [4, 5].

Table 6 Material properties of (CSG) [12, 14, 17]

Characteristic	Lyapichev	Dam Cindere	Dam Kahir	Bulletin 117	Japanese engineers
E MPa	10,000	10,000	5000	10,000	2000
ν	0,25	0,22	0,25		0,25
γ (T/M^3)	2,3	2,4	2,2	2,3	2,2
C MPa	0,01–0,5	0,8	–	0,3	–
φ (degree)	45	45	–	22	–
σ_c MPa	–	6	5	5	5
σ_T MPa	–	0,6	0,6	–	–
Content	30	80 ± 10	50	50	60–80

7 Material Properties (CSG)

CSG (Cemented Sand and Gravel) is a new material for dam construction works and it is produced by adding cement and water to rock-based material such as riverbed gravel or excavation muck that can be easily obtained near dam sites and mixing them with a simple device.

The physical properties of CSG are affected by gradation curves of raw materials, unit water content, unit cement content, and so on. Presented in Table 6 are the results of previously carried out experimental studies, such as Yu. P. Lyapicheva, Bulletin 117 of the Cindere dam, Kahir dam [10, 13], showed that, depending on the particle size distribution and water-cement ratio, the properties of especially lean rolled concrete (CSG) may differ.

8 Methods

In the studies, 2 variants of dams were considered, the profiles of which were obtained on the basis of solving the optimization problem.

Option 1: Trapezoid-shaped CSG profile (laying slopes 0.4 ÷ 0.6) - a dam, the main volume of the body, which is lean concrete Fig. 2a.

Option 2: Profile with a vertical pressure face in the upper part of the profile and with a 0.05 inclination in the lower part. The lower face of a variable profile with a laying of 0.82 and 1.0 is a dam made of plastic rolled concrete, Fig. 2b.

Estimated seismicity of the construction area - 9 points. The obtained profiles of the considered dams are shown in Figs. 2a, 2b. The aim of the work is to compare the stress–strain state and the cost of sections of dam profiles erected using lean and ductile rolled concrete. The initial data for the design of dams are given in tabular form 7, 8.

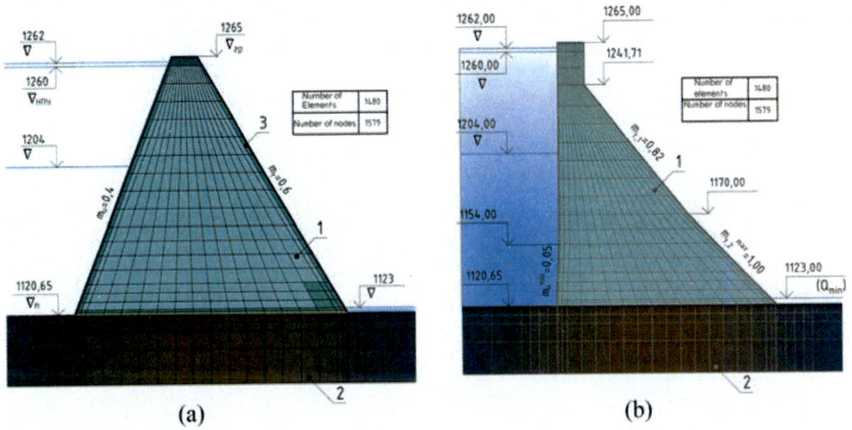

Fig. 2 Profiles of the dam M 1:1000 1 - Rolled concrete, 2 - Base (serpentinites), 3 - Vibrated concrete

Table 7 Material properties

Value	Roller compacted concrete	Foundation	Vibrated concrete
Unit mass kn/m^3	23	–	24
Modulus of Elasticity E, (MPa)	10 000	12 000	25 000
Compressive strength (MPa)	5.00	7.50	11.50
Tensile strength (MPa)	0,25	0,40	0,90
Poisson's ratio	0,22	0,26	0,18

Table 8 Properties of the simulated contact section

Value	Meaning
C, MPa	0,20
φ, °	38,7
σ_T (MPa)	0,35
kn, MPa /m	0.1,105
ks, MPa /m	0.25,105

9 Results

Calculation of the stress–strain state of various options for the profiles of a dam made of CSG material and a dam made of rolled concrete material. Calculations for the main and special combination of loads were carried out using the CRACK software package [5, 21], which implements the finite element method in a strength setting and allows modeling various discontinuities in seams, cracks, and contacts. Let us set the

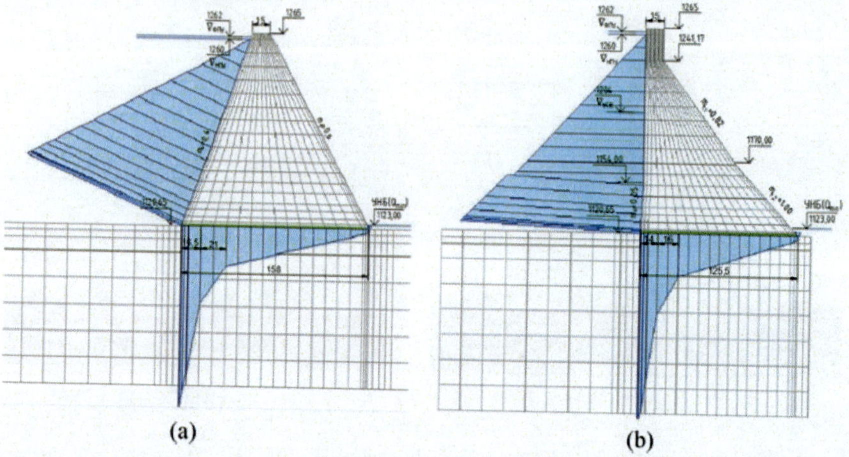

Fig. 3 Design schemes for the calculation of the stress–strain state

initial data and grid layout parameters (the number of elements is 1480, the number of nodes is 1579), after which we obtain the design schemes shown in Fig. 3a, 3b.

10 Comparative Calculations of the Stress–strain State and Stability of the Dam

Comparative calculations of the stress–strain state and stability of a concrete gravity dam of a traditional profile made of roller compacted concrete (RCC) and a gravity dam of the Trapezoid-shaped profile of (CSG). To analyze the stress–strain state, a profile was chosen with the following characteristics: Dam height H = 140 m; Ridge width b = 15 m; (laying slopes 0.4 ÷ 0.6). The results of the stress–strain state calculation are presented in Fig. 4a, 4b and Fig. 5a, 5b. Calculations were performed for the action of the main static (own weight of the structure, hydrostatic pressure from the upstream and downstream, filtration and weighing pressures on the dam foot, taken in accordance with STO 17,330,282.27.140.002-2008) loads of the operational period.

Let's summarize the obtained data in Table 9. The normative values of the stability coefficient and the absolute value of the maximum compressive stress were found from the relevant normative literature [2, 3]. An approximate estimate of the cost of structures is given in Table 10. The cost of 1 m^3 of material was selected from the recommendations [2, 3].

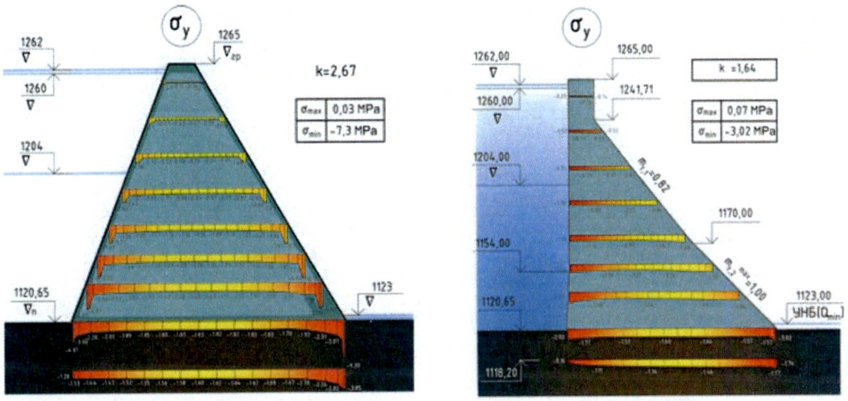

(a) (b)

Fig. 4 The main combination of loads distribution of normal stresses

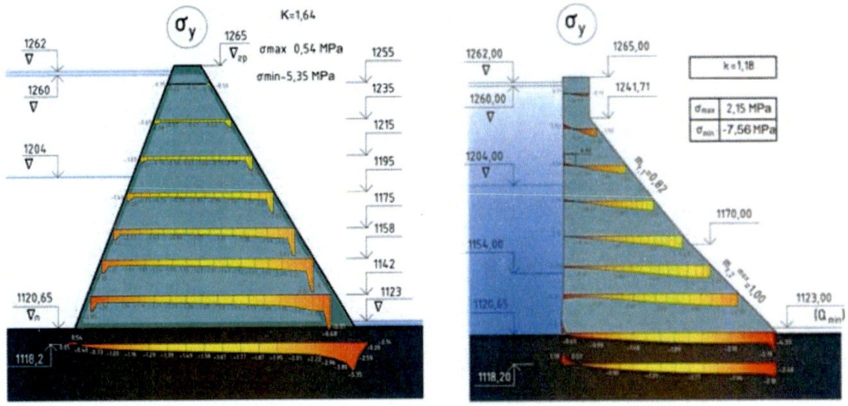

(a) (b)

Fig. 5 Special combination of loads distribution of normal stresses

Table 9 Comparison of the results of the calculation of the stress–strain state of gravity dams

The main combination of loads			
	Ks	σy max, MPa	σy min, MPa
CSG (option 1)	2,67	0.03	−7,3
Roller Compacted Concrete (option2)	1,64	0,07	−3,02
Standard value	1,316	–	8,84
Special combination of loads			
CSG (option 1)	1,64	0,54	−5,35
Roller Compacted Concrete (option2)	1,185	2,15	−7,56
Standard value	1,184	–	10,28

Table 10 Estimated cost of the dam section

Kind of concrete			$ м^3 material	
	DAM of CSG	Dam of RCC	Iran, 2009	Lyapichev 1995
CSG	11,168	0	$ 20	$ 21
RCC	0	5476	$ 31	$ 35
Vibrated concrete	1407	3051	$ 60	
The total cost of the section	$ 307,780	$ 351,173	–	–

11 Conclusions

1) For the Trapezoid-shaped profile dam (CSG) with a height of 140 m, the nature of the stress distribution in the body of the dam and in the foundation is more favorable than for classic profile dams made of (RCC).
2) The coefficients of stability for the main and special combination of loads in a lean concrete dam are higher than in a rolled concrete dam.
3) Reducing construction time by –1.5 times.
4) Symmetrical dams made of extra lean rolled concrete have better seismic resistance and technical and economic efficiency compared to conventional rolled concrete dams.

Acknowledgements The author expresses his deepest gratitude to his supervisor, Doctor of technical sciences Tolstikov, V.V, who helped with his invaluable guidance in difficult moments of reflection on the solution of the assigned tasks.

References

1. Sudakov VB (2011) Construction of dams from rolled concrete. Prospects and tasks, JSC VNIIG named after B.E. Vedeneev, St. Petersburg, p 42
2. Ahab Ghods MG (2008) Price List of Dam, Consulting Engeneering, Iran
3. Lyapichev YP (2004) Design and Construction of Modern High Dams. Publishing House of the Peoples' Friendship University of Russia, Moscow
4. Tolstikov VV (1993) Mathematical modeling of the static operation of concrete dams, taking into account discontinuities and elastic-plastic work of the material. Diss.kan.tekh.nauk: 05.23.07-M., Cc 11, 19, 65.67, p 86.231
5. Tolstikov VV (2006) Simulation of the work of joints and cracks in the calculations of the stress-strain state of concrete dams Vestnik MGSU, no 2
6. Orekhov VG (1984) Calculation of the bearing capacity of concrete dams on rocky foundations. In: Orekhov VG (ed.) Materials of Conferences and Meetings on Hydraulic Engineering. L.: Energoatomizdat, pp 42–49
7. Ghaemian M, Ghobarah A (1999) Nonlinear seismic response of concrete gravity damswith dam-reservoir interaction. J Eng Struct 306–315
8. Rueda F, Camprubi N, Garcia G, Pardo JM (2006) Thermo mechanical analysis of La Breña II Dam during its construction process: evaluation of potential thermal cracking. In: Proceedings of 22th Congress on Large Dams, Q.84, R.31, Barcelona, Spain, June 2006

9. Berton R (2004) Estimation of Nonlinear Parameters Roller Compacted Concrete Using Experimental Results for Dams. Sharif University of Technology, Tehran-Iran
10. Lyapichev YP (2009) Design and construction of modern high dams: Textbook/Yu.P. Lyapichev. Publishing house of RUDN University
11. Tolstikov VV, Nguyen Dyk Ngia DN (2011) Numerical study of possible schemes of destruction of the system Concrete gravity dam-rock foundation. Bulletin of MGSU. 2011, no 5
12. Torkamani HM (2010) Evaluation and Comparing of Faced Symmetrical Hardfill Dams (FSHD) and Concrete Faced Rock-Fill Dams (CFRD), University of Kurdistan, Sanandaj, Iran, May 2010
13. Hiros T, Fujisawa T, (2007) Design concept of trapezoidal shapped CSG dams. In: Proceeding of the Fifth International Symposium on Roller Compacted Concrete Dams, Guiyand, China, 2007, pp 457–464
14. Hitoshi Yoshida HY, Daisuke Hirayama DH, Noboru Ogiya NO, Hagiwaza J (2003) Effect of the amount of paste on CSG material properties. Civ Eng J 45–5:60–65. Japan
15. ICOLD-Bulletin 126: Roller-Compacted Concrete Dams. State of the art and case histories. – Ed. International Comission on Large Dam, Paris, France, 2003
16. Jia Jinsheng JJ (2013) New progress for better development of dams and reservoirs. In: Annual Meeting of Committee on Integrated Operation of Hydropower Stations and Reservoirs, Washington
17. Kabir MD, Mustang Rifath MR, Mahzuz MHMA (2010) Comparative study on compressive strenhth of medium grade cement concrete using various types of coarse aggregates. World Appl Sci J 8(29):206–209
18. Official site [Electronic resource] (2016). https://docplayer.ru/67297308-2-5-modul-uprugosti-polzuchest-i-vodopronicaemost-ukatannogo-i-vibrirovannogo-betona-prochnost-na-sdvig-ukatannogo-betona-mezhdu-shvami-i.html
19. Okamura Ouchi HM (2003) Self-compacting concrete. J Adv Concrete Technol 1(1):5–15
20. Hirose T, Fujisawa T, Nagayama NI, Yoshida YH, Sasaki T (2001) Design criteria for trapezoid-shaped CSG dams. In: Proceedings of the 69th ICOLD Annual Meeting; 9–15 Sep 2001; Dresden, Germany
21. Zertsalov MG, Tolstikov VV (1988) Accounting for the elastic-plastic work of concrete dams and rock foundations in calculations using the FEM. Hydrotechnical construction, no 8
22. Romero F, Sandoval A, Ibáñez de Aldecoa R, Noriega G (2007) Plans for the construction of La Breña II Dam in Spain. In: Proceedings of 5th International Symposium on RCC dams, Guiyang, China, November 2007

Mode Shapes of Hysteresis Type Elastic Dissipative Characteristic Plate Protected from Vibrations

M. M. Mirsaidov, O. M. Dusmatov, and M. U. Khodjabekov

Abstract In this study, hysteresis type elastic dissipative characteristic plate under the influence of kinematic excitations is devoted to reducing the values of low frequency transverse mode shapes in conjunction with a liquid section dynamic absorber. A mathematical model and a solution method were developed to estimate the mode shapes of an elastic dissipative characteristic plate of the hysteresis type, which is protected from vibrations under the influence of kinematic motions. In the differential equations of motion, the elastic dissipative characteristics of the plate material of the hysteresis type are taken into account by means of harmonic linearity coefficients based on the Pisarenko-Boginich hypothesis. The efficiency of the liquid section dynamic absorber in the reduction of harmful vibrations of the plate at low frequencies is evaluated on the basis of the results of numerical calculations, and it is shown that the mode shapes can achieve minimum values. In this case, the boundary conditions are taken for the plate where the two opposite sides are fixed and the other two sides are free. As a dynamic condition, the force exerted on the plate by a liquid section dynamic absorber is equal to the shear force at the point where the plate is connected to the liquid section dynamic absorber. Numerical calculations were performed and conclusions were made for the case where the liquid section dynamic absorber was mounted in the geometric center of the plate. Optimal values of elastic and damping elements of the liquid section dynamic absorber, where the mode shapes reach the minimum values, have been determined.

M. M. Mirsaidov
National Research University - Tashkent Institute of Irrigation and Agricultural Mechanization Engineers, 39 Kori Niyoziy str., Tashkent 100000, Uzbekistan
e-mail: mirsaidov1948@mail.ru

Institute of Mechanics and Seismic Stability of Structures of the Academy of Sciences of the Republic of Uzbekistan, 33 Durmon str., Tashkent 100125, Uzbekistan

O. M. Dusmatov
Samarkand State University, 15 University blv., Samarkand 140104, Uzbekistan
e-mail: dusmatov62@bk.ru

M. U. Khodjabekov (✉)
Samarkand State Architectural and Civil Engineering Institute, 70 Lolazor str., Samarkand 140147, Uzbekistan
e-mail: uzedu@inbox.ru

P. Akimov et al. (eds.), *Proceedings of FORM 2022*, Lecture Notes in Civil Engineering 282, https://doi.org/10.1007/978-3-031-10853-2_12

Keywords Vibrations · Plate · Liquid section dynamic absorber · Hysteresis · Mode shapes · Boundary and dynamic conditions

1 Introduction

Taking into account the hysteresis type elastic dissipative characteristics of mechanical systems under the influence of various excitations used in modern techniques and technologies, the study of the dynamics and exploration of stability using the mode shapes is one of the urgent problems. To date, many theoretical and experimental studies have been carried out to study the dynamics and stability of mechanical systems. However, there is a need to conduct research to improve research methods using mode shapes, taking into account the elastic dissipative characteristics of hysteresis type of mechanical systems.

In the works [1–3], the vibrations of a rectangular plate with edges mounted on elastic springs were studied using converging rows. The natural frequency was determined depending on the system parameters. Mode shapes were analyzed numerically and the effect of system parameters on their change was evaluated as a result of numerical calculations. Recommendations are given for the identification of mode shapes that meet the boundary conditions.

The works [4, 5] focuse on the optimal design of liquid section dynamic absorbers to control the seismic impact of structures. Algorithms were used in the optimization process to achieve the optimal solution.

The problem of structural modeling of distributed systems with linear elastic characteristics by the method of bond graph is considered in the works [6–10]. The motion equations are expressed by first-order derivatives, and the motion stability is explored from the system matrix. Nonlinear free and forced vibrations of discrete and continuous systems have also been studied. The systems are mathematically modeled using the Lagrange equation.

In the works [11, 12] have shown that semi-active systems cover the properties of passive and active systems. A dynamic absorber with a passive liquid section converted to a semi-active system with variable damping properties was analyzed.

The works [13, 14] have studied the dynamics of complex mechanical systems and solved the problems of bending of systems consisting of several layers of distributed parameters.

In the works [15, 16] the mathematical model of transverse vibrations of systems of elastic dissipative characteristic of hysteresis type protected from vibrations is given and the stability is explored. In this case, a conventional dynamic absorber is taken as an object that protects the plate from harmful vibrations.

Mathematical model of transverse vibrations under the influence of harmonic and random excitations of an elastic dissipative characteristic plate of the hysteresis type, protected from vibrations, is presented and studied in the work [17]. The elastic characteristics of the plate material of the hysteresis type are obtained on the basis of the Pisarenko-Boginich hypothesis. Quantitative calculations were performed and analyzed on the obtained results, conclusions were drawn and recommendations were developed.

The work [18] considered the problem of determining the mode shapes of rod with liquid section dynamic absorber when it is used as an object to reduce vibrations.

The analysis of the above works shows that the study of the dynamics of transverse vibrations of the plate with an elastic dissipative characteristic in conjunction with the liquid section dynamic absorber using mode shapes is one of the current problems of mechanics.

This work is devoted to solving the problem of reducing the mode shapes of low frequency transverse vibrations of the plate under the influence of kinematic excitations in conjunction with the liquid section dynamic absorber and determining the optimal parameters of the liquid section dynamic absorber.

2 Materials and Methods

We consider the vibration of the rectangular hysteresis type elastic dissipative characteristic plate with sides *a*, *b*, and thickness *h* under the influence of kinematic excitations in conjunction with the liquid section dynamic absorber (Fig. 1).

m_1 and q_3 are the mass and displacement of the outer body of the dynamic absorber surrounding the liquid, respectively; m_2 and q_4 are the mass and displacement of the solid of the dynamic absorber, respectively; m_3 is mass of liquid; m_4 is mass of liquid adhering to the body with mass m_2; b_F is the coefficient of resistance of the dynamic absorber; c_1 and c_2 are coefficients of stiffnesses.When the kinematic excitations consist of forces $F_L(t)$ and $F_R(t)$ placed between the opposite sides of the plate, the motion of the elastic dissipative characteristic plate of the hysteresis type protected from the vibrations under consideration can be mathematically modeled as follows:

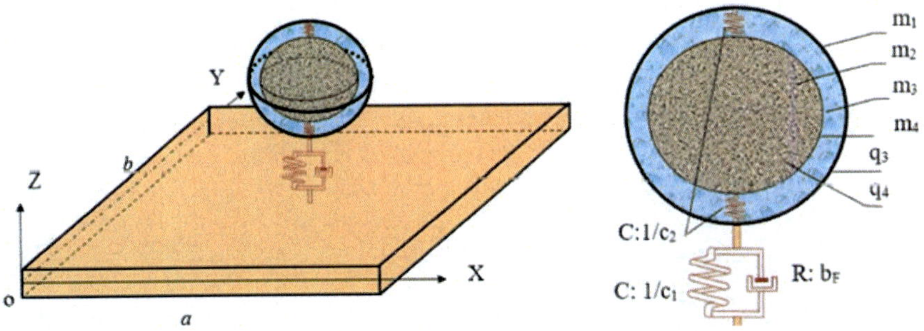

Fig. 1 Scheme of plate and liquid section dynamic absorbing system

$$A\ddot{Q} + B\dot{Q} + CQ = F, \tag{1}$$

where

$$\ddot{Q} = \begin{bmatrix} \ddot{q}_{km} \\ \ddot{q}_3 \\ \ddot{q}_4 \end{bmatrix};\ \dot{Q} = \begin{bmatrix} \dot{q}_{km} \\ \dot{q}_3 \\ \dot{q}_4 \end{bmatrix};\ Q = \begin{bmatrix} q_{km} \\ q_3 \\ q_4 \end{bmatrix};\ F = \begin{bmatrix} u_{km0}F_L + u_{kma}F_R \\ 0 \\ 0 \end{bmatrix};$$

$$A = \begin{bmatrix} m_{km} & 0 & 0 \\ (m_{13*} + m_2)u_{km1} & m_{13*} + m_2 & m_2 + m_v \\ (m_2 - m_v)u_{km1} & m_2 - m_v & m_2 + m_4 \end{bmatrix};\ B = \begin{bmatrix} 0 & -u_{km1}b_F & 0 \\ 0 & b_F & 0 \\ 0 & 0 & b_S \end{bmatrix};$$

$$C = \begin{bmatrix} c_{km} & -u_{km1}c_1 & 0 \\ 0 & c_1 & 0 \\ 0 & 0 & 2c_2 \end{bmatrix};$$

q_{km} are displacements of plate points; $m_{13*} = m_1 + m_3$; m_v is mass of liquid squeezed by the solid with mass m_2; b_S is the viscosity coefficient of the liquid; m_{km} and c_{km} are modal masses and stiffnesses, expressed as (k, $m = 1 \ldots n$):

$$m_{km} = \int_0^a\int_0^b \rho h u_{km}^2 dxdy;$$

$$c_{km} = \Bigg[(1 + D_0(-\eta_1 + j\eta_2))\int_0^a\int_0^b \rho h u_{km}^2 dxdy$$

$$+\frac{3D}{\omega_{km}^2}(-\eta_1 + j\eta_2)\sum_{R=1}^{S_1} D_R q_{kma}^R \frac{h^R}{2^R(R+3)}$$

$$\times \int_0^a\int_0^b u_{km}\Big[\frac{\partial^2}{\partial x^2}\Big(\Big(\frac{\partial^2 u_{km}}{\partial x^2} + \mu_n\frac{\partial^2 u_{km}}{\partial y^2}\Big)\Big|\frac{\partial^2 u_{km}}{\partial x^2} + \mu_n\frac{\partial^2 u_{km}}{\partial y^2}\Big|^R\Big) + \frac{\partial^2}{\partial y^2}\Big(\Big(\frac{\partial^2 u_{km}}{\partial y^2} + \mu_n\frac{\partial^2 u_{km}}{\partial x^2}\Big)$$

$$\times\Big|\frac{\partial^2 u_{km}}{\partial y^2} + \mu_n\frac{\partial^2 u_{km}}{\partial x^2}\Big|^R\Big)\Big]dxdy + \frac{6D}{\omega_{km}^2}(1 - \mu_n)(\nu_1 - j\nu_2)\sum_{N=1}^{S_2} K_N q_{kma}^N \frac{h^N}{2^N(N+3)}$$

$$\times \int_0^a\int_0^b u_{km}\frac{\partial^2}{\partial x\partial y}\left(\left(\frac{\partial^2 u_{km}}{\partial x\partial y}\right)\left|\frac{\partial^2 u_{km}}{\partial x\partial y}\right|^N\right)dxdy$$

$$+\frac{2D}{\omega_{km}^2}(1 - \mu_n)(\nu_1 - j\nu_2)K_0\int_0^a\int_0^b u_{km}\frac{\partial^4 u_{km}}{\partial x^2\partial y^2}dxdy\Bigg]\omega_{km}^2;$$

ρ is the density of the plate material; $u_{km} = u_{km}(x, y)$ are mode shapes; $D_0, D_1, \ldots, D_{S_1}$, $K_0, K_1, \ldots, K_{S_2}$ are parameters determined experimentally [19, 20]; $D = \frac{Eh^3}{12(1-\mu_n^2)}$ is cylindrical stiffness; E is Young's module; μ_n is Poisson's ratio; q_{kma} are amplitude values of plate vibration forms; ω_{km} are natural frequencies of the plate; $\eta_1, \eta_2 = \eta_{22} sign(\omega)$, $\nu_1, \nu_2 = \nu_{22} sign(\omega)$ are constant coefficients depending on the dissipative properties of the plate material, determined experimentally [17]; ω is system frequency; $j^2 = -1$; $u_{km0} = u_{km}\left(\frac{x}{2}, 0\right)$, $u_{kma} = u_{km}\left(\frac{x}{2}, b\right)$

and $u_{km1} = u_{km}(x_1, y_1)$ are the values of the mode shapes at the points where the forces are applied and at the point where the liquid section dynamic absorber is installed.

We express the bending of the plate.

$$w_{km}(x, y, t) = \sum_{k=1}^{\infty}\sum_{m=1}^{\infty} u_{km}(x, y)q_{km}(t). \tag{2}$$

We get the function $q_{km}(t)$ in the following form:

$$q_{km}(t) = q_{kma}(t)\cos(\omega t + \beta_{km}),$$

where ω and β_{km} are the frequencies and phases of vibrations.

From the given system of differential Eqs. (1) and the expression of the bending of the plate (2) it can be seen that the dynamic nature of the system under consideration depends on its mode shapes. With this in mind, we define the mode shapes.

3 Results and Discussion

Suppose that the two opposite sides of the plate are fixed and the other two sides are free, the liquid section dynamic absorber is installed in the middle of the plate. In this case, it is expedient to obtain the mode shapes that satisfies these boundary conditions as follows:

$$\begin{aligned} u_{km}(x, y) = {} & B_1 cosh\frac{\pi kx}{a} cosh\frac{\pi my}{b} + B_2 sinh\frac{\pi kx}{a} sinh\frac{\pi my}{b} + B_3 cos\frac{\pi kx}{a} cos\frac{\pi my}{b} \\ & + B_4 sin\frac{\pi kx}{a} sin\frac{\pi my}{b}, \end{aligned} \tag{3}$$

where B_1, B_2, B_3, B_4 are the coefficients.

We determine the coefficients B_1, B_2, B_3, B_4 from the boundary conditions. To do this, we write the following boundary conditions:

The following conditions apply to the fixed sides of the plate:

$$\left.\begin{matrix} x = 0 \\ y = 0 \end{matrix}\right\}, \rightarrow w_{km} = w_0, \frac{\partial w_{km}}{\partial x} = 0, \tag{4}$$

$$\left.\begin{matrix} x = a \\ y = b \end{matrix}\right\}, \rightarrow w_{km} = w_0, \frac{\partial w_{km}}{\partial x} = 0, \tag{5}$$

where w_0 is the displacement of the base.

The following conditions are appropriate for free sides:

$$\left.\begin{array}{l} x=0 \\ y=0 \end{array}\right\}, \rightarrow \frac{\partial^2 w_{km}}{\partial y^2}+\mu_n \frac{\partial^2 w_{km}}{\partial x^2}=0,$$

$$\frac{\partial^3 w_{km}}{\partial y^3}+(2-\mu_n) \frac{\partial^3 w_{km}}{\partial x^2 \partial y}=0, \tag{6}$$

$$\left.\begin{array}{l} x=a \\ y=b \end{array}\right\}, \rightarrow \frac{\partial^2 w_{km}}{\partial y^2}+\mu_n \frac{\partial^2 w_{km}}{\partial x^2}=0,$$

$$\frac{\partial^3 w_{km}}{\partial y^3}+(2-\mu_n) \frac{\partial^3 w_{km}}{\partial x^2 \partial y}=0. \tag{7}$$

The following dynamic condition is appropriate at the point where the liquid section dynamic absorber is installed:

$$D \int_0^a \int_0^b \left(\frac{\partial^4 w_{km}}{\partial x^4}+2 \frac{\partial^4 w_{km}}{\partial x^2 \partial y^2}+\frac{\partial^4 w_{km}}{\partial y^4}\right) dx dy=-F_0, \tag{8}$$

where F_0 is the force representing the effect of the liquid section dynamic absorber on the plate.

Taking the displacement of the base in the form $w_0=-\frac{\varepsilon p_*}{\omega^2} \cos \omega t$ (εp_* is amplitude value of the acceleration of the base), we analyze the boundary and dynamic conditions (4)–(8).

We write the following from the boundary condition $\left.\begin{array}{l} x=0 \\ y=0 \end{array}\right\}, w_{km}=w_0$:

$$w_{km}=u_{km}(0,0)\, q_{kma} \cos(\omega t+\beta_{km})=w_0=-\frac{\varepsilon p_*}{\omega^2} \cos \omega t. \tag{9}$$

We receive the following from Eq. (9):

$$u_{km}(0,0)=\pm \frac{\varepsilon p_*}{q_{kma} \omega^2}. \tag{10}$$

We put (3) mode shapes in (10) condition. As a result, we obtain the following relationship between the coefficients B_1 and B_3:

$$B_1+B_3=\pm \frac{\varepsilon p_*}{q_{kma} \omega^2}. \tag{11}$$

Given the mode shapes (3), the condition $\left.\begin{array}{l} x=0 \\ y=0 \end{array}\right\}, \frac{\partial w_{km}}{\partial x}=0$ of the boundary conditions (4) becomes an identity.

We can write the following relation from the boundary condition $\left.\begin{array}{l} x=a \\ y=b \end{array}\right\}, w_{km}=w_0$:

$$w_{km} = u_{km}(a, b)q_{kma}\cos(\omega t + \beta_{km}) = w_0 = -\frac{\varepsilon p_*}{\omega^2}\cos\omega t. \tag{12}$$

We have the following relation from Eq. (12):

$$u_{km}(a, b) = \pm\frac{\varepsilon p_*}{q_{kma}\omega^2}. \tag{13}$$

If we put mode shapes (3) in Eq. (13) then:
when one of the *k, m* is odd and the other is even,

$$B_1\cosh(\pi k)\cosh(\pi m) + B_2\sinh(\pi k)\sinh(\pi m) - B_3 = \pm\frac{\varepsilon p_*}{q_{kma}\omega^2}. \tag{14}$$

when both *k, m* are even or odd,

$$B_1\cosh(\pi k)\cosh(\pi m) + B_2\sinh(\pi k)\sinh(\pi m) + B_3 = \pm\frac{\varepsilon p_*}{q_{kma}\omega^2}. \tag{15}$$

We obtain the following relation from the boundary condition $\left.\begin{matrix} x = a \\ y = b \end{matrix}\right\}, \frac{\partial w_{km}}{\partial x} = 0:$

$$B_1\sinh(\pi k)\cosh(\pi m) + B_2\cosh(\pi k)\sinh(\pi m) = 0. \tag{16}$$

We write from the boundary condition $\left.\begin{matrix} x = 0 \\ y = 0 \end{matrix}\right\}, \frac{\partial^2 w_{km}}{\partial y^2} + \mu_n\frac{\partial^2 w_{km}}{\partial x^2} = 0.$

$$B_1 - B_3 = 0. \tag{17}$$

The boundary condition $\left.\begin{matrix} x = 0 \\ y = 0 \end{matrix}\right\}, \frac{\partial^3 w_{km}}{\partial y^3} + (2 - \mu_n)\frac{\partial^3 w_{km}}{\partial x^2 \partial y} = 0$ becomes an identity.

We write following from the boundary condition $\left.\begin{matrix} x = a \\ y = b \end{matrix}\right\}, \frac{\partial^2 w_{km}}{\partial y^2} + \mu_n\frac{\partial^2 w_{km}}{\partial x^2} = 0:$
when one of the *k, m* is odd and the other is even,

$$B_1\cosh(\pi k)\cosh(\pi m) + B_2\sinh(\pi k)\sinh(\pi m) + B_3 = 0, \tag{18}$$

when both *k, m* are even or odd,

$$B_1\cosh(\pi k)\cosh(\pi m) + B_2\sinh(\pi k)\sinh(\pi m) - B_3 = 0. \tag{19}$$

We write following from the boundary condition $\left.\begin{matrix} x = a \\ y = b \end{matrix}\right\}, \frac{\partial^3 w_{km}}{\partial y^3} + (2 - \mu_n)\frac{\partial^3 w_{km}}{\partial x^2 \partial y} = 0:$

$$B_1\cosh(\pi k)\sinh(\pi m) + B_2\sinh(\pi k)\cosh(\pi m) = 0. \tag{20}$$

The following can be written from the dynamic condition (8):
when *k, m* are both odd,

$$\frac{4D\pi^2\left((mb)^2 + (ka)^2\right)^2}{km(ab)^3}(B_1\sinh(\pi k)\sinh(\pi m) + B_2(\cosh(\pi k) - 1)(\cosh(\pi m) - 1) + 4B_4) = \pm\frac{q_3}{q_{kma}}\sqrt{c_1^2 + (b_F\omega)^2}. \tag{21}$$

when both *k, m* are not odd,

$$\frac{4D\pi^2\left((mb)^2 + (ka)^2\right)^2}{km(ab)^3}(B_1\sinh(\pi k)\sinh(\pi m) + B_2(\cosh(\pi k) - 1)(\cosh(\pi m) - 1)) = \pm\frac{q_3}{q_{kma}}\sqrt{c_1^2 + (b_F\omega)^2}. \tag{22}$$

The ratio of amplitudes $\frac{q_3}{q_{kma}}$ in conditions (21) and (22) can be formed from the differential equations of motion (1) by the differential operator $S = \frac{d}{dt}$ as follows:

$$\frac{q_3}{q_{kma}} = \omega^2 u_{km1}\sqrt{\frac{\left(2c_2M_1 - \omega^2\Delta\right)^2 + (M_1b_S\omega)^2}{P_0^2 + P_1^2}}, \tag{23}$$

where

$$P_0 = p_0 + p_1\omega + p_2\omega^2 + p_3\omega^3 + p_4\omega^4;\ P_1 = p_5 + p_6\omega + p_7\omega^2 + p_8\omega^3;$$
$$p_0 = 2c_1c_2;\ p_1 = 0;$$
$$p_2 = -(c_1M_4 + b_Fb_S + 2c_{2*}M_1);\ p_3 = 0;\ p_4 = \Delta = M_1M_4 - M_2M_3;$$
$$p_5 = 0;\ p_6 = 2b_Fc_2 + b_Sc_1;\ p_7 = 0;$$
$$p_8 = -(b_FM_4 + b_SM_1);\ M_1 = m_{13*} + m_2;\ M_2 = m_2 + m_v;$$
$$M_3 = m_2 - m_v;\ M_4 = m_2 + m_4.$$

Given that $x_1 = \frac{a}{2}$, $y_1 = \frac{b}{2}$, we put the amplitude ratio (23) in conditions (21) and (22).

$$
\begin{aligned}
&(\tfrac{4D\pi^2\left((mb)^2+(ka)^2\right)^2}{km(ab)^3}\sinh(\pi k)\sinh(\pi m)\\
&\pm\omega^2\sqrt{\tfrac{((2c_2M_1-\omega^2\Delta)^2+(M_1b_S\omega)^2)\left(c_1^2+(b_F\omega)^2\right)}{P_0^2+P_1^2}}\\
&\times cosh\tfrac{\pi k}{2}cosh\tfrac{\pi m}{2})B_1+(\tfrac{4D\pi^2\left((mb)^2+(ka)^2\right)^2}{km(ab)^3}(\cosh(\pi k)-1)(\cosh(\pi m)-1)\\
&\pm\omega^2\sqrt{\tfrac{((2c_2M_1-\omega^2\Delta)^2+(M_1b_S\omega)^2)\left(c_1^2+(b_F\omega)^2\right)}{P_0^2+P_1^2}}sinh\tfrac{\pi k}{2}sinh\tfrac{\pi m}{2})B_2\\
&+(\tfrac{16D\pi^2((mb)^2+(ka)^2)^2}{km(ab)^3}\\
&\pm\omega^2\sqrt{\tfrac{((2c_2M_1-\omega^2\Delta)^2+(M_1b_S\omega)^2)(c_1^2+(b_F\omega)^2)}{P_0^2+P_1^2}})B_4=0.
\end{aligned}
\tag{24}
$$

$$
\begin{aligned}
&(\tfrac{4D\pi^2\left((mb)^2+(ka)^2\right)^2}{km(ab)^3}\sinh(\pi k)\sinh(\pi m)\\
&\pm\omega^2\sqrt{\tfrac{((2c_2M_1-\omega^2\Delta)^2+(M_1b_S\omega)^2)\left(c_1^2+(b_F\omega)^2\right)}{P_0^2+P_1^2}}\\
&\times cosh\tfrac{\pi k}{2}cosh\tfrac{\pi m}{2})B_1+(\tfrac{4D\pi^2\left((mb)^2+(ka)^2\right)^2}{km(ab)^3}(\cosh(\pi k)-1)(\cosh(\pi m)-1)\\
&\pm\omega^2\sqrt{\tfrac{((2c_2M_1-\omega^2\Delta)^2+(M_1b_S\omega)^2)\left(c_1^2+(b_F\omega)^2\right)}{P_0^2+P_1^2}}sinh\tfrac{\pi k}{2}sinh\tfrac{\pi m}{2})B_2=0.
\end{aligned}
\tag{25}
$$

We analyze the equations obtained on the basis of boundary and dynamic conditions. It follows from conditions (16) and (20) that $k = m$ is required for both of these equations to be satisfied. When both k, m are not odd, conditions (22) and (25) show that they can be obtained arbitrarily, since the conditions B_3 and B_4 do not depend on the coefficients. In this case, when $k = m$, the following equation is valid:

$$
\begin{aligned}
&\frac{8Dk^2\pi^2\left(a^2+b^2\right)^2}{(ab)^3}(\cosh(\pi k)-1))\\
&\quad=\mp\omega^2\sqrt{\frac{((2c_2M_1-\omega^2\Delta)^2+(M_1b_S\omega)^2)\left(c_1^2+(b_F\omega)^2\right)}{P_0^2+P_1^2}}.
\end{aligned}
$$

Hence, the coefficients sought are as follows:
when $m = k$ is odd,

$$
\begin{aligned}
&B_1=-B_2=B_3=\pm\tfrac{\varepsilon p_*}{2q_{kka}\omega^2},\\
&B_4=\left(\pm\tfrac{\varepsilon p_0}{2q_{kka}\omega^2}\right)(\tfrac{16Dk^2\pi^2\left(a^2+b^2\right)^2}{(ab)^3}\\
&\pm\omega^2\sqrt{\tfrac{((2c_2M_1-\omega^2\Delta)^2+(M_1b_S\omega)^2)\left(c_1^2+(b_F\omega)^2\right)}{P_0^2+P_1^2}})^{-1}\\
&\times(\tfrac{8Dk^2\pi^2\left(a^2+b^2\right)^2}{(ab)^3}(\cosh(\pi k)-1))\\
&\pm\omega^2\sqrt{\tfrac{((2c_2M_1-\omega^2\Delta)^2+(M_1b_S\omega)^2)\left(c_1^2+(b_F\omega)^2\right)}{P_0^2+P_1^2}}).
\end{aligned}
\tag{26}
$$

when $m = k$ is even,

$$B_1 = -B_2 = B_3 = B_4 = \pm \frac{\varepsilon p_*}{2q_{kka}\omega^2}. \tag{27}$$

When the determined coefficients and (3) the mode shapes are considered together, it is possible to determine the vibration forms of the plate protected from vibrations at different values of the system parameters.

In order to perform numerical calculations on the obtained results, we obtain the aluminum alloy AL19 for the plate material: $E = 6964119 \cdot 10^4 \frac{N}{m^2}$, $G = 26487 \cdot 10^6 \frac{N}{m^2}$; $\rho = 2780 \frac{kg}{m^3}$; $\mu_n = 0.32$.

Let the sides of the plate be $a = b = 2.6\,m$ and the thickness $h = 3 \cdot 10^{-3}\,m$. In that case the natural frequency of the plate for $k = m = 1$ will be $\omega_{11} = 13.4s^{-1}$.

Based on the results obtained in the work [17], we obtain the parameters of the liquid section dynamic absorber as follows:

$$b_S = 264.7 \cdot 10^3 \frac{N \cdot s}{m};\ m_v = 0.524\,kg;\ m_2 = 1.456\,kg;\ m_3 = 3.665\,kg;$$
$$m_4 = 0.314\,kg.$$

We select the remaining parameters as follows: $m_1 = 6.987kg$; $c_1 = c_2 = 10^4 \frac{N}{m}$; $b_F = 10^4 \frac{N \cdot s}{m}$.

Based on the given values of these variables and parameters, we analyze the mode shapes.

We draw a graph of the mode shapes.

In Fig. 2, the effect of the liquid section dynamic absorber on the mode shapes of the plate can be seen. In the mode shapes of the plate without the liquid section dynamic absorber (line 1, brown), the dimensionless value of the point at its geometric center is 22.1839, and after the installation of the liquid section dynamic absorber at this point the dimensionless value of the mode shapes at this point decreases to 15.1651 (line 2, blue).

This means that the use of the liquid section dynamic absorber as an object to protect the plate from harmful vibrations reduces the values of the mode shapes of the plate by 1.5 times. This shows that the liquid section dynamic absorber is effective in reducing low frequency harmful vibrations of the plate.

We analyze the change in the mode shapes depending on the elastic and damping elements of the liquid section dynamic absorber.

The graphs in Fig. 3. show the change in the mode shapes of the plate depending on the elastic c_1 and damping elements b_F of the liquid section dynamic absorber ($c_1 = 2,5 \cdot 10^3 \frac{N}{m}$, $b_F = 10^3 \frac{N \cdot s}{m}$ (line 1, blue), $c_1 = 10^2 \frac{N}{m}$; $b_F = 10^2 \frac{N \cdot s}{m}$ (line 2, yellow), $c_1 = 0$; $b_F = 0$ (line 3, red)). From this it can be said that at the values of $c_1 \geq 2,5 \cdot 10^3 \frac{N}{m}$, $b_F \geq 10^3 \frac{N \cdot s}{m}$ the mode shapes are very close to 15.1651 and at values $c_1 < 2,5 \cdot 10^3 \frac{N}{m}$, $b_F < 10^3 \frac{N \cdot s}{m}$ greater than 15.1651. Therefore, the values of $c_1 = 2,5 \cdot 10^3 \frac{N}{m}$, $b_F = 10^3 \frac{N \cdot s}{m}$ are the optimal values of the elastic and damping elements of the liquid section dynamic absorber.

The graphs in Fig. 4. show the change in the mode shapes of the plate depending on the elastic element c_2 of the liquid section dynamic absorber (($c_2 = 10^4 \frac{N}{m}$ (line

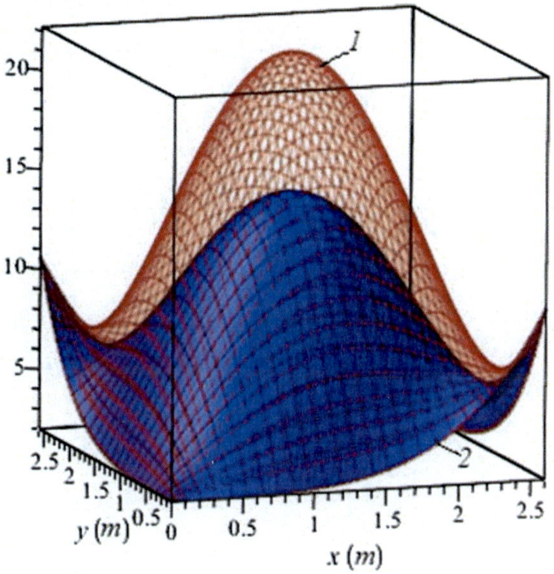

Fig. 2 Graph of the mode shapes

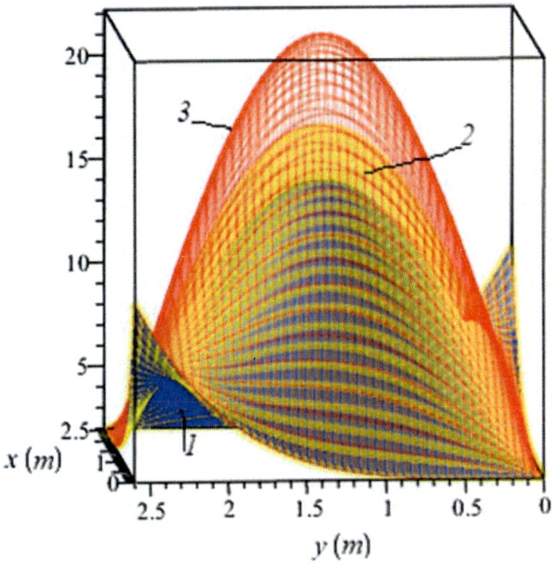

Fig. 3 Changing of the mode shapes depending on the elastic and damping elements of the liquid section dynamic absorber

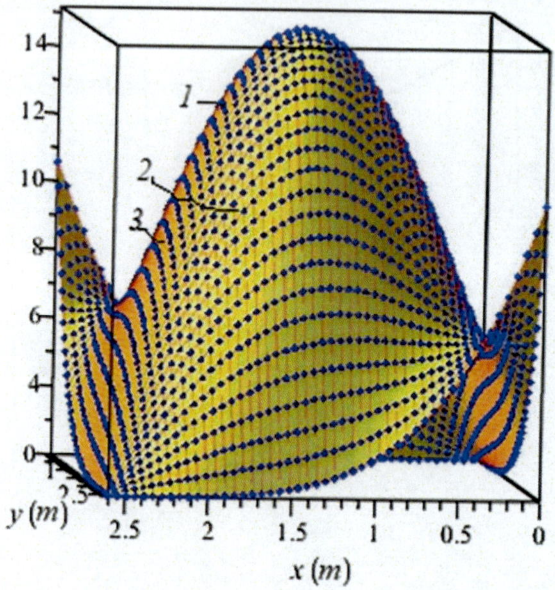

Fig. 4 Changing of the mode shapes depending on the elastic element of the liquid section dynamic absorber

1, blue), $c_2 = 10^6 \frac{N}{m}$ (line 2, yellow), $c_2 = 10^8 \frac{N}{m}$ (line 3, red)) when $c_1 = 2,5 \cdot 10^3 \frac{N}{m}$, $b_F = 10^3 \frac{N \cdot s}{m}$. From this it can be seen that the mode shapes of the plate change slowly with respect to the elastic element c_2. This means that the elastic element c_2 is the passive element relative to the elastic and damping elements c_1, b_F. This is of practical importance in the design of this system.

4 Conclusion

1. A mathematical model and solution was developed to estimate the mode shapes of the plate with elastic dissipative characteristics of the hysteresis type, protected from vibrations under the influence of kinematic excitations.
2. The expression of mode shapes allows to select the system parameters that achieve the minimum value of the displacement of the plate points protected from vibrations.
3. The use of the liquid section dynamic absorber as an object to protect the plate from harmful vibrations has been found to reduce the mode shapes values of the AL 19 aluminum alloy plate by 1.5 times.
4. The optimum values of the elastic and damping elements of the liquid section dynamic absorber for the plate of aluminum alloy AL 19 material were determined.

5. It was found that the elastic element inside the liquid of the liquid section dynamic absorber is a passive element relative to the elastic and damping elements outside the liquid.

References

1. Dal H, Morgül ÖK (2011) Sci Res Essays 6(34):6811–6816. https://doi.org/10.5897/SRE11.356
2. Changjian J, Linquan Y, Jiping S, Tonghao H, Cheng L (2019) Glob J Eng Sci 2(3). GJES.MS.ID.000537. https://doi.org/10.33552/GJES.2019.02.000537
3. Zhong Y, Zhao X, Liu H (2014) Vibration of plate on foundation with four edges free by finite cosine integral transform method. Lat Am J Solids Struct 854–863
4. Linsheng H, Wenhe S, Hongnan L, Yaowen Z (2013) Smart control algorithms and technology in civil infrastruct. Article ID 191279. https://doi.org/10.1155/2013/191279
5. Tanveer M, Usman M, Khan IU, Ahmad S, Hanif A, Farooq SH (2019) PLoS ONE 14(10):e0224436. https://doi.org/10.1371/journal.pone.0224436
6. Stian S, Eilif P (2016) The theory of bond graphs in distributed systems and simulations. In: Proceedings of the International Conference on Bond Graph Modeling and Simulation, ACM Proceedings, (Society for Computer Simulation International, San Diego, CA, United States, 2016), pp 147–156
7. Cezary O, Rafał H (2011) Modelling and analysis of beam/bar structure by application of bond graphs. J Theor Appl Mech 49(4):1003–1017
8. Migdalovici M, Baran D (2007) Theoretical research regarding any stability theorems with applications. In: Proceedings of the 14th International Congress on Sound and Vibration, (Cairns, Australia, 2007), pp 575–581
9. Jans V, Līga G (2015) Overview of tuned liquid dampers and possible ways of oscillation damping properties improvement. In: Proceedings of the 10th International Scientific and Practical Conference, (Rezekne, Latvia, 2015), pp 233–238. https://doi.org/10.17770/etr2015vol1.230
10. Alberto DM, Mario DP, Antonina P (2016) Smart Struct Syst 18(1):117–138. https://doi.org/10.12989/sss.2016.18.1.117
11. Swaroop KY, Ahsan K, Jeffrey CK (2001) Semi-active tuned liquid column dampers for vibration control of structures. Eng Struct 23:1469–1479
12. Swaroop KY, Ahsan K (2003) Semiactive tuned liquid column dampers: experimental study. J Struct Eng 129(7):960–971
13. Ahmedov O, Mirsaidov M (2021) Finite element of rotating wheelset and its natural frequencies determination. Mater Phy Mech 47(5):706–719
14. Mirsaidov M, Mamasoliev O (2020) Contact problems of multilayer slabs interaction on an elastic foundation. IOP Conf Ser Earth Environ Sci 614(1):012089. https://doi.org/10.1088/1755-1315/614/1/012089
15. Mirsaidov MM, Dusmatov OM, Khodjabekov MU (2021) Stability of nonlinear vibrations of plate protected from vibrations. J Phys Conf Ser 1921. https://doi.org/10.1088/1742-6596/1921/1/012097
16. Mirsaidov MM, Dusmatov OM, Khodjabekov MU (2020) The problem of mathematical modeling of a vibration protected rod under kinematic excitations. In: Proceedings of VII International Scientific Conference Integration, Partnership and Innovation in Construction Science and Education, 11–14 November 2020, Tashkent. https://doi.org/10.1088/1757-899X/1030/1/012069
17. Pavlovsky MA, Ryzhkov LM, Yakovenko VB, Dusmatov OM (1997) Nonlinear problems of the dynamics of vibration protection systems (Technique, Kiev, 1997), p 204

18. Mirsaidov MM, Dusmatov OM, Khodjabekov MU (2022) Lect Notes Civ Eng 170:217–227
19. Pisarenko GS, Yakovlev AP, Matveev VV (1971) Vibration-damping properties of construction materials (Nauk. Dumka, Kiev, 1971), p 327
20. Pisarenko GS, Boginich OE (1981) Vibrations of kinematically excited mechanical systems taking into account energy dissipation (Nauk. Dumka, Kiev, 1981), p 219

Dynamics of FRC Slabs on Elastic–Plastic Supports Under Blast Loading

Ashot Tamrazyan, Anatoly Alekseytsev, and Svetlana Sazonova

Abstract The article is devoted to the solution of the actual problem of the analysis of the stress–strain state of fiber-reinforced concrete (FRC) structures under emergency dynamic effects. A prefabricated roof slab of a protective structure buried in the ground is considered. Shock effects on the slab are considered to be blast impulse ones. The cause of their occurrence is believed to be an explosion at the surface of the earth or another impulse transmitted through the soil of the undisturbed structure to the structure. The nature of the change for the dynamic effect in the concrete and reinforcement when the supporting conditions are changed is investigated. These conditions are understood as a possibility of rigid or elastoplastic deformation of supports under impact action. By means of a FRC example the efficiency of the suggested approach is demonstrated under symmetrical loading by impulse loading taking into account the variable area of impact spot. Fields of normal stress distribution in time for fiber concrete, stress distribution in a compliant support in the form of a round electric-welded tube are presented. Numerical simulations have shown that both under static and dynamic limit loading, the slab collapses over the concrete with the formation of a rebar-cable mechanism, while maintaining the survivability property while ensuring the strength of the reinforcement. The main research perspectives and conclusions about the correctness of the assumed approach to the modeling of the dynamic behavior of the structure have been formed. It has been proved that the use of compliant supports plays an essential role in damping of vibrations and makes it possible to regulate the level of mechanical safety of the structure and its material capacity. The significance, necessity and relevance of determining the actual damping properties of the structure has been established.

Keywords FRC slabs · Blast load · Elastic–plastic supports · Direct transient analysis · Structural safety

A. Tamrazyan · A. Alekseytsev (✉) · S. Sazonova
Moscow State University of Civil Engineering, Moscow 129337, Russian Federation
e-mail: aalexw@mail.ru

P. Akimov et al. (eds.), *Proceedings of FORM 2022*, Lecture Notes in Civil Engineering 282, https://doi.org/10.1007/978-3-031-10853-2_13

1 Introduction

The calculation and design of load-bearing structures of protective structures is an urgent engineering and research task. At the same time to the used reinforced concrete, FRC, steel and other structures the main requirement is to provide mechanical safety. In scientific researches it is of interest to consider blast load influences [1–7], including pulses of various forms [8–16].

In a number of cases, ensuring those or other performance qualities of the structure under such influences requires a significant overconsumption of material. To solve this problem, supports can be used to dampen dynamic impacts, which can significantly reduce the dynamic overloading in structures [17–20]. Approaches that implement the selection of rational object parameters on the basis of metaheuristic search procedures are also promising [21–23].

To realize such an approach to design, the dynamic load should be represented in the form of some static equivalent, which can be implemented on the basis of the energy approach [24]. This approach can be applied in case of qualitative coincidence of the structure's own form of vibrations with the deformed scheme from the load action. Thus, one of the main problems in the design of load-bearing reinforced concrete structures of protective structures is the assessment of the stress–strain state under dynamic influences, including the determination of the limit load.

Another no less important task is dynamic calculation of structures under emergency loads of relatively low velocity impacts for civil engineering structures, for example, falling of mounted structures on the set in the designed position, impact effects of man-made origin, implemented during the operation of structures.

There are known works [25–27], in which the approaches to dynamic calculation of structures on deformed supports are represented by rather complicated models. Their application in engineering practice is somewhat constrained by high labor intensity and adaptation to a certain type of structures and support links. In this relation, the development of numerical models for reproducing the stress–strain state of such structures under various forms of momentum and time of dynamic loads seems relevant.

This paper presents an approach to the numerical modeling of FRC slabs taking into account the presence of rigid and deformable supports. The influence of damping effects on the values of stresses, deformations and displacements of these structures is evaluated. Proposals on the method of searching rational parameters of supports stiffness for dynamic impacts of a given type and intensity in order to obtain the least material-intensive structures are given.

2 Methods

We consider an explosive detonation impact on the ground surface. At the same time, accurate modeling of the pulse shock load acting on the structure buried in the ground is a difficult and poorly studied problem, however, the current level of research, in particular, the provisions of works [17, 28, 29] allow an approximate estimate of the intensity of the dynamic load as a pulse transmitted through the ground to the structure and its force equivalent q_{eq}. For an impulse load at a detonation explosion (Fig. 1,a), the influence of the rarefaction phase on the structure when the explosion epicenter is on the ground surface is insignificant, so we consider only the compression phase. We assume a dense ground, and the graph of momentum intensity variation in time with the depth of the ground has the form shown in Fig. 1,b. The character of propagation of the blast wave front in the ground is taken on the basis of the data of numerical analysis of the work [29]. For the structure we are considering, we will consider two possible variants of impulse load transfer to the structure (Fig. 1,c, d).

To switch from momentum i [29] to equivalent force impact q_{ed}, it is used the condition of equality of momentum transmitted by the explosion by the ground, perceived by the structure, to the momentum of the force q_{ed}:

$$i = \mu P \tau_1 / 2 = q_{ed} t, \tag{1}$$

where μ is the coefficient taking into account the burial of the structure (see Fig. 1,b), t is the time of the dynamic load acting on the structure. The momentum intensity P can be determined based on the Jones-Wilkins-Lee (JWL) model using the equation:

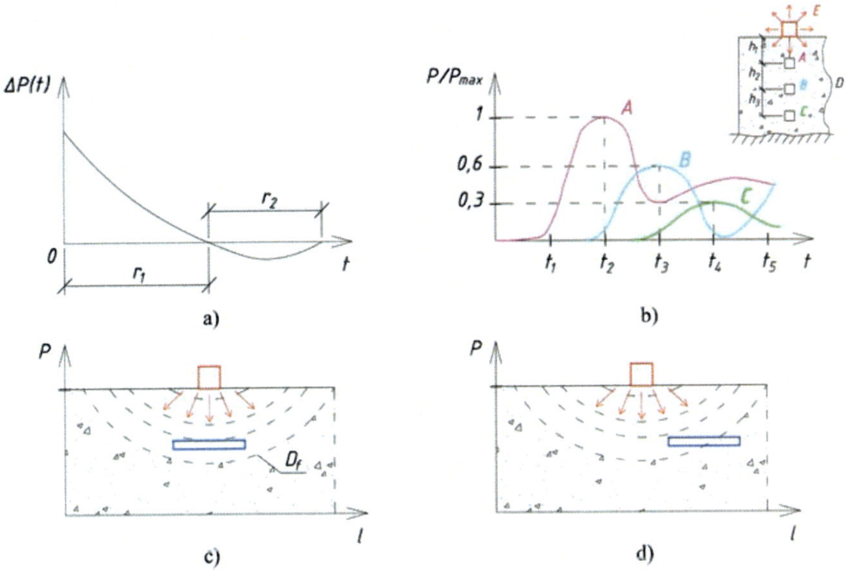

Fig. 1 To model the load on the structure: **a** change in time t the momentum Δp during a detonation explosion: τ_1, τ_2 are times of discharge and time of compression with the formation of waves in the ground; **b** change in the intensity of the blast impulse momentum in the ground by depth: A, B, C are points by depth, D is a ground base, E is an explosive charge; **c-d** location of the structure relative to the epicenter of explosion: Df is the distribution of the wave front in the ground

$$P = A\left(1 - \frac{\omega}{R_1 V}\right)e^{-R_1 V} + B\left(1 - \frac{\omega}{R_2 V}\right)e^{-R_2 V} + \frac{\omega E}{V}, \tag{2}$$

where the quantities included in this equation are described in detail in [30].

The solution of the problem within the framework of the finite element method is realized on the basis of the equation

$$[M]\ddot{y}(t) + [C]\dot{y}(t) + [K_\tau(t)]y(t) = F(t) + G\chi(t), \tag{3}$$

where $[M]$, $[C]$ are global mass and damping matrices for the finite-element model, $[K_\tau(t)]$ is the global tangent stiffness matrix for the deformed state of the system; $y(t)$, $\dot{y}(t)$, $\ddot{y}(t)$ are vectors of displacements, velocities and accelerations, $F(t)$ is the vector of external load reduced to nodes, $G\chi(t)$ is the Heaviside function, t is the integration time represented by a set of discrete values with a certain step. The matrix $[K_\tau(t)]$ for the problem we consider is defined by the formula:

$$[K_\tau(t)] = [K_{bo}(t)] + [K_{so}(t)] + \left[K_{po}(t)\right] + [K_{b\sigma}(t)] + [K_{s\sigma}(t)] + \left[K_{p\sigma}(t)\right], \tag{4}$$

where $[K_{bo}(t)]$, $[K_{so}(t)]$, $\left[K_{po}(t)\right]$ are matrices of deformations for finite elements of fiber concrete, reinforcement and steel shells describing deformations of supports, respectively; $[K_{b\sigma}(t)]$, $[K_{s\sigma}(t)]$, $\left[K_{p\sigma}(t)\right]$ are geometric matrices for, respectively, elements of fiber concrete, reinforcement and shells.

Within the framework of the finite element method, the representation of a time-varying impulse load with a varying impact spot is realized on the basis of the equations

$$F(t) = q_{st} Q_{st}(t) + \sum_{i=1}^{S} q_{i,eq} Q_i(t), \tag{5}$$

where q_{st}, $q_{i,eq}$ are constants, values of loads acting during operation and emergency dynamic overloading respectively, S is the number of sections (areas) where the load acts, $Q_{st}(t)$, $Q_i(t)$ are functions of load changes in time, normalized by the value of the corresponding loads.

Deformation modeling of FRC was performed on the basis of Drucker-Prager plasticity theory, the material of reinforcement and pipe was taken as steel, taking into account the bilinear diagram of deformation without strengthening.

3 Results

Let us consider the results of modeling of the stress–strain state of the building slab having a protective functional purpose and located at some depth from the surface of the ground. The operational load $q_{st} = 15$ kPa includes the plate's own weight of 3 kPa and the weight of the ground of 12 kPa. The geometric and design diagrams of the slab are shown in Fig. 2.

The structure of the typical solution from the series U 01-01-/80 "Unified prefabricated monolithic structures of buried premises with beam-type floor slabs" was taken as a basis. Fibrous concrete with tensile strength of 4 MPa and compression resistance of 22 MPa was used. Initially, verification of the model was performed by estimating the ultimate static loading. In the nonlinear calculation, the value of limiting static loading by uniformly distributed load $q_{st,\lim} = 41,5$ kPa was obtained. The deflection of the structure was 8.3 cm. The stress–strain state is shown in Fig. 3.

As can be seen from the calculation at the selected mechanical characteristics, the failure of the structure occurs along the normal sections. Stresses in the transverse reinforcement in the supported sections do not exceed 170 MPa. Until the level of load in 20 kPa is reached, the slab and beams work together, after which the fiber concrete of the slab over the entire height of its cross section begins to work in tension, then the slab collapses, the reinforcement of its lower grid begins to work according to the scheme of the cable-stayed mechanism.

Let's consider dynamic loading by impulse load with variable impact spot. For this purpose, let's divide the slab by its area into 5 sections (Fig. 4,a), describing the pulse propagation scenario, and we will use the time functions (Fig. 4,b) for each of the loads $q_1 - q_3$. The operating load q_4 increases for 0.5 s. and remains constant over the entire integration time interval. The distribution of the load in time is shown in Fig. 4,c. It corresponds to the loading variant shown in Fig. 1,c.

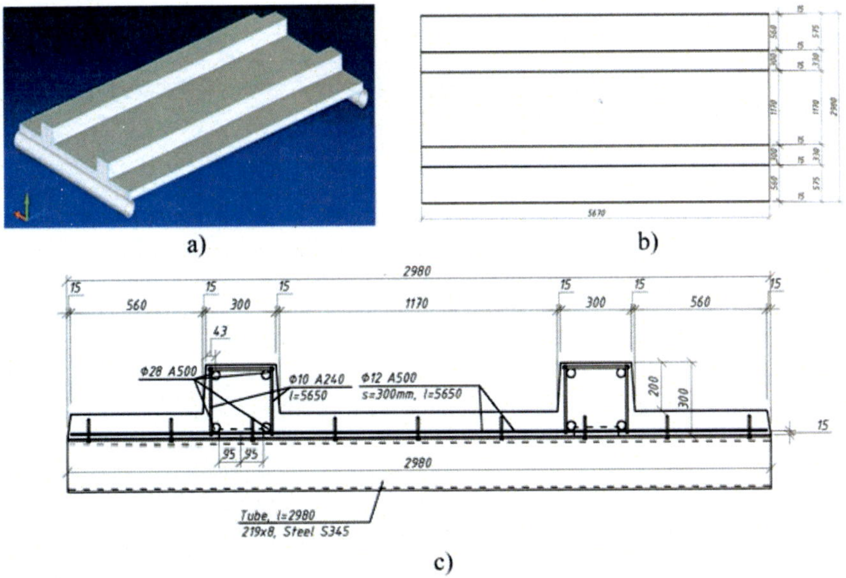

Fig. 2 Cover slab of a buried structure: **a** solid model; **b** plan; **c** cross-section

a)

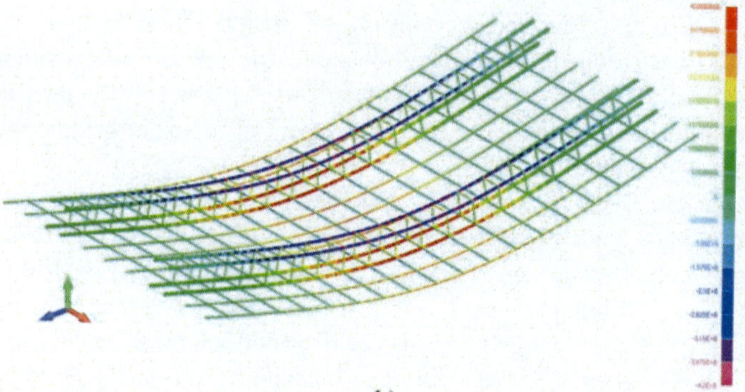

b)

Fig. 3 Stresses in fiber concrete **a** and in reinforcement **b** in ultimate static loading

Dynamic analysis was performed for two cases of supports. The first with the slab resting on a pliable support in the form of a steel tube, the second with the support on a rigid support. The calculation showed that at the level of the maximum impulse of the dynamic load of 28 kPa · sec., the system with rigid supports has collapsed, while the structure with deformable supports it still has a carrying capacity.

The residual load level of 0.3 $q\forall\{q_1, q_2, q_3\}$ approximates the additional load from the soil of the disturbed structure as a result of dynamic impact. The qualitative situation for the distribution of stresses in fiber concrete in time as a result of dynamic loading is shown in Fig. 5, 6. The stress scale is the same as in Fig. 3,a. It can be noted that after a time moment of 1.4 s, the stresses in the upper fibers of the compressed fiber concrete begin to decrease, which can be interpreted as a failure. The compressed zone shifts downward in the vertical direction. Analysis of the work of the suppressed support in the form of a tube has shown that at a pulse level of 28 kPa · s the stresses in the walls reach the yield strength. At the same time, it seems promising to study the influence of tube stiffness on oscillations damping.

It should be noted that in case of large vertical displacements of the tube, it is necessary to model the support with the use of GAP finite elements, determining

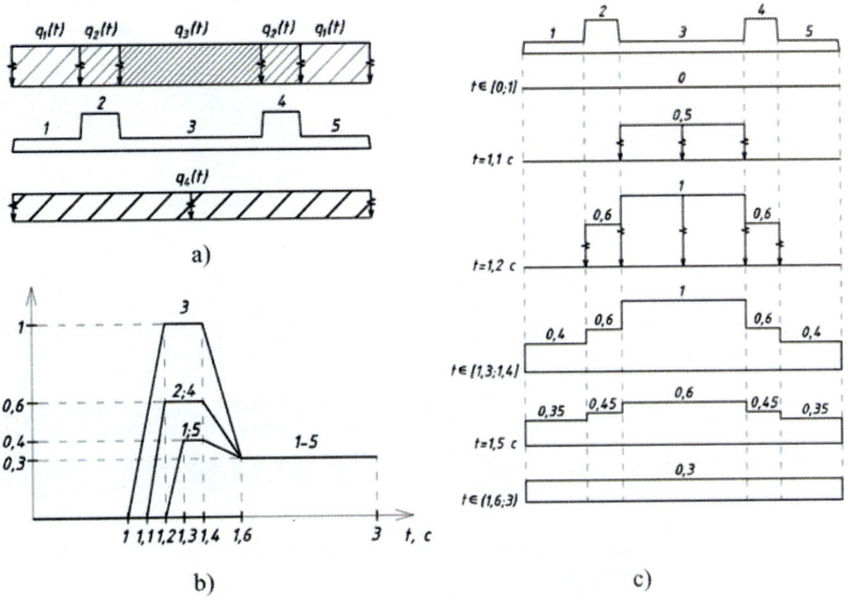

Fig. 4 Pulse load application diagram

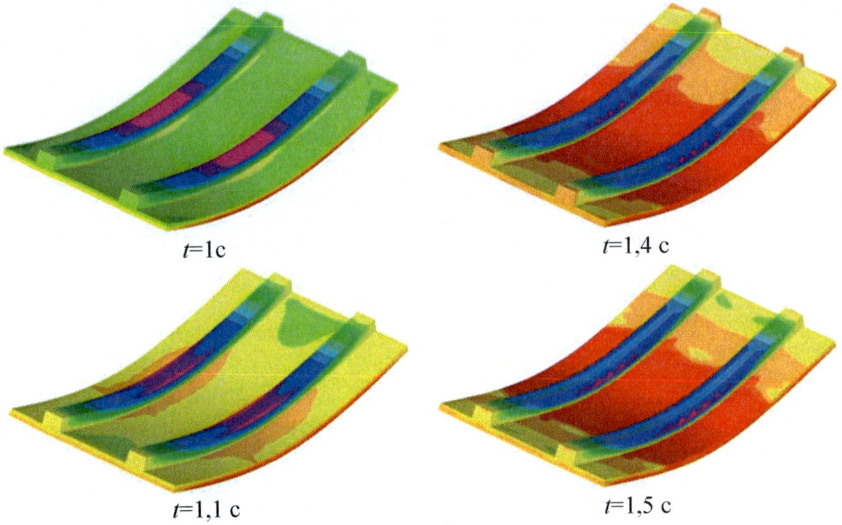

Fig. 5 The minimum principial stresses in time (begin)

Fig. 6 The minimum principial stresses in time (end)

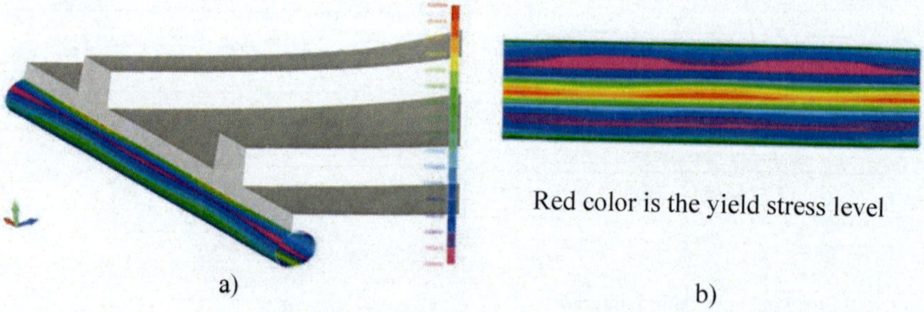

Fig. 7 Distribution of von Mises stresses along the pipe length at time t = 1.2 s **a**, image with 3 times increased cross-sectional scale **b**

deformation phase of the support, which is similar to the absolutely rigid one. Stress distribution along the length for a compliant support is shown in Fig. 7.

4 Discussion

In the calculations under consideration, the total damping coefficient, which takes into account the structural and inertial damping, was used. Its value was assumed to be 0.1. The quantitative estimation of the components of the stress–strain state of the structure when solving problems of this type depends significantly on both the damping properties of the structural materials and the damping of the medium. However, no experimental data are given for such problems. Besides, the damping of vibrations in the problem under consideration essentially depends on the suppleness of supports and their elastoplastic properties. Therefore, in the process of designing, it is advisable to solve the problem of searching for the rational rigidities of supports. This can be done using the deformable polyhedron method or other metaheuristic algorithms [21, 22]. When solving nonlinear problems, a modified Newton–Raphson method was used in combination with the method of dividing the segment in half with the value of the unconstraint by forces equal to 0.001. At the same time, up to 25 iterations were used at each step of the dynamic calculation.

5 Conclusion

An approach to numerical modeling of dynamic behavior of FRC slabs taking into account the time-varying spot of impulse loading has been developed. It has been established that under such load the introduction of deformable supports contributes to the improvement of mechanical safety and reduction of material capacity of structures. The effectiveness of introduction of a yielding support in the form of a round tube has been shown, which led to an increase in the ultimate impulse load by approximately 5% in comparison with a rigid support. In a particular case it is established that the critical damage of the structure at the considered impulse action occurs in fibrous concrete, despite the presence of the working rebars in the compressed concrete zone.

Acknowledgements This work was carried out with the financial support of the Ministry of Science and Higher Education of the Russian Federation (project "Theoretical and experimental design of new composite materials to ensure safety during the operation of buildings and structures in conditions of man-made and biogenic threats" No. FSWG-2020-0007).

References

1. Chen HL, Xia ZC, Zhou JN, Fan HL, Jin FN (2013) Dynamic responses of underground arch structures subjected to conventional blast loads: Curvature effects. Arch CivMech Eng 13:322–333
2. Kerber A, Gargano A, Pingkarawat K, Mouritz AP (2017) Explosive blast damage resistance of three-dimensional textile composites. Compos A Appl Sci Manuf 100:170–182
3. Malak SA, Krstulovic-Opara N (2020) American Concrete Institute, ACI Special Publication SP-340, pp 114–136
4. Nurick GN, Martin JB (1989) Deformation of thin plates subjected to impulsive loading—A review. Int J Impact Eng 8(2):159–170
5. Senthil K, Singhala A, Shailjab B (2019) Damage mechanism and stress response of reinforced concrete slab under blast loading. Coupled Syst Mech 8(4):315–338
6. Choi H, Kim S-J, Hong J-HJ, Choi K-N (2018) Evaluation of blast resistance and failure behavior of prestressed concrete under blast loading. Constr Build Mater 173:550–572
7. Park G-K, Kwak H-G, Filippou FC (2017) Blast analysis of RC beams based on moment-curvature relationship considering fixed-end rotation. Eng Fail Anal 93:268–288
8. Tamrazyan AG (2016) The assessment of reliability of punching reinforced concrete beamless slabs under the influence of a concentrated force at high temperatures. Procedia Eng 153:715–720
9. Tamrazyan A, Alekseytsev A (2019) Evolutionary optimization of reinforced concrete beams, taking into account design reliability, safety and risks during the emergency loss of supports. E3S Web Conf 97:04005

10. Fujikake K, Li B, Soeun S (2009) Impact response of reinforced concrete beam and its analytical evaluation. J Struct Eng 135(8):938–950
11. Zeng X, Xu B (2012) Modelling of damage patterns of RC concrete columns under demolition by blasting. Chin Civil Eng J 45(9):63–73
12. Kou J, Wang H (2020) J Vib Shock 39(11):239–247
13. Savinykh AS, Garkushin GV, Kanel GI, Razorenov SV (2019) Dynamic strength of a eutectic bismuth–lead alloy in the solid and liquid states. Int J Fract 215(1–2):138
14. Geng F, Ding Y, Wu H, Yang K (2020) Seismic risk assessment of a novel self-centering precast concrete frame under near-fault ground motions. Appl Sci (Switzerland) 10(18):6510
15. Zhang C, Abedini M, Mehrmashhadi J (2020) Dynamic vulnerability assessment and damage prediction of RC columns subjected to severe impulsive loading. Eng Struct 224:111219
16. Mousavi T, Shafei E (2019) Impact response of hybrid FRP-steel reinforced concrete slabs. Structures 19:436–448
17. Pham TM, Hao Y, Hao H (2018) Sensitivity of impact behaviour of RC beams to contact stiffness. Int J Impact Eng 112:155–164
18. Kumpyak OG, Galyautdinov ZR, Kokorin DN (2016) AIP conference proceedings. In: Proceedings of the II All-Russian Scientific Conference of Young Scientists Advanced Materials in Technology and Construction, p 070006
19. Kim T-Y, Chang Y-S (2021) Numerical investigation on postulated aircraft crash to tokamak building Fusion. Eng Des 173:112958
20. Serpik I, Alekseytsev A (2018) Optimization of steel frame building systems in terms of parameters and reliability requirements. IOP Conf Ser 365:052003
21. Alekseytsev AV, Al Ali M (2018) Optimization of hybrid I-beams using modified particle swarm method. Mag Civ Eng 7(83):75–185
22. Alekseytsev AV, Gaile L, Drukis P (2019) Optimization of steel beam structures for frame buildings subject to their safety requirements. Mag Civ Eng 7(91):3–15
23. Alekseytsev A, Botagovsky M, Kurchenko N (2019) Cost minimization for safety enhancing of timber beam structures in historical buildings. E3S Web Conf 97:03002
24. Geniev GA (2000) Soil Mech Found Eng 37(6):179–183
25. Radchenko A, Radchenko P, Batuev S, Plevkov V (2019) Archit Eng 4(3):22–29
26. Fialko YS, Kabantsev OV, Perelmuter AV (2021) Elasto-plastic progressive collapse analysis based on the integration of the equations of motion Mag Civ Eng 2(102):10214
27. Kabantsev OV, Mitrovitch B (2018) Justification of the special limit state characteristics for monolithic reinforced concrete bearing systems in the progressive collapse mode. IOP Conf Ser Mater Sci Eng 456:012002
28. Mkrtychev OV, Savenkov AY (2019) Int J Comput Civ Struct Eng 15(4):111–122
29. Chernukha NA (2014) Mag Civil Eng 1:12–22
30. Castedo R et al (2018) Estimation of Jones-Wilkins-Lee parameters of emulsion explosives using cylinder tests and their numerical validation. Int J Rock Mech Min Sci 112:290–301

Concrete Radiation Resistance Dependency on Phase Composition of HCP with Superplasticizers

Vyacheslav Medvedev

Abstract The use of various admixtures in concrete today has become a standard, since this makes it possible to obtain high-quality concrete with the properties required for a particular design. Of greatest interest in recent years is the use of functional admixtures in the construction of modern nuclear power plants, since this directly affects the quality and construction time. The impact of admixtures is observed mainly on the chemical and mineral composition of concrete. Superplasticizers are of the greatest interest in this area, since their introduction into the composition of the concrete mixture can have a significant effect on the radiation resistance of concrete. The cause of this phenomenon is supposed to be a change in phase composition. In this paper, the mechanism of the influence of polycarboxylate superplasticizers on the process of phase formation in hardened cement paste (HCP) and its effect on the stability of the cement matrix during radiation-thermal changes is researched in more detail. The method of X-ray diffraction analysis (XRD) was chosen for the research. The main task of XRD is the identification of various phases in a mixture of crystalline substances based on the diffraction pattern they create.

Keywords Concrete · Radiation resistance · Hardened cement paste · Phase composition · Superplasticizers

1 Introduction

The use of various admixtures in concrete today has become a standard, since this makes it possible to obtain high-quality concrete with the properties required for a particular design. Of greatest interest in recent years is the use of functional admixtures in the construction of modern nuclear power plants, since this directly affects the quality and construction time [1–3]. At the same time, in contrast to civil engineering, in the operating conditions of nuclear power plants, concrete is subjected to loads that differ significantly from those considered in most existing researches. The impact of

V. Medvedev (✉)
Moscow State University of Civil Engineering, Moscow 129337, Russian Federation
e-mail: MedvedevVV@mgsu.ru

P. Akimov et al. (eds.), *Proceedings of FORM 2022*, Lecture Notes in Civil Engineering 282, https://doi.org/10.1007/978-3-031-10853-2_14

admixtures is observed mainly on the chemical and mineral composition of concrete. The complexity of conducting field tests and the enormous variety of different types of admixtures is a significant obstacle in the way of modern researchers in this field, as evidenced by the works [1, 3–9].

As shown in works [1, 3, 4, 7], radiation exposure, which is unique for nuclear power plants, can be simulated in laboratory conditions by using short-term thermal exposure due to the similarity of thermal and radiation-thermal changes in concrete [1, 3, 7], which significantly reduces the time of experiments and resource costs.

In many domestic and foreign works [4–18], an analysis was made of the effect of various admixtures on the radiation resistance of concrete in terms of changes in their physical and mechanical properties. At the same time, as proved in [1, 3, 15, 18], superplasticizers are of the greatest interest in this area, since their introduction into the composition of the concrete mixture can have a significant effect on the radiation resistance of concrete. The cause of this phenomenon is supposed to be a change in phase composition.

In [3], using the DSC method, the influence of the hydration process with the introduction of polycarboxylate superplasticizers and the change in the degree of hydration during thermal tests simulating radiation exposure were researched. According to the data of [6, 7, 9, 12, 18], changes in the phase composition can have a significant effect on the radiation resistance.

In this paper, the mechanism of the influence of polycarboxylate superplasticizers on the process of phase formation in hardened cement paste (HCP) and its effect on the stability of the cement matrix during radiation-thermal changes is researched in more detail.

2 Methods

The method of X-ray diffraction analysis (XRD) was chosen for the research. The main task of XRD is the identification of various phases in a mixture of crystalline substances based on the diffraction pattern they create. The most commonly used practical method of X-ray phase analysis is the powder diffraction method [19].

XRD was carried out on a full-size powder diffractometer with θ-θ geometry with a goniometer radius of 260 mm and a source - a narrow-focus tube with a power of 2200 W (Cu anodes). This device uses an energy-dispersive solid-state detector with a Peltier cooler, which makes it possible to exclude passive elements (beta filters/monochromators) from the optical scheme of the device due to the software separation of $K\beta$ and fluorescent radiation. Instrumental resolution of the device is 0.04° 2θ while maintaining a high signal-to-noise ratio.

The registration parameters of diffraction patterns are shown in Table 1:

Based on the results of [3], the research in this work was carried out for superplasticizers with long side chains and a high steric effect at the maximum concentration, since they have the greatest effect on the change in the radiation resistance of concrete.

Table 1 Diffraction patterns registration parameters

Parameters	Value
Sample rotation	60 rpm
Voltage	40 kV
Current	40 mA
DS	1.5 mm
Soller slit on the primary beam	1.3°
SS1	3 mm
SS2	0.7 mm
Soller slit on the secondary beam	1.3°
RS	0.3 mm
Capture mode	step-by-step
Step	0.02° 2θ
Scan speed	1.2° 2θ/min
Range 2θ	4.00–70.00°

Just as in [3], the studies were carried out on HCP samples made in accordance with the procedure described in [1, 3, 6]. For the subsequent correlation of the obtained data with the results of work [3], a similar Portland cement was chosen with the following content of the main minerals: C3S - 64.8%; C2S - 11.1%; C3A - 4.4%; C4AF - 15.5%; Bassanite-2.3%; Gypsum-1.9%.

For testing, two groups of samples were made - without admixtures and with the use of the superplasticizer used in [3] at the maximum concentration.

For XRD the samples were preliminarily ground in an agate mortar to a powder state, the sample was homogenized. To exclude the interaction of heat-treated samples with moisture from the surrounding air, the crushed samples were placed in a desiccator.

3 Results

To study the process of phase formation and its change under the influence of radiation-thermal irradiation, samples were examined before heat treatment (after hardening), and at each of the heating stages, according to [3, 16].

The results are shown in Figs. 1, 2, 3 and 4:

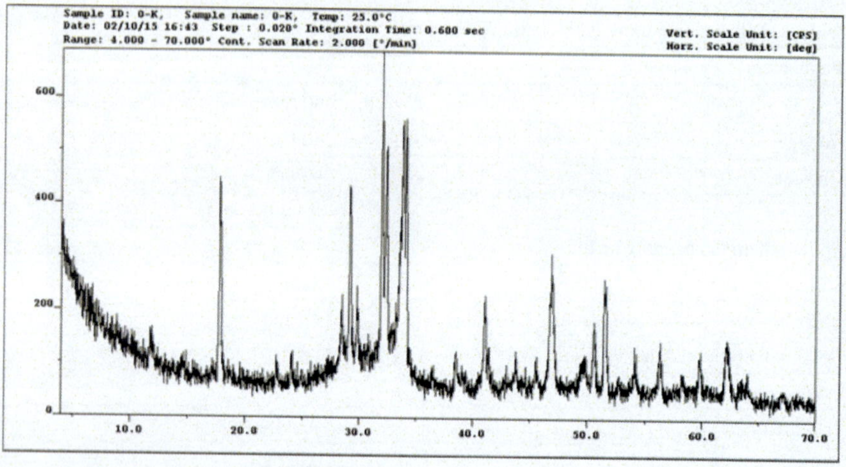

Fig. 1 Registered diffraction pattern of the sample without admixtures before thermal testing

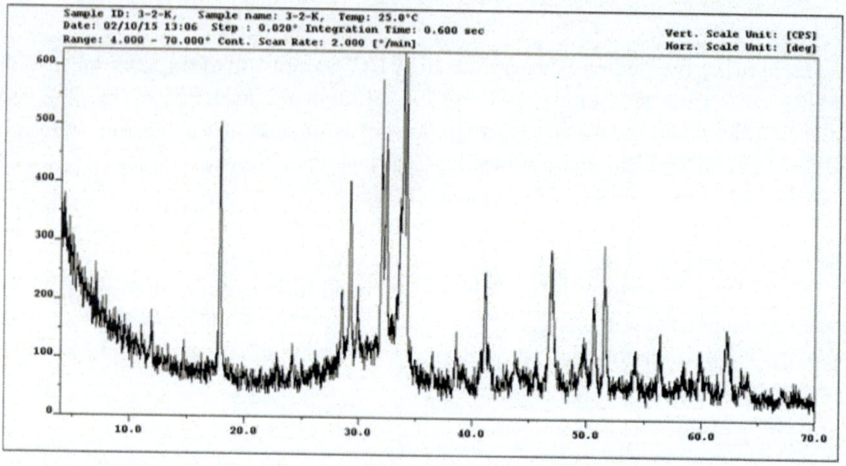

Fig. 2 Registered diffraction pattern of the sample with the investigated superplasticizer before thermal testing

4 Discussion

After thermal testing at 150 ° C, ettringite dissociates, which is clearly visible on the diffraction patterns of the samples (Figs. 3(a) and 4(a)). The mineral composition of the sample with the researched superplasticizer does not change in comparison with the control sample.

After thermal testing at 350 ° C, slight changes in the phase composition are also observed (Figs. 3(b) and 4(b)). In the composition of the sample with the use of the

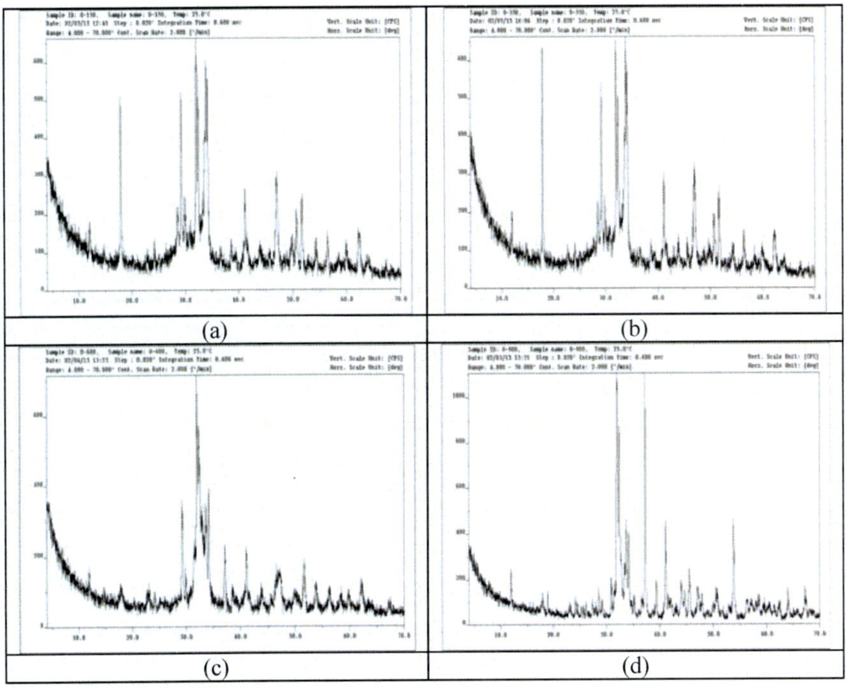

Fig. 3 Recorded diffraction patterns of the sample without admixtures after each stage of thermal testing: **a** –150 ° C; **b** –350 ° C; **c** –600 ° C; **d** –900 ° C

studied superplasticizer at this stage, a lower content of portlandite is observed in comparison with the control sample. The mineral composition of the sample with the researched superplasticizer does not change in comparison with the control sample.

After carrying out thermal tests at 600 ° C, complete decomposition of portlandite with the formation of CaO is observed (Figs. 3(c) and 4(c)). The amount of the amorphous part is also much less. In addition, at this stage, the dissociation of the C-S-H gel occurs with the formation of anhydrous calcium silicates. A separate diffraction reflection of portlandite is observed, which indicates partial hydration of CaO formed during thermal decomposition. The mineral composition of the sample with the researched superplasticizer does not change in comparison with the control sample.

After thermal testing at 900 ° C, the mineral composition remains unchanged (Figs. 3d and 4d). In addition, in the course of sample preparation $CaCO_3$ was formed as a result of CaO carbonation. At this stage, no increase in background intensity is observed, which indicates complete decomposition followed by crystallization of the amorphous C-S-H gel. The mineral composition of the sample with the researched superplasticizer does not change in comparison with the control sample.

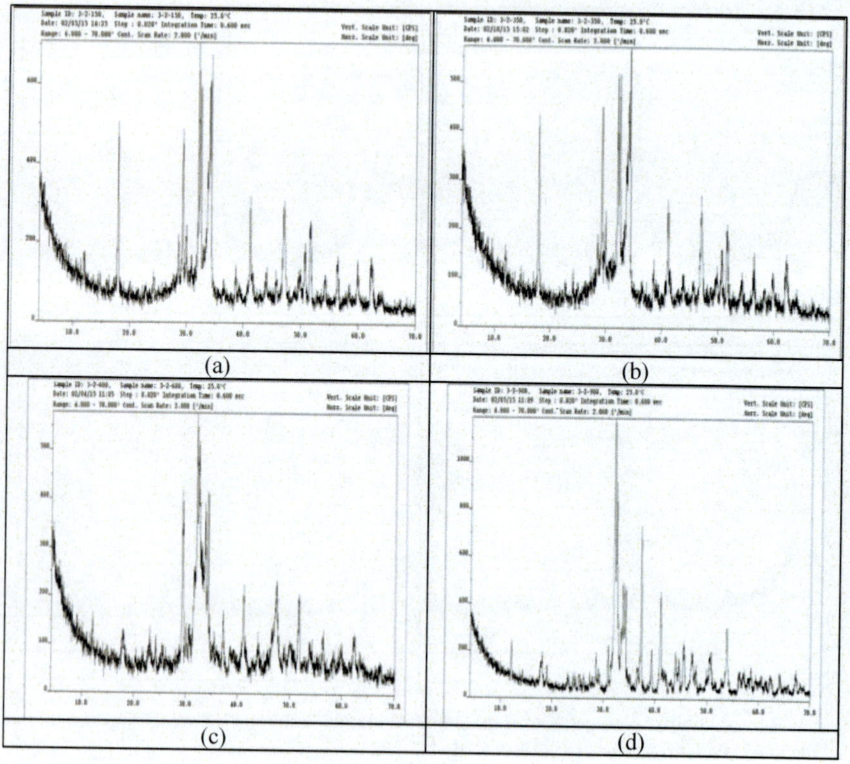

Fig. 4 Recorded diffraction patterns of samples with the investigated superplasticizer after each stage of thermal tests: **a** –150 ° C; **b** –350 ° C; **c** –600° C; **d** –900 ° C

5 Conclusion

The results obtained under laboratory conditions correlate with the results of [3]. According to the data of [20], during direct irradiation of HCP, structural radiation-thermal changes lead to amorphization of calcium oxide $Ca(OH)_2$ and calcium hydroaluminate C4CH13, as well as partial recrystallization of calcium hydrosilicates. Subsequently, amorphous $Ca(OH)_2$ reacts with atmospheric carbon dioxide to form calcium carbonate $CaCO_3$. The amount of the non-hydrated phase does not change in HCP samples, but decreases after irradiation due to additional hydration [3, 7].

In the case of laboratory tests, the mineral composition of the sample with the introduction of a superplasticizer based on polycarboxylate ether with long side chains and a high steric effect does not change compared to the control sample without admixtures. The content of calcite and portlandite with the introduction of the admixture is slightly lower than that of the control sample, which also confirms the data from [3]. At the same time, an increase in the radiation resistance of concrete can be justified by the formation of a larger amount of hydrated phases. It should be noted that this process can cause more intense outgassing during irradiation, however, this phenomenon requires a separate research and was not considered within the framework of the current research.

Acknowledgements This work was financially supported by the Ministry of Science and Higher Education of the Russian Federation (grant # 075-15-2021-686). All tests were carried out using research equipment of The Head Regional Shared Research Facilities of the National Research Moscow State University of Civil Engineering.

References

1. Ershov VY (1992) Radiation resistance of hardened Portland cement paste with chemical and mineral additives, The dissertation for the degree of candidate of technical sciences, Moscow
2. Nithya R, Barathan S, Govindarajan D, Raghu K, Anandhan N (2010) A thermal analysis study on blended ternary cement paste. Int J Chem 2(1):121–127
3. Medvedev V (2020) Radiation resistance evaluation of HCP with superplasticizers using DSC. IOP Conf. Ser Mater Sci Eng 869:032033
4. Dubrovsky VB, Korenevsky VV, Muzalevsky LL, Pergametsik BK, Sugak EB (1980) Radiation resistant concretes for reactor containment (in Russian). Radiat Saf Prot Nucl Power Plants 4:240
5. Dubrovsky VB, Korenevsky VV, Pospelov VP, Sugak EB (1985) Investigation of a binder for protective concretes with increased plasticizing properties (in Russian). In: Proceedings of the third all-union scientific conference on protection against ionizing radiation of nuclear engineering installations, vol 5. 27–29 October 1985, Tbilisi, Georgia, pp 62–69
6. Sugak EB, Denisov AV, Korenevsky VV, Muzalevsky LL, Pergametsik BK (1978) Radiation changes in hardened cement paste with the addition of fly ash. Energy Constr 9:11–13
7. Ershov VY, Dubrovsky VB, Muzalevsky LL, Kolesnikov NA (1988) Behavior of hardened cement paste with admixtures under conditions of prolonged exposure to high temperatures. Quest At Sci Technol (Design and construction series) 2:120–129
8. Hilsdorf HK, Kropp J, Koch HJ (2014) The effects of nuclear radiation on the mechanical properties of concrete. Int Comm Irradiat Concr (ICIC) Inf Exch Framew Meet 1:223–251
9. Batrakov VG, Shanov FM, Silina BS, Falikman VR (1988) The use of superplasticizers in concrete. Constr Mater Prod 1.2(7):59
10. Field KG, Remec I, Le Pape L (2015) Radiation effects in concrete for nuclear power plants – part I: quantification of radiation exposure and radiation effects. Nuclear Eng Des 282:126–143
11. Batrakov VG (1998) Modified concrete - Theory and practice, Construction materials and products, p 768
12. El-Gamal SMA, Al-Nowaiser FM, Al-Baity AO (2012) Effect of superplasticizers on the hydration kinetic and mechanical properties of Portland cement pastes. J Adv Res 3(2):119–124
13. Thomas DR (1965) Temperature and thermal stress distributions in concrete primary shields for nuclear reactors. Nucl Struct Eng 1:368–384
14. Nowak-Michta A (2015) Influence of superplasticizer on porosity structures in hardened concretes. Procedia Eng 108:262–269
15. Nkinamubanzi PC, Mantellato S, Flatt RJ (2016) Science and Technology of Concrete Admixtures, pp 353–377. Woodhead Publishing, Cambridge
16. Medvedev V, Pustovgar A (2015) Influence of chemical additives on radiation stability of concrete - theoretical basis and evaluation method. Appl Mech Mater 725–726:337–382

17. Lowinska-Kluge A, Piszora P (2008) Effect of gamma irradiation on cement composites observed with XRD and SEM methods in the range of radiation dose 0-1409 MGy. In: Proceedings of the 7-th national meeting of synchrotron radiation users, vol 114, pp 399–411
18. Ohta A, Sugiyama T, Tanaka Y (1997) Fluidizing mechanism and application of polycarboxylate-based superplasticizers. In: Proceedings of the fifth CANMET ACI international conference, Rome, Italy, pp 173–19
19. Kowba LM, Trunov VK (1976) X-ray diffraction analysis, Moscow State University, p 232
20. Pospelov VP, Mirenkov AF, Pokrovsky SG (2006) Radiation shielding concretes for nuclear power plants, AugustBorg, p 652

Operational Durability of Facade Systems

Elina Gorbunova, Ilya Govryakov, Igor Bessonov, Olga Lyapidevskaya, and Boris Efimov

Abstract Plaster coatings applied to facade surfaces using reinforcing meshes can be considered as a kind of textile reinforced concrete, a material consisting of a mineral binder and reinforcing components. Such coatings are used in facade thermal insulating composite systems, as well as on a concrete base (without wall insulation), or on any surface during the reconstruction of building facades.

In addition to decorative characteristics and external expressiveness, facade systems and materials must meet the requirements for durability and operational durability under climatic influences: solar radiation, precipitation, alternating and negative temperatures.

The purpose of the research presented in the article was to assess the resistance of facade cladding and decorative systems on a concrete base and on ceramic tiles to climatic temperature and humidity cyclic influences.

The test procedure was based on creating conditions for artificial aging of samples under cyclic temperature and humidity effects. Modeled: solar irradiation, atmospheric precipitation, as well as the impact of negative temperatures, while ensuring freezing and thawing of the outer surface of the facade system by 20–25 mm.

The decrease in adhesion strength after cyclic temperature and humidity exposures was 9–13%. Upon completion of cyclic exposures, no external changes were found on the front surface of the samples (color, cracks, chips, peeling).

Additional protection systems for the facade cladding and decorative building system, which is a special case of textile-reinforced concrete material, have fully passed comprehensive climatic tests. Facade protection systems on a concrete base and on ceramic tiles withstood temperature and humidity cyclic effects equivalent to 30 conditional years of operation.

E. Gorbunova (✉) · I. Govryakov · O. Lyapidevskaya · B. Efimov
Moscow State University of Civil Engineering, Moscow 129337, Russian Federation
e-mail: eg15082000@mail.ru

E. Gorbunova · I. Govryakov · I. Bessonov
Research Institute of Building Physics Russian Academy of Architecture and Construction Sciences (RAACS, Transliterating the Russian Acronym, RIBPh RAACN), Lokomotivny pr. 21, Moscow 127238, Russian Federation

P. Akimov et al. (eds.), *Proceedings of FORM 2022*, Lecture Notes in Civil Engineering 282, https://doi.org/10.1007/978-3-031-10853-2_15

Keywords Plaster coating · Textile reinforced concrete · Climatic tests · Adhesion to the substrate

1 Introduction

Textile reinforced concrete is a group of modern natural materials, consisting of two components: mineral binders, hydraulic raw materials and natural resources: mineral or synthetic, dispersed in the material or woven in meshes with volumetric cell sizes [1–3]. In all cases, the use of hybrid fibers is allowed, that is, fibers of various nature: for example, mineral and carbon, glass and cellulose, etc. [4–6].

The classic is textile-reinforced concrete, in which layers of fine-grained concrete are reinforced with flat meshes of glass or polymer fibers, geotextile and its analogues [7–9]. Fine-grained concretes reinforced with individual fibers dispersed in the concrete mix - fiber-reinforced concretes - are widely used. Fiber concretes can have a dense structure and high strength; such materials can be used in loaded structures, including thin-walled structures and shells. If the fiber-reinforced material has a porous structure, then piece products are made on its basis: cellular concrete blocks that have higher performance compared to non-reinforced counterparts [10–12].

The so-called concrete canvas is becoming popular. This type of reinforced concrete textile consists of an outer shell made of two webs of non-woven fabric and a core consisting of mineral binder and filler. As a filler, either quartz sand with a particle size of 0.5 to 1 mm, or a mixture of quartz sand and reinforcing fiber (with a consumption of up to 1.5%) is used. These products are delivered in rolls to the object, the rolls are rolled out, moistened, the core concrete sets and, as a result, a finished structure element is obtained [13–15]. This method is technologically advanced, quick to implement and is used both in new construction and in reconstruction.

A variety of concrete textiles can also be considered plaster coatings applied to the surfaces of facades using reinforcing meshes. Such coatings are used in facade heat-insulating composite systems, as well as on a concrete base (without wall insulation), or on any surface during the reconstruction of building facades [16–18].

The requirements for the facade of the building, in addition to decorative characteristics and external expressiveness, also include the requirement for durability, taking into account the fact that the facade is exposed to the whole complex of atmospheric factors. The materials used in facade systems must withstand such influences and maintain structural integrity, that is good adhesion to the substrate [19–21].

The study of the properties of such systems was the goal of the research presented in the article. An assessment was made of the resistance of the facade cladding and decorative system on a concrete base and on ceramic tiles to climatic temperature and humidity cyclic influences.

2 Methods

The test procedure is based on creating conditions for artificial aging of samples under cyclic temperature and humidity effects. Heating the surface of a structural fragment by means of the radiant action of a heating device simulates solar irradiation of the infrared spectrum, humidification simulates atmospheric precipitation, and freezing simulates the impact of negative temperatures, while freezing and thawing of the outer surface of the facade system by 20–25 mm is ensured.

According to the results of processing long-term meteorological data of the Moscow State University Observatory, the RIMC (Research Institute of Moscow Construction) laboratory and the laboratory of Research Institute of Building Physics, the number of generalized freeze–thaw cycles for facades in the climatic conditions of Moscow is 14 on average. 420 cycles of exposure were carried out with a temperature difference from minus 20 °C to plus 60 °C, and 300 cycles were carried out in parallel with a temperature difference from minus 40 °C to plus 70 °C.

Testing of fragments of concrete slabs with applied compositions of facing and decorative systems on a concrete base and on ceramic tiles was carried out in a cooling and sprinkling installation KhDU-02 "Thermoisolation" (Fig. 1). The cycle in the XDU-02 installation included humidification, freezing at minus 25 °C, heating to plus 60° C under IR and UV irradiation. The number of test cycles was 420.

The sequence of preparation and testing of fragments of concrete slabs with applied compositions of facing and decorative systems in the climatic thermal shock installation TSK300 is shown in Fig. 2. Sample preparation consisted in moistening

(a) (b)

Fig. 1 Testing of fragments of concrete slabs with applied compositions of cladding and decorative systems on a concrete base **a** and on ceramic tiles **b**, in a refrigeration and sprinkling installation XDU-02 "Thermoisolation"

(a) (b) (c)

Fig. 2 Preparation **a** and testing of fragments in a thermal shock chamber TSK300 **b, c**

the outer surface of the sample by immersing it in water for 24 h. The test cycle consisted of the following stages: freezing to minus 40 °C for 3 h, heating to plus 70 °C for 3 h. The number of cycles was assumed to be 300.

In the experiment, plaster systems of four compositions were studied.

System №1 (on a concrete base) included a primer Ceresit ST 16, a plaster-reinforcing layer with a mesh made of a plaster-glue mixture Ceresit ST 190, a decorative layer made of mineral decorative plaster Ceresit ST 137, painted with acrylic paint Ceresit ST 42.

System №2 (for ceramic tiles) included a primer Ceresit ST 16, a plaster-reinforcing layer with a mesh made of a plaster-adhesive mixture Ceresit ST 190, a decorative layer made of acrylic decorative plaster Ceresit CT 60.

System №3 (for ceramic tiles) included a primer Ceresit ST 16, a plaster layer without a mesh, made from the Ceresit ST 190 plaster-adhesive mixture, a decorative layer made from Ceresit ST 137 mineral decorative plaster, painted with Ceresit ST 42 acrylic paint.

System №4 (for ceramic tiles) included a primer Ceresit ST 16, a plaster-reinforcing layer with a mesh, made from a plaster-adhesive mixture Ceresit ST 190, a decorative layer made from mineral decorative plaster Ceresit ST 137, painted with acrylic paint Ceresit ST 42.

Determination of adhesion of the coating to the base was carried out according to the standard method. The sequence of operations implementation is shown in Fig. 3.

3 Results

Upon completion of cyclic exposures, a visual inspection of the samples was carried out and experimental determinations of the adhesion strength of the layers were carried out. External changes on the front surface of the samples (color, cracks, chips, peeling) were not found. The averaged results of experimental determinations of the adhesion strength of layers on the main and control samples with cuts to

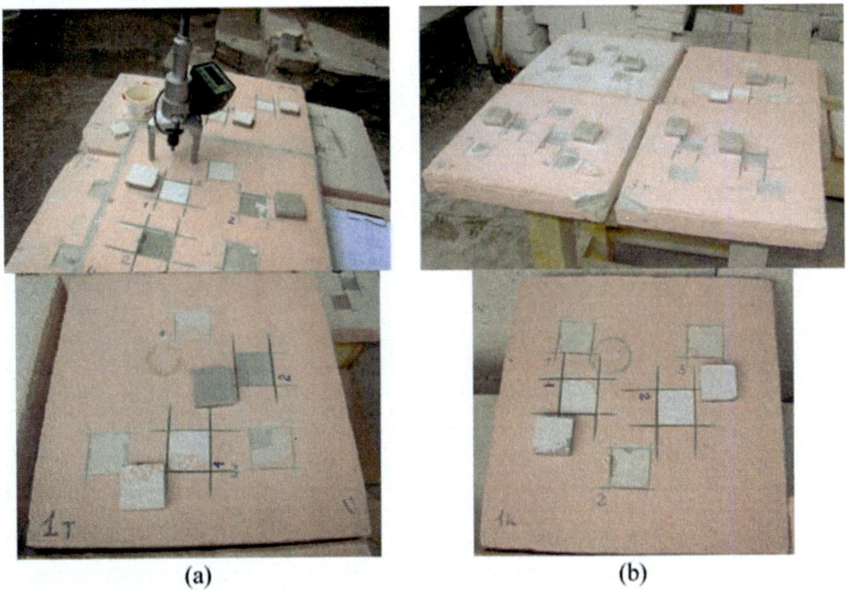

Fig. 3 Carrying out experimental determinations of the adhesion strength of layers. Samples: main **a** and control **b**

the concrete base (along the perimeter of the glued metal plates) are presented in Table 1.

Visual inspection of the samples showed no changes in appearance, color, cracks, cavities, etc. The decrease in the strength of adhesion was, respectively: in system 2–9%, in system 3–12%, in system 4–13%. In all main series, the destruction occurred with the separation of the tile, which confirms the high adhesion of the protective coating to the base.

In series 1, there was an increase in adhesion strength after cyclic climatic influences, and the destruction of the system occurred in the main series along the grid in the plaster-reinforcing layer, and in the control series - along the layer of decorative plaster. This also confirmed the high adhesion of the protective coating to the concrete base. The increase in adhesion strength as a result of climatic influences, apparently, was associated with a deeper hydration of the binder in the coating under conditions of a humid environment and a cyclic increase in temperature.

4 Discussions

Expansion of the scope of application of building materials and systems similar in structure and composition to textile-concrete is a promising direction in the development of the construction industry. The result is an increase in the durability and

Table 1 The characteristics of tearing off the protective coating in various systems

System number	Characteristics	Sample series type	
		Basic	Control
1	Adhesion strength, MPa	1,20	0,93
	The nature of the separation	On the grid in the plaster-reinforcing layer	On a layer of decorative plaster
2	Adhesion strength, MPa	0,98	1,08
	The nature of the separation	Tile detachment	Tile detachment
3	Adhesion strength, MPa	0,77	0,88
	The nature of the separation	Tile detachment	On a layer of decorative plaster
4	Adhesion strength, MPa	0,86	0,99
	The nature of the separation	Tile detachment	On a layer of decorative plaster

operational stability of facade systems, which is one of the factors for increasing their energy efficiency. An increase in the service life of the structure reduces the frequency of major repairs, and, consequently, the cost of operating the structure.

The second component, which determines the increase in energy efficiency, is the possibility of using the considered solutions in facade heat-insulating composite systems, which provide for the installation of a heat-insulating layer and a finishing protective and decorative coating on a reinforcing mesh. Non-combustible stone wool boards or extruded polystyrene foam boards can be used as thermal insulation.

Expanded polystyrene plates are used on the basement of the building and in areas of the facade, where penetration (including due to capillary phenomena) of ground moisture is possible. In the event that slabs of extruded polystyrene foam are mounted along the surface of the wall, fire cuts are required at the level of interfloor ceilings, as well as along the perimeter of window and door openings. The cuts are made from strips of non-combustible thermal insulation.

Regardless of the thermal insulation used, the system of plaster coatings based on mineral plasters reinforced with meshes performs protective functions in relation to the insulating layers. Firstly, it is weather protection, secondly, it is protection against vandalism, and thirdly, it is protection against possible fire exposure, which is especially important in the case of combustible thermal insulation.

5 Conclusions

Additional protection systems for the facade cladding and decorative building system, which is a special case of textile-concrete material, have fully passed comprehensive climatic tests. Facade protection systems on a concrete base and on ceramic tiles withstood temperature and humidity cyclic effects equivalent to 30 conditional years of operation. This makes it possible to recommend similar design solutions for the reconstruction of building facades and, in particular cases, for new construction, including in facade heat-insulating composite systems.

Acknowledgements The research presented in the article was carried out within the framework of the plan of fundamental scientific research of the Ministry of Construction of Russia and the Russian Academy of Architecture and Civil Sciences for 2022-2023 on the topic 3.1.2. "Development of scientific foundations for the creation of fine-grained concrete reinforced with textile material (textile concrete)".

References

1. Lesovik VS (2017) Construction materials. Present and future. Vestnik MGSU 12(1(100)):9–16. (In Russian)
2. Kulagin AV, Bessonov IV, Bogomolova LK, Zhukov AD (2020) Investigation of coatings reinforced with fiberglass reinforcement during the restoration of reinforced concrete panels. Bull Constr Mach (BST) 6(6):32–35 (In Russian)
3. Zhukov A, Medvedev A, Poserenin A, Efimov B (2019) Ecological and energy efficiency of insulating systems. In: E3S Web of Conference, vol 135, p 03070, 04 December 2019. (ITESE-2019). https://doi.org/10.1051/e3sconf/201913503070
4. Schladitz F, Lorenz E, Jesse F, Curbach M (2009) Verstärkung einer denkmalgeschätzten Tonnenschale mit Textilbeton. Beton- und Stahlbetonbau 104(2009):432–437. Heft 7, S
5. Hankers C, Matzdorf D: Verstärkung von Stahlbetonbauteilen mit textilbewehrtem Spritzbeton. Fachausatz, S. 10. http://www.torkret.de/leistungsvielfalt/kernkompetenzen/textilbeton-carbonbet-on.html
6. Gelbrich S (2012) Organisch geformter Hybridwerkstoff aus textil-bewehrtem Beton und glasfaserverstrktem Kunststoff. Leichter bauen—Zukunft formen. TUDALIT 7. S. 9
7. Scherer S, Michler H, Curbach M (2014) Brückenaus Textilbeton. Handbuch Brücken: Entwerfen, Konstruieren, Berechnen, Bauen und Erhalten, pp 118–129
8. Curbach M, Graf W, Jesse D, Sickert JU, Weiland S (2007) Segmentbräcke aus textilbewehrtem Beton - Konstruktion, Fertigung, numerische Berechnung. Beton- und Stahlbetonbau, no 102, Heft 6, pp 342–352
9. Hegger J, Goralksi C, Kulas S (2011). Schlanke Fußgängerbrücke aus Textilbeton — Sechsfeldrige Fuägängerbrücke mit einer Gesamtlänge von 97 m. Beton- und Stahlbetonbau, no 106, Heft 2, pp 64–71
10. Tuchaev D, Zarmanyan E, Petrovskiy E, Zemlyanko A, Ivanov K, Zhukov A (2018) Thermal insulation systems for the Arctic. In: FORM 2018. IOP Conference Series: Materials Science and Engineering. https://doi.org/10.1088/1757-899X/365/3/032015032041
11. Zhukov AD, Bessonov IV, Bobrova EY, Gorbunova EA, Demissie BA (2021) Materials based on modified gypsum for facade systems. Nanotechnol Constr Sci J 13(3):144–149. https://doi.org/10.15828/2075-8545-2021-13-3-144-149

12. Bobrova E, Pilipenko A, Zhukov A (2019) Insulating sheath system and energy efficiency of buildings, p 02019. TPACEE 2018, 02 April 2019. https://doi.org/10.1051/e3sconf/20199102019
13. Pyataev ER, Pilipenko ES, Burtseva MA, Mednikova EA, Zhukov AD (2018) Composite material based on recycled concrete. In: FORM 2018. IOP Conference Series: Materials Science and Engineering, p 03201. https://doi.org/10.1088/1757-899X/365/3/032041
14. Zhukov AD, Bessonov IV, Demissi Bekele A, Zinoveva EA (2021) Analytical optimization of the dispersion-reinforced fine-grained concrete composition. In: CATPID 2020. IOP Conference Series: Materials Science and Engineering, vol 1083, p 012037. https://doi.org/10.1088/1757-899X/1083/1/012037
15. Tolstoy AD, Lesovik VS, Kovaleva IA (2016) Kompozitsionnye vyazhushchie dlya poroshkovykh betonov s promyshlennymi otkhodami. Vestnik BGTUimeni V.G. Shukhova, no 1, pp 6–9. (In Russian)
16. Zhukov A, Ushakov A (2020) Isolation systems based on energy efficient basalt fiber wares. In: XXIII International Scientific Conference on Advance in Civil Engineering: FORM-2020, vol 869, p 032001, 23–26 September 2020, Hanoi, Vietnam, 09 July. https://doi.org/10.1088/1757-899X/869/3/032001. Accessed 20 May 2020
17. Bessonov IV, Bulgakov BI, Zhukov AD, Gradov VA, Ivanova NA, Kodzoev MBK (2021) Lightweight concrete based on crushed foam glass aggregate. In: CATPID 2020. IOP Conference Series: Materials Science and Engineering, vol 1083, p 012038. https://doi.org/10.1088/1757-899X/1083/1/012038
18. Bessonov I, Zhukov A, Efimov B, Gorbunova E, Govryakov I (2021) Gypsum polymer materials in construction. In: E3S Web of Conferences, vol 258, p 09087. https://doi.org/10.1051/e3sconf/202125809087
19. Pukharenko YV, Panteleev DA, Morozov VI, Magdeev UK (2016) Durability and deformativnost of the polyreinforced fibrobeton with application of an amorphous metal fiber. Academia. Arkhitektura i stroitel'stvo. 2016, no 1, pp 107–111. (In Russian)
20. Shitikova MV, Bobrova EY, Popov II, Zhukov AD (2019) Energy efficiency technical thermal insulation, no 8934917. In: 2019 International Multi-Conference on Industrial Engineering and Modern Technologies, FarEastCon. 2019, October 2019. Номер категории CFP19M35-ART, код 156113. https://ieeexplore.ieee.org/stamp/stamp.jsp?arnumber=8934917. https://doi.org/10.1109/FarEastCon.2019.8934917
21. Efimov B, Isachenko S, Kodzoev MB, Dosanova G, Bobrova E (2019) Dispersed reinforcement in concrete technology, p 01032, 09 August 2019. https://doi.org/10.1051/e3sconf/201911001032

Scanning Electron Microscopy of Plasticized Cement Stone After Heat-Moisture Treatment

Andrei Leshkanov, Lev Dobshits, and Sergey Anisimov

Abstract The use of superplasticizers (SPs) in concrete and mortar mixtures makes it possible to obtain materials with high strength, frost resistance, and water resistance by reducing the water-cement ratio. The most widely used SPs are based on polycarboxylate esters (PCE) and sulfonated naphthalene formaldehydes (SNF). Under the conditions of heat-moisture treatment (HMT), temperature deformations and pores inevitably form in the body of the hardening cement stone. The use of SPs allows you to reduce the size and number of pores, leading to some relaxation of internal stresses in the stone. In particular, the HMT mode is considered without preliminary holding according to the mode (0–6–2) h at a steaming temperature of 60 °C. It is important at the same time to understand which phases are formed during the hardening of cement systems with SP at HMT. Scanning electron microscopy (SEM) analysis allowed us to draw several conclusions. The size of calcium hydrosulfaluminate crystals is much smaller in systems with SP. In particular, in cement stone with SNF, AFt crystals after HMT are significantly shorter, up to 500 nm long, indicating the stone's loose structure in this period. The length of ettringite crystals in compositions without additives reaches 3 microns. Zones of unreacted $CaSO_4 \cdot 2H_2O$ minerals were recorded in cement stone with the addition of PCE, and the gradual ettringite formation was shown in the contact zone with tricalcium aluminate. The results obtained can be used to develop HMT modes to get high early strength with a shortened steaming duration.

Keywords Concrete · Cement · Superplasticizer · Sulfonated naphthalene formaldehyde · Polycarboxylate ester · Heat-moisture treatment · Ettringite · Portlandite · Gypsum

A. Leshkanov (✉) · S. Anisimov
Department of Construction Technologies and Highways, Volga State University of Technology, 3 Lenin square, Yoshkar-Ola 424000, Russia
e-mail: andruxalesh@gmail.com

L. Dobshits
Department of Construction Materials and Technologies, Russian University of Transport (MIIT), 9, b 9 Obrazcova street, Moscow 127994, Russia

P. Akimov et al. (eds.), *Proceedings of FORM 2022*, Lecture Notes in Civil Engineering 282, https://doi.org/10.1007/978-3-031-10853-2_16

1 Introduction

Currently, most cement-containing concretes and mortars are manufactured using chemical additives. The most widely used superplasticizers are, in particular, based on sulfonated naphthalene formaldehydes (SNF) and polycarboxylate esters (PCE). The principle of operation of the SNF is mainly associated with a decrease in the water-cement ratio due to adsorption on cement clinker ("electrostatic repulsion"). At the same time, the PCE has a so-called "steric repulsion" [1–3]. A distinctive feature of PCE is that its molecule is selective in terms of adsorption on clinker minerals. At the initial stage, it mainly occurs on the surface of positive-potential minerals C_3A and C_4AF, as a result of which less ettringite (AFt) and monosulfate (AFm) are formed at the initial stage of hydration [4, 5].

It is known that the use of PCE prevents the coagulation of ettringite with minerals C_3S and CSH phases. In the hardening cement dough, the carboxylate groups are adsorbed directly on the ettringite surface [6]. Along with a change in morphology, the strength of cement compositions with superplasticizers increases due to a change in the nature of hydrated cement stone neoplasms [7, 8].

In the presence of both PCE and SNF additives, ettringite crystals are much smaller. In addition, the morphology changes from long and thin to short and compact. Most likely, changes in the size and morphology of crystals caused by superplasticizers are associated either with a higher rate of embryo formation or with preferred adsorption on certain crystal faces, which prevents normal crystal growth [4, 9–12]. Studies have shown that superplasticizers can inhibit the growth of crystals and, consequently promote their nucleation, which leads to smaller crystals in larger quantities [13, 14]. When the PCE is adsorbed on ettringite and stabilizes it, the transformation rate of ettringite into monosulfate slows down [15].

Some scientists note that in cement systems with an increased amount of $Ca(OH)_2$ and a high concentration of sulfates, AFt is formed in the form of short prismatic crystals; on the contrary, elongated needle-like AFt crystals are formed in unsaturated solutions. The first crystals of calcium hydrosulfoaluminate formed in the supersaturated $Ca(OH)_2$ liquid phase of cement dough have the form of short prisms. At later stages of hydration, elongated needle-like crystals develop. [16, 17].

Nowadays, the problem of accelerating the strength gain of concrete, both on construction sites and on factory lines, is particularly relevant. It should be noted that the main way to intensify the hardening of concrete is HMT. The essence of the HMT is that the concrete mixture hardens at elevated temperatures (up to 90–95° C) and atmospheric pressure.

At the same time, the pores between the formed crystals are well filled with hydration products. With an increase in the heating time, cement clinker particles continuously dissolve, hydration products gradually precipitate, and the pore structure is crushed [18]. It is also noted that a higher steam curing temperature (70–100 °C) leads to a lower final strength; therefore, the steam curing temperature is a compromise factor between the rate of strength gain at an early age and the ultimate strength [19, 20]. In addition, elevated temperatures and a long warm-up period lead to a decrease in the durability of concrete, which is especially important for products operated in direct contact with water and the atmosphere [21].

The authors of this article previously conducted a study of the influence of low-temperature and reduced modes of HMT on the strength of concrete and the phase composition of cement stone. Experiments have confirmed the possibility of obtaining concrete with PCE with high early strength at a warm-up time of only 6 h and a steaming temperature of 60° C. In addition, it is shown that the pre-aging of concrete with PCE can be reduced due to the insignificant effect on the strength after the HMT [22]. The reduced content of AFt and portlandite in cement stone with PCE at the HMT mode (0–6–2) h at a temperature of 60° C was found to be 3% and 8% in comparison with the sample with SNF and without additives, respectively. Also, a greater number of unreacted C_3A and C_4AF grains were detected in the compositions with PCE [22, 23].

Of particular interest is the microstructure of cement stone after low-temperature modes of HMT. It is important to understand what changes occur when PCE and SNF additives are introduced into the formulation of cement systems. Therefore, the authors carried out a further study of cement stone by scanning electron microscopy.

2 Materials and Methods

Cement produced by "Asia Cement" (Penza, Russia) was used in this study. The mineralogical composition of clinker and some properties are presented in Table 1.

Table 1 Some properties of the cement used

№	Description of characteristics	Unit of measurement	Actual readings
Physical and mechanical properties of cement			
1	Normal consistency of cement paste	%	0.304
2	Compressive strength after HMT according to the Russian State Standard (GOST) 31,108-2020	MPa	36.89
3	Efficiency group after steaming according to the Russian State Standard (GOST) 31,108-2020		I
Mineral composition of the cement clinker			
4	Tricalcium silicate ($3CaO \cdot SiO_2$)	%	64.1
5	Dicalcium silicate ($2CaO \cdot SiO_2$)	%	13.1
6	Tricalcium aluminate ($3CaO \cdot Al_2O_3$)	%	7.3
7	Tetracalcium aluminoferrite ($4CaO \cdot Al_2O_3 \cdot Fe_2O_3$)	%	11.5

Table 2 Compositions for the preparation of cement pastes

№	Admixture	Dosage of the admixture, in % by cement mass	w/c
1	Without admixtures	–	0.309
2	S-3 (SNF)	0.4%	0.302
3	Sika ViscoCrete 24HE (PCE)	0.4%	0.268

Superplasticizers based on PCE and SNF were used as an additive. The PCE additive, Sika ViscoCrete 24HE, was produced in the Swiss concern "Sika" (Lobnya, Russia). The superplasticizer SNF is represented by the additive C-3 produced by "Polyplast" (Novomoskovsk, Russia). The dosage of the additives was 0.4%, following the technical descriptions. The superplasticizer is introduced into the cement mixture with the last quarter of the mixing water. As a result, equally mobile cement pastes were prepared. The addition of PCE allowed reducing the w/c ratio by 13.3%, while SNF by 2.3%. The mobility of cement pastes was controlled by lowering the Vic device's pestle according to the Russian standard (GOST) 310.3–76 until the normal density was reached.

The compositions of the prepared samples are shown in Table 2.

Next, cubes with a rib size of 20 mm were poured and placed in a steaming chamber. The samples were hardened according to the HMT (0–6–2) h regime at a temperature of 60° C.

According to the above regime, the samples were immediately prepared for further SEM analysis after steaming and cooling.

Studies of the obtained cement stone samples were carried out using an electron microscope equipped with an Aztec X-MAX energy dispersion spectrometer. The resolution of the spectrometer is 127 eV. Cement stone was sprayed with gold alloy on a turbomolecular pump sprayer T150 ES (Quorum Technologies, UK) under vacuum. SEM of samples was carried out on an auto-emission scanning electron microscope Merlin of CARL ZEISS company.

3 Results and Discussion

Scanning electron microscopy at 100× magnification revealed that the number and pore size of the sample with PCE is smaller, which is consistent with the previous studies of the authors of this article (Figs. 1, 2 and 3) [22].

Fig. 1 SEM-image of a sample without additives at 100× magnification

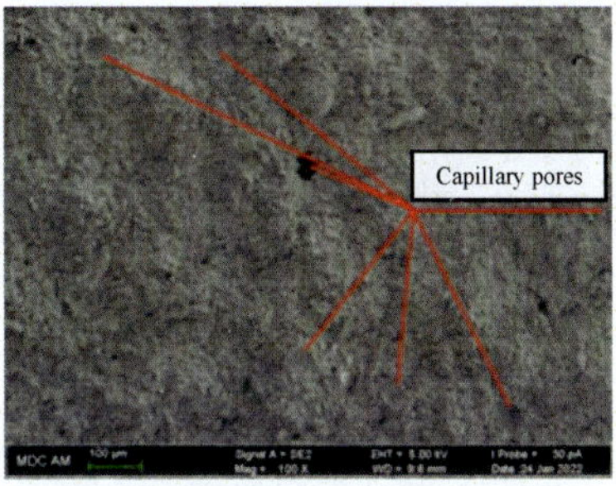

Fig. 2 SEM-image of a sample with SNF at 100× magnification

Fig. 3 SEM-image of a sample from a PCE at 100× magnification

It has been revealed that the introduction of both SNF and PCE additives into cement systems leads to a compaction of the structure, and the pores in compositions with superplasticizers are more homogeneous and have a smaller size.

Thermal cracks were found inside the pores in the sample without additives at 1000× magnification (Fig. 4).

Fig. 4 SEM-image of a sample without additives at 1000× magnification. Illustration of cracks in pore

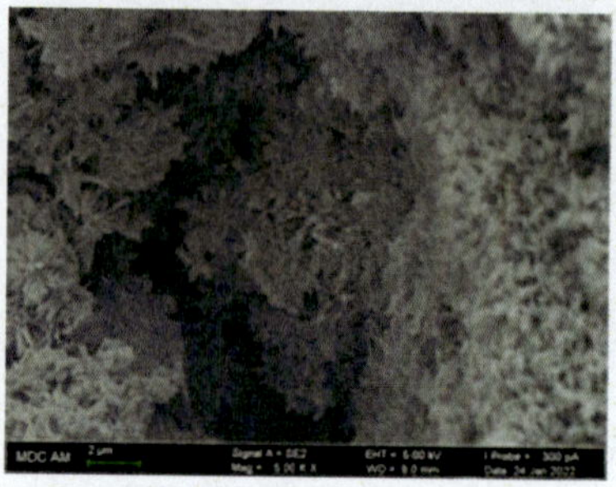

Fig. 5 SEM-image of a sample with PCE at 5000× magnification. Illustration of AFt growth in a pore

Fig. 6 SEM-image of a sample without additives at 1000× magnification

In addition, crystal formation at the edges of capillary pores has been studied. Active formation of needle-like AFt was detected in the contact zone. The compaction and curing of the structure at this stage occur since hydrated calcium sulfoaluminate crystallizes with an increase in volume in the pore spaces of the cement stone (see Fig. 5).

The microstructure of the samples at a magnification of 1000× is shown in Figs. 6, 7 and 8. It was revealed that unreacted clinker minerals are present in the stone

Fig. 7 SEM-image of a sample with SNF at 1000× magnification

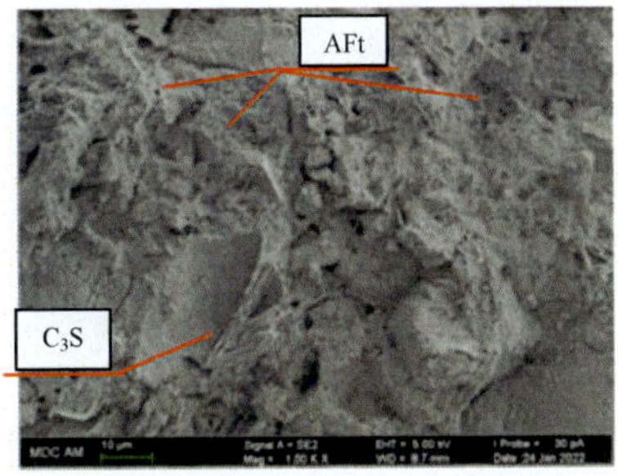

Fig. 8 SEM-image of a sample with PCE at 1000× magnification

with the addition of PCE. AFt crystals are deposited as separate or compacted and agglomerated mass, where individual crystals are barely distinguishable [6].

For a more specific inspection of the formed phases, a survey was carried out at a resolution of 5000× in Figs. 9, 10 and 11. It was revealed that homogeneous long needle-shaped calcium hydrosulfoaluminates, up to 3 microns long, were formed in the cement stone after HMT according to the regime (0–6–2) h. At the same time, in a stone with SNF, the hydration reaction significantly slowed down. The length

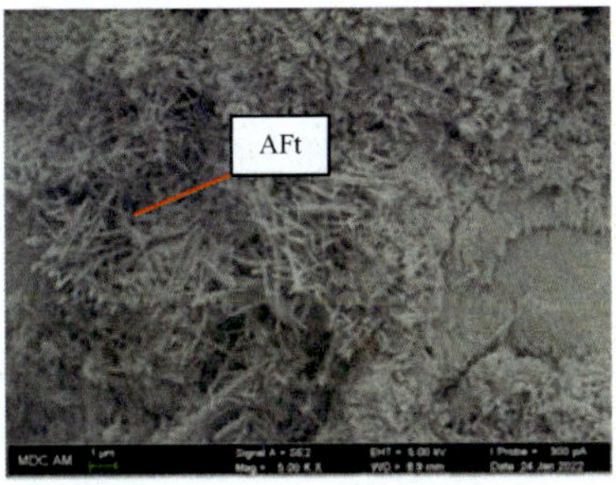

Fig. 9 SEM-image of a sample without additives at 5000× magnification

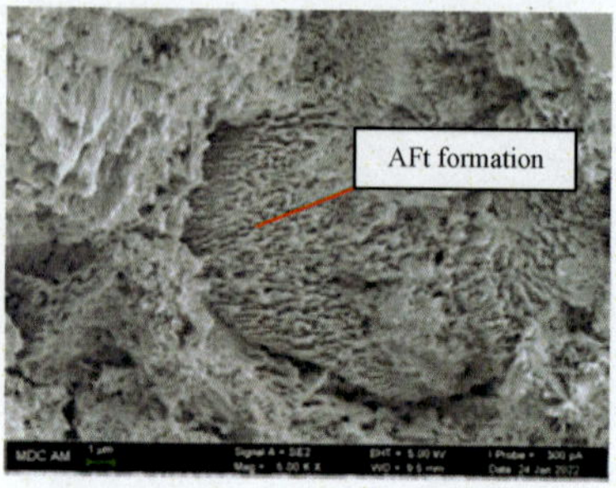

Fig. 10 SEM-image of a sample with SNF at 5000× magnification

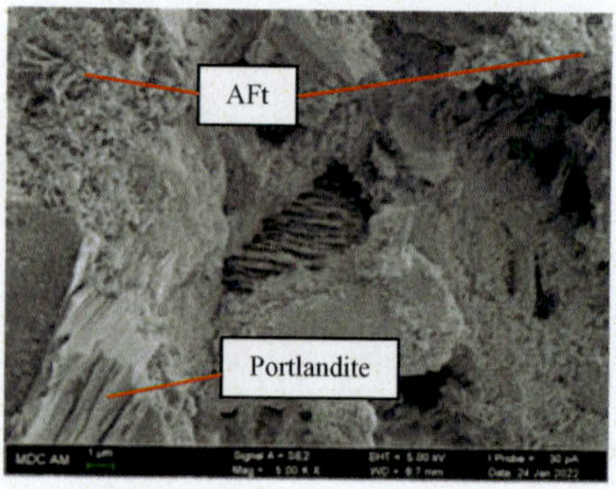

Fig. 11 SEM-image of a sample from a PCE at 5000× magnification

of the AFt minerals is from 100 to 500 nm, which explains the low strength of the composition at this time.

The slowing down of hydration in systems with PCE is associated with the high adsorption capacity of the superplasticizer on cement particles and, in particular, its interaction with ettringite. It leads to a delay in its deposition and a significant change in morphology. Also, the compositions with PCE and SNF contain a relatively large amount of non-hydrated cement particles [11, 23].

Further research was carried out at 10,000× magnification. Micrographs are shown in Figs. 12, 13 and 14. The structure of the sample without additives is represented by long needle-like calcium hydrosulfoaluminate, as well as clusters of C-S-H phases, which converges with previous studies [23]. The amount of clinker minerals in the composition with PCE is also higher than in the sample without additives and with SNF. Since the adsorption component of the PCE molecule has a comb shape [1], and SNF has a flat rigid rod shape; thus, SNF can be adsorbed on the surface of gypsum particles layer by layer, which leads to an increase in the amount of adsorption, and hence a significant slowdown in hydration [24]. The distribution of hydration products of steamed cement stone is more uniform and orderly [13].

Unreacted $CaSO_4 \cdot 2H_2O$ minerals were also detected in cement stone samples, which may indicate selective adsorption of PCE on minerals, in particular on C_3A (Fig. 15). It is known that ettringite is formed by the interaction of tricalcium alumi-

Fig. 12 SEM-image of a sample without additives at 10,000-fold magnification

Fig. 13 SEM-image of a sample with SNF at 10,000-fold magnification

Fig. 14 SEM-image of a sample from a PCE at 10,000-fold magnification

nate with the formation of a trisulfate form of calcium hydrosulfoaluminate. Ettringite, which forms short crystals up to several hundred nanometers long, is clearly manifested relative to C_3A grains and unreacted gypsum minerals [25].

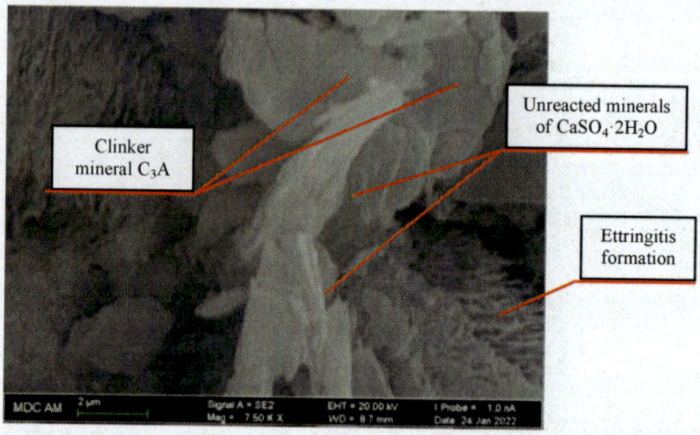

Fig. 15 SEM-image of a sample from a PCE at 7500× magnification

4 Conclusions

As a result of the performed SEM analysis, it was revealed that using both SNF and PCE-type additives contributes to the compaction of the cement stone structure, reduction and more uniform formation of capillary pores.

Calcium hydrosulfoaluminate crystals are much smaller in a system with superplasticizers. The sites in the compositions with SNF were identified, in which the hydration of C_3A was greatly slowed down, AFt crystals are just beginning to form, as evidenced by their length (up to 100–500 nm). At the same time, the length of ettringite crystals in compositions without additives reaches 3 microns.

Analysis of the pair's contact zone with the cement stone's surface showed that the pores are covered with multiple formations of ettringite, which strengthens its structure. A significant number of temperature cracks were also detected after HMT in the pores of the composition without additives.

Zones of unreacted $CaSO_4 \cdot 2H_2O$ minerals were detected in cement stone with the addition of PCE at high magnification. A gradual formation of ettringite was recorded in the contact zone with tricalcium aluminate, and this fact indicates selective adsorption of PCE on clinker minerals, particularly on C_3A.

The results obtained can be used to develop low-temperature HMT modes with a shortened steaming time.

Acknowledgements The research was supported by the Ministry of Science and Higher Education of the Russian Federation (Grant № 075-15-2021-674) and Core Facility Centre «Ecology, biotechnologies and processes for obtaining environmentally friendly energy carriers» of Volga State University of Technology, Yoshkar-Ola.

References

1. Yamada K, Ogawa S, Hanehara S (2000) In: Proceedings of the 6th CANMET/ACI International Conference on Superplasticizers (Nice, 2000), pp 351–366
2. Flatt RJ, Houst YF (2001) Cem Concr Res 31:1169
3. Yoshioka K, Sakai E, Daimon M, Kitahara J (2005) J Am Ceram Soc 80:2667
4. Plank J, Hirsch C (2007) Cem Concr Res 37:537
5. Schmidt W, Brouwers HJH, Kühne HC, Meng B (2014) Cem Concr Compos 49:111
6. Zingg A, Winnefeld F, Holzer L, Pakusch J, Becker S, Gauckler L (2008) J Colloid Interface Sci 323:301
7. Ibragimov R, Fediuk R (2019) Constr Build Mater 226:839
8. Khuzin A, Ibragimov R (2021) J Build Eng 35:102030
9. Barneoud-Chapelier A, Saout Le G, Azéma N, El Bitouri Y (2022) Constr Build Mater 322:126483
10. El-Gamal SMA, Al-Nowaiser FM, Al-Baity AO (2012) J Adv Res 3:119
11. Tian H, Kong X, Cui Y, Wang Q, Wang D (2019) Constr Build Mater 228:116711
12. Li S et al (2019) Colloids Surfaces A Physicochem Eng Asp 564:78
13. Yu Z, Ma J, Shi H, Shen X, Ye G (2016) In: International Conference on the Durability of Concrete Structures ICDCS 2016 (Purdue University, 2016), pp 171–173
14. Niu M, Li G, Li Q, Zhang G (2021) Constr Build Mater 312:125374
15. Liu M, Gao Y, Zhang L, Jiang G, Zeng C, Wang P (2021) Thermochim Acta 696:178821
16. Winnefeld F, Zingg A, Holzer L, Pakusch J (2009) In: 9th ACI International Conference on Superplasticizers and Other Chemical Admixtures Concrete (Sevilla, Spain, 2009), pp 1–17
17. Rößler C, Eberhardt A, Kučerová H, Möser B (2008) Cem Concr Res 38:897
18. Shi J, Liu B, Wu X, Qin J, Jiang J, He Z (2020) J Build Eng 32:101796
19. Zeyad AM et al (2022) Clean Mater 3:100042
20. Mironov SA (1966) In: Symposium on Structure of Portland Cement Paste and Concrete (Washington, 1966), pp 465–474
21. Hanson JA (1963) ACI J 60:75
22. Leshkanov AY, Dobshits LM, Anisimov SN (2020) IOP Conference Series: Materials Science and Engineering 896:1
23. Leshkanov A, Dobshits L, Anisimov S (2022) Lect Notes Civ Eng 170:53
24. Cao W, Yi W, Yin S, Peng J, Li J (2021) Constr Build Mater 278:122421
25. Kirchheim AP, Fernàndez-Altable V, Monteiro PJM, Dal Molin DCC, Casanova I (2009) J Mater Sci 44:2038

Flexural Capacities of Cold-Formed Steel Channel Sections with Perforations

Ngoc Hieu Pham

Abstract Cold-formed steel beams with perforations are commonly applied to meet the requirements of technical system installations. The presence of holes leads to the capacity reduction of this structure; and their influence on the flexural capacities of cold-formed steel sections has been regulated in AISI S100-16 using a new design method named the Direct Strength Method (DSM). This method provides predictions of the flexural capacities of cold-formed steel sections based on elastic buckling analyses. This paper, therefore, investigates the effects of variations of web hole dimensions on the flexural capacities of cold-formed steel channel sections. Based on the investigated results, the analyses are subsequently provided for the selection of proper web hole dimensions to optimise the flexural capacities of the investigated channel sections.

Keywords Web holes · Flexural capacities · Cold-formed steel channel sections · Perforations

1 Introduction

Cold-formed steel channel beams are commonly pre-punched in the web to allocate technical systems. The presence of these web holes has been illustrated to have significant impacts on the sectional capacities under bending [1], and have been considered in the design of cold-formed steel structures as regulated in American Specification AISI S100 [2] with the variety of hole shapes such as circular, rectangular holes or slotted holes. The flexural capacities of channel sections can be determined using the Direct Strength Method (DSM) proposed by Schafer [3–5] that provides the predictions of the sectional capacities of the investigated sections based on sectional elastic buckling loads. These elastic buckling loads are obtained with the support of a module software program (CUFSM v0.1 – Hole Effect Analysis Tool) developed

N. H. Pham (✉)
Faculty of Civil Engineering, Hanoi Architectural University, Hanoi, Vietnam
e-mail: hieupn@hau.edu.vn

P. Akimov et al. (eds.), *Proceedings of FORM 2022*, Lecture Notes in Civil Engineering 282, https://doi.org/10.1007/978-3-031-10853-2_17

by the American Iron and Steel Institute [6, 7] based on the research results of Moen and Schafer [8–10].

In the literature, a large number of studies on cold-formed steel sections with perforations have been available. A variety of hole shapes have been investigated to determine their effects on the capacities of cold-formed stub columns [1, 11–13]. The influence of web hole locations or hole lengths on the capacities of stub columns was also investigated [14–17]. Moen and Schafer carried out a large number of research studies [8–10, 18–21] to investigate the strength and behavior of cold-formed steel sections with perforations. The design of perforated sections was subsequently proposed and included in the Specification AISI S100-16 [2] using the Direct Strength Method that will be used in this investigation. This paper, therefore, is aimed to investigate the flexural capacities of cold-formed perforated steel channel sections using the DSM design regulated in AISI S100-16 with the variation of rectangular hole sizes. Based on the investigated results, recommendations are given for the selection of hole sizes of channel sections with perforations under bending.

2 Determination of Flexural Capacities of Cold-Formed Steel Channel Sections

The flexural capacity of the cold-formed steel channel section with perforations is determined using DSM design specified in AISI S100-16 [2] and is the least of the following moment values including yield moment of the net cross-section (M_{ynet}), local buckling moment (M_{nl}), and distortional buckling moment (M_{nd}).

- Yield moment of the net cross-section:

$$M_{ynet} = S_{fnet} F_y \tag{1}$$

- Local buckling moment:

$$M_{nl} = M_y \quad \text{for} \quad \lambda_l \leq 0.776 \tag{2}$$

$$M_{nl} = \left[1 - 0.15\left(\frac{M_{crl}}{M_y}\right)^{0.4}\right]\left(\frac{M_{crl}}{M_y}\right)^{0.4} M_y \quad \text{for} \quad \lambda_l > 0.776 \tag{3}$$

- Distortional buckling moment:

If $\lambda_d \leq \lambda_{d2}$:

$$M_{nd} = M_{ynet} \quad \text{for} \quad \lambda_d \leq \lambda_{d1} \tag{4}$$

$$M_{nd} = M_{ynet} - \left(\frac{M_{ynet} - M_{d2}}{\lambda_{d2} - \lambda_{d1}}\right)(\lambda_d - \lambda_{d1}) \text{ for } \lambda_{d1} < \lambda_d \leq \lambda_{d2} \tag{5}$$

If $\lambda_d > \lambda_{d2}$:

$$M_{nd} = M_y \text{ for } \lambda_d \leq 0.673 \tag{6}$$

$$M_{nd} = \left[1 - 0.22\left(\frac{M_{crd}}{M_y}\right)^{0.5}\right]\left(\frac{M_{crd}}{M_y}\right)^{0.5} M_y \text{ for } \lambda_d > 0.673 \tag{7}$$

where:

S_{fnet} is the net section modulus referenced to the extreme fiber at first yield;

F_y is the yield stress;

λ_l, λ_d are slenderness factors of local buckling or distortional buckling for a beam, respectively;

M_y, M_{ynet} are yield moments of the gross section and net section;

M_{crl}, M_{crd} are the critical elastic moments of local and distortional buckling, can be determined using elastic buckling analyses presented in Sect. 3.1;

λ_{d1}, λ_{d2} are slenderness factors of distortional buckling for a beam;

$$\lambda_{d1} = 0.673\left(\frac{M_{ynet}}{M_y}\right)^3;$$

$$\lambda_{d2} = 0.673\left[1.7\left(\frac{M_y}{M_{ynet}}\right)^{2.7} - 0.7\right];$$

$$M_{d2} = \left[1 - 0.22\left(\frac{1}{\lambda_{d2}}\right)\right]\left(\frac{1}{\lambda_{d2}}\right)M_y$$

3 Effects of Web Hole Dimensions on the Flexural Capacities of Cold-Formed Steel Channel Sections with Perforations

Sections of C25015 and C40019 are used for this investigation with the nomenclature illustrated in Fig. 1 and their dimensions listed in Table 1. The hole heights (h_{hole}) are varied from 0.2 to 0.8D whereas the hole lengths (L_{hole}) are from 0.5 to 3.0D, where D is the depth of the investigated section. The material properties are used for the investigation including the yield stress $F_y = 345$ MPa and Young modulus $E =$ 203,400 MPa.

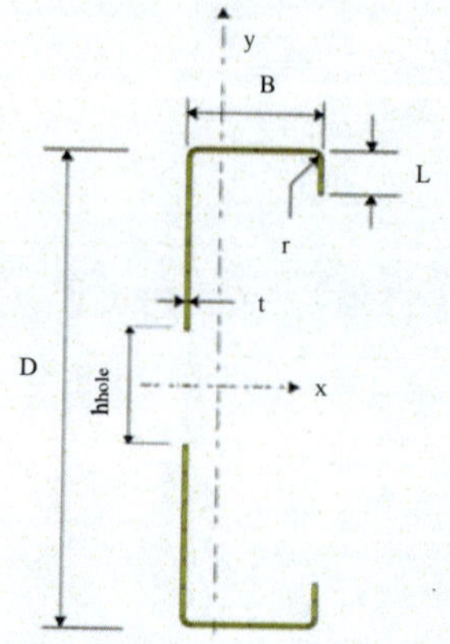

Fig. 1 Nomenclature

Table 1 The dimensions of investigated sections (Unit: mm)

Sections	t	D	B	L
C25015	1.5	254	76	21.5
C40019	1.9	400	125	30

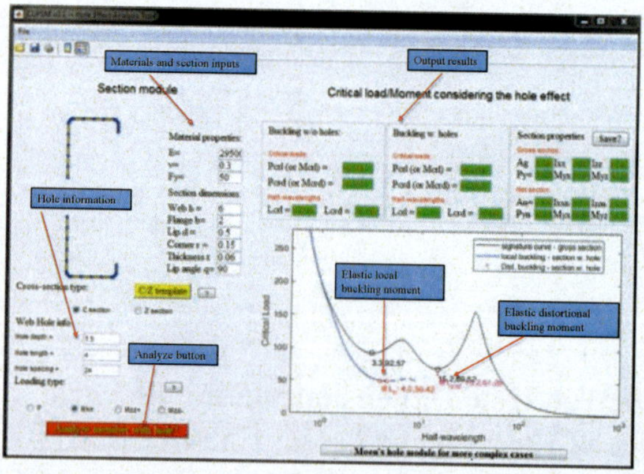

Fig. 2 CUFSM software program for elastic buckling analysis

3.1 Elastic Buckling Analysis of Cold-Formed Steel Channel Sections with Perforations

Elastic buckling analyses for the investigated channel sections with perforations are carried out using the CUFSM software program [22], as illustrated in Fig. 2. Based on Pham's research report [23], it was found that the elastic local buckling moments

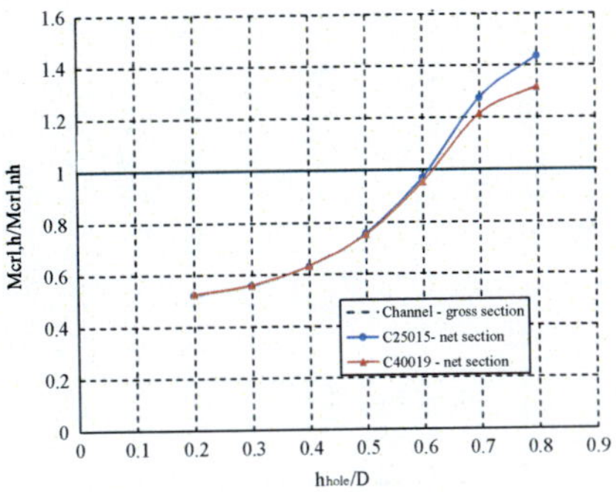

(a) Elastic local buckling

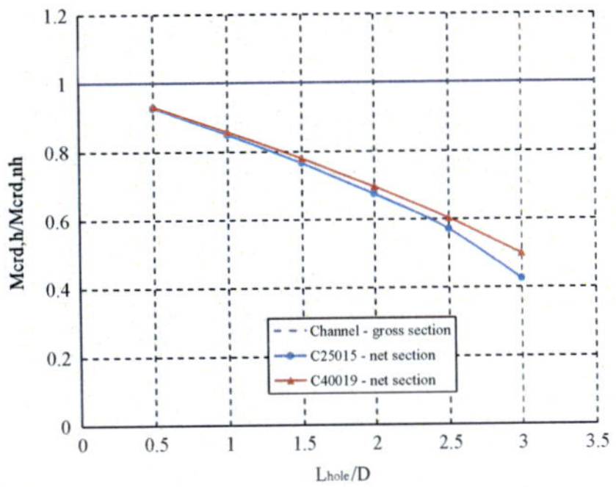

(b) Elastic distortional buckling

Fig. 3 Elastic local and distortional buckling moments of the investigated channel sections

only depended on the ratios of the hole heights and sectional depths (h_{hole}/D) whereas elastic distortional buckling moments depended on the ratios of the hole lengths and sectional depths (L_{hole}/D). The elastic local and distortional buckling moments, therefore, are determined with the variations of the hole heights and the hole lengths, respectively, as shown in Fig. 3, where ($M_{crl,h}$, $M_{crd,h}$) and ($M_{crl,nh}$, $M_{crd,nh}$) are the pairs of elastic local and distortional buckling moments for the net section and the gross section, respectively.

The investigated results show that when the hole sizes increase, an increasing trend is observed for elastic local buckling moments of the net sections, but it is a decreasing trend for elastic distortional buckling moments. The elastic local buckling moments of the net sections are seen to be reduced by 40% compared to those of the gross sections with the small hole heights. The local buckling, therefore, occurred at the net section areas. If the ratios of h_{hole}/D are greater than 0.6, the elastic local buckling moments of the net cross-sections are larger than those of the gross sections; this leads to the occurrence of local buckling at the gross section areas between holes. The elastic local buckling moments of perforated sections in these cases can be taken as those of the gross sections. This can be explained that the flat areas at the top and bottom of the holes in the web sections are reduced with the increase of the hole heights (h_{hole}); this leads to the reduction of the slenderness of these flat areas and the webs at the net sections become more stable. Meanwhile, the elastic distortional buckling moments undergo a decreasing trend for the net sections, reaching 50% when the hole lengths reach 3D, where D is the depth of the investigated section.

3.2 *Flexural Capacities of Cold-Formed Steel Channel with Perforations*

The flexural capacities of the investigated perforated channel sections are determined according to AISI S100 [2], as presented in Sect. 2. The investigated results include the local buckling moments and the distortional buckling moments, as demonstrated in Figs. 4(a) and (4b), where the former moments only depended on the hole heights (h_{hole}), but the latter moments are determined with the variations of both hole lengths (L_{hole}) and hole heights (h_{hole}). The flexural capacities of the investigated perforated

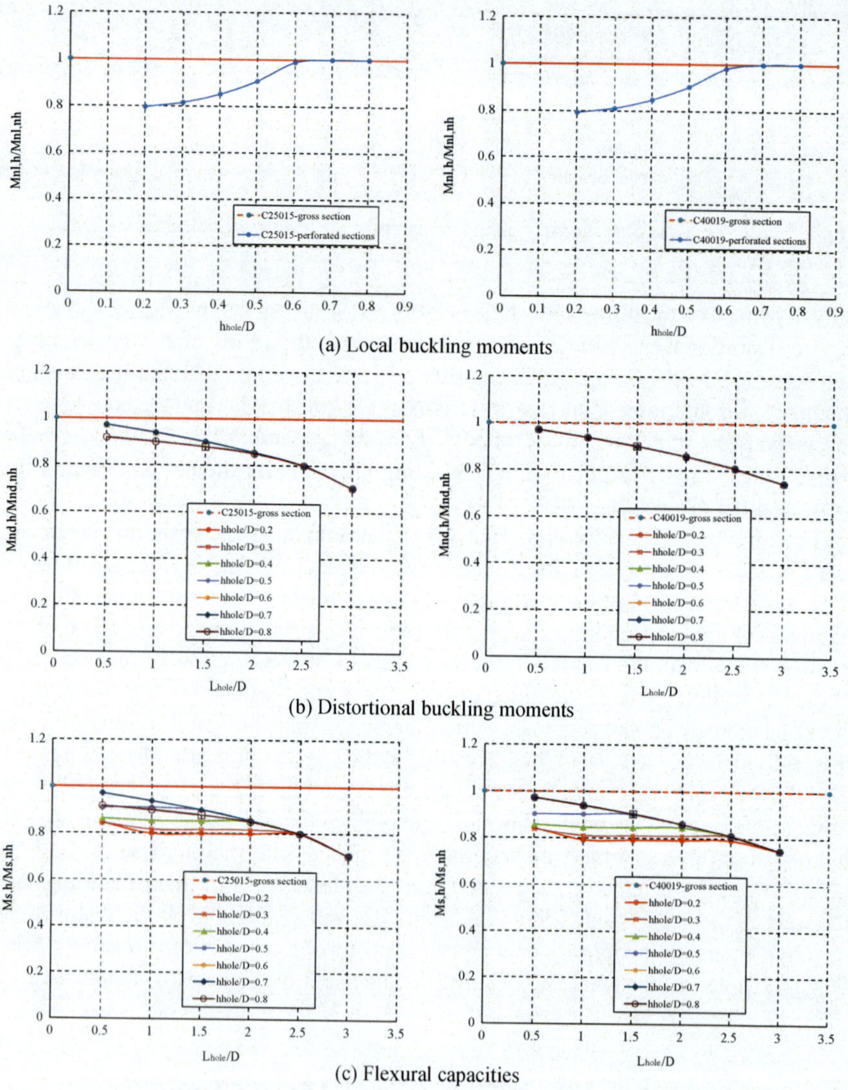

Fig. 4 Flexural capacities of the investigated perforated channel sections Note: ($M_{nl,nh}$, $M_{nd,nh}$, $M_{s,nh}$) and ($M_{nl,h}$, $M_{nd,h}$, $M_{s,h}$) are the local bucking, distortional buckling moments and flexural capacities of gross cross-sections and net cross-sections, respectively

channel sections are the lesser of the component moments in Figs. 4(a) and 4(b), as illustrated in Fig. 4(c).

Figure 4(a) shows that it is a reduction of 20% for local buckling moments of perforated channel sections in comparison with those of gross channel sections with the ratios h_{hole}/D of 0.2. These moments of perforated sections subsequently undergo an increasing trend and reach those of the gross sections when the ratios h_{hole}/D vary from 0.2 to 0.6. If these ratios are higher than 0.6, the local buckling moments of perforated sections are equal to those of gross sections due to the occurrence of local buckling at the gross section areas between holes as discussed in Sect. 3.1.

The distortional buckling moments of perforated sections are seen as a decreasing trend if the hole lengths increase; these reach a reduction of 30% compared to those of the gross cross-sections. These moments of the perforated sections are nearly unchanged with the variations of the hole heights. This can be illustrated in Fig. 4(b) for both investigated sections although there is a slightly lower for section C25015 with the ratio (h_{hole}/D) of 0.8.

The flexural capacities of the investigated perforated sections demonstrated in Fig. 4(c) are found to be governed by local buckling modes if the ratios (h_{hole}/D) vary from 0.2 to 0.6; this can be seen as the significant changes of Fig. 4(c) compared to Fig. 4(b) in these ranges. Meanwhile, these flexural capacities are governed by distortional buckling when the ratios (h_{hole}/D) of 0.7 and 0.8, as the unchanged curves between Fig. 4(c) and Fig. 4(b).

The investigated moment results in Fig. 4(c) demonstrated that the smaller holes are not better than the bigger holes in terms of flexural capacities. This is illustrated when the ratios (L_{hole}/D) vary from 1.0 to 2.0, the flexural capacities of the investigated sections are seen as an increasing trend with the increase of ratios (h_{hole}/D) from 0.2 to 0.7. This can be explained due to the increasing trend of local buckling moments as discussed in Fig. 4(a).

Based on the investigated results in Fig. 4(c), a recommendation is given as follows: the optimum hole sizes should be larger heights (h_{hole}) and shorter lengths (L_{hole}) in terms of a specified hole area. This is explained that this trend results in the increase of both local and distortional buckling moments, but the opposite trend leads to the reduction of both these component moments.

4 Conclusions

The paper investigates the flexural capacities of the cold-formed steel perforated channel sections with the variation of the hole sizes. The flexural capacities of the investigated sections were determined using the Direct Strength Method according to American Specification AISI S100-16. Based on the investigated results, the following conclusions can be given:

- The local buckling moments of perforated channel sections undergo an increasing trend with the increase of hole height. The local buckling modes, therefore, were occurred at the net sections with the small hole heights, but were at the gross section areas between holes with the large hole heights.

- When the hole lengths increase, the distortional buckling moments of perforated sections are seen as a reduction trend.
- For a specified hole area, the selection of larger height and shorter length of the hole sizes are recommended to obtain the optimum sections in terms of flexural capacities.

These conclusions provide deep understandings of the flexural capacities of cold-formed steel perforated channel sections to help the designers to use perforated channel sections more properly in the design.

References

1. Ortiz-Colberg RA (1981) The load carrying capacity of perforated cold-formed steel columns, Ph.D. thesis, Cornell University, Ithaca, New York
2. American Iron and Steel Institute (2016): North American Specification for the Design of Cold-formed Steel Structural Members. Washington DC: American Iron and Steel Institute
3. Schafer BW, Peköz T (1998) Direct strength prediction of cold-formed members using numerical elastic buckling solutions. In: Fourteenth International Specialty Conference on Cold-Formed Steel Structures
4. Schafer BW (2002) Local, distortional, and Euler buckling of thin-walled columns. J Struct Eng 128(3):289–299
5. Schafer BW (2008) Review: the direct strength method of cold-formed steel member design. J Constr Steel Res 64(7–8):766–778
6. American Iron and Steel Institute (2021): Development of CUFSM Hole Module and Design Tables for the Cold-formed Steel Cross-sections with Typical Web Holes, in AISI D100. Research Report RP21-01
7. American Iron and Steel Institute (2021): Development of CUFSM Hole Module and Design Tables for the Cold-formed Steel Cross-sections with Typical Web Holes, in AISI D100. Research Report RP21-02
8. Moen CD (2008) Direct Strength Design for cold-formed steel members with perforations, Ph.D. thesis, Johns Hopkins University, Baltimore
9. Moen CD, Schafer BW (2008) Experiments on cold-formed steel columns with holes. Thin-Walled Struct 46(10):1164–1182
10. Moen CD, Schafer BW (2009) Elastic buckling of cold-formed steel columns and beams with holes. Eng Struct 31(12):2812–2824
11. Sivakumanran KS (1987) Load capacity of uniformly compressed cold-formed steel section with punched web. Can J Civil Eng 14(4):8
12. Banwait AS (1987) Axial load behaviour of thin-walled steel sections with openings, Master dissertation, Hamilton, Ontario
13. Abdel-Rahman N (1997) Cold-formed steel compression members with perforations, Master dissertation, Hamilton, Ontario
14. Rhodes J, Schneider FD (1994) The compressional behaviour of perforated elements. In: Twelfth International Specialty Conference on Cold-Formed Steel Structures, pp 11–28
15. Loov R (1984) Local buckling capacity of C-shaped cold-formed steel sections with punched webs. Can J Civil Eng 11(1):1–7
16. Pu Y, Godley MHR, Beale RG, Lau HH (1999) Prediction of ultimate capacity of perforated lipped channels. J Struct Eng 125(5):4

17. Rhodes J, Macdonald M (1996) The effects of perforation length on the behaviour of perforated elements in compression. In: Thirteenth International Specialty Conference on Cold-Formed Steel Structures, pp 91–101
18. Moen CD, Schafer BW (2006) Impact of holes on the elastic buckling of cold-formed steel columns. In: International Specialty Conference on Cold-Formed Steel Structures, pp 269–283
19. Moen CD, Schafer BW (2010) Extending direct strength design to cold-formed steel beams with holes. In: 20th International Specialty Conference on Cold-Formed Steel Structures - Recent Research and Developments in Cold-Formed Steel Design and Construction, pp 171–183
20. Cai J, Moen CD (2016) Elastic buckling analysis of thin-walled structural members with rectangular holes using generalized beam theory. Thin-Walled Struct 107:274–286
21. Moen CD, Schafer BW (2009) Elastic buckling of thin plates with holes in compression or bending. Thin-Walled Struct 47(12):1597–1607
22. Li Z, Schafer BW (2010) Buckling analysis of cold-formed steel members with general boundary conditions using CUFSM: Conventional and constrained finite strip methods, Saint Louis, Missouri, USA
23. Pham NH (2021) Investigation of web hole effects on the elastic buckling loads of cold-formed steel members, Research Report, Hanoi Architectural University

Influence of Sectional Imperfections on Strength and Behavior of Cold-Rolled Aluminium Alloy Channel Stub Columns

Ngoc Hieu Pham

Abstract Geometric imperfections have significant effects on the buckling and capacities of cold-formed thin-walled structures, and should be considered in structural analysis. Cold-rolled aluminium alloy sections are new structural products of BlueScope Permalite that can be seen as thin-walled structures to be prone to bucklings. This paper investigates the effects of sectional imperfections on the strength and behavior of cold-rolled aluminium alloy channel sections under compression using finite element simulations. The compression calibrated models presented in publications of Pham et al. on cold-rolled aluminium structures are utilized in this investigation. Sectional measured imperfection data is also collected from Pham et al., and is the basis to generate random samples for numerical investigation using the Latin Hypercube technique. The random imperfection values are incorporated into the simulations with a variety of imperfection shape modes. The simulation results are thoroughly analysed to demonstrate the imperfection effects on the capacities and behaviors of cold-rolled aluminium alloy channel sections. Recommendations are given to the selections of imperfection shape modes for further studies.

Keywords Sectional imperfections · Strength and behavior · Cold-rolled aluminium alloy channel · Stub columns

1 Introduction

Aluminium is a relatively new material in the structural engineering point with a variety of applications in bridges and buildings [1]. The advantageous properties of this material are excellent corrosion resistance, high strength to weight ratio and convenience in transportation and assembly. Aluminium sections are normally produced by two methods. Extrusion is the conventional method commonly used around the world whereas cold-rolling is a new method recently used by BlueScope Permalite in Australia to successfully produce aluminium alloy sections. The new

N. H. Pham (✉)
Faculty of Civil Engineering, Hanoi Architectural University, Hanoi, Vietnam
e-mail: hieupn@hau.edu.vn

P. Akimov et al. (eds.), *Proceedings of FORM 2022*, Lecture Notes in Civil Engineering 282, https://doi.org/10.1007/978-3-031-10853-2_18

products are found to be more cost-effective in comparison with the extruded ones, and should be commercialized in the worldwide market. A large number of recent research studies are on extruded aluminium alloy structures in the literature, whereas research studies on cold-rolled aluminium alloy structures remain scarce.

The actual sections appear to be not "perfect" geometries as nominal dimensions known as geometric imperfections. They are caused by unavoidable disturbances during assembly, transportation or production processes. Aluminium is a lightweight and soft material; this leads to significantly large imperfections are contained in cold-rolled aluminium alloy members. Geometric imperfections are determined using a variety of measuring methods, as reviewed and presented in Pham et al. [2]. The effects of geometric imperfections on the instabilities of cold-formed steel members have been studied in past investigations [3–6]. Cold-rolled aluminium alloy sections in the form of thin-walled sections are sensitive to instabilities that are significantly affected by geometric imperfections. The imperfections are classified into global and sectional imperfections. The global imperfections are twists or deflections of members (G_1, G_2, G_3) whereas displacements of plate elements are sectional imperfections including local imperfections (d_1) and distortional imperfections (d_2), as illustrated in Fig. 1.

There were a large number of studies on geometric imperfections and their effects on the strengths and behaviors of cold-formed steel members. Pi and Trahair [4, 7] investigated the effects of initial twist imperfections on the lateral buckling strengths of cold-formed steel channel and zed section beams. Dubina and Ungureanu [6] also pointed out that the combination of initial twist and deflection imperfections have considerable impacts on the ultimate strengths of flexural members, whereas the effects of sectional imperfections are negligible and can be ignored. Dinis et al. [8] examined the effects of pure local and pure distortional imperfections on the post-buckling behavior of cold-formed steel channel columns. Dominik et al. [9] developed a random field approach of modeling geometric and thickness imperfections to investigate the buckling behavior of thin-walled I section beam-columns. Andrei et al. [10] conducted an imperfection sensitivity analysis to identify the most critical imperfection modes which have impacts on the practical erosion of cold-formed steel perforated sections in compression. Chao and Yong-Lin [11] studied

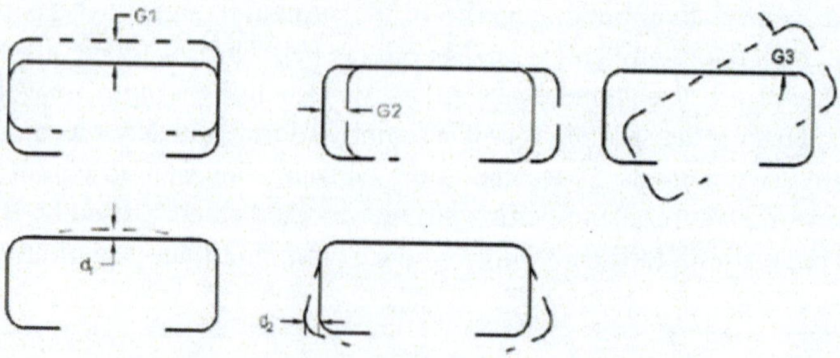

Fig. 1 Representatives of imperfection components

the effects of deflection imperfections on the inelastic flexural bucking resistance of box columns. Radom imperfections were generated based on existing measured data, and were then incorporated into finite element models to obtain the corresponding ultimate loads. A critical geometric imperfection shape was proposed to get a rational buckling resistance result. Dinis et al. [12] investigated the behavior and strength of cold-formed steel channel columns affected by global and distortional imperfection modes. This investigation was the basis to identify the most detrimental imperfections that were subsequently used in a parametric study. In terms of aluminium structures, studies on the effects of geometric imperfections on the strength and behavior of aluminium sections or members remain scarce. Pham [13] investigated the influence of global imperfections on the member strengths of cold-rolled aluminium alloy members under compression or bending. Pham's research showed the effects of initial twist and flexural imperfection modes on the behavior and strengths of cold-rolled aluminium alloy members. Detrimental directions of global imperfection modes were then proposed to develop parametric studies based on the investigated results.

This paper presents a FE model to investigate the effects of sectional imperfections on the behavior and strength of cold-rolled aluminium alloy channel stub columns. The amplitudes of these imperfections were collected from the measured imperfection data in Pham's publications [2, 14] and are then used to determine the probability density functions of sectional imperfections. Random amplitude samples of sectional imperfections are generated based on these functions by using the Latin Hypercube sampling technique [15]. Huynh et al. [16] pointed out that sectional capacities of cold-rolled alumnium alloy columns are very sensitive to sectional imperfections that should be included in the analysis. Combinations of local and distortional imperfections, therefore, are included in the analysis with a variety of amplitudes and imperfection shape modes. The investigated model is the calibrated compression model presented in Pham and Huynh's publications [17, 18] to ensure the reliability and accuracy of investigated results. The sectional behaviors are thoroughly analysed to illustrate the influence of geometric imperfections. The investigated results are the basis to identify the most detrimental modes of imperfections corresponding to the lowest column strengths that are then proposed for further research.

2 Sectional Imperfections and Nonlinear Finite Element Models

The procedures to measure the geometric imperfections and to incorporate them into the FE models were presented in Pham et al. [2]. Figure 2 shows an imperfect specimen after incorporating initial imperfections with the presence of initial twist (G_3) at two end sections, initial local imperfections (d_1) in the web and initial distortional imperfections (d_2) along the flanges. Sections used for the investigation are C10030 and C25025 with the nomenclature illustrated in Fig. 3 and their dimensions listed in Table 1.

Amplitudes of each sectional imperfection component are collected from the measured imperfections as fully reported in Pham et al. [2, 14], which allows determining their probability density functions with the support of the Easy-fit software program [19] as presented in Pham [13]. These obtained functions are subsequently used to generate random samples of sectional imperfections using the Latin Hypercube sampling technique [15]. This technique allows minimising the number of

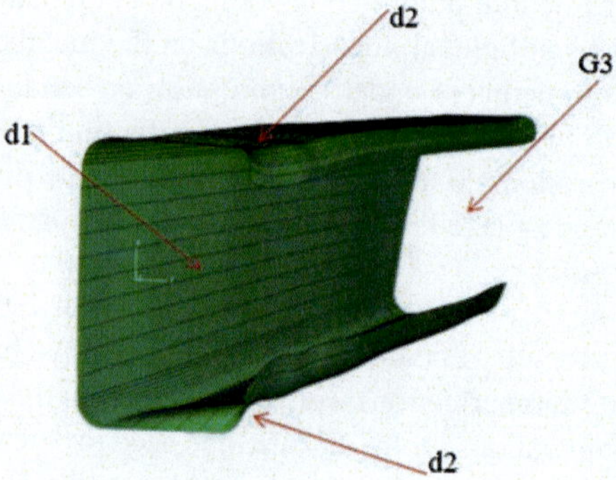

Fig. 2 An imperfect specimen

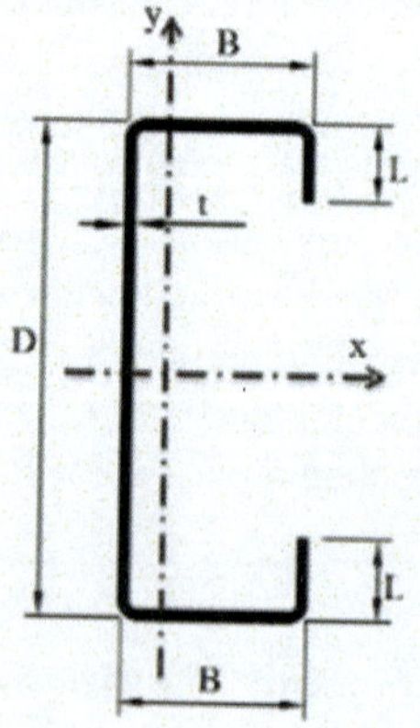

Fig. 3 Nomenclature

Table 1 The dimensions of investigated sections and elastic bucking analysis results

Sections	t mm	D mm	B mm	L mm	A mm^2	f_{ol} MPa	L_{ol} mm	f_{od} MPa	L_{od} mm
C10030	3.0	103	60.5	16	703.7	302.8	80	240.6	350
C25025	2.5	255	76	25.5	1093	36.21	200	66.03	800

random samples based on the determined probability density functions. The random samples of each sectional imperfection are generated and listed in Table 2 that will be used to incorporate into finite element (FE) models.

The nonlinear finite element model was developed and validated against experiments, as fully presented in [17, 18], and boundary conditions of this model are shown in Fig. 4(b). Two end boundary conditions were fixed with a uniform displacement at one end. Pattern stone was poured at two ends to ensure load uniformly transfer across the section, to prevent two end sections from sliding while testing, and to prevent local failure adjacent to the end sections. Material properties for this model are based on the coupon test results as presented in Huynh et al. [20] on cold-rolled aluminium alloy sections for both flat and corner parts in the cross-section. As discussed in Pham [13], assumed imperfections used for the investigation are performed as simple sinusoidal functions, as follows:

$$f(x) = A\sin(n\pi z/l) \tag{1}$$

Table 2 The random values of sectional imperfection amplitudes

Number	Sections			
	C10030	C25025		
	d_1 (mm)	d_2 (mm)	d_1 (mm)	d_2 (mm)
1	0.2901	0.9642	0.7409	1.1967
2	0.2372	0.7809	0.6351	1.9602
3	0.4451	1.1248	0.8336	0.9183
4	0.3369	1.3565	0.9673	1.334
5	0.1927	0.3541	0.3888	1.5599
Mean	0.3004	0.9161	0.7131	1.3938

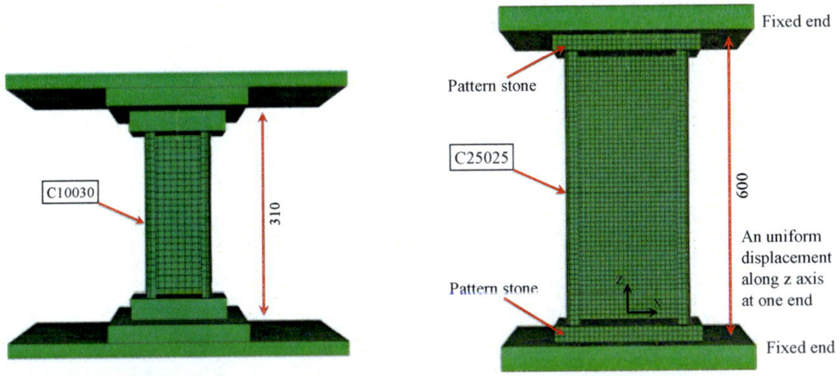

(a) Model 1: C10030

(b) Model 2: C25025

Fig. 4 The investigated finite element models

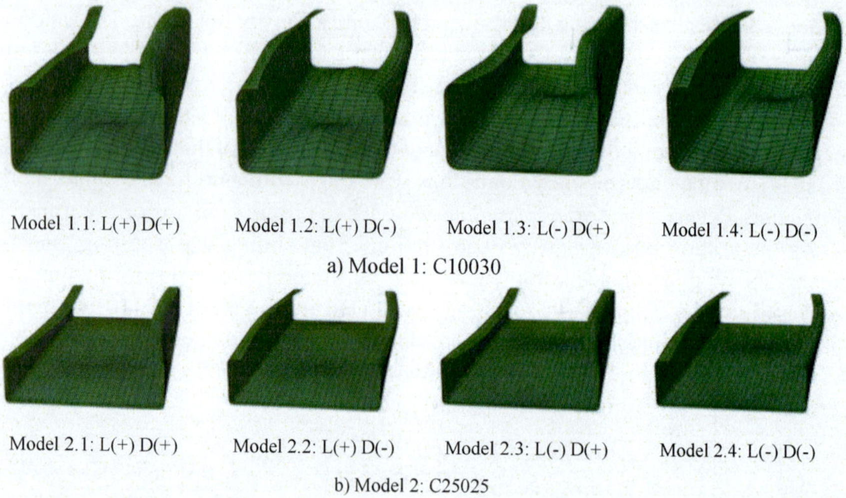

Fig. 5 Sign conventions of sectional imperfections and imperfection shape modes of the investigated FE models Note: L, D stand for local and distortional imperfections

where A is the amplitude of the imperfection component; n is the number of half-wavelengths; z is the coordinate along the specimen length, and; l is the length of the specimen. The amplitude imperfections (A) are taken from the generated random values in Table 2. According to the Guide to Stability Design Criteria for Metal Structures [21], the specimen lengths are selected as three times of the half-wavelengths corresponding to the local or distortional buckling stresses. These half-wavelengths are determined on the basis of the elastic buckling analysis with the support of the THIN-WALLED-2 software program [22]. Buckling analyses for the investigated sections are carried out to obtain the elastic sectional buckling stresses and corresponding half-wave lengths, as listed in Table 1, where A is the gross-section area; (f_{ol}; f_{od}) and (L_{ol}; L_{od}) are the elastic bucking stresses and half-wave lengths for local and distortional bucklings respectively. Therefore, the investigated specimen lengths for C10030 and 25,025 sections are 310 and 600 mm to study the local buckling behaviors, as demonstrated in Fig. 4. Imperfection shape modes for each model are illustrated in Fig. 5 with the adopted sign conventions for local and distortional imperfections.

3 Numerical Investigations

Sectional imperfections including both local and distortional components are incorporated into the FE models. In order to estimate the effects of each component, one component is taken as the average value, and another one is varied using the random values in Table 2 and vice versa. FE models are developed by using the ABAQUS software program [23], the behaviors and strength results are obtained that can be used for later discussion.

3.1 Local Imperfection Investigations

The compression strengths for each configuration model are listed in Tables 3 and 4, and the behaviors of each imperfection shape mode are demonstrated in Fig. 6.

Tables 3 and 4 show that the strength results in Models (1.2; 1.3) and (2.2 and 2.3) are higher than those in the rest group models (Models (1.1; 1.4) and (2.1; 2.4) for both investigated sections. The later group models have the same sign conventions

Table 3 The compression strengths of C10030 models with the variation of local imperfection amplitudes (Unit: kN)

Random	Sign conventions of distortional imperfections			
Local	(+)		(–)	
Values	Model 1.1	Model 1.3	Model 1.2	Model 1.4
0.2901	143.00	152.33	153.92	142.83
0.2372	143.31	152.14	154.67	143.39
0.4451	141.92	148.27	153.24	141.76
0.3369	142.68	149.15	153.74	142.40
0.1927	147.78	151.79	155.58	143.79
μ	143.74	150.74	154.23	142.83
σ	0.01613	0.01253	0.00592	0.00562

Note μ, σ are the mean and coefficient of variation values

Table 4 The compression strengths of C25025 models with the variation of local imperfection amplitudes (Unit: kN)

Random	Sign conventions of distortional imperfections			
Local	(+)		(–)	
Values	Model 2.1	Model 2.3	Model 2.2	Model 2.4
0.7409	128.95	132.61	137.50	126.39
0.6351	128.99	132.03	138.74	126.46
0.8336	128.92	132.80	137.43	126.33
0.9673	128.89	132.66	138.47	126.24
0.3888	129.26	132.56	137.34	126.67
μ	129.00	132.53	137.90	126.42
σ	0.001168	0.002232	0.004762	0.001284

Note μ, σ are the mean and coefficient of variation values

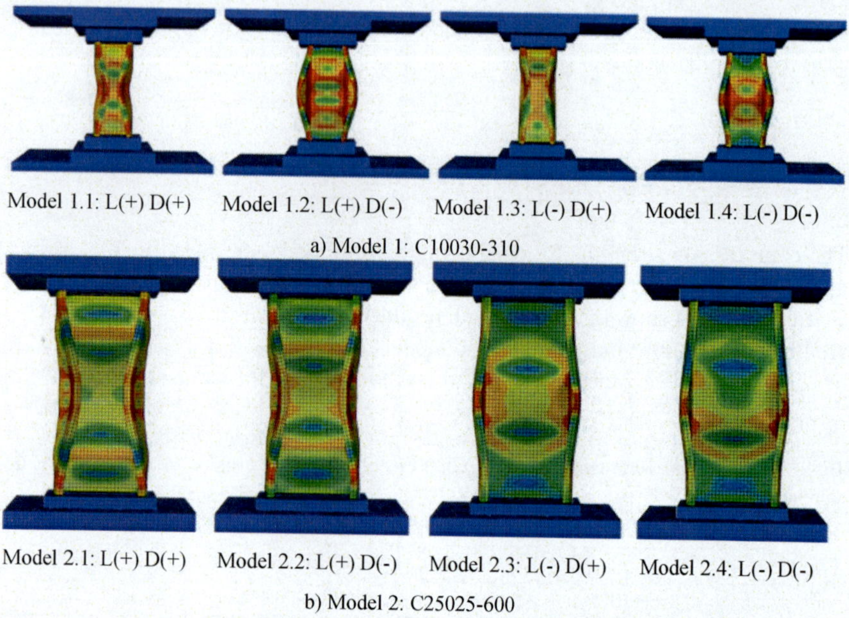

Fig. 6 Behaviors of the investigated FE models Note: L, D stand for local and distortional imperfections

of local and distortional imperfections (the same positive or negative) whereas the sectional imperfections of the former group models have the opposite sign conventions. The strength deviations between the two group models reach 8% for both C10030 and C25025 models. This illustrated that the detrimental strength results do not depend on the single sign conventions of sectional imperfections, but depend on the combinations of their sign conventions.

Figure 6 shows that similar sectional behaviors are observed with the same sign conventions of local imperfections for C25025 models, but the same distortional imperfection signs for C10030 models. It means that the behaviors of C25025 models are governed by the sign conventions of local imperfections whereas they are governed by distortional imperfection signs for C10030 models.

The coefficients of variation of both models are insignificant with less than 0.02. It illustrated the strength results are very consistent despite the variation of random local imperfection values. It can be concluded that local imperfections have insignificant impacts on the strengths of the investigated FE models.

Table 5 The compression strengths of C10030 models with the variation of distortional imperfection amplitudes (Unit: kN)

Random distortional values	Model 1.1	Model 1.4
0.9642	142.85	142.94
0.7809	143.34	143.55
1.1248	142.43	142.45
1.3565	141.67	141.66
0.3541	149.19	146.18
μ	143.89	143.36
σ	0.02100	0.01202

3.2 *Distortional Imperfection Investigations*

The detrimental group models (Models (1.1; 1.4) and (2.1; 2.4)) are used to investigate the distortional imperfection effects. The compressive strengths are listed in Tables 5 and 6 with the average strengths and coefficients of variation included.

The coefficients of variation are also very small for both two models although they are slightly higher for C10030 models. That illustrates the insignificant effects of distortional imperfections on the strengths of both C10030 and C25025 models. Models 1.4 and 2.4 provide the least strengths for both investigated models of local or distortional imperfection effects, and can be selected to develop parametric studies for further research.

Local buckling failure modes are observed in these C25025 models. To better understand the behaviors of these models, the progressive deformation behaviors of a typical channel column undergoing local buckling at various stages of the loading process corresponding to these points on the load versus end-shortening graphs are illustrated in Fig. 7.

Table 6 The compression strengths of C25025 models with the variation of distortional imperfection amplitudes (Unit: kN)

Random distortional values	Model 2.1	Model 2.4
1.1967	128.83	126.73
1.9602	127.64	125.48
0.9183	129.96	127.25
1.3340	128.67	126.49
1.5599	128.23	126.14
μ	128.666	126.418
σ	0.006689	0.005235

Note μ, σ are the mean and coefficient of variation values

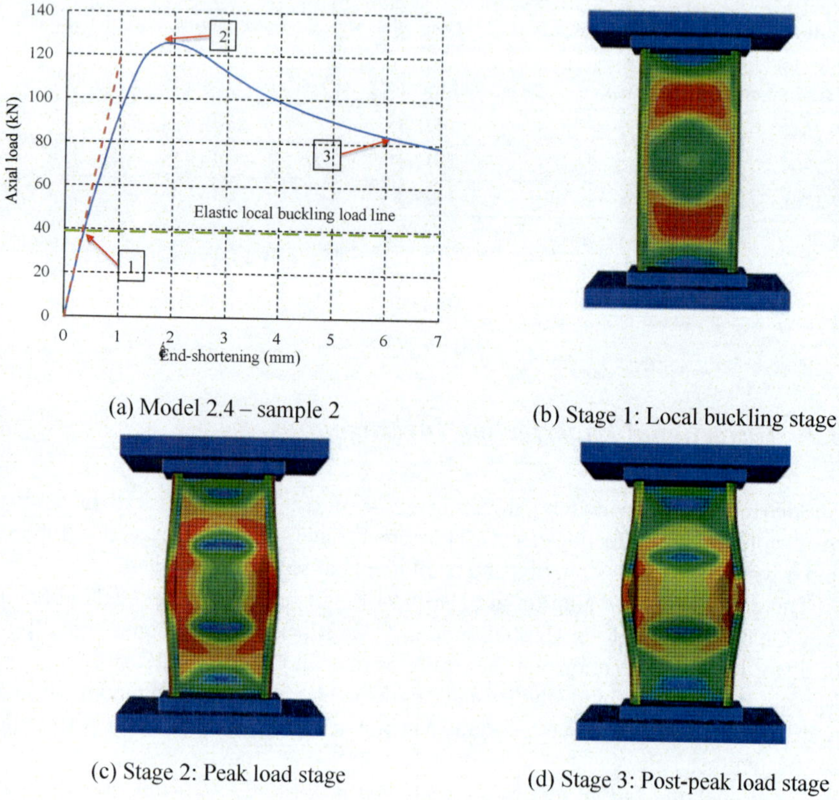

Fig. 7 Loading process stages of Model 2.4- sample 2 Note: Points (1), (2) and (3) correspond to stages 1, 2 and 3 in Fig. 7; sample 2 corresponds to the second imperfection random value in Table 6

Figure 7(b) shows that applied load reaches approximately 40 kN, local half-wave lengths are observed in the web. This load value is also in good agreement with the elastic local buckling load (see the horizontal green line in Fig. 7(a)) as shown in Table 1 (the elastic local buckling load $= A \times f_{ol} = 39.6$ kN). The load versus end-shortening graph also shows a change in slope at this load value due to the effect of the local buckling (see point 1 in Fig. 7(a)). Beyond the local buckling point, the applied load increases to the peak loads at point 2 in Fig. 7(a), followed by the significant increase of end-shortening versus the decreasing trend of the axial load. The specimen is failed due to the occurrence of the yield mechanism at the corners between the flanges and the lips (see point 3 in Fig. 7(a)).

4 Conclusions

This paper presents a numerical investigation to study the sectional imperfection effects on the strength and behavior of cold-rolled aluminium alloy channel stub columns. Two model simulations including C10030 and C25025 are investigated with the consideration of both local and distortional imperfections and a variety of imperfection shape modes and amplitudes. Amplitudes of sectional imperfections are generated on the basis of collected sectional imperfection data in publications of Pham et al. by using the Latin Hypercube technique. There are following conclusions can be drawn based on the numerical investigation results:

- The amplitudes of sectional imperfections have negligible effects on the sectional capacities of cold-rolled aluminium alloy channel sections. This is illustrated on the basis of very small values of coefficients of variation.
- The imperfection shape modes have significant impacts on both strengths and behaviors of the investigated section. The capacity deviations can reach 8% with the change of imperfection shape modes for both C10030 and C25025 models.
- The detrimental imperfection shape mode is not governed by a single sign convention of a sectional imperfection, but depends on combined sign conventions of sectional imperfection components.
- The sectional behaviors can be governed by the local imperfection as observed in C25025 models, but also can be governed by the distortional imperfection as illustrated in C10030 models.

Recommendations are provided to consider the sectional imperfections in finite element simulations, as follows: the amplitudes of sectional imperfections can be taken as the average values due to the high consistency of investigated strength results, but imperfection shape modes should be carefully analysed to identify the most detrimental shape mode for further studies.

References

1. Szumigała M, Polus Ł (2015) Applications of aluminium and concrete composite structures. Procedia Eng 108(61):544–549
2. Pham NH, Pham CH, Rasmussen KJR (2017) Incorporation of measured geometric imperfections into finite element models for cold-rolled aluminium sections. In: Proceeding of 4th Congres International de Geotechnique-Ouvrages-Structures, pp 161–171
3. Pit YL, Trahair NS (1997) Lateral-distortional buckling of hollow flange beams. J Struct Eng 123(6):695–702
4. Pi YL, Put BM, Trahair NS (1998) Lateral buckling strengths of cold-formed channel section beams. J Struct Eng 10(124):1182–1191
5. Wilkinson T, Hancock GJ (2002) Predicting the rotation capacity of cold-formed RHS beams using finite element analysis. J Constr Steel Res 58(11):1455–1471
6. Dubina D, Ungureanu V (2002) Effect of imperfections on numerical simulation of instability behaviour of cold-formed steel members. Thin-Walled Struct 40(3):239–262

7. Put BM, Pi YL, Trahair NS (1999) Lateral buckling tests on cold-formed channel beams. J Struct Eng 125(5):532–539
8. Borges Dinis P, Camotim D, Silvestre N (2007) FEM-based analysis of the local-plate/distortional mode interaction in cold-formed steel lipped channel columns. Comput Struct 85(19–20):1461–1474
9. Schillinger D, Papadopoulos V, Bischoff M, Papadrakakis M (2010) Buckling analysis of imperfect I-section beam-columns with stochastic shell finite elements. Comput Mech 46:495–510
10. Crisan A, Ungureanu V, Dubina D (2012) Behaviour of cold-formed steel perforated sections in compression: Part 2-numerical investigations and design considerations. Thin-Walled Struct 61:97–105
11. Dou C, Pi YL (2016) Effects of geometric imperfections on flexural buckling resistance of laterally braced columns. J Struct Eng 142(9):10, 04016048
12. Dinis PB, Young B, Camotim D (2016) Local-distortional-global interaction in cold-formed steel lipped channel columns: behavior, strength and DSM design. In: Structural Stability Research Council Annual Stability Conference 2016, pp 654–687
13. Pham NH (2019) Strength and behaviour of cold-rolled aluminium members, Ph.D. thesis, The University of Sydney
14. Pham NH, Pham, CH, Rasmussen KJR (2019) Numerical investigation of the member buckling of cold-rolled aluminium alloy channel beams. In: 9th International Conference on Steel and Aluminium Structures, p 11
15. McKay MD, Beckman RJ, Conover WJ (2000) A comparison of three methods for selecting values of input variables in the analysis of output from a computer code. Technometrics 42(1):55–61
16. Huynh LAT, Pham CH, Rasmussen KJR (2016) Stub column tests and finite element modelling of cold-rolled aluminium alloy 5052 channel sections. In: 8th International Conference on Steel and Aluminium Structures
17. Pham NH, Pham CH, Rasmussen KJR (2020) Finite element simulation of member buckling of cold-rolled aluminium alloy 5052 channel columns. In: Ha-Minh C, Dao D, Benboudjema F, Derrible S, Huynh D, Tang A (eds.) CIGOS 2019, Innovation for Sustainable Infrastructure. LNCE, vol 54, pp 263–268. Springer, Singapore. https://doi.org/10.1007/978-981-15-0802-8_39
18. Huynh LAT, Pham CH, Rasmussen KJR (2019) Finite Element analyses of cold-rolled aluminium alloy 5052-H36 beams subjected to distortional buckling. In: 9th International Conference on Steel and Aluminium Structures 2019
19. Mathforum.org, EasyFit (1994): Distribution-Fitting Software. Park City Mathematics Institute, Chicago
20. Huynh LAT, Pham CH, Rasmussen KJR (2016) Mechanical properties of cold-rolled aluminium alloy 5052 channel sections. In: 8th International Conference on Steeland Aluminium Structures
21. Ziemian RD (2010) Guide to Stability Design Criteria for Metal Structures, 6th edn. Wiley, New York, NY
22. Nguyen VV, Hancock GJ, Pham CH (2015) Development of the thin-wall-2 for buckling analysis of thin-walled sections under generalised loading. In: 8th International Conference on Advances in Steel Structures
23. Dassault Systemes Simulia Corp (2014) ABAQUS/CAE User's Manual. Providence, RI, USA

Crystallization of Alkanes at Different Substrates: Computer Simulation

Sergey Matseevich and Tatiana Matseevich

Abstract The mechanism of crystallization of tetradecane and polycaprolactone on silicon and graphite substrates has been studied. The polymer compounds near the substrates behave differently with the temperature decreasing. The crystallization process can proceed either due to heterogeneous nucleation, or due to the prefreezing process, when a crystalline layer of finite thickness is formed on the entire surface of the substrate. Since it is not always possible to determine which scenario is realized during cooling, the computer simulation should be made first. Two valence-force fields were studied by a dissipative particle dynamic to model systems, which represents various models of coarse-grained polymers. The simplest systems are a melt of tetradecane ($C_{14}H_{30}$) and a solution of polycaprolactone in toluene. Graphite and silicon were chosen as substrate materials, due to the frequent use of them in precision microscopy studies.

Keywords Polyethylene · Computer simulation · Prefreezing · Coarse-grained force fields

1 Therory and Modeling

Tatradecane and polycaprolactone were chosen as objects for modeling the crystallization processes of polyolefins and polyesters. Tatradecane is a model of low molecular weight polyethylene, the chemical structure of which is described by the formula CH_3-$(CH_2)_{12}$-CH_3. As is known, polyethylene is widely used in construction as coatings on various substrates, as well as moldings. Polycaprolactone, the structure of which is represented by the formula -$(CH_2)_5$-CO-O-, is widely used to obtain special-purpose polyurethanes, which are widely used in construction. Modeling the

S. Matseevich
Lomonosov Moscow State University, Physical Faculty, Leninskie gory, 1, Moscow 119991, Russian Federation
e-mail: cvmac@mail.ru

T. Matseevich (✉)
Moscow State University of Civil Engineering, Moscow 129337, Russian Federation
e-mail: MatseevichTA@mgsu.ru

P. Akimov et al. (eds.), *Proceedings of FORM 2022*, Lecture Notes in Civil Engineering 282, https://doi.org/10.1007/978-3-031-10853-2_19

behavior of tetradecane and polycaprolactone during their crystallization and the influence of the type of substrate on this process, but which is coated, is an urgent task of research: the results obtained make it possible to evaluate the mechanism of crystallization and, as a result, obtain coatings with improved structure and properties.

Nucleation and Prefreezing

The process of crystallization of polymers on solid substrates can occur according to two scenarios (Fig. 1) – due to heterogeneous nucleation [1] or due to prefreezing [2, 3]. During heterogeneous nucleation, a nonequilibrium process occurring at a temperature below the melting point T_m, individual crystalline nuclei are formed on the substrate and then their growth begins. Prefreezing, on the other hand, is a less studied phenomenon that occurs at a temperature above the melting point T_m, with the formation of a macroscopic layer of finite thickness (about 5 nm) on the substrate, which completely covers it. Prefreezing is commonly seen near highly attractive and/or structured surfaces. Type of crystallization scenario depends on the polymer and on the substrate. Thus, for example [4, 5], the crystallization of polycaprolactone near the graphite substrate proceeds according to the prefreezing type, while its heterogeneous nucleation is observed near the silicon substrate.

Prefreezing is based on the process of prewetting close to a solid substrate. It is possible to construct an analogy for the crystal-melt system. Although a theoretical description of this process already exists [6–10], the process has been experimentally demonstrated only in recent works [4, 5]. The resulting diagram (Fig. 2) describes the prefreezing process.

The red dotted line corresponds to the experiment in which cooling occurs at atmospheric pressure. The process running in the region between the prefreezing line and the coexistence line is the prefreezing process. It can be seen from the diagram that below a certain temperature T_{max} and at a pressure above the critical value (corresponding to the critical temperature Tc), the process begins to occur in the system. Upon cooling, the thickness of the crystalline layer near the substrate, which was formed at a temperature T_{max}, continues to grow, acting as one large center of bulk crystallization.

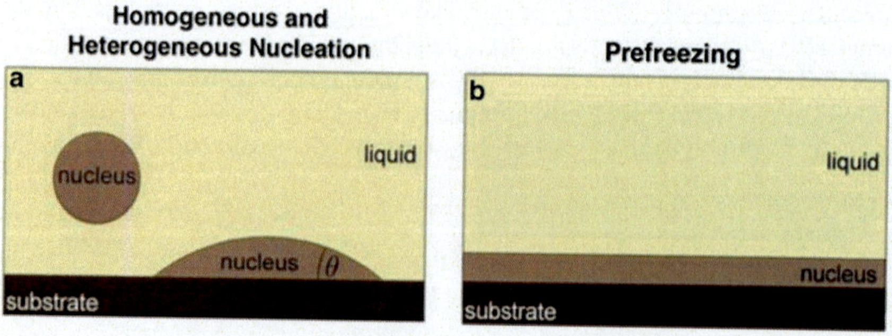

Fig. 1 Phase transitions near the substrate. Figure by T. Thurn-Albrecht

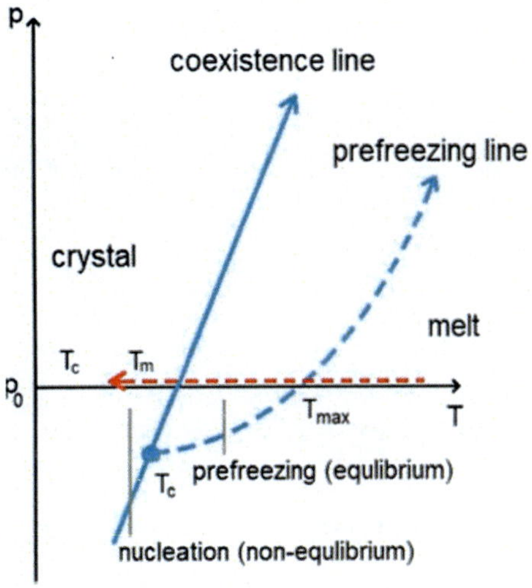

Fig. 2 Phase diagram [4]

2 Methods and Models

In this paper, we study two modified atomic-scale force-fields: a low coarse-grained one, represented by the UA GROMOS FF, and a high coarse-grained one, represented by the Martini FF. The evolution of systems was carried out by the methods of molecular dynamics: numerical solution of Newton's equations. The program for implementing molecular dynamics methods in this work is the GROMACS 2020.1 [11]. Snapshot images were obtained using the VMD program [12].

Valence-force fields (VFF or FF) - a way of representing the molecules under study in the form of interaction potentials that look like:

$$U(R) = U_b(R) + U_{LJ}(R) + U_{el}(R), \tag{1}$$

Here U_b is the bonded atom interaction potential, U_{LJ} is the Lennard–Jones interaction potential, and U_{el} is the electrostatic interaction potential. Both VFFs represent a class of atomistic force fields. GROMOS 54a7 FF [13–15] is a united atom (UA) force field (Fig. 3), while Martini 3.0 [16, 17] is high coarse-grained.

The most important difference between the two VFF is the number of atoms in the molecular models. For GROMOS 54a7 FF, the molecules of both studied substances (tetradecane and polycaprolactone) are represented by a more complete number of atoms (Fig. 4).

Studied Systems

In this paper, 4 different systems are studied (Fig. 5). All systems have a size of 3.58 × 3.58 × 3.58 nm and periodic boundary conditions in all 3rd coordinates. Each

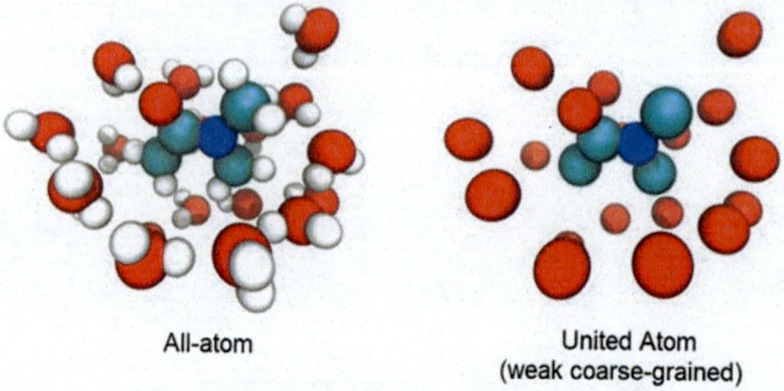

Fig. 3 Difference between full-atomic VFF and VFF UA. White spheres represent hydrogen atoms

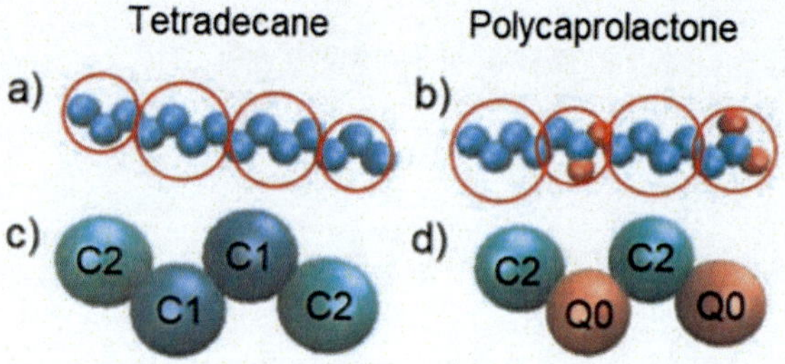

Fig. 4 Comparison of tetradecane and polycaprolactone molecules in the representation of GROMOS 54a7 FF (**a**), **b**)) and Martini 3.0 FF (**c**), **d**))

face of the cell is twice the length of the extended chain (1.63 nm), which excludes interaction with itself.

In the center of the cell there is a substrate made of α-graphite (thickness 1.02 nm) or silicon (thickness 1.03 nm). The density of the substance in the cell corresponds to the density of the test substance, and melting points for both substances are also known. In addition, for polycaprolactone, the maximum temperature value is known, below which prefreezing occurs: $T_{max} \approx 355.5\ K$ (82.3 °C).

The phase transition in the system occurs due to successive passage of several NVT ensembles with decreasing temperature. The state of a system can be estimated from its potential energy. In the case of a phase transition, a leap in this quantity should be observed. However, to give a more correct answer, cluster analysis is carried out for systems. 4 layers of atoms near the surface are combined into one cluster, and all the rest - into another. For the particles that are part of their cluster, the Steinhardt bond order parameters [18] are considered. The calculation algorithm is based on spherical harmonics, which makes it possible to quite well distinguish "liquid" (internal structure according to the type of liquid) clusters from crystalline ones (with a dense internal structure with crystalline symmetry). The determination of the crystal structure can be improved by calculating the average Steinhardt bond order parameters:

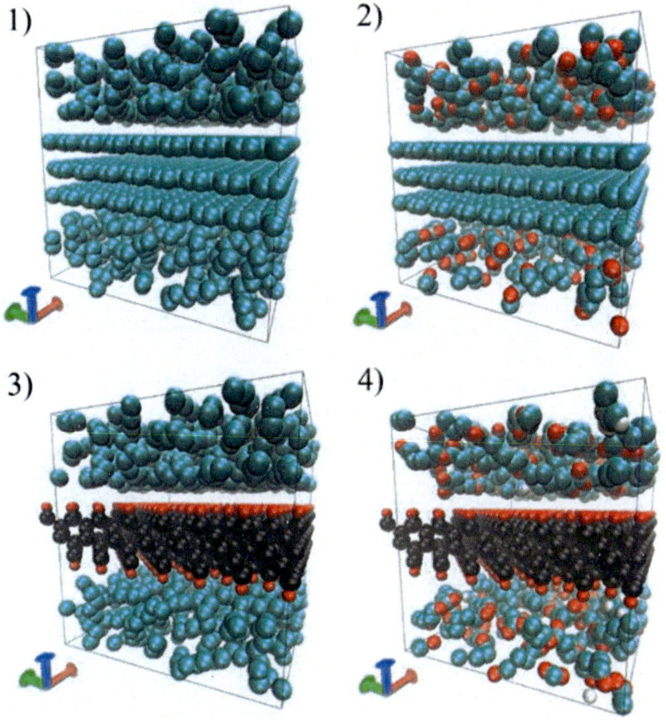

Fig. 5 Visualization of systems using the VMD program in GROMOS 54a7 FF: 1) tetradecane melt on a graphite substrate, 2) polycaprolactone solution in toluene ($\varphi = 10\%$) on a graphite substrate, 3) tetradecane melt on a silicon substrate and 4) polycaprolactone solution in toluene ($\varphi = 10\%$) on a silicon substrate

$$\overline{q_l}(i) = \sqrt{\frac{4\pi}{2l+1}\sum_{m=-l}^{l}\left|\overline{q}_{lm}(i)\right|^2}, \qquad \overline{q}_{lm}(i) = \frac{1}{\tilde{N}_b(i)}\sum_{k=0}^{\tilde{N}_b(i)}\frac{1}{N(i)}\sum_{j=1}^{N_b(i)}Y_{lm}\left(\vec{r}_{ij}\right), \tag{2}$$

The advantage of these parameters is that they visualize the results better (Fig. 6).

3 Results and Analysis

During the computer experiment, 10 systems of each of the 4 types were investigated. As a result, graphs of the change in the potential energy of molecules and the Steinhardt parameters for atoms in the layer near the substrate were plotted.

GROMOS 54a7 FF. Tetradecane and Polycaprolactone on Different Substrates For GROMOS 54a7 FF, in systems with a graphite substrate, both substances (tetradecane and polycaprolactone) exhibited cases of potential energy leaps. The graphs

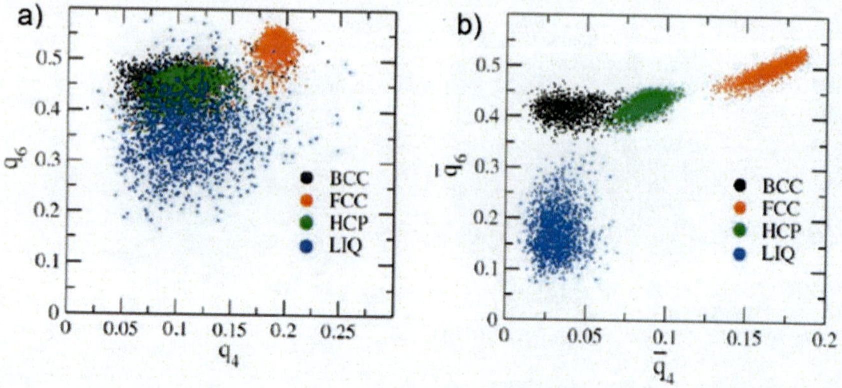

Fig. 6 Steinhardt bond order parameters **a** and mean bond order parameters **b** for body-centered cube (BCC), face-centered cube (FCC), hexagonal close packing (HCP), and liquid (LIQ) [18]

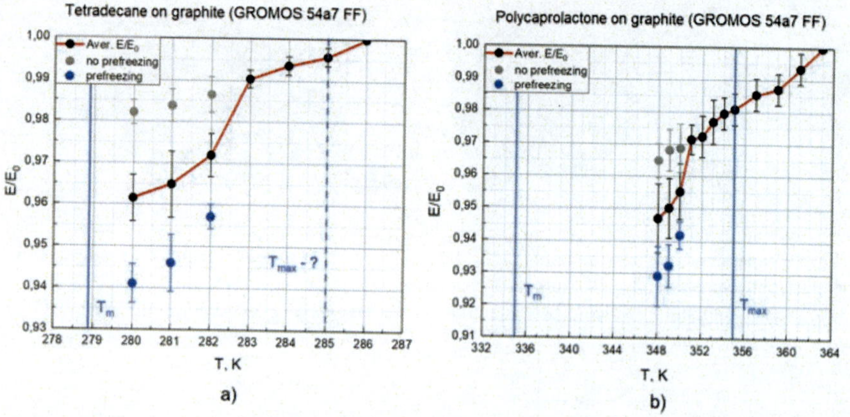

Fig. 7 Potential energy leaps for tetradecane **a** and polycaprolactone **b** on a graphite substrate in GROMOS 54a7 FF (above the melting point)

show the values of the leaps averaged over ten systems in the case of its complete absence, and by the number of representatives of the subtype, if it is present, i.e., a separate average for systems with a leap, and a separate average for systems without a leap (Fig. 7). In the case of a silicon substrate, the potential energy leap was recorded only at a temperature below the melting temperature (Fig. 8). The Steinhardt order parameters of these systems also tend towards the crystal structure near the substrate, which may indicate the onset of the nucleation process.

The graphs (Fig. 9) show the results of calculating the Steinhardt bond order parameters. It can be seen that in a system with an energy leap for systems with graphite substrate, the bond order parameters of the cluster atoms near the substrate tend to the values of the crystal lattices.

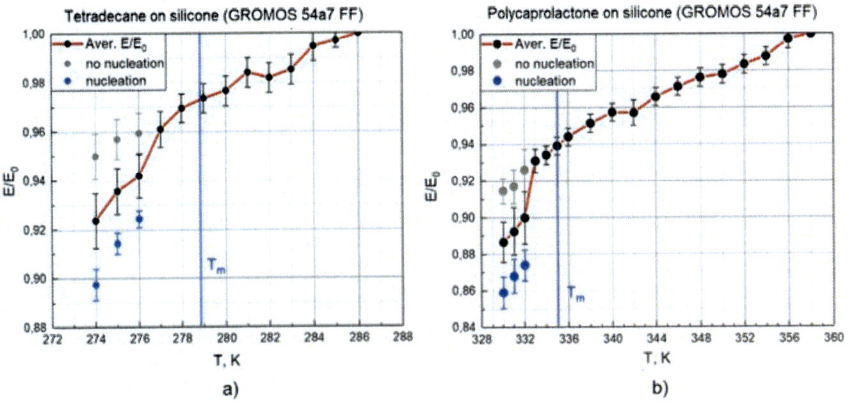

Fig. 8 Potential energy leaps for tetradecane **a** and polycaprolactone **b** on a silicon substrate in GROMOS 54a7 FF below the melting point

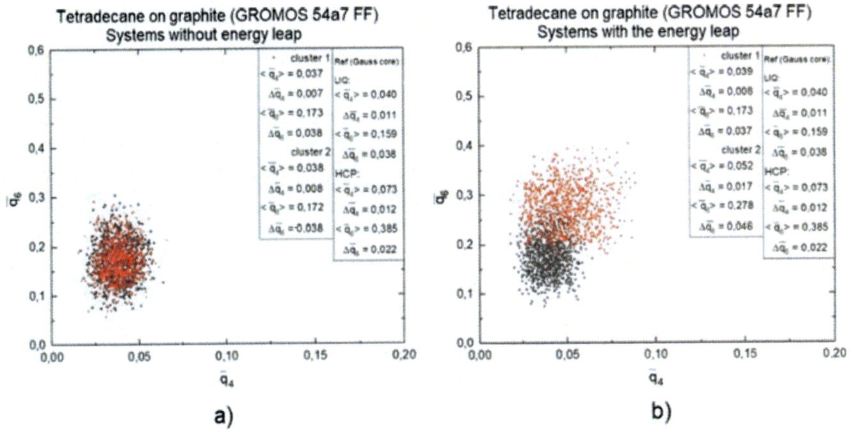

Fig. 9 The average Steinhardt bond order parameters for systems without a potential energy leap **a** and systems with a potential energy leap **b** at the same temperature

Martini 3.0. Tetradecane and Polycaprolactone on Different Substrates

In the case of Martini 3.0 FF, it was not possible to find a leap in potential energy above the melting temperature in any of the systems (Fig. 10).

The Steinhardt bond order parameters are in the range of values corresponding to the liquid, and only at a temperature below the melting point tend to the values of the crystal lattices.

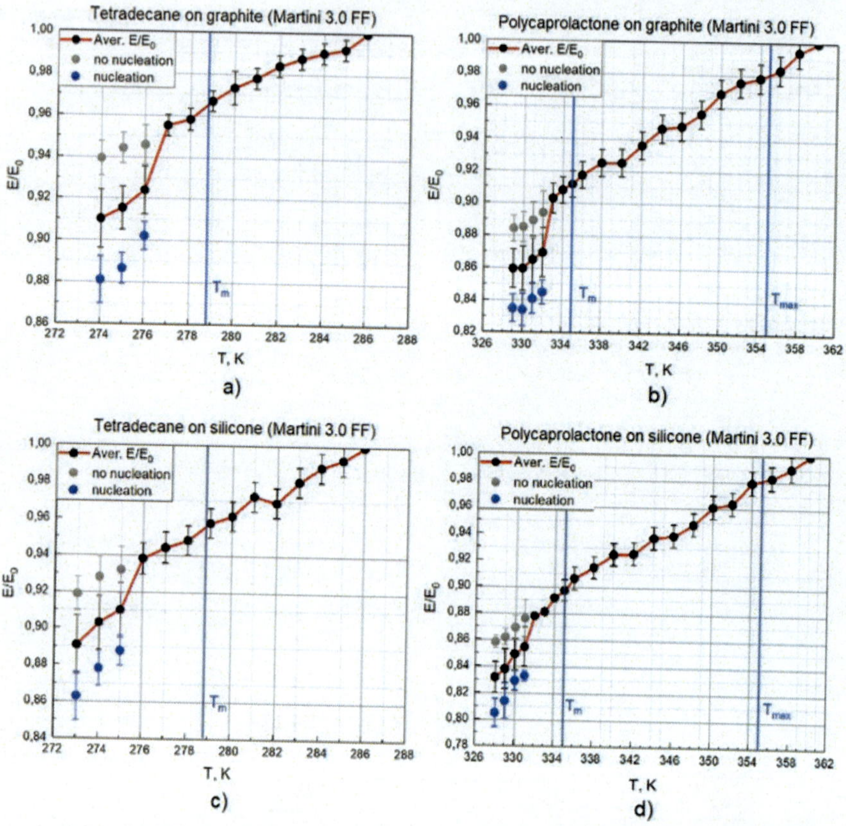

Fig. 10 Potential energy leaps for tetradecane **a** and polycaprolactone **b** on a graphite substrate and **c** and **d** on silicon substrate at temperatures below the melting point

4 Conclusion

The analysis of the obtained data gave the following results:

- GROMOS 54a7 FF is able to reproduce the process of prefreezing on a graphite substrate. This is evidenced by the leap in potential energy at a temperature above the melting point Tm and the closeness of the values of the average Steinhardt order parameters $\overline{q^4}$ and $\overline{q^6}$ to the values of the crystal lattice.
- Absence of the prefreezing process on the silicon substrate for both VFFs: there is no potential energy leap at temperatures above the melting point Tm, the Steinhardt parameters describing the appearance of a crystalline order with different symmetries have values corresponding to the structure of the liquid.
- Martini 3.0 FF with the set of parameters used in this paper does not correctly reproduce the prefreezing process: there is no potential energy leap, the order parameters are close to the liquid phase.

- Both VFFs are capable of reproducing a phase transition at temperatures below the melting point Tm: the presence of a potential energy leap in systems of all four types, the tendency of the order parameters to the values of the crystal lattice.

References

1. Sear RP (2007) Nucleation: theory and applications to protein solutions and colloidal suspensions. J Phys Condens Matter 19:033101. https://doi.org/10.1088/0953-8984/19/3/033101
2. Iwamatsu M (2011) Heterogeneous critical nucleation on a completely wettable substrate. J Chem Phys 134:234709. https://doi.org/10.1063/1.3599710
3. Heni M, Löwen H (2000) Phys Rev Lett 85:3668–3671. https://doi.org/10.1103/PhysRevLett.85.3668
4. Flieger A-K, Schulz M, Thurn-Albrecht T (2018) Macromolecules 51(1):189–194. https://doi.org/10.1021/acs.macromol.7b02113
5. Löhmann A-K, Henze T, Thurn-Albrecht T (2014) Proc Natl Acad Sci USA 111(49):17368. https://doi.org/10.1073/pnas.1408492111
6. Pereira GG (1997) Theoretical study of the prewetting transition in polymer blends. J Chem Phys 106:4282–4290. https://doi.org/10.1063/1.473130
7. Bonn D, Ross D (2001) Rep Prog Phys 64:1085. https://doi.org/10.1088/0034-4885/64/9/202
8. Archer AJ, Malijevsky A (2016) J Phys Condens Matter 28:244017. https://doi.org/10.1088/0953-8984/28/24/244017
9. Dolynchuk O, Tariq M, Thurn-Albrecht T (2019) J Phys Chem Lett 10:1942–1946. https://doi.org/10.1021/acs.jpclett.9b00608
10. Page AJ, Sear RP (2009) Phys Rev E 80:031605. https://doi.org/10.1103/PhysRevE.80.031605
11. Berendsen HJC, van der Spoel D, Van Drunen R (1995) Comput Phys Commun 91(1):43–56. https://doi.org/10.1016/0010-4655(95)00042-E
12. William H, Andrew D, Klaus S (1996) Vmd: Visual molecular dynamics. J Mol Graph 14(1):33–38. https://doi.org/10.1016/0263-7855(96)00018-5
13. Van Gunsteren WF, Berendsen HJC (1987) Groningen Molecular Simulation (GROMOS) Library Manual. Biomos, Groningen, The Netherlands, pp 1–221
14. Gunsteren WF et al (1996) Biomolecular Simulation: The GROMOS96 Manual and User Guide, pp 1–1042. Vdf Hochschulverlag AG an der ETH Zürich Zürich, Switzerland
15. Poger D, Van Gunsteren WF, Mark AE (2010) J Comput Chem 31(6):1117–25. PMID: 19827145. https://doi.org/10.1002/jcc.21396
16. Marrink SJ, Tieleman DP (2013) Chem Soc Rev 42:6801–6822. https://doi.org/10.1039/c3cs60093a
17. Souza PCT et al (2021) Nat Methods 18:382–388. https://doi.org/10.1038/s41592-021-01098-3
18. Lechner W, Dellago C (2008) Accurate determination of crystal structures based on averaged local bond order parameters. J Chem Phys 129:114707 https://doi.org/10.1063/1.2977970

Spectrum of Transverse Vibrations of a Layered Viscoelastic Composite

Tatyana Bobyleva and Alexey Shamaev

Abstract The article deals with the problems of constructing effective characteristics of a layered composite material, the layers of which are viscoelastic material. In this case, the hereditary property of viscosity is modeled using the Rabotnov functions, which have singularities at zero. It is proved that the average properties of a material can be analytically described in the form of a homogeneous viscoelastic material with memory, which is also specified using kernels, which are also specified analytically by the Rabotnov functions. It is of interest to study the natural vibrations of such a material. The easiest way is to obtain analytical formulas for the natural frequencies of oscillations transverse with respect to the layers. Such analytical expressions for natural frequencies are obtained in the present work. The spectra are compared for the case of composites made of elastic materials, materials in which nonlocal terms are specified with the help of exponential functions and with the help of Rabotnov functions.

Keywords Natural frequencies · Transverse vibrations · Viscoelastic media · Rabotnov fractional exponential kernels · Homogenization

1 Introduction

During the last years, a special type of material, multilayered composites has become quite popular in construction, automotive and aerospace industries. These materials

T. Bobyleva (✉)
Moscow State University of Civil Engineering, Moscow 129337, Russian Federation
e-mail: tatyana2211@outlook.com

A. Shamaev
Ishlinsky Institute for Problems in Mechanics of the Russian Academy of Sciences, Pr. Vernadskogo, 101-1, Moscow 119526, Russian Federation
e-mail: sham@rambler.ru

Lomonosov Moscow State University, GSP-1, Leninskie Gory, Moscow 119991, Russian Federation

P. Akimov et al. (eds.), *Proceedings of FORM 2022*, Lecture Notes in Civil Engineering 282, https://doi.org/10.1007/978-3-031-10853-2_20

have excellent characteristics such as light weight, high fatigue strength and good corrosion resistance.

The state of any system often depends on the processes that occurred in the past, so the appropriate mathematical models must take into account the hereditary properties of the system [1–3]. In the theory of viscoelasticity, integro-differential equations in partial derivatives arise. Mathematical models in these cases contain integro-differential equations in partial derivatives. Such equations include terms corresponding to the elastic and viscous parts. Monographs [4, 5] consider the homogenization problems of the equations of elasticity theory with rapidly changing coefficients in perforated domains with different boundary conditions, and finding effective characteristics. In the paper [6] a model has been constructed that makes it possible to quickly and efficiently calculate the average characteristics of heterogeneous layered soils consisting of layers of elastic-creeping materials. The creep kernels of an averaged medium are described as a sum of decreasing exponential functions. A similar method can be used to investigate the natural vibrations of a layered composite when each layer is a material with fractional Kelvin-Voigt friction [7]. In work [8] the spectrum of one- dimensional vibrations along the axis of periodically alternating elastic and viscoelastic horizontal layers is found and its asymptotic behavior is analyzed.

The subject of our article also is a layered viscoelastic material. Figure 1 shows an example of a layered array. The hereditary properties (the memory of the material) are described in integral terms using the Rabotnov function. The models using fraction-exponential functions are simpler than the complex ones obtained from combination of dashpots and springs and, furthermore, are suitable for fitting experimental data with good approximation allowing, at the same time, obtaining inverse Laplace transform in closed form. In 1948 Rabotnov Yu. N. [9] suggested to use fraction-exponential operators that can describe experimental data of real materials with sufficient accuracy and allows one to obtain inverse Laplace transforms analytically. Various applications of kernels with power singularities in hereditary mechanics are described in the papers [10, 11]. In the paper [12] the properties of operators with an integral aftereffect are investigated without connection with the theory of composite

Fig. 1 Layered array in nature

materials and the theory of homoganization. The structure of the spectrum of operators with integro-differential terms of the convolution type is studied. The next works [13, 14] are also devoted to mathematical problems that arise when considering the behavior of hereditary viscoelastic bodies. Fractional calculus is successfully used in viscoelasticity based on the analysis of experimental results for elastic-viscous bodies [15–17]. In the work [18] the authors consider the issue of constructing effective (averaged) characteristics of a layered composite with layers of viscous-creep material with non-local terms in the form of convolutions with exponential kernels.

In this paper, we show that the results of the previous paper can be generalized to the case of kernels of the type of Rabotnov functions, and the parameters of the Rabotnov functions in the averaged model can be determined through the solution of algebraic equations. It is shown that one can simply calculate the natural vibration frequencies of a sample of such a layered composite for vibrations in the direction perpendicular to the layers of the composite. A layered array in nature is shown in Fig. 1.

2 Problem Specification and Decision

Let Ω be a multilayer composite structure composed of the superposition of alternating layers of two materials with different physical properties. Each layer is made of an isotropic material, i.e., a material with uniform behaviour in all directions. Let a be the total thickness of two adjacent layers, and ε be the length ratio of two adjacent layers to the characteristic size of the sample H, $\varepsilon = \frac{a}{H}$, and the relative thicknesses of each individual layer are respectively equal to: εh and $\varepsilon(1-h)$, $0 \le h \le 1$. The density of layers of thickness ah is equal to ρ_1, and, accordingly, the density of layers of thickness $a(1-h)$ is equal to ρ_2. Cartesian coordinate system $Ox_1x_2x_3$ is used. All the layers are parallel to the coordinate plane Ox_1x_2 (Fig. 2). The Ox_2 axis, perpendicular to the plane of the figure, is omitted.

In this article, viscoelastic properties of composite are under consideration. The origin of modern viscoelasticity and, more generally, of hereditary theory traces

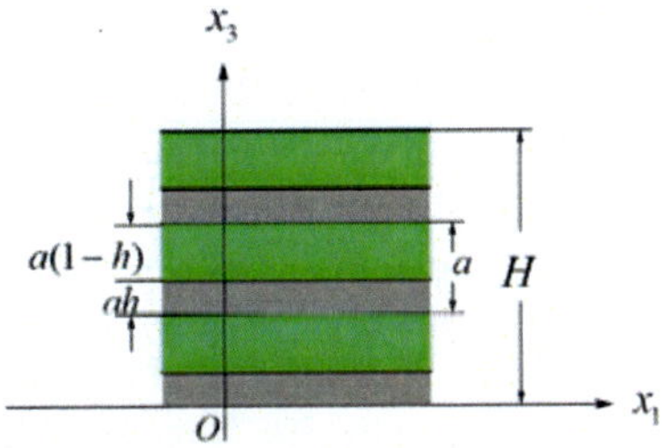

Fig. 2 Layered material

back to the works of Ludwig Boltzmann and Vito Volterra. They first introduced the concept of memory in connection with the analysis of elastic materials.

The key assumption of the hereditary theory of elasticity can be formulated as follows. For a viscoelastic body the deformation at any point x is a function both of the instantaneous stress and of all the past stresses at this point. For such materials the stress at the moment t are governed not only by the strain at the given moment, but by the entire pre-history of the deformation. These are materials with memory. Many physical processes are accompanied by such after-effect. For it describing Volterra suggested to use linear integral equations.

Components of the stress $\sigma_{ij}^{(s)}$ $(i, j = 1, 2, 3)$ and strain $e_{ij}^{(s)}$ tensors are connected by the following equations of state (here $s = 1, 2$ is the layer number) for each layer:

$$\sigma_{ij}^{(s)} = b_{ijkh}^{(s)} * e_{kh}^{(s)}, \tag{1}$$

where $b_{ijkh} = c_{ijkh}\delta(t) + d_{ijkh}$, and $e_{kh}^{(s)} = \frac{1}{2}\left(\frac{\partial u_k^{(s)}}{\partial x_h} + \frac{\partial u_h^{(s)}}{\partial x_k}\right)$, $(k, h = 1, 2, 3)$, u_k are components of the displacement vector, $c_{ijkh}^{(s)}$ are components of the elastic modulus tensor, $d_{ijkh}^{(s)}(t, \tau)$ are Volterra integral operators, characterizing hereditary properties of materials, namely

$$d_{ijkh} * e_{kh} = \int_0^t d_{ijkh}(t-\tau)e_{kh}(\tau)\,d\tau, \tag{2}$$

and t is time variable. (Einstein convention for repeated indices is used.)

Relaxation kernels $d_{ijkh}(t-\tau)$ depend on the difference $t-\tau$. This follows from the condition of invariance of σ_{ij} with respect to the origin of time.

The following kernel functions are widely used in the static and dynamic problems for materials with memory and viscoelasticity: $P(t) = e^{-\alpha t}$, $\alpha > 0$ and

$$e_\alpha(-\beta, t) = t^{\alpha-1}\sum_{n=0}^{\infty}\frac{(-\beta)^n t^{\alpha n}}{\Gamma\left[(n+1)\alpha\right]}. \tag{3}$$

Exponential kernels can be recommended for analysis of long-term deformation processes. The function $e_\alpha(-\beta, t)$ is fractional-exponential function, where $0 < \alpha \leq 1$, $\Gamma(\cdot)$ is Euler gamma function, $e_\alpha(-\beta, t)$ is called the Rabotnov function.

Yu.N. Rabotnov constructed a class of fractional-exponential functions with an integral singularity at the initial moment of the time (weakly singular functions) and showed the efficiency of using functions of this type as kernels of integral operators in the hereditary theory of elasticity. This direction in the construction of models of viscoelastic media seems to be quite effective. The theoretical results are validated against experimental data.

Questions concerning the construction of rheological models for viscoelastic materials under various types of loading using fractional derivatives are based on the assumption that they are equivalent to weakly singular functions.

The ideal contact conditions are assumed on the horizontal surfaces of the layers, namely: components of the displacement and the components of stress parallel the x_3-axis are continuous, i.e. $[u_i] = 0,\ [\sigma_{i3}] = 0,\ (i = 1 \div 3)$.

Equations of motion in the theory of elasticity have the form

$$\rho \frac{\partial^2 u}{\partial t^2} = \frac{\partial \sigma_{ij}(x,t)}{\partial x_j} + f_i(x,t). \tag{4}$$

In (4) we designated: $x = (x_1, x_2, x_3)$, $f_i(x,t)$ are components of a vector of external forces.

We consider isotropic materials, therefore, the components of the elastic tensors and relaxation kernel tensors in (1) have the form [2]:

$$\begin{aligned} c_{ijkh} &= \lambda\delta_{ij}\delta_{kh} + \mu(\delta_{ik}\delta_{jh} + \delta_{ih}\delta_{jk}),\\ d_{ijkh} &= -(D_v(t) - \tfrac{1}{3}D_{sh}(t))\delta_{ij}\delta_{kh} - \tfrac{1}{2}D_{sh}(t)(\delta_{ik}\delta_{jh} + \delta_{ih}\delta_{jk}). \end{aligned}$$

We denote here by D_{sh} and D_v the regular part of the shear and the bulk relaxation respectively, by δ_{ij} Kronecker symbol. Suppose that the amplitude of a bulk relaxation kernel is proportional to the amplitude of the shear relaxation kernel with a proportionality coefficient k_s for each layer, that is:

$(D_v)_s = k_s(D_{sh})_s, k_s = \text{const}, k_s > 0, (s = 1, 2)$. Further, D_{sh} is denoted by D.

According to the condition of this problem, all elastic modulus and relaxation kernels are periodic functions of the coordinate $y = \frac{x_3}{\varepsilon}$ and are piecewise constant functions of this variable, i.e., elastic modulus, volume force and relaxation kernels have the form [6]:

$$\lambda(y) = \begin{cases} \lambda_1, & y \in [0; h] \\ \lambda_2, & y \in [1-h; 1] \end{cases}, \quad \mu(y) = \begin{cases} \mu_1, & y \in [0; h] \\ \mu_2, & y \in [1-h; 1] \end{cases},$$
$$D(y,t) = \begin{cases} D_1(t), & y \in [0; h] \\ D_2(t), & y \in [1-h; 1]. \end{cases}$$

As usual, we denote here by λ_s, μ_s $(s = 1, 2)$ Lame parameters for each layer.

We apply the Laplace transform in the time domain to the Eqs. (4) with the assumption (1)

$$\tilde{f}(p) = \int_0^\infty f(t)e^{-pt}dt. \tag{5}$$

The result is the system of elasticity theory with a complex parameter p. We apply homogenization method described in [4, 5] to this system. In accordance with the classical theory of two-scale expansions the solution of the problem should be sought in the next form

$$\tilde{u}_i(x, y, p) = \tilde{u}_i^{(0)}(x, p) + \varepsilon \tilde{u}_i^{(1)}(x, y, p) + \cdots . \tag{6}$$

Variable p is the complex parameter.

$\hat{\varphi} = \langle \varphi \rangle = \int\limits_0^1 \varphi(\xi)\, d\xi$ is the operation of homogenization over a variable ξ. In addition, the vector-function of the displacements of the medium $(\tilde{u}_1^{(0)}(x, p), \tilde{u}_2^{(0)}(x, p), \tilde{u}_3^{(0)}(x, p))$ must satisfy to appropriate boundary conditions on the boundary of the region Ω.

$$\begin{aligned} b_{1111} = b_{2222} = \frac{(1-h)h}{L_z} &\left\{ \sum_{n=1}^{2} (-1)^n \left[\frac{e_\alpha(q_1, t) T_n^2}{(\beta_1 + q_n)(\beta_2 + q_n)(q_2 - q_1)} \right.\right. \\ &\left.\left. - \frac{e_\alpha(\beta_n, t) j_n^2 (\beta_1 - \beta_2)}{(\beta_n + q_1)(\beta_n + q_2)} \right] + (\lambda_1 - \lambda_2)^2 \delta(t) \right\} - e_\alpha(q_1, t) G_1 h \\ &- e_\alpha(q_2, t) G_2 (1-h) + L\delta(t), \end{aligned}$$

$$\begin{aligned} b_{1122} = \frac{(1-h)h}{L_z} &\left\{ \sum_{n=1}^{2} (-1)^n \left[\frac{e_\alpha(\beta_n, t) T_n^2}{(\beta_1 + q_n)(\beta_2 + q_n)(q_2 - q_1)} \right.\right. \\ &\left.\left. - \frac{e_\alpha(\beta_n, t) j_n^2 (\beta_1 - \beta_2)}{(\beta_n + q_1)(\beta_n + q_2)} \right] + (\lambda_1 - \lambda_2)^2 \delta(t) \right\} - e_\alpha(q_1, t) j_1 h \\ &- e_\alpha(q_2, t) j_2 (1-h) + l\delta(t), \end{aligned}$$

$$b_{1212} = -\tfrac{1}{2}[e_\alpha(q_1, t) g_1 h + e_\alpha(q_2, t) g_2 (1-h)] + M\delta(t),$$
$$b_{3333} = \tfrac{1}{L_z} \left\{ \sum_{n=1}^{2} \tfrac{(-1)^{n+1} e_\alpha(q_n, t) B_1^{(n)} B_2^{(n)}}{q_1 - q_2} + L_1 L_2 \delta(t) \right\},$$

In these formulas q_1, q_2 are the roots of the following quadratic equation:

$$\begin{aligned} &L_z q^2 + [L_z(\beta_1 + \beta_2) - ((1-h)G_1 + hG_2)]q + [L_z \beta_1 \beta_2 - G_2 h \beta_1 - G_1 (1-h) \beta_2] = 0 \\ &(L_z = L_1(1-h) + L_2 h) \end{aligned} , \tag{7}$$

q_1, q_2 are real numbers, since the discriminant *DISC* is positive:

$$DISC = [G_1(1-h) + L_z(\beta_1 - \beta_2) - G_2 h]^2 + 4G_1 G_2 h(1-h) > 0.$$

$$b_{1313} = b_{2323} = \frac{4\mu_1 \mu_2 \delta(t)}{M_z} + \frac{1}{M_z} \sum_{n=1}^{2} \frac{(-1)^{n+1} e_\alpha(q_n, t)[g_1 - 2\mu_1(\beta_1 + q_n)][g_2 - 2\mu_2(\beta_2 + q_n)]}{q_2 - q_1}.$$

In this formula q_1, q_2 are roots of the following quadratic equation:

$$M_z q^2 + [M_z(\alpha_1+\alpha_2) - 2((1-h)g_1 + hg_2)]q + [M_z\alpha_1\alpha_2 - 2g_2h\alpha_1 - 2g_1(1-h)\alpha_2] = 0, \quad (8)$$

which also has a positive discriminant, consequently, q_1, q_2 will be real numbers as well.

In the preceding formulas we have designated:

$$L_s = \lambda_s + 2\mu_s,\ G_s = g_s\left(k_s + \tfrac{2}{3}\right),\ j_s = g_s\left(k_s - \tfrac{1}{3}\right), M_z = 4[\mu_1(1-h) + \mu_2 h],$$
$$H_s^1 = \begin{cases} 1-h, & \text{if } s=1 \\ h, & \text{if } s=2 \end{cases},$$
$$L_z = L_1(1-h) + L_2h,\ A_s^{(n)} = \lambda_s(\alpha_s + q_n) - j_s,$$
$$T_n = [\lambda_1(\alpha_1+q_n) - j_1](\alpha_2+q_n) - [\lambda_2(\alpha_2+q_n) - j_2](\alpha_1+q_n),$$
$$B_1^{(n)} = L_1(\beta_1+q_n) - G_1,\ B_2^{(n)} = L_2(\beta_2+q_n) - G_2.$$

The following relation is valid $b_{1212} = \frac{1}{2}(b_{1111} - b_{1122})$.

If Eq. (4) does not contain the second derivative with respect to time, then such a system of equations with the indicated effect coefficients b_{ijkl}, $(ijkl = 1, 2, 3)$ describes elastic-creeping small displacements of the layered composite.

In general, the three-dimensional case, the study of the natural vibration frequencies of a layered composite is a very difficult problem. Therefore, in this paper, we consider a special case of transverse vibrations, which simply reduces to finding the roots of some polynomials in the complex plane.

Let us consider a particular case of transverse vibrations of a layered viscoelastic composite, i.e. oscillations along the x_3 axis. We have the following equation of motion

$$L\frac{\partial^2 u}{\partial x_3^2} + d_{33} * \frac{\partial^2 u}{\partial x_3^2} = \rho\frac{\partial^2 u}{\partial t^2}, \quad (9)$$

here we have designated $L = \lambda + 2\mu$.

We choose the kernel in the second term in the form of a fractional exponential Rabotnov function for each layer: $D_s = g_s e_\alpha(-\beta_s, t)$, where α, β_s, g_s are the constants, $0 < \alpha \le 1$, $\beta_s > 0$,$g_s > 0$, t is the variable that specifies time.

Let's perform the Laplace transform (5) for Eq. (9).

The Laplace transform of the Rabotnov function has the following form [1]:

$$\tilde{e}_\alpha(-\beta, t) = \frac{1}{p^\alpha + \beta}. \quad (10)$$

Further we apply homogenization method described in [4] to the resulting equation in Laplace images. As a result, we obtain the following equation for each layer

$$< \tilde{c}_{33} > \tilde{u}''_{x_3x_3} + < \rho > p^2\tilde{u} = 0. \quad (11)$$

Here $< \cdot >$ is the notation for the homogenization operation, that is $< \tilde{c}_{33} >$ is the effective coefficient after homogenization, and $< \rho > = \rho_1(1-h) + \rho_2 h$ is the average density of the array.

We introduce the following notation $q = p^\alpha$, then we get

$$< \tilde{c}_{33} > = \left\langle \left(L - \frac{G}{q+\beta} \right)^{-1} \right\rangle^{-1}. \tag{12}$$

After calculations the coefficient $< \tilde{c}_{33} >$ looks like

$$< \tilde{c}_{33} > = \frac{[L_1(q+\beta_1) - G_1][L_2(q+\beta_2) - G_2]}{[L_2(q+\beta_2) - G_2](q+\beta_1)h + [L_1(q+\beta_1) - G_1](q+\beta_2)(1-h)}.$$

In the preceding formula we have designated: $L_s = \lambda_s + 2\mu_s$, $G_s = g_s\left(k_s + \frac{2}{3}\right)$, $(s = 1, 2)$, s is the layer number.

In the formula for $< \tilde{c}_{33} >$ and further, q_1, q_2 are the roots of the Eq. (7).

The effective coefficient $< \tilde{c}_{33} >$ can be represented as

$$< \tilde{c}_{33} > = \frac{L_1 L_2}{L_z} + \frac{N_1}{q - q_1} + \frac{N_2}{q - q_2}, \tag{13}$$

where $L_z = L_1(1-h) + L_2 h$, N_1, N_2 are the next constants

$$N_1 = \frac{L_z G_1 G_2 - G_2(1-h)L_1^2(\beta_1+q_1) - G_1 h L_2^2(\beta_2+q_1)}{q_1 - q_2},$$
$$N_2 = \frac{L_z G_1 G_2 - G_2(1-h)L_1^2(\beta_1+q_2) - G_1 h L_2^2(\beta_2+q_2)}{q_1 - q_2}$$

The oscillation spectrum can be found from the following equation

$$\frac{< \rho > p^2}{\frac{L_1 L_2}{L_z} + \frac{N_1}{q-q_1} + \frac{N_2}{q-q_2}} = \frac{(in)^2 \pi^2}{H^2}, \quad (n = 1, 2, \ldots). \tag{14}$$

Spectrum is the union of solutions to the Eq. (14) with respect to $n = 1, 2, \ldots$ (relatively p, $(q = p^\alpha)$).

If $N_1 = N_2 = 0$ and $\alpha = 1$, then for the spectrum of the problem with Dirichlet conditions, the well-known expressions are obtained .

$$p_{\pm n} = (\pm in)\frac{\pi}{H}\sqrt{\frac{< \tilde{c}_{33} >}{< \rho >}} = (\pm in)\frac{\pi}{H}\sqrt{\frac{L_1 L_2}{< \rho > L_z}}, \quad n = 1, 2, \ldots .$$

Spectrum is shown on the Fig. 3.

If $\alpha = 1$, $N_1 \neq 0$, $N_2 \neq 0$, then the functions $e_\alpha(-\beta, t)$ turn into exponential functions. The spectrum of such a problem was studied earlier [12]. In this case, the $p_{\pm n}$ have a vertical asymptote and accumulation point on the real axis (Fig. 4).

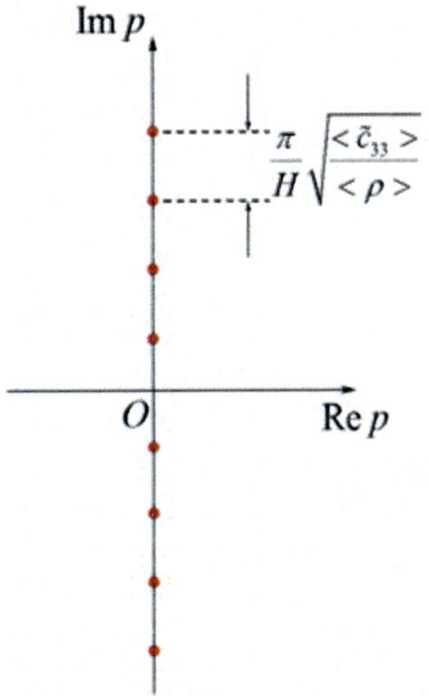

Fig. 3 Frequency spectrum in the purely elastic case

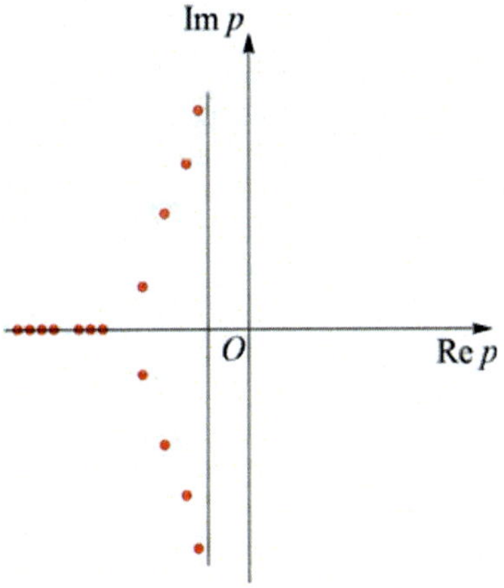

Fig. 4 Frequency spectrum in the case of exponential viscosity kernels

Now suppose $0 < \alpha < 1$, then the oscillations perpendicular to the layers will satisfy the average, or "effective" equation .

$$< \rho > \frac{\partial^2 u}{\partial t^2} = \frac{L_1 L_2}{L_z} \frac{\partial^2 u}{\partial x_3^2} + [N_1 e_\alpha(q_1, t) + N_2 e_\alpha(q_2, t)] * \frac{\partial^2 u}{\partial x_3^2}.$$

It is easy to see from the Eqs. (14) that for large n will be

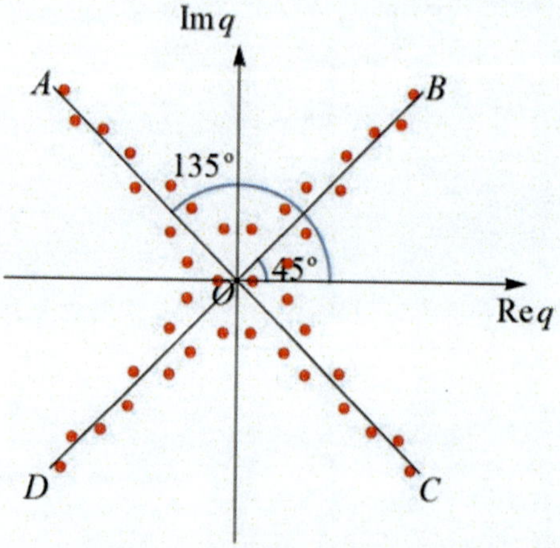

Fig. 5 Frequency spectrum for the case of viscosity kernels in the form of Rabotnov functions

$$q_n \sim (\pm in)^{\alpha} \left[\frac{\pi L_1 L_2}{L_z < \rho >} \right]^{\alpha}. \tag{15}$$

The roots of the complex numbers $(\pm in)$ must lie in the left half-plane, otherwise the solution will be devoid of physical meaning. Let $\alpha = {}^1/_2$, then the quantity q_n has six values, since the Eq. (14) with respect to the variable q $(p = q^2)$ is an equation of the sixth degree. The four roots of this equation are complex, and two are real. For large n we have $q_n \sim \sqrt{\pm(in)\left[\frac{\pi L_1 L_2}{L_z<\rho>}\right]}$. The points of the spectrum asymptotically approach the rays *OA, OB, OC, OD* (Fig. 5).

Let us give an example of calculating the values of natural frequencies in this case. Let the numerical values of the input parameters be the following:

$$h = 0.8,\ L_1 = 673 \text{ MPa},\ L_2 = 4711 \text{ MPa},\ G_1 = 3.31 \frac{\text{n}}{\text{m}^2 \cdot \text{sec}},$$

$$G_2 = 2.27 \frac{\text{n}}{\text{m}^2 \cdot \text{sec}},\ \beta_1 = 0.0015 \text{ sec}^{-1},\ \beta_2 = 0.003 \text{ sec}^{-1}.$$

Using the software system Wolfram Mathematics, we have the following results:

$q_1 = -0.002,\ q_2 = -0.00056$ are the roots of Eq. (7), and the roots of Eq. (14) are:

for $n = 1 \rightarrow q^{(1,2,3,4)} = \pm 1.26613 \pm 1.26551i$, for $n = 2 \rightarrow q^{(1,2,3,4)} = \pm 1.79033 \pm 1.7897i$,

for $n = 3 \rightarrow q^{(1,2,3,4)} = \pm 2.19255 \pm 2.19193i$, for $n = 10 \rightarrow q^{(1,2,3,4)} = \pm 4.00252 \pm 4.00189i$ and so on.

In addition for any $n = 1, 2, \ldots$ we have the following two real roots $q^{(5)} = -0.000797,\ q^{(6)} = 0.000522$ (Fig. 5).

Two real roots $q^{(5)}, q^{(6)}$ can be written analytically as follows

$$q^{(5,6)} = \frac{1}{2L_z < \rho >}\Big[(\beta_1 + \beta_2)L_z \pm \sqrt{Z}\Big],$$

here $Z = (\beta_1\beta_2 L_z - G_z)^2 - 4 < \rho > (G_1 - \beta_1 L_1)(G_2 - \beta_2 L_2)L_z$, and $G_z = G_1(1-h) + G_2 h$.

According to the physical interpretation, the number $p = q^2$ must lie in the left half-plane. Therefore, from the set of solutions (14) with respect to the variable q, one must choose such that $\mathrm{Re} q^2 \leq 0$.

3 Conclusion

We see that in order to calculate the natural frequencies of the transverse vibrations of the composite, we need to solve a countable sequence of algebraic equations. For the case when the kernels that determine the hereditary properties of the material are given by exponential functions, these are quadratic equations. If the kernels are given by Rabotnov functions with a rational exponent α, then the degree of the equations will be more than two. In this case, this circumstance is not an obstacle to the accurate calculation of the frequencies of natural oscillations. So, if $\alpha = {}^1\!/_2$, then the calculation of natural frequencies will be reduced to the calculation of a sequence of fourth-order algebraic equations, which is easy to do using standard software packages. The use of Rabotnov kernels to describe viscoelastic properties is experimentally justified in hereditary mechanics. Fractional-exponential functions (Rabotnov functions) as kernels that determine the hereditary properties of a material, according to a number of researchers in the field of hereditary mechanics, describe the hereditary properties of a material better than exponential kernels or Abelian-type kernels [10]. Of course, we have given a calculation of the natural frequencies of the simplest partial motion for the obtained model of a layered composite. It is very easy to implement using standard packages. Cases of higher dimension can also be investigated. The problem will be reduced to the problem of finding the eigenvalues of some polynomial operator pencils, which is much more difficult than the problems of transverse and longitudinal vibrations of the composite, but can also be studied.

Acknowledgements The work was supported by the Russian Science Foundation, project № 21-11-00151.

References

1. Rabotnov YuN (1980) Elements of hereditary solid mechanics. Mir, Moscow
2. Ilyushin AA, Pobedrya BE (1970) Foundations of the mathematical theory of thermovisco-elasticity. Science, Moscow
3. Christensen RM (1971) Theory of viscoelasticity: an introduction. Academic Press, New York and London

4. Oleynik OA, Shamaev AS, Yosifian GA (1992) Mathematical problems in elasticity and homogenization. Elsevier, North-Holland
5. Bardzokas DI, Zobnin AI (2003) Mathematical modelling of physical processes in composite materials of periodical structures. URSS, Moscow
6. Bobyleva TN, Shamaev AS (2017) Grounds. Found Mech Soils 4:7–12
7. Shamaev AS, Shumilova VV (2016) Proc Steklov Inst Math 295(1):202–212
8. Shamaev AS, Shumilova VV (2016) Proc Steklov Inst Math 295(1):213–224
9. Rabotnov YuN (1948) Prikladnaya Matematika i Mekhanika 12(1):53–62 (1948), (in Russian). Reprinted: Fract Calculus Appl Anal 3:684–696 (2014). (Translated in English by: Marina Shitikova)
10. Zhuravkov MA, Romanova NS (2013) On the prospects of using fractional calculus in mechanics. BSU, Minsk
11. Konstantinova SA, Pestrenin VM, Pestrenina IV (2007) Proceedings of higher educational institutions. Min J 4:92–98
12. Vlasov VV, Rautian NA, Shamaev AS (2012) Modern mathematics. Fund Direct 45:43–61
13. Orlik J (2000) Transmission and homogenization in hereditary viscoelasticity with aging and shrinkage. Shaker, Berlin
14. Orlik J (2000) Abstr Appl Anal 5:1–19
15. Kilbas AA, Saigo M, Saxena AA (2004) Fractional calculus operators. Integral Transf Spec Funct 15(1):31–49
16. Rossikhin YA, Shitikova MV (2018) J Phys Conf Ser 991(1):012069
17. Yang XJ (2019) General fractional derivatives: theory, methods and applications. CRC Press, Boca Raton
18. Bobyleva TN, Shamaev AS (2018) IFAC PapersOnLine 51(2)

Exact Controllability of the Distributed System Governed by the Wave Equation with Memory

Igor Romanov and Alexey Shamaev

Abstract We will consider the exact controllability of the distributed system governed by the wave equation with memory. The theorem will be formulated that this mechanical system can be driven to rest in finite time, the absolute value of the distributed control function being bounded. In this case the memory kernel is a linear combination of decreasing exponentials.

Keywords Integrodifferential systems · Distributed controllability · Bounded control

1 Introduction

In this article we will consider the problem of the exact controllability of a system governed by the integro-differential equation

$$\theta_{tt}(t,x) - K(0)\Delta\theta(t,x) - \int_0^t K'(t-s)\Delta\theta(s,x)ds = u(t,x), \quad x \in \Omega, \quad t > 0. \tag{1}$$

$$\theta|_{t=0} = \varphi_0(x), \quad \theta_t|_{t=0} = \varphi_1(x), \tag{2}$$

$$\theta|_{\partial\Omega} = 0, \tag{3}$$

here

$$K(t) = \sum_{j=1}^{N} \frac{c_j}{\gamma_j} e^{-\gamma_j t}, \quad N \geq 2,$$

I. Romanov (✉) · A. Shamaev
Institute for Problems in Mechanics RAS, 101 Prosp. Vernadskogo, Block 1, Moscow 119526, Russia
e-mail: romm1@mail.ru

P. Akimov et al. (eds.), *Proceedings of FORM 2022*, Lecture Notes in Civil Engineering 282, https://doi.org/10.1007/978-3-031-10853-2_21

where c_j, γ_j are given positive constant numbers such that

$$0 < \gamma_1 < \gamma_2 < ... < \gamma_N,$$

$u(t, x)$ is a control supported (in x) on a bounded domain Ω and

$$|u(t, x)| \leq M,$$

$M > 0$ is a given constant number. The goal of the control is to drive this mechanical system to rest in finite time. We say that the system is *controllable to rest* when for every pair of initial conditions φ_0, φ_1 it is possible to find a control $u(t, x)$ and a time $T > 0$ such that $u(t, x)$ is equal to zero for any $t > T$ and the corresponding solution $\theta(t, x, u)$ of problem (1)–(3) equals zero for any $t > T$ too.

Note that exponent rates and exponential coefficients in the representation for the convolution kernel are usually derived by approximating the experimentally obtained relaxation kernel function by the sum of a finite number of exponents using, for example, the least squares method. In some cases, for example, for suspension acoustics models, these values are obtained by calculating using the homogenization method.

Similar controllability problems for membranes and plates were studied earlier in [1]. It was proved that vibrations of these mechanical systems could be driven to rest by applying bounded (in absolute value) and volume-distributed control functions. An overview of the results concerning the boundary controllability of distributed systems can be found in [2]. Problems of controllability of systems similar to (1) were considered in [3]. A condition was formulated under which a solution to the heat equation with memory could not be driven to rest in a finite time. This condition is that there is a null of some analytic function of complex variable in the domain of holomorphism.

Problems similar to (1)–(3) for integro-differential equations were studied earlier in many articles. Equation (1) was firstly derived in [4]. The solvability and asymptotic behaviour for an abstract equation of this type were investigated, for example in [5] and [6]. In [7], it was proved that the energy for some dissipative system decays polynomially when the memory kernel decays exponentially. Problems of the solvability of system (1)–(3) were considered in [8]. It was proved that the solution belongs to some Sobolev space on the semi-axis (in t) when the kernel $K(t)$ is the series of exponentials, each exponential function tending to zero when $t \to +\infty$. The explicit formulae for the solution of (1)–(3) were obtained in [9]. In this case kernel $K(t)$ is also the series of decreasing exponentials. It follows from these formulae that solutions tend to zero when $t \to +\infty$. In all these articles kernels in integral summands of the equation are suggested to be none-increasing. The problem of controllability to rest for one-dimensional string equation is considered in [10]. In this case, the kernel in the integral term of the equation is identical to 1 and control is focused on a compact (part of a string) that moves at a constant speed.

In this article we also consider the so-called "null controllability". It means that for every initial conditions φ_0 and φ_1 there is a control $u(t, x)$ and a time $T > 0$ such that the corresponding solution $\theta(t, x, u)$ and its first derivative, with respect to t equals zero for $t = T$. Null controllability and controllability to rest are not the same for systems with memory. In many cases, controllability to rest is impossible. Let us consider for example the one-dimensional case (Ω is an interval $(0, \pi)$). We prove now that the system (1) is uncontrollable to rest if $u(t, x) \in L_2((0, \infty), L_2(0, \pi))$ is supported (in x), as well as in [3], on an interval $[a, b]$ which is properly contained in $(0, \pi)$. It means that $u(t, x) \equiv 0$ outside $[a, b]$. It is clear that the Eq. (1) can be written in the following form

$$\frac{\partial}{\partial t}\left(\theta_t(t, x) - \int_0^t K(t-s)\theta_{xx}(s, x)ds - \int_0^t u(s, x)ds\right) = 0.$$

Obviously function $\theta(t, x)$ is a solution of the Eq. (1) if and only if this function is a solution of the following equation

$$\theta_t(t, x) - \int_0^t K(t-s)\theta_{xx}(s, x)ds - \int_0^t u(s, x)ds = f(x), \tag{4}$$

where $f(x)$ is an arbitrary function. Let t in (4) be equal to zero then we obtain

$$f(x) = \varphi_1(x).$$

Let $\varphi_1(x) \equiv 0$. We introduce

$$P(t, x) = \int_0^t u(s, x)ds.$$

Thus the problem (1)–(3) reduces to the form

$$\theta_t(t, x) - \int_0^t K(t-s)\theta_{xx}(s, x)ds = P(t, x), \quad x \in (0; \pi), \quad t > 0. \tag{5}$$

$$\theta|_{t=0} = \varphi_0(x), \tag{6}$$

$$\theta|_{x=0} = 0, \quad \theta|_{x=\pi} = 0. \tag{7}$$

Note that $P(t, x) \equiv 0$ outside $[a, b]$ and $P(t, x)$ can be considered as a new control function. Similar problem was considered in [3] (in a more general case). If $K(t)$ is a linear combination of two exponentials then the system (5)–(7) is uncontrollable to rest. It means that there is an initial condition φ_0 such that, for any control $P(t, x)$, where $P(t, x)$ belongs to the corresponding space, the solution of (5)–(7) can not be driven to rest. Using arguments similar to the above it can be proved that the system (1)–(3) is uncontrollable to rest if $K(t)$ is a linear combination of N exponentials, where $N \geq 2$. In [11] this result was generalized to the multidimensional case. More accurately, controllability to rest is impossible if $N \geq 2$ and control function is applied only to the subdomain. But, as it will be shown in this article, the system (1)–(3) is controllable to rest when the control is distributed on the whole domain Ω.

2 Preliminaries

Let $A := -\Delta$ be an operator acting on a space

$$D(A) := H^2(\Omega) \cap H_0^1(\Omega),$$

$\Omega \subset R^s$ ($s \in \mathbf{N}$) be a bounded simply connected domain with an infinitely smooth boundary. Let also $\{\psi_n(x)\}_{n=1}^{+\infty}$ be a corresponding orthonormal system of eigenfunctions and $\{\alpha_n^2\}_{n=1}^{+\infty}$ are corresponding eigenvalues such as

$$-\Delta\psi_n(x) = \alpha_n^2\psi_n(x).$$

We denote $W_{2,\gamma}^2(R_+, A)$ the linear space of functions $f : R_+ = (0, +\infty) \to D(A)$ equipped with the norm

$$\|\theta\|_{W_{2,\gamma}^2(R_+,A)} = \left(\int_0^{+\infty} e^{-2\gamma t}\left(\left\|\theta^{(2)}(t)\right\|_{L_2(\Omega)}^2 + \|A\theta(t)\|_{L_2(\Omega)}^2\right)dt\right)^{\frac{1}{2}}, \quad \gamma \geq 0.$$

Definition 1 *A function $\theta(t, x)$ is called a strong solution of the problem (1)–(3) if for some $\gamma \geq 0$ this function belongs to the space $W_{2,\gamma}^2(R_+, A)$, satisfies the Eq. (1) nearly everywhere (in t) on the positive semiaxis R_+ and satisfies the initial conditions (2).*

Let us consider the space l_β of sequences $\{c_n\}_{n=1}^{+\infty}$ such that the series

$$\sum_{n=1}^{+\infty} |c_n|^2\alpha_n^{2\beta}$$

converges. Then we define the space

$$D(A^{\frac{\beta}{2}}) = \left\{f(x) = \sum_{n=1}^{+\infty} f_n\psi_n(x) : \{f_n\}_{n=1}^{+\infty} \in l_\beta\right\}.$$

3 Main Results

The section is devoted to the formulation of the main theorem which states that the solution and its first derivative with respect to time can be driven to the null state during finite time. Then it shows that the control function which is constructed in the proof drives the system to rest actually.

It is the following theorem, which presents the main result of the article.

Theorem 3.1 *Let $\varphi_0 \in D(A^{\beta+\frac{1}{2}})$ and $\varphi_1 \in D(A^{\beta})$, where $\beta > \frac{s}{2}$, $M > 0$ be a certain constant. Then, there are, depending on the value M, control $u(t,x) \in C([0,T] \times \Omega)$ and the time $T > 0$, such that the solution of the problem (1)–(3) has the equalities.*

$$\theta(T,x) = \theta_t(T,x) = 0, \tag{8}$$

and the restriction

$$|u(t,x)| \leq M,$$

for any $t \in (0,T]$, $x \in \Omega$ is done.

If we extend the constructed control function $u(t,x)$ by zero when $t > T$ then the system (1)–(3) will stays in the null state for $t > T$.

A detailed proof of the theorem is given in the preprint [12].

4 Discussion

In this paper we consider the problem of controllability to rest of the system described by the integrodifferential equation with a convolution kernel representing the sum of a finite number of decreasing exponentials. At the same time, the transition to kernels of a more general kind, for example, to sums of convergent series from an infinite number of exponentials, is far from a trivial task.

For the kernel in the form of the sum of an infinite number of exponentials the results of this work do not take place, with the proper choice of exponent rates the problem lacks the property of complete controllability. To obtain this, the rates of the exponential functions should slowly tend to minus infinity. A close result is proved in the article [13]. This result indicates the "stability" of the controllability property in the close problem. Namely, the "tail" of the series, with the help of which the convolution kernel is set, can be arbitrarily small and can decrease arbitrarily quickly. If we discard this "tail", then the system will become controllable to rest, and its restoration leads to an uncontrollable system. This is the property of the "stability" of the task. It is a "dissonance" with the problem of controllability of linear finite-dimensional systems. According to the classical Kalman criterion of controllability of linear finite-dimensional systems, full controllability is equivalent to the full rank of some rectangular matrix composed of the problem data, which in turn is equivalent to the difference from zero of several determinants from the elements of this matrix. It is clear that with a sufficiently small arbitrary perturbation of the problem data, this property of difference from zero determinants is preserved, which indicates the "roughness" of the controllability property.

In general, the properties of controllability problems for integrodifferential systems are radically different from the properties of similar controllability problems for differential systems. So, if the boundary controllability problems for differential systems are usually solvable (while of course there are certain solvability conditions), then the solvability of similar problems for integrodifferential systems are exceptional cases. To illustrate this fact we can consider the boundary controllability problem for the one-dimensional Gurtin-Pipkin equation. It turns out that an obstacle to boundary controllability is, for example, that the density of the spectrum of the problem under consideration is equal to infinity, where density is understood in the sense of some numerical characteristic (see [14]). Often the spectra of integrodifferential problems have an accumulation endpoint and therefore the density characteristics are equal to infinity, which implies the absence of complete controllability. So, in the one-dimensional case for the Gurtin-Pipkin equation, the boundary controllability problem is solvable, for example, in the case when the kernel consists of a one decreasing exponential function and the corresponding exponent rate is close enough to zero, in cases when the kernel is the sum of two or more decreasing exponentials it is unsolvable due to the presence of accumulation endpoints in the spectra.

One important feature should be pointed out in the controllability problems for integrodifferential equations when the control force is applied to the subdomain. For these problems in most cases there is no complete controllability (see [11]). Besides, in [11] it is proved that if even an arbitrarily small neighborhood inside the domain does not belong to the region of the application of the control force, then the system is not controllable, i.e. there is an initial condition that we cannot drive to complete rest in a finite time, no matter what control action we apply that satisfies the conditions of the problem.

Recently, research has been quite actively conducted on the problem of controllability for an integrodifferential system when the control is applied on an open subset of the one-dimensional torus and this open subset moves with a constant velocity. In [10] controllability of this kind is established for some cases. Such a statement of the problem is, in some sense, intermediate between the problem of stopping vibrations using a force applied to a fixed part of the domain (or the one-dimensional torus as in [10]) and the controllability problem in which the force is applied to the entire domain. This formulation leads to interesting spectral problems about the existence of a biorthogonal system of functions for a system of exponentials on a segment and about estimates on the elements of this system, if it exists. At the same time, the study of the problem of controllability for a moving open subset is significantly more complicated compared to the case of a stationary one, and positive results are obtained only for special cases. So, in [10], the equation of string vibrations with a convolution type term is considered, and the convolution kernel is equal to 1. The transition to an exponential function as the convolution kernel is not at all obvious. The paper also considers the boundary conditions of periodicity, and the transition to Dirichlet conditions is also not obvious. This class of interesting problems is still waiting for new research.

5 Conclusion

The article is devoted to the problem of distributed (throughout the domain) controllability of multi-dimensional wave equation with integral memory. The aim is to drive the system to rest by means of control function which is bounded by its absolute value. The additional difficulties deal with the fact that the driving of a solution (and its first derivative, with respect to time) to null is not equivalent to controllability to rest. More accurately, not every control function, which drives the solution and its time derivative to null, leaves them in this state in the future.

Note that when control function is applied only to the subdomain, controllability to rest is (generally speaking) impossible. This can be proved using methods of [11] which were applied to a similar problem.

Now suppose that control function $u(t, x)$ is identically equal to zero. If the second initial condition φ_1 is also null in the problem (1)–(3), then, as it is proved in Y [9], the solution of (1) tends to zero exponentially at $t \to +\infty$. In some sense, this obstacle "helps" the process of control and it can be shown that the upper estimate of time T_* of the control process (when $\varphi_1 = 0$) has the order $\ln \varepsilon^{-1}$, if the absolute value of $u(t, x)$ is bounded by the close to zero parameter ε, i. e. $|u(t, x)| \leq \varepsilon$. If $\varphi_1 \neq 0$, then the order T_* is ε^{-1}.

Note that, in case if $\varphi_1 \neq 0$ and $u(t, x) \equiv 0$, the solution $\theta(t, x)$ tends (at $t \to +\infty$) not to zero, but to a function (limiting state of the system):

$$\frac{1}{\sqrt{2\pi}} \sum_{n=1}^{\infty} \frac{\varphi_{1n}}{l_n^{(1)}(0)} \psi_n(x).$$

Acknowledgements This work was supported by the Russian Science Foundation, project no. № 21-11-00151.

References

1. Chernousko FL (1992) Bounded control in distributed-parameter systems. J Appl Math Mech 56(5):707–723

2. Lions JL (1988) Exact controllability, stabilization and perturbations for distributed systems. SIAM Rev 30(1):1–68
3. Ivanov S, Pandolfi L (2009) Heat equations with memory: lack of controllability to rest. J Math Anal Appl 355(1):1–11
4. Gurtin ME, Pipkin AC (1968) A general theory of heat conduction with finite wave speeds. Arch Ration Mech Anal 31:113–126
5. Dafermos CM (1970) Asimptotic stability in viscoelasticity. Arch Ration Mech Anal 37:297–308
6. Desch W, Miller RK (1987) Exponential stabilization of volterra integrodifferential equations in hilbert space. J Differ Eqn 70:366–389
7. Rivera JEM, Naso MG, Vegni FM (2003) Asymptotic behavior of the energy for a class of weakly dissipative second-order systems with memor. J Math Anal Appl 286:692–704
8. Vlasov VV, Rautian NA, Shamaev AS (2011) Spectral analysis and correct solvability of abstract integro-differential equations arising in thermophysics and acoustics. Contemp Math Fund Direc 39:36–65
9. Rautian NA (2011) On the structure and properties of solutions of integro-differential equations arising in thermal physics and acoustics. Math Notes 90(3):455–459
10. Biccari U, Micu S (2019) Null-controllability properties of the wave equation with a second order memory term. J Differ Eqn 267:1376–1422
11. Romanov I, Shamaev A (2016) Noncontrollability to rest of the two-dimensional distributed system governed by the integrodifferential equation. J Optim Theory Appl 170(3):772–782
12. Romanov IV, Shamaev AS (2015) Exact controllability of the distributed system governed by the wave equation with memory. arXiv:1503.04461
13. Romanov IV (2022) Study of controllability for some dynamical systems with distributed parameters described by integrodifferential equations (in Russian). Izvestiya Akademii Nauk, Teoriya i Sistemy Upravleniya 2:58–61
14. Romanov IV, Shamaev AS (2018) Some problems of distributed and boundary control for systems with integral aftereffect. J Math Sci 234(4):470–484

Usage of Foamed Construction Materials in Wastewater Treatment Technology

Nina Zaletova and Sergey Zaletov

Abstract A typical feature of traditional biological waste watwer treatment with suspended activated sludge is big size of structures and high energy consumption at treatment plants, especially nowadays, when it is necessary to meet high requirements for residual content of nitrogen compounds in treated water. By present time a number of processes for nitrogen removal by means of activated sludge were elaborated. In order to intensify the process of treatment it was proposed to use an microflora immobilized on different carriers made of different materials. Preference for carrier was given to synthetic foamed materials with a density < 1, and surface of the materials with micro mesh structure. ***Methods*** The researches were conducted with foamed polystyrene. The elements of the carrier were in the form of small pieces and cords. The researches were carried out on an experimental installation. The specific aspects of biomass carriers' work in the biological wastewater treatment process were studied. Chemical and microbiological studies were conducted on real waste water. ***Results*** The experimental data showed that the application of the chosen type of carrier allows to hasten the biological process of elimination of organic matter and at the same time allows to increase the depth of purification of organic matters. Almost complete removal of ammonium salt has been achieved. It is shown the ability of the immobilized microflora regenerated under certain technological parameters of biological treatment. It is shown, that the flow of alkaline wastewater or acidic wastewater leads to a fast interruption of ammonium removal. ***Conclusions*** Foamed synthetic elements in the form of pieces and cords can be used as effective carriers of immobilized microflora. It provide an intensive treatment of wastewater from organic matter and ammonium. It is necessary to avoid the flow of alkaline or acidic wastewater into aeration tank.

Keywords Biological treatment · Organic matters · Ammonium · Aeration tank · Carrier · Microflora · Intensification · Stability · Regeneration

N. Zaletova (✉) · S. Zaletov
Moscow State University of Civil Engineering, Moscow 129337, Russian Federation
e-mail: ZaletovaNA@mgsu.ru

P. Akimov et al. (eds.), *Proceedings of FORM 2022*, Lecture Notes in Civil Engineering 282, https://doi.org/10.1007/978-3-031-10853-2_22

1 Introduction

According to modern requirements for the composition of treated waste water discharged into water objects, it is necessary to remove nitrogen compounds up to the level prescribed in regulations [1, 13–15]. The main method of nitrogen removal is nitrification–denitrification. A feature of the biological treatment of nitrogen with suspended microflora is to use large volumes of tanks, and, therefore, outsized areas of occupied lands and obvious increase in the cost of treatment.

There are various technologies to intensify the biological process, in particular, the use of immobilized microflora [2–8]. The materials for carrier should have required properties: physical strength, chemical and biological resistance. In order to work effectively, the necessary technical characteristics of the media must be: a large specific surface of the medium. The carrier must be regenerated without troubles. Many variants of materials are known: these are voluminous blocks made of solid plastic materials (cell, sheet, mesh), framed structures with ruff media on them, mesh boxes filled with large grainy material (such as ceramist) [9–12, 18]. Another type of carrier of immobilized microflora in the aeration tank is the use volume synthetic structured free-floating elements with a density close to 1 g/l [14–16].

In our opinion, the attractive solution of this problem is the use of synthetic materials with a density of much less than one [17].

The aim of this work is to assess the stability of the biological system with immobilized microflora on synthetic media with a density less than one. Certain interest was the biological system's ability to accelerate the process and improve the depth of purification from organic matter and nitrogen compounds.

2 Materials and Method

The studies were conducted on three models of laboratory aeration tanks with a capacity of four liters each. They are made of plexiglas and equipped with small-bubble air diffusor for air supply. Two models were equipped with carriers of foamed synthetic material, which has a different form of elements - the shape of pieces (carrier type 1-st) and the shape of cords (carrier type 2-nd). The total geometric surface area of the elements placed in the model was roughly the same. The third model served as a control - it carried out biological treatment of sewage with suspended activated sludge.

Every day, the sample of sewage waste were supplied to each model. The amount of air did not limit the biological process. The effectiveness of biological purification was judged by the change in the composition of treated water during the treatment. The effectiveness of cleaning was assessed by definition of amount of BOD and NH_4, as well as by comparison with indicators of treatment in the third column. The ability for regenerate the surface of the carrier was determined. The effect of sludge stability in resistance to treatment processes by immobilized sludge when acid or alkaline has been used. It was determined that the attached microflora could be restored during long (for several months) interruptions in the supply of wastewater. The influence of prolonged drying or freezing effects on samples with immobilized biofilm on the ability and effectiveness of the biological treatment system was assessed.

3 Results and Discussion

Visual observations showed that after the introduction of the experimental installation in a stable mode of cleaning, the surface of the media elements was covered with a thin regularly distributed layer of light gray-brown microflora. The more detailed nature of the growth is tested during microscopy on the electron microscope. All microorganisms are in secretion, which characterized the dominant factor of retaining the microflora on the carrier. A dense bacterial growth is formed on the surface of the carrier. Various types of microorganisms are found: sticks, cocci, growth of coquettes, etc. Figure 1 photo 1 and photo2 are snapshots of samples of an active media in the established cleaning mode. From this it is clear that the micro-cellular structure of the surface of the media material, filled with bacterial cells, indicates the excess of the area of attachment of microorganisms over its geometric sizes of the media elements.

In the models with carriers, the removal of organic substances, estimated by the BOD indicator, reached 70% for three days day. By 10–12 days the removal of

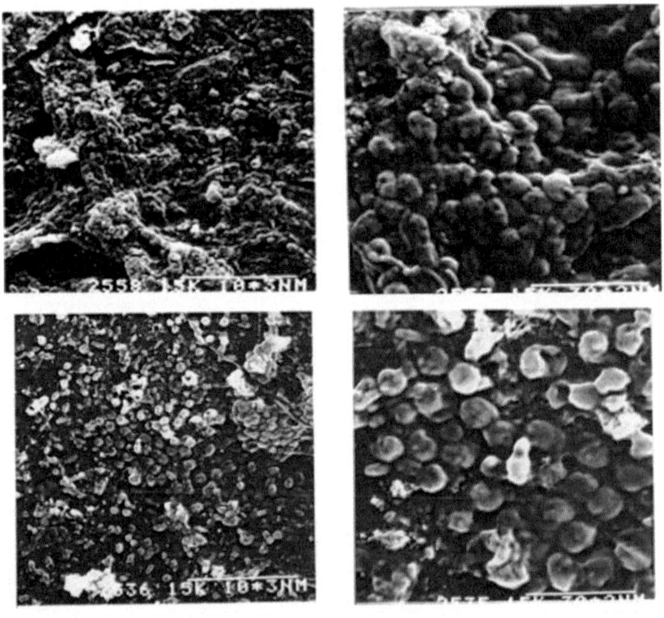

Fig. 1 Microphotography of immobilized microflora with varying degrees of enlargement

organic substances exceeded up to 90% in both models (Fig. 2 curves 1 and 2). At the same time, in the control model, the effect of organic removal was well (significantly) below, (curve 3). The difference in the effectiveness of the reduction in BOD indicates that the immobilization of the microflora really allows to intensify, namely, to hasten removal of contaminants and increase the depth of the removal of organic matter from wastewater.

The dynamic of the removal of ammonium salts from wastewater differed from the dynamics of the removal of organic substances (Fig. 3). Within the first one

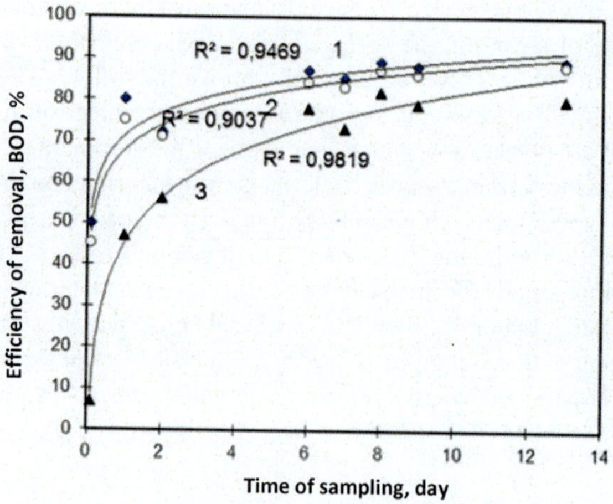

Fig. 2 Effect of sampling time on removals of organic matters: 1st - 1st type of media; 2 - 2nd type of media, 3 - suspended active sludge

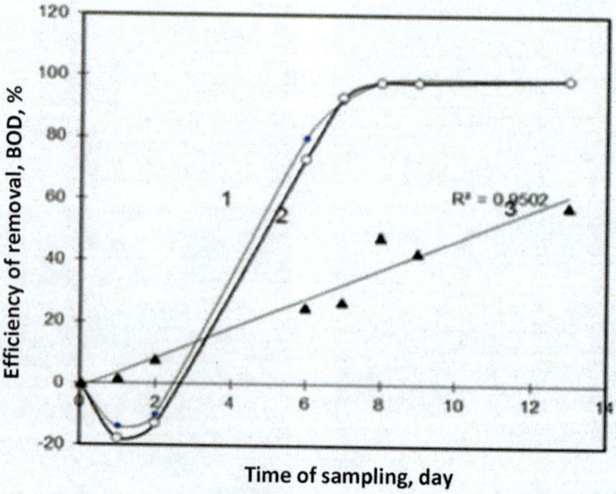

Fig. 3 Effect of sampling time on the removal of ammonium salt: 1 - 1st type of carrier; 2 - 2nd type of carrier, 3 - suspended activated sludge

or two days in the models with carriers (curves 1 and 2) concentration of ammonium increased in purified water. This result is probably connected with ammonium formation.

During first period, there is no microflora which capable to oxidize ammonium. At the same time oxidizing ammonium microflora began to grow in the immobilized biofilm. As a result after a third day the effectiveness of ammonium removal begins to raise. By 8–10 days of the experiment there was a build-up of microflora, which allowed to ensure stable removal of ammonium salt at the level close to full withdrawal. Ammonium oxidation was accompanied by the development of the microflora, which is attended by a change in the bacterial composition of immobilized microflora from fast-growing forms to the development of a long-term microflora with a powerful destructive apparatus that provides the removal of contaminants from low-concentrated solutions.

In the model 3 with the traditional build-up of suspended microflora (curve 3), variations in ammonium concentrations in purified water remained consistently positive. However, the growth rate in the efficiency of ammonium removals was small compared to models with immobilized microflora. By the time NH_4 is almost completely removed in models 1 and 2 with immobilized microflora, the efficiency of removing it in model 3 s increased to only 40%.

Studies have observed that the prolonged (for some days) absence of the supply by new portions of wastewater in the model leads to the gradual lightening of the surface of the loading carrier. Checking the condition of the film under a microscope confirmed the reduction of the area of growth of the elements, reducing the amount of immobilized microflora. This allowed us to consider the temporary ending of supply the purified water (with the filled capacity) as a way of regeneration, which significantly reduces the labor costs of replenishing the necessary and sufficient amount of working microflora. (Fig. 4).

A series of experiments in which an immobilized microflora adapted to wastewater treatment was exposed to extreme conditions: wastewater with pH 11–12 or pH 3.5–4. In experiments, such wastewater was created artificially by the introduction of alkali or acid (depending on the conditions of the experiment) in the original sewage.

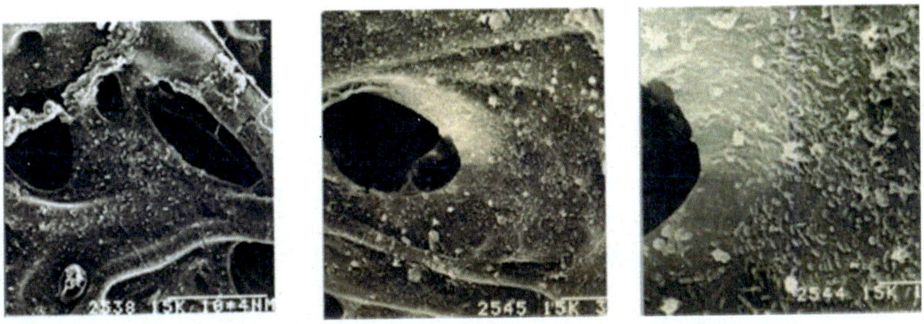

Fig. 4 Microphotography of immobilized microflora in process of regeneration

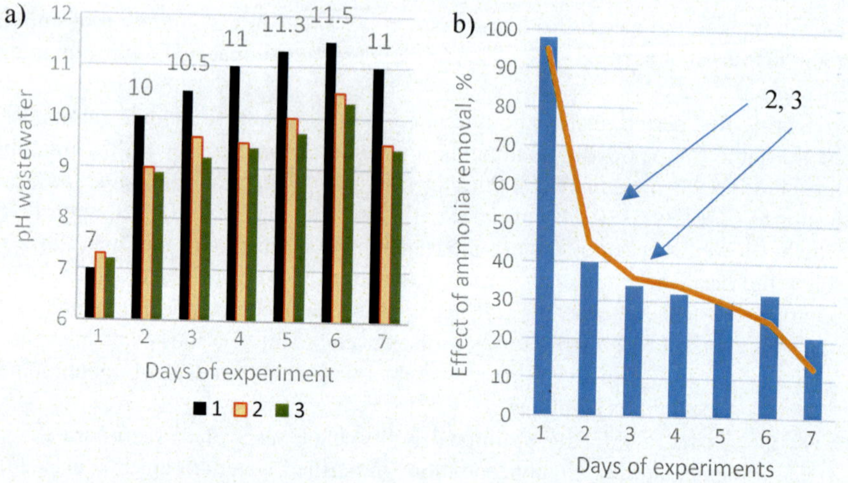

Fig. 5 Effect of alkaline wastewater on the alteration in the pH of waste water **a** and the effectiveness of ammonium removal **b** 1 - original wastewater; 2–1-st type of carrier; 3 - 2nd type of carrier

The results of the experience with the introduction of alkali are shown on Fig. 5. Sewage c pH – 10–11 during a week daily served to the aeration tank once a day. In the aeration tanks for 5 days the liquid phase with immobilized microflora increased to pH 9–10, i.e. there was some neutralization of the injected runoff, on the 6th day the pH of the sewage rose above 10. At the same time, the microflora retained the ability to consume organic matters. However, the removal of ammonium immediately dropped to 30%, and on the 7th day–to 20%. (Fig. 5b).

When acidic wastewater was served to models (i.e. wastewater with pH = 5) for 3 days and further with pH 3.5 (for 2 days), the pH of treated water was maintained at the level of 7 - 8. During this period, the immobilized microflora ensured that the pH was restored to a level favorable for biological purification. With a longer supply of acidic water, the rate of purified water began to decrease intensely. (Fig. 6a), on the 9th day it was below 6.5. The removal of ammonium salt stopped almost immediately (Fig. 6b).

Microbiological assessment shows that the reasons for the survival of biofilm in extreme conditions of pH disturbance, especially in the alkaline side, is the natural more priority development of acid-forming bacteria, which allows more successfully resist the effects of alkaline water. In addition, in all extreme conditions, it seems that biofilm helps to survive the presence of dissolved oxygen and water vapor in closed media pores. Microorganisms consume oxygen through the cell wall, which represents the thinnest film, which causes the stability of the biological system.

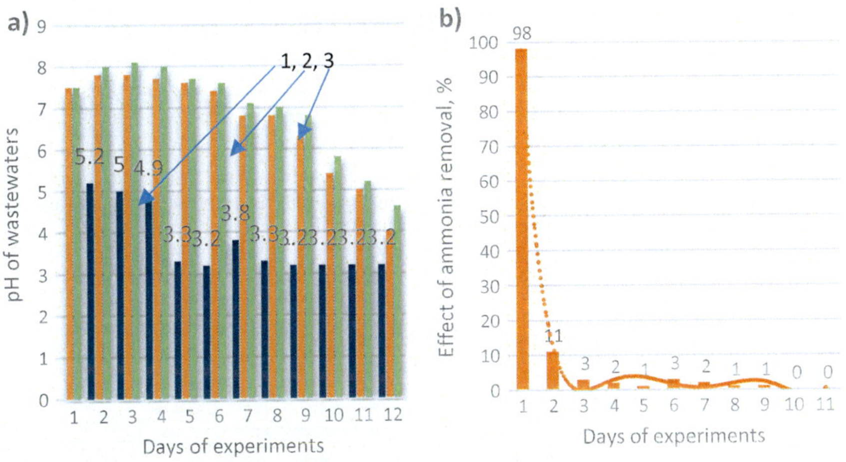

Fig. 6 Effect of acid wastewater on the alteration in the pH of purified wastewater **a** and the effectiveness of ammonium removal **b**: 1- original wastewater; 2 - 1 type of carrier; 3rd - 2nd type of carrier

Cases of wastewater in the field with extreme pH levels in wastewater treatment can be classified as extreme situations in the operating conditions of the facilities. However, in order to ensure the stable operation of the biological system with the immobilized microflora, the requirements for the response of the cleaned wastewater environment in working mode should be the same as is customary in the for suspended sludge.

4 Conclusions

Foamed synthetic materials in the form of pieces or cords can be successfully used as carriers of immobilized microflora. They provide an intensive regime of wastewater treatment from organic substances and ammonium.

The microporous structure of the surface of elements of the carrier can significantly increases the surface of attaching of microorganisms and rises the efficiency of biological wastewater treatment.

Due to the buoyancy of foam materials, various design solutions of carrier blocks are possible.

The shape of the media elements has no effect on the cleaning efficiency when it is equal geometric surface area of the carriers.

The selected carrier material allowed to conduct regeneration without removing the carrier elements from the aeration tank.

Immobilized microflora keep on the surface of the dried carrier for a long time, it makes possible to use such a load for the operation of structures with periodic sewage incoming flow.

Acknowledgements This work was financially supported by the Ministry of Science and Higher Education of the Russian Federation (grant # 075-15-2021-686). All tests were carried out using research equipment of The Head Regional Shared Research Facilities of the National Research Moscow State University of Civil Engineering.

References

1. Shvetsov VN, Morozova KM, Smirnova II, Semenov MY (2007) Technological efficiency of bioloading production. Techvodpolimer Water supply San Technica (2):33–37
2. Gerasimenko IA, Kiselyova YuA, Musinova NL (2008) Modular wastewater treatment station. Water Supply San Technica (2):17
3. Lessel TH (1991) First practical experiences with submerged rope-tipe biofilm reactors for upgrading and nitrification. Wat/Sci/Tech/Val 23:824–834
4. Loading for biological cleaning. Pat 4678567, USA. Accessed 07 July 1987
5. A method and device for biological wastewater treatment. Germany's bid 4329239. Accessed 2 Mar 1995
6. Kletsev ES, Kulikov NI (2015) Device for biological wastewater treatment.A.c. 42:1437356. Accessed 08 Nov 2015
7. Lubchenko OA, Mogilevevich NF, Gvozdyak PI (1996) The effect of the fibrous nozzle on the activity of nitrification in water purification. Chem Water Technol 18(3):323–328
8. Narkevic J (1992) Studies of pressure losses in loading with carrier. Water Supply San Technica 11(12):35–37
9. Sposob and installing wastewater (2–231407). Accessed 27 Apr 94
10. Installation to clean contaminated water using nozzle with immobilized microorganisms. (62–188050). 27 Apr 1994
11. A.C. № 4114583/26. Aeration tank with porous carrier (8). Accessed 20 Mar 1996
12. Ebner L, Reinmann H (1987) Schaumstoffwürfel reinigen häusliche Abwässer. Umvelt 4:173–175
13. Kozlov MN, Harkina OV et al (2014) Nitrification in tertiary treatment of treated wastewater with floating carrier. Water Supply San Technica 14(3):50–54
14. Sirotkin AS, Shaginurova GI, Ippolitov KG (2007) Microorganisms Aggregation: flocules, biofilms, microbial pellets. Academy of Sciences RT Kazan
15. Henze M (2006) Wastewater treatment. World, Moscow
16. Sinitsin AP, Rainina EI, Lozinskiy VI, Spasov SD (1994) Immobilized cells of microorganisms. Publishing House of Moscow University, Moscow
17. Zaletova NA (1999) Treatment of civil waste water (nitrogen and phosphorus compounds) Disseration, Doctor Science, Moscow
18. Pukemo MM (2015) Modern technology in the construction of treatment plants. Water Purif 4(88):26–28

Use of White Quartz Sand as Fine Aggregate in Concrete for Hydraulic Structures

Sergey Fedosov, Olga Aleksandrova, Nguyen Duc Vinh Quang, Boris Bulgakov, and Nadezhda Galtseva

Abstract This paper presents experimental results on mechanical and durability properties such as chloride ion permeability tests, water absorption of modified concrete containing whiter sand used as fine aggregate to partial or complete replacement of river sand. At the same time, the influence of mineral admixtures upto 35% contained in the mixture with and without nanosilica was used at 1 and 1,5% by wt. of cement as a partial replacement of cement on the properties of concrete was also studied. The results indicated that increase of white sand content replacement improves the chloride permeability and water absorption resistance of concrete. It was shown that when the replacement rate of white sand upto 100%, the water-absorption and chloride ion permeability values got been values of 0.374% and 72.4 C, respectively. Meanwhile, the reference sample containing 100% river sand obtained values of 0.442% and 284.2 C, respectively. The lowest water absorption and chloride ion permeability value in these concrete samples studied belongs to the nanosilica series with 66.2 C for № NS2 mixture, decreased by 5.97 and 6.89% compared to the control sample № NS1 and № NR1 (containing 1% Nano-SiO2), respectively. Also, the test results revealed that the optimum content of the white sand was 60% to produce the maximum increase in strength, in which the compressive strength value reached 95.5 MPa, flexural strength attained 9.78 MPa, and splitting tensile strength achieved 7.26 MPa at 28 days curing age.

Keywords Reinforced concrete structure · Concrete corrosion · Water absorption · Chloride ions · Natural river sand · Natural white sand · Nanosilica

S. Fedosov · O. Aleksandrova (✉) · N. D. V. Quang · B. Bulgakov · N. Galtseva
Moscow State University of Civil Engineering, Moscow 129337, Russian Federation
e-mail: aleks_olvl@mail.ru

S. Fedosov
Volga State Technological University, Lenin Square, 3, Yoshkar-Ola, Russian Federation

N. D. V. Quang
Hue Industrial College, Hue City, Vietnam
e-mail: ndvquang@hueic.edu.vn

P. Akimov et al. (eds.), *Proceedings of FORM 2022*, Lecture Notes in Civil Engineering 282, https://doi.org/10.1007/978-3-031-10853-2_23

1 Introduction

According to the United Nations Environment Programme, sand accounts for about 85% of the extraction of natural resources in the world annually, which is equivalent to more than 40 billion tons. In Asia and Africa, the construction boom has tripled demand over the past two decades. Globally, extraction of sand and gravel is anticipated to increase to 82 billion tons by 2060 [1]. Sand is the main raw material of the construction industry and an indispensable material in areas such as oil drilling, electronic chip production, glass production, cosmetics and many other manufacturing fields, the demand for sand has been growing at breakneck speed in recent years, especially in developing countries such as Vietnam [2]. While desert sand with abundant reserves, but its particle size is too fine to use as a building material, so most sand is obtained from quarries or from coastal beaches, or from the dredging of river beds, lead to damage to the underground aquifer, fishery, and nature reserve. Forecasting demand for sand is expected to increase in the coming years, especially in developing countries that are facing population growth, urbanization, and economic growth. Nearly two-thirds of global cement production is increasing sharply in China and India, and Vietnam ranks sixth in the world in terms of output, its reflected the rapid urbanization rate. Sand is the second most-consumed natural resource in the world and as cities grow rapidly like today, the demand for sand is enormous and it is inadvertently draining the world's supply of sand. It is estimated that humans are using about 50 billion tons of sand every year for all activities. This amount of sand is enough to cover the entire UK. The demand for socio-economic development, construction of works, and urban construction require a huge amount of sand. Therefore, sand and gravel gradually become scarce resources. However, large-scale sand and gravel mining activities will damage the environment, affect people's lives and consume resources.

In Vietnam, yellow sand, also known as natural river sand, is a building material that plays a key role in the construction industry for the manufacture of concrete. At present, the construction industry in Vietnam uses only natural river sand extracted from rivers and streams used as fine aggregate for the production of concrete mixtures or building materials. Natural river sand is not only a building material, but also plays an important role in the tectonics and construction delta, stabilization of riverbeds and riverbanks, as well as in creating a habitat for aquatic species of flora and fauna. Excessive extraction of sand from rivers will lead to coastal erosion, which affects the livelihoods of people and aquatic ecosystems. River sand resources could be exhausted in the near future. The excessive extraction of sand from rivers has caused adverse effects to the environment, will cause the river bed to drop, making the riverbank unstable and leading to erosion (Fig. 1).

In fact, some places along the Red River and other rivers stretching from the North to the South of Vietnam to lower their water levels and reduce their water volume. In Hanoi, for example, over-exploitation of sand has caused the bottom of the Red River Delta in Northern Vietnam to be much lower than the bottoms of the small rivers that connecting it. The result that the water of the Red River could not flow

(a) nateral river sand (b) riverbank erosion (c) nateral white sand

Fig. 1 River sand mining causes landslides and gradual loss of agricultural land

into these small rivers was the cause of the death of the Nhue Giang River which feeds almost two million inhabitants in its basin of the Nhue Giang River [3].

The excessive exploitation of river sand for a long time in many localities the whole of Vietnam has been leaving heavy consequences. That is the situation of upsetting the environment around rivers, subsidence, landslides, seriously affecting the environment, reducing aquatic resources. The bottom of large rivers is lowered as "facilitation" for saltwater to easily overflow. With the current rapid urban development, the increasing demand for sand raw materials to service of construction, leading to over-exploitation of sand mining out of control in many riverbeds and streams that have been drained and it is inadvertently draining this natural resource. From apartments, office buildings and to shopping malls, everything is built with concrete – a material created from sand and gravel mixed with cement and water. Every meter of concrete or asphalt road connecting all the buildings, every window frame is also composed of sand. That shows that sand is an indispensable material in the modern life of Vietnamese people. To serve the needs of sand mining for the construction industry, the sand in most rivers and streams has been sucked clean, even in order to get the sand layer deep below the alluvial soil, thousands of hectares of forest were excavated and wiped out mercilessly. River sand mining is also gradually contributing to the disappearance of the Mekong Delta in Vietnam. The risk of "disintegration" of the Mekong Delta seems to be clear, as saltwater intrusion is happening every day regardless of the rainy season or summer. The region is home to more than 20 million people and supplies half of Vietnam's food and most of the world's rice. Construction sand gradually became a scarce resource. Therefore, Vietnam needs to cherish every grain of sand. Because of, "Sand is part of the territory of the Vietnamese national, losing sand means the national territory area is being narrowed." According to a report by the Ministry of Natural Resources and Environment of Vietnam, currently, the total sand resource is just over 2 billion cubic meters, but in 2020 alone, the demand for sand for construction has reached 182 - 197 million cubic meters. By 2030 this figure will increase to 326 - 477 million cubic meters. Without planning and finding other materials to replace natural river sand, Vietnam is at risk of running out of sand and having to import it within 10 years [4]. In the context of the fact that natural river sand is gradually becoming a scarce resource in Vietnam at the present time, it is important to search for other alternative sources of raw materials, the solution to this problem may be the use of natural white sand in the construction industry. In Vietnam, natural sources of white sand are found in most places, with endless reserves, so the research and use of natural white sand to replace natural river sand as aggregate for concrete is meaningful. huge for the concrete industry, thereby reducing construction costs, acceleration construction progress of construction works, reduce difficulties in exploitation and transportation for deep-lying and remote areas; at the same time contribute to reducing the depletion of environmental resources. Natural white sand has abundant reserves, its deposits extend along the coastline of Vietnam with a length of 3260 km, the area of natural white sand in Vietnam is hundreds of thousands of hectares with reserves of up to billions of m3, and high quality among the world's best (Figs. 2). Compared to other countries in the world, Vietnam's white sand in its original form has the same quality as the processed sand of other countries. However, Vietnam does not know how to take advantage of this material to put it into use for the construction industry. So, the research and use of natural white sand to replace natural river sand as aggregate for concrete is meaningful. huge for the concrete industry, thereby reducing construction costs, acceleration construction progress of construction works, reduce difficulties in exploitation and transportation for deep-lying and remote areas; at the same time contribute to reducing the depletion of environmental resources.

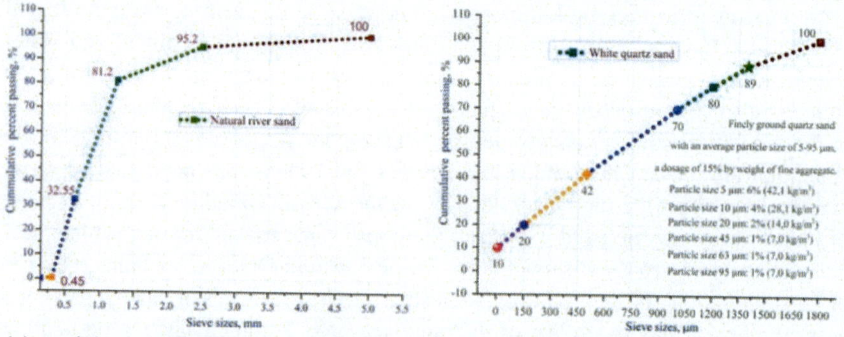

(a) particle size distribution curve of river sand

(b) particle size distribution curve of white sand

Fig. 2 Size distribution of fine aggregate particles

2 Materials and Methods

2.1 Materials

Fine aggregate consists of natural river sand (R) of the Huong river (Vietnam), compliant with TCVN 7570–2006 [5] standards, and white quartz sand (S) meeting the requirements of TCVN 10,796:2015 [6]. The granulometric composition and main characteristics of white quartz sand and river sand are presented in Table 1 and Fig. 2.

Finely ground quartz powder (Qp) obtained by grinding white quartz sand (with an average particle size of 5–95 microns and a SiO_2 content = 99.7%) was used as a micro filler that compaction the structure of the cement matrix [7], as well as improving the mechanical properties and increasing the durability of concrete exposed to chloride and sulfate media [8, 9].

Coarse aggregates used in this research is crushed granite rock from Ga Loi quarry (Vietnam), in the form of a mixture of two fractions of 5–10 and 10–20 mm with a ratio of 40:60, conforms to the standard TCVN 7570:2006 [10]. Characteristics of coarse aggregate are given in Table 2.

The studies used sulfate-resisting Portland cement -PC_{SR}40 - Kim Dinh, produced by company Luks Cement (Vietnam) Limited in accordance with TCVN 6067:2018 [11], to have compressive strength at 3 days (35.0 ± 1 MPa) and 28 days (50.0 ± 1 MPa), initial setting time 120 ± 10 min and final setting time 160 ± 10 min,

Table 1 Physical properties of fine aggregate

Material names	pH value	Solubility in HCl, %	Clay content, %	Bulk density, kg/m^3	Density, kg/m^3	Fineness modulus
Natural river sand	6.2 ÷ 7.3	0.52	0.5 ÷ 1.1	1510 ÷ 1610	2670	3.0
White quartz sand	6.5 ÷ 7.5	0.48	0 ÷ 0.2	1400 ÷ 1500	2240	2.25

Table 2 Main characteristics of the crushed granite aggregate from the Ga Loi quarry

Characteristics	Value
Moisture, %	0,0
Specific gravity, kg/m^3	2680
Bulk density, kg/m^3	1390
Voidness, %	46,9
Silt and Dust content, %	0,73
Water absorption, %	0,2
Average compressive strength, MPa	105,6

Table 3 Physical–mechanical properties of the mineral admixture

Index	Unit	Finely ground quartz powder	Silica fume Sikacrete® PP1	Fly ash (class F)	Nano SiO_2
Specific gravity	kg/m^3	2600	2150	2260	2330
Bulk density	kg/m^3	1400 ÷ 1500	500 ÷ 700	890	<0.15·10^3
Surface area	m^2/kg	0.25 × 10^3	16.3 × 10^3	1830	(150 ± 20) × 10^3
Average particle size	μm	5–95	<0.1	3.05	(10 ÷ 50) × 10^{-3}
Loss on ignition	%	0.044	1.68	0.95	–

content C_3A is 2.2 ± 0.5% and content ($C_4AF + 2C_3A$) of 18,48%, specific gravity of 3,15 g/cm^3.

Supplementary cementing materials consist of condensed microsilica (silica fume - SF) manufactured by Sikacrete®PP1 - Sika Limited (Vietnam) with particle size < 0.1 mm and specific gravity approx. 2.15 g/cm^3, complies with TCVN 8827:2013 [12]; Class F fly ash of thermal power stations "Pha Lai" conforms to the standard TCVN 10,302:2014 [13]. The properties of mineral additives are presented in Table 3.

In the present study, silica (SiO_2) nanoparticles (N) with an average particle size of 10 to 50 nm were synthesized through the decomposition of silica sand in sodium hydroxide (NaOH) molten salt at 500 °C under ultrasound waves [14].

Chemical admixture - polycarboxylate superplasticizer (SP) - Sika®Viscocrete® -151 with density range of 1.075 to 1.095 kg/l and pH value range between 4.0 and 6.0, meets the requirements of TCVN 8826:2011 [15].

2.2 *Methods*

2.2.1 Design of Concrete Composition

Concrete mixtures were developed with using existing composition design methods, namely TCVN 10,306:2014 [16]. At the beginning of the experiment, the concrete mixture compositions for concrete class ≥B65 (grade - M900, average strength 80 MPa) were obtained without taking into account the addition of any mineral and chemical additives and were considered as a reference composition. The composition of the reference concrete mixture is given in Table 4.

On experience, as seen in a number of other authors research [17], for modified concrete, the ratio of coarse-to-fine aggregate (CA/FA) used in practice vary in the range of 1.45 to 1.8. In this research, the optimal ratio of coarse to fine aggregate is CA/FA = 1,55. Therefore, weight of fine aggregate is 1088/1,55 = 701.9 kg/m^3.

Table 4 The composition of the concrete mixture

Material	Value	Absolute Volune m^3	Mix proportions
Water, kg/m^3	156	$156/10^3$	0.284
Cement, kg/m^3	550	$550/3{,}15 = 175/10^3$	1.0
Coase aggregate, kg/m^3	1088	$1088/2{,}68 = 406/10^3$	1.98
Fine aggregate, kg/m^3	654	$654/2{,}6 = 252/10^3$	1.19
Entrapped air, %	2	$2/10^3$	–
Superplasticizer Sika®Viscocrete®-151, lit/m^3	9,9	$9{,}9/1{,}075 = 9{,}2/10^3$	0.018
	Total	*$1000/10^3$*	*4.472*
Mix proportion received is: 0,28:1:1,12:1,98			

2.2.2 Properties of Concrete Mixtures

Eight compositions of concrete mixtures were prepared. A control mixture (C) containing 100% natural river sand and binder material was 100% cement PCsr40. The reference mixture (S0) containing mineral admixtures including fly ash and micro-silica were used for partially replacing Cement PCsr40 in high-performance concrete determined in previous studies by Nguyen D.V.Q. and et al. [7, 18, 19] was 35% by weight of cement consist of 5% micro-silica combined with 30% class F fly ash.

In this research, deals with the possibility of natural river sand to be replaced by white quartz sand in various proportions from 40 to 100% in increments of 20%. The influence of white natural sand on the workability, mechanical properties, corrosion resistance of concrete and reinforcing steel in concrete was investigated [18].

Mixture designations for concrete mixtures containing white quartz sand in the amount of 40, 60, 80 and 100% by weight of fine aggregate, namely № S1, S2, S3 and S4 respectively. Followed by № NS1 and NS2 - mixtures containing 100% white quartz sand and nano-SiO2 content of 1 and 1.5%, respectively. The compositions of concrete mixtures are presented in Table 5.

2.2.3 Determination of the Mechanical Characteristics of Hardened Concrete

After mixing, freshly cast specimens were left in the molds until 24 h at room temperature 28 ± 2 °C, then removed from the mold and submerged in freshwater at temperatures 22 ± 2 °C for curing until the time of testing.

The 100^3 mm cube specimens were used to test the compressive strength of concrete at the age of 3, 7, and 28 days according to TCVN 3118:1993 [20] and water absorption conforms to TCVN 3113:1993 [21] standards. The 100 × 200 mm cylinder specimens were used for tensile splitting strength tests and prism specimens

Table 5 Mix proportions of concrete mixtures, kg/m³

Mixture №	Binding materials (Bm), kg/m³					Aggregate concrete, kg/m³					Water	Super plasticizer	W/Bm
	Cement	Microsilica	Fly ash	Nano SiO_2		Coarse aggregate	Fine aggregate						
							Rive sand	White sand		Quartz powder			
	kg	kg	kg	%	kg	kg	kg	%	kg	kg	kg	kg	-
C	550	–	–	–	–	1088	701.9	–	–	105.3	156	10.6	0.28
S0	357.5	27.5	165	-	–	1088	596.6	–	–	105.3	156	10.6	0.28
S1	357.5	27.5	165	–	–	1088	358.0	40	238.7	105.3	156	10.6	0.28
S2	357.5	27.5	165	–	–	1088	238.7	60	358.0	105.3	156	10.6	0.28
S3	357.5	27.5	165	–	–	1088	119.3	80	477.3	105.3	156	10.6	0.28
S4	357.5	27.5	165	–	–	1088	–	100	596.6	105.3	156	10.6	0.28
NS1	352.0	27.5	165	1.0	5.5	1088	–	100	596.6	105.3	156	10.6	0.28
NS2	349.3	27.5	165	1.5	8.25	1088	–	100	596.6	105.3	156	10.6	0.28
NR1	352.0	27.5	165	1.0	5.5	1088	596.6	–	–	105.3	156	10.6	0.28

Fig. 3 Determination of water absorption of concrete specimens

of 100 × 100 × 400 mm to test flexural strength of concrete according to TCVN 3119:1993 [22] at 28 days curing age.

The 100 × 50 mm cylinders samples were used to test for the Rapid Chloride Permeability Test (RCPT) according to TCVN 9337:2012 [23]. The water absorption capacity and resistance to chloride ion penetration in concretes were tested and compared.

2.2.4 Determination of Water Absorption of Concrete Samples

Water absorption of concrete specimens was determined in accordance with TCVN 3113:1993 [21] standards. The tests were carried out on 100^3 mm^3 cube samples after 28-days of curing. The samples were immersed in water tank (Fig. 2) at temperature of 22 ± 2 °C and kept for 24 h until they were completely saturated. After that, the samples were removed from the water and placed for 1 min on a wire mesh, at the moment, the wet mass was weighed (m_1). Subsequent to samples were placed in a ventilated oven at temperature of 105 ± 5 °C for at least 24 h while allowing for two successive weightings at intervals of 2 h to determine increment of mass losses with an error of no more than 0.1% of the last value of the sample mass, the dry mass was weighed (m0). The water absorption was being determined formula (1) (Fig. 3):

$$W_a(\%) = \frac{m_1 - m_0}{m_0} \cdot 100 \tag{1}$$

2.2.5 Determination of the Density of the Concrete Structure

The density of the modified concrete structure was assessed by determining its permeability to chloride ions according to the TCVN 9337:2012 standard using concrete disk samples with a diameter of 100 ± 2 mm and a thickness of 50 ± 3 mm.

Table 6 Rating of chloride permeability of concrete based on charge passed according to TCVN 9337:2012

The value of the total passed charge Q, Coulombs (C)	The chloride permeability level of the concrete concrete structure
>4000	High
2000 ÷ 4000	Moderate
1000 ÷ 2000	Low
100 ÷ 1000	Very low
<100	Negligible

After 28 days of hardening, concrete samples were tested for the permeability of their structure to chloride ions by impressing a voltage of 60 V between two containers filled, respectively, with 3% NaCl solution and 0.3 N NaOH. The electric current strength was measured every 30 min for during a 6-h period. The total electric charge in coulombs that passed through the concrete sample during the tests, calculated according to the formula (2), was measured as an indicator of the resistance of concrete to the penetration of chloride ions:

$$Q = (\frac{95}{x})^2 \cdot 900 \cdot (I_0 + 2I_{30} + 2I_{60} + \ldots 2I_{300} + 2I_{330} + I_{360}) \quad (2)$$

where, Q is the total electric charge passed through the concrete sample for 6 h, in coulombs; I0, I30, I60 … I330, I360 is the current strength, in amperes (A), for a test duration of 0, 30, 60 … 330, 360 min after voltage is applied; x is the actual diameter of the test disk samples, mm.

It was found that the transmitted electric charge Q has a close relationship with the resistance of the concrete structure to the penetration of chloride ions, and therefore TCVN 9337:2012 recommends a qualitative criterion – "the level of permeability of the concrete structure for chloride ions" according to the range of values of the transmitted electric charge, as shown in Table 6.

3 Results and Discussion

3.1 Properties of Fresh Concrete

In this study, the workability of all fresh concrete mixtures was checked meets the requirements of TCVN 12,209–2018 [24]. Value slump flow of the respective concrete mixtures containing 100% white sand varied in the range of 620–670 mm, while № S0 reference sample obtained at 590 mm and control sample (№ C) attained at 550 mm. The L-box passing ratio for the (№ C) control mixes is 0.7. Conversely, the passing ratio is between 0.85 and 0.89 for all other mixes. The V-funnel flow time (Tv) of the fresh mixes varied in the range of 10–13 s.

3.2 Mechanical Properties of HPC

Compressive, flexural and splitting tensile strength of concrete increased with age for all mixtures, as shown both in Figs. 4 and 5 and Table 4. From the results in Figs. 5 and 6, it was shown that when using white sand to replace for river sand, increased the slump flow of the concrete mix. The strength of the concrete at the test ages gradually increase with the percentage of white sand replaced compared with the № S0 reference samples. This shows that the white sand combined with finely ground sand powder increased in the density of particle packing in the mix, thereby increasing the strength as well as other physical and mechanical properties of the concrete.

Of all the concrete mixtures containing white sand and without nano-SiO2, the strength of concrete reached the highest value. The compressive strength value reached 95.5 MPa, flexural strength attained 9.78 MPa, and splitting tensile strength achieved 7.26 MPa at 28 days curing age, while № S0 reference sample obtained at 89.1, 9.71 and 7.22 MPa, respectively, and control sample (№ C) got at 79.9, 8.4 and 5.7 MPa, respectively.

A comparison of the river sand and white sand nanosilica concrete series in Figs. 5 and 6 shows that the strength of the concrete mixtures for the river sand series are lower than those of the corresponding mixture of the white sand series. The compressive strength of № NS1 sample value won at 106 MPa, flexural strength

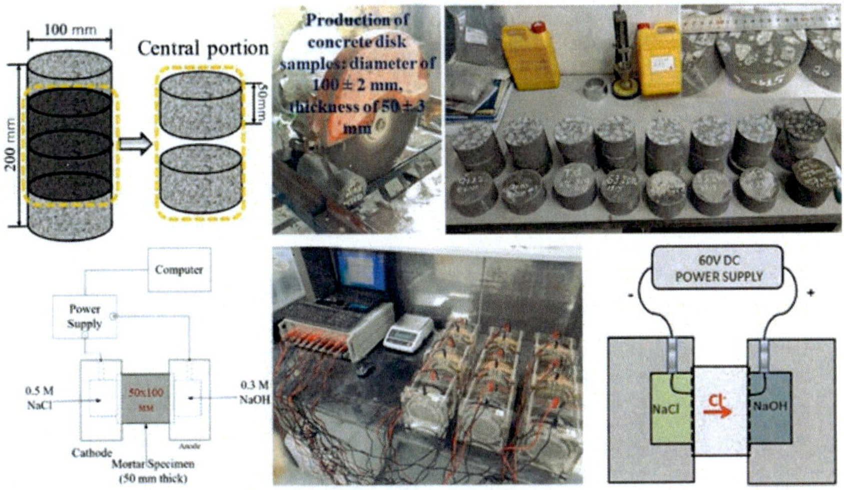

Fig. 4 Diagram of experimental setup for determining rapid chloride permeability of concrete

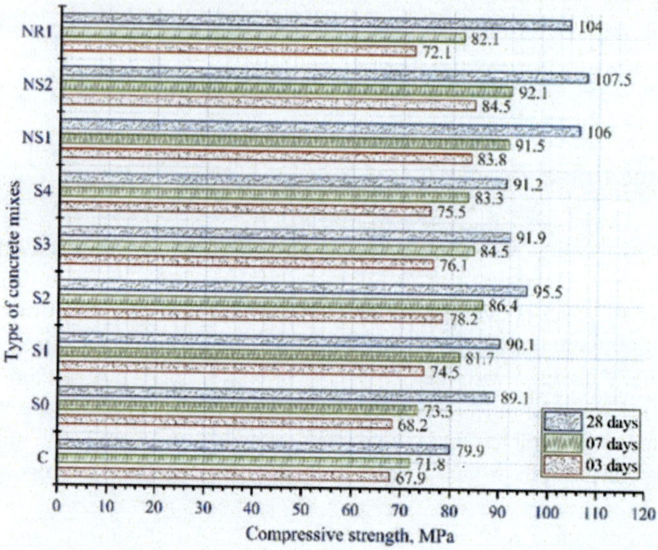

Fig. 5 Compressive strength of concrete containing various content of natural white sand

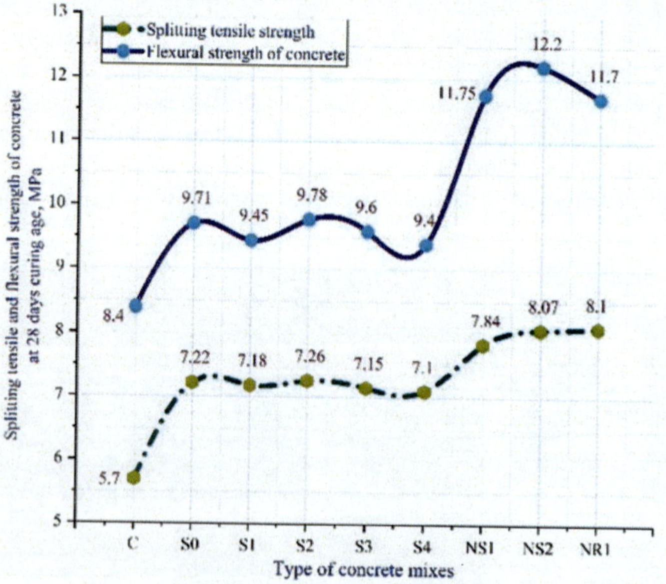

Fig. 6 Splitting tensile and flexural strength of concrete at 28 days

Table 7 Effect of white sand content and nano-SiO2 on concrete water absorption

Mix №	Cube specimen 1			Cube specimen 2			Cube specimen 3			Average Wa, %
	m_1, g	m_0, g	W_{a1}, %	m_1, g	m_0, g	W_{a2}, %	m_1, g	m_0, g	W_{a3}, %	
C	2514.5	2495	0.782	2534	2516.5	0.695	2531.5	2512	0.776	0.751
S0	2480	2468	0.486	2521	2511	0.398	2497	2486	0.442	0.442
S1	2524	2514	0.398	2515	2505	0.399	2510	2499	0.440	0.412
S2	2515	2504	0.439	2504	2494	0.401	2501	2492	0.361	0.400
S3	2500	2491	0.361	2509	2499	0.400	2502	2492	0.401	0.388
S4	2510	2502	0.320	2500	2490	0.402	2508	2498	0.400	0.374
NS1	2519	2513	0.239	2527	2523	0.159	2508	2503	0.200	0.199
NS2	2500	2496	0.160	2496	2492	0.161	2510	2506	0.160	0.160
NR1	2475	2465	0.406	2500	2490	0.402	2494	2485	0.362	0.390

achieved 11.75 MPa, and splitting tensile strength reached 8.78 MPa at 28 days curing age, while № NR1 reference sample obtained at 104, 11.7 and 8.22 MPa, respectively.

3.3 Determination of Water Absorption

The average of the three samples was reported as the water absorption (Wa) for each mixture. The test results are presented in Table 7. The water absorption of concrete decreases with an increase in the proportion of white sand in concrete mixtures. The decrease in water absorption may be due to fineness modulus of white sand. The modulus of fineness of white sand (Mk < 2.25) is smaller than river sand (Mk = 3.0), which leads to a decrease in porosity of concrete, thereby increasing the density and strength, which leads to a decrease in water absorption of concrete.

3.4 Determination of Chloride Ion Penetration in Concrete

The penetration resistance of chloride ions was measured by determining the amount of charge transferred to the sample was monitored for 6 h. The rapid chloride penetrability test results are presented in Table 8, which show a decrease dramatically in thc total charge passed through № S0 concrete specimens compared to № C (control) specimen after 28 days of hardening, by substitution of 35% weight of cement PCsr 40 with mineral admixture consisting of 5% silica fume and 30% fly ash, combined with the use of finely ground quartz powder to replace fine aggregate (river sand) with a content of 15% by weight of river sand in the composition of concrete mixtures. Similarly, the S4 concrete mix containing white sand was used as fine aggregate to

Table 8 Average value of current strength and total electrical charge passing through concrete specimens during 6 h

Mix №	T, min	0	30	60	90	120	150	180	210	240	270	300	330	360	Q, Coulombs	The level of permeability of chloride ions
	t, ^0C	28	28	27	27	27	27	27	28	28	28	28	28	28		
C	I, mA	0	57,8	55,6	54	52,4	50,8	51	50,8	50,8	51,0	51,3	51,3	50	1086	Low
S0	I, mA	0	12,9	13,4	13,9	14,2	14,2	13,9	13,9	14,2	13,9	13,1	13,3	13,1	284,2	Very low
S1	I, mA	0	3,1	3,3	3,5	3,5	3,5	3,5	3,5	3,6	3,6	3,6	3,6	3,6	72,4	Negligible
NS1	I, mA	0	3,1	3,3	3,4	3,5	3,5	3,4	3,5	3,4	3,4	3,4	3,4	3,4	70,4	Negligible
NS2	I, mA	0	3,0	3,0	3,0	3,2	3,2	3,2	3,4	3,2	3,3	3,3	3,3	3,1	66,2	Negligible
NR1	I, mA	0	3,05	3,25	3,35	3,4	3,35	3,5	3,5	3,5	3,55	3,6	3,6	3,5	71,1	Negligible

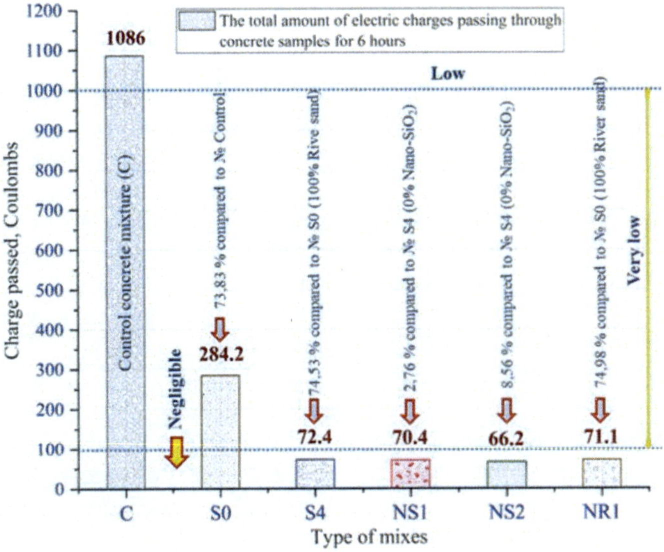

Fig. 7 Results of tests chloride ions penetration into concrete

completely replace river sand. The chloride ion permeability test results show that the chloride ion permeability of № S4 concrete sample obtained of 72.4 Coulombs, than decreased by 74.5% compared to control sample № S0 (containing 100% river sand as fine aggregate). Additionally, as the white sand weight is increased, there is a dropped considerably in the permeability of chloride ions.

The results are presented in Fig. 7 have been evaluated according to the limit values shown in Table 8. Figure 7 also shows that the use of nano-silica to partial replacement of cement in concrete mixtures containing 100% white quartz sand leads to has decreased significantly in the permeability of chloride ions. The content of nano-silica replaces at 1% by weight of cement improves the permeability for chloride ions of concretes containing both white quartz sand and river sand (compositions mixes № NS1 and NR1). The chloride ion permeability of № NS1 concrete mixes attained at 70.4 Coulombs has dropped by 2.76% compared to reference specimens № S4, and 1.0% compared to № NR1 sample (containing 100% river sand as fine aggregate). The chloride ion permeability values of NS1 and NS2 samples at 28-days curing age were achieved 70.4 and 66.2 Coulombs, respectively, while the control composition NR1 reached a value of 71.1 Coulombs, which is only 1.8% lower than that of № S4 samples.

A factor affecting the depth of penetration of chloride ions may be the content of clay particles and plant waste contained in the sand when they are not washed, which causes the formation of air voids in concrete. As a rule, penetration occurs through capillary or open pores. The formation of this continuous pore system promotes the penetration of chloride ions into concrete structure.

4 Discussions

The water absorption of concrete decreases with an increase in the proportion of white sand in concrete mixtures. The decrease in water absorption may be due to the fineness modulus of white sand. The modulus of fineness of white sand (M_k < 2.5) is smaller than river sand ($M_k = 3.0$), which leads to a decrease in the porosity of concrete, thereby increasing the density and strength, which leads to a decrease in water absorption of concrete. The experimental results showed that the reference concrete mixture № S0, containing natural river sand as fine aggregate, shows the value of rapid penetration of chloride ions of 284.2 Coulombs, while the mixture № S4, containing natural white sand as fine aggregate 72.4 Coulombs.

This indicates a better resistance of concrete samples containing white sand to the penetration of chloride ions from aggressive environments. The value of the permeability of chloride ions in sample № S4 is 292.5% lower than that of the reference specimens № S0, but 1.83% higher than that of sample № NR1. The addition of 1 and 1.5% of nano-silica (by weight of cement) in the concrete mixture containing white sand as fine aggregate contributed to an increase in resistance to the penetration of chloride ions. Samples № NS1 and № NS2 showed values of 70.4 and 66.2 Coulombs, respectively. However, the degree of improvement is insignificant. It can be said that the developed composition of № S4 is the optimal mix composition for concrete structures exposed to chloride aggressive environments.

5 Conclusions

In general, the water absorption and chloride ion permeability values of all concrete series using nanosilica were lower than the without nano-silica samples. The longer the curing time, the better the chemical reactions in the cement and the decrease in the chlorine permeability values. The optimum percent replacement is 60% white sand, obtained the maximum strength value among all white sand replacement rates, in which the compressive strength value reached 95.5 MPa, flexural strength attained 9.78 MPa, and splitting tensile strength achieved 7.26 MPa at 28-days curing age.

The use of natural white sand as fine aggregate leads to an increase in the density of concrete, which leads to a decrease in the water absorption of concrete, reduces the penetration rate of chloride ions, and, accordingly, increases the resistance in aggressive environments. Research results confirm that natural white sand can be used as fine aggregate instead of river sand for underground concrete structures exposed to chlorides environments.

Acknowledgements This work was financially supported by the Ministry of Science and Higher Education of the Russian Federation (grant # 075-15-2021-686). All tests were carried out using research equipment of The Head Regional Shared Research Facilities of the National Research Moscow State University of Civil Engineering.

References

1. Filho WK, Hunt J, Lingos A, Platje J, Vieira L et al (2021) The unsustainable use of sand: reporting on a global problem. Sustainability 13(6):3356. https://doi.org/10.3390/su13063356
2. Nam CPV (2020) Nghị định số 23/2020/NĐ-CP của Chính phủ: Quy định về quản lý cát, sỏi lòng sông và bảo vệ lòng, bờ, bãi song, 20 tr. Accessed 24 Feb 2020
3. Pilarczyk KW (2016) Bank erosion Mekong Delta and Red River, Vietnam, p 157
4. Tài B, Và N, Trường M, Nam V (2017) Nguồn cát tự nhiên sẽ cạn kiệt trong vòng 10 năm tới. Quản Lý Tài Nguyên Thiên Nhiên, 168 tr
5. TCVN 7570–2006 (2006) Cốt liệu cho bê tông và vữa - Yêu cầu kỹ thuật. Hà Nội, 94 tr
6. TCVN 10796:2015 (2015) Cát mịn cho bê tông và vữa. Hà Nội, 35 tr
7. Quang NDV, Bazhenov YM, Aleksandrovna OV (2019) Effect of quartz powder and mineral admixtures on the properties of high-performance concrete. Vestnik MGSU 14(1):102–117 (2019). https://doi.org/10.22227/1997-0935.2019.1.102-117
8. Quang NDV, Aleksandrova O, Bulgakov B (2020) Mechanical and durability properties of high-performance concrete in corrosive medium of vietnam. In: Akimov P, Vatin N (eds) Proceedings of FORM 2021, Lecture Notes in Civil Engineering, vol. 170, pp 29–43 (2022). https://doi.org/10.1007/978-3-030-79983-0_4
9. Bazhenov Yuv, Aleksandrovna OV, Quang NDV, Bulgakov BI, Larseh OA, Golotenko DS (2020) High-performance concrete produced with locally available materials in Vietnam. Constr Mater 3:32–38. (In Russian). https://doi.org/10.31659/0585-430X-2020-779-3-32-38
10. TCVN 7570:2006 (2006) Cốt liệu cho bê tông và vữa - Yêu cầu kỹ thuật. Hà Nội, 94 tr
11. TCVN 6067:2018 (2018) Xi măng poóc lăng bền sulfat. Hà nội, p 7.
12. TCVN 8827:2011 (2011) Phụ gia khoáng hoạt tính cao dùng cho bê tông và vữa - Silicafume và tro trấu nghiền mịn. Hà Nội, 17 tr
13. TCVN 10302–2014 (2014) Phụ gia hoạt tính tro bay dùng cho bê tông, Vữa xây và xi măng. Hà Nội, 14 tr
14. Quang NDV, Le DV, Alexandrov OV, Bulgakov BI (2019) Synthesis and characterization of silica nanoparticles from Vietnam. Nanomater Energy 8(1):73–77. https://doi.org/10.1680/jnaen.18.00012
15. TCVN 8826:2011 (2011) Phụ gia hóa học cho bê tông. Hà Nội, 22 tr
16. TCVN 10306–2014 (2014) Bê tông cường độ cao - Thiết kế thành phần mẫu hình trụ. Hà Nội, 24tr
17. Aitcin PC, Shirlaw M, Fines E (1992) High performance concrete: removing the myths. In: Concrescere, newsletter of the high-performance concrete network of centres of excellence, Canada, 6 March 1992
18. Quang NDV, Aleksandrova OV, Samchenko CV (2020) Combined effect of mineral admixtures and fine aggregate on the mechanical properties of ultrahigh performance concrete. IOP Conf Ser Mater Sci Eng 869:032034. https://doi.org/10.1088/1757-899X/869/3/032034
19. Нгуен Дык Винь Куанг ОВ, Александрова БИ, Булгаков ВФ, Коровяков МБ (2021) Каддо. Влияние золы-уноса в многокомпонентном вяжущем на прочность бетонов. Техника и технология силикатов. Том 28(3):110–116
20. TCVN 3118: 1993 (1993) Bê tông nặng - phương pháp xác định cường độ nén. Hà Nội, 7 tr
21. TCVN 3113–1993 (1993) Bê tông nặng - Phương pháp xác định độ hút nước. Hà Nội, 3 tr
22. TCVN 3119:1993 (1993) Bê tông nặng - Phương pháp xác định cường độ kéo khi uốn. Hà Nội
23. TCVN 9337–2012 (2012) Bê tông nặng - Xác định độ thấm ion clo bằng phương pháp đo điện lượng. Hà Nội, 14 tr
24. TCVN 12209-2018 (2018) Bê tông tự lèn - Yêu cầu kỹ thuật và phương pháp thử. Hà Nội, 22 tr

The Fulfilment of the Zhucowski's Theorem Conditions in Research of the Stationary Wind Exposure to the Lattice Structure

Lyudmila Frishter and Maxim Lukin

Abstract The article deals with an experimental simulation of the way an industrial stack behaves when exposed to wind blowing at various angles of attack. In construction of high-rise buildings and facilities, such lattice structures are used for placing of the process equipment. The problem of aeroelasticity for a lattice structure has been demonstrated, and indications of strain gauges installed at its base have been analyzed. Then, data obtained were used for verifying the fulfilment of the aeroelasticity (Zhucowski's) theorem conditions. At first, Zhucowski's theorem deals with a plane profile lattice, and later it takes up with the designated lattice structure exposed to the wind that is treated as a regular system of prismatic elements. The work is aimed at establishing a correspondence between the conditions of experiment performance (finding the stationary mode zone) and the solution of the aeroelasticity problem through the use of the Zhucowski's theorem. Verification of the Zhucowski's theorem conditions for regular lattice of prismatic bodies affords finding the area of wind stream-to-lattice structure stationary action. Simplified mathematical model based on the experiment data is treated as a test model. It affords both improvement of the experimental procedure and correction of experimental conditions through close inspection look at the lattice structure exposed to the wind.

Keywords Lattice structure · Wind exposure · Stream stationary action area · Experimental modelling · Plane profiles lattice · Zhucowski's theorem

1 Introduction

In construction of high-rise buildings and facilitites, lattice external structures are used for storage, placing of the process equipment and supporting of the work process.

L. Frishter (✉)
Moscow State University of Civil Engineering, Moscow 129337, Russian Federation
e-mail: lfrishter@mail.ru

M. Lukin
Construction and Assembling Department No. 1 (OOO SMU No.1), Nauchnii proyezd, d. 17, Moscow 117246, Russian Federation

P. Akimov et al. (eds.), *Proceedings of FORM 2022*, Lecture Notes in Civil Engineering 282, https://doi.org/10.1007/978-3-031-10853-2_24

This work studies wind exposure to an open, detached, rectangular-plan industrial stack. The problem of the constructions aerodynamics is solved, numerically and theoretically, by methods of structural mechanics [1–8], and experimentally [9–15]. Design of lattice structures exposed to the wind is governed by general normative documents [16–18], that ignore design features. According to norms, design process of such structures includes wind-tunnel testing of mock-ups [16].

The work describes an experimental research of the lattice structure exposed to the wind. The angle between the speeddirection of the approach wind and the face area of the structure, i.e. the angle of attack, varies. For experiment, the mock-up is positioned in the part of the wind-tunnel, where it should be exposed to the stationary air stream [9–12, 14–16]. Mathematical test model of such exposure agrees both with indications of strain gauges installed at the base, and with provisions of the Zhucowski's theorem [19, 20].

The importance of the work consists in the need for correction of experimental studies methods that will afford the use of the mathematical model adequate to data obtained from experiments.

Practically, the work is valuable because correlation of the wind exposure data obtained from experiments and conditions of the Zhucowski's theorem defines both the mode of stationary action towards the lattice structure, and areas where this mode is violated, i.e. areas of turbulence. These areas will require closer study and interpretation of the data obtained from experiment to draw recommendations for future experiments.

The work establishes a correspondence between the conditions of experiment performance (finding the stationary mode zone) and the solution of the aeroelasticity problem under the Zhucowski's theorem.

The work pursues the following tasks: 1. Experimental aerodynamic studies of wind exposure to an open, rectangular-plan lattice structure for various angles of the wind attack and obtaining of internal forces. 2. Verification of the Zhucowski's theorem conditions in application to open lattice structures, finding the area of wind stream stationary action with regard to variation of angle of attack to the lattice structure.

At first, Zhucowski's theorem deals with a plane profile lattice [19], and later it takes up with the particular lattice structure exposed to the wind.

2 Experiment Performed and Data Obtained

The experiment has been performed in the special wind-tunnel of the Training, Research and Production Laboratory for Aerodynamics and Aeroacoustics of Moscow State University of Civil Engineering. The construction has a length of $l = \xi$, a height of $H = 3\xi/4$, and a width of $B = 5\xi/8$, where ξ means a characteristic dimension, mm. The regular pitch of the lattice makes $l_0 = l/4 = \xi/4$ in length, $h_0 = h/5 = \xi/8$ in height, and $b_0 = b/5 = \xi/8$ in width. Mock-up industrial stack (Fig. 1) was designed and made for experimental research [13].

Fig. 1 Mock-up industrial stack, side view

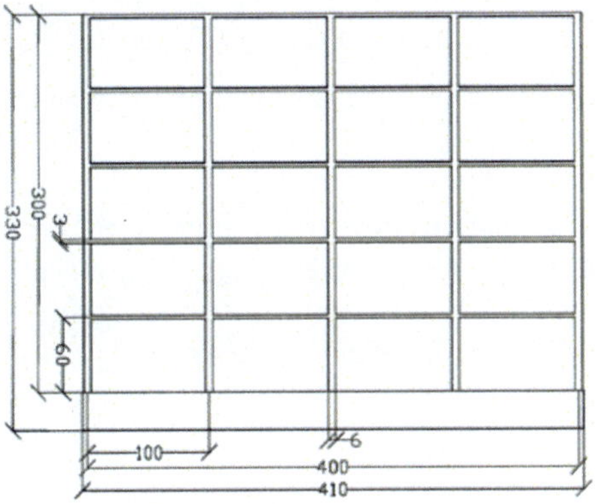

Fig. 2 Schematic mock-up stack, side view

Mocked-up construction with rectangular-plan lattice of columns (Fig. 2) has size of 250 × 400 mm in plan and height of 300 mm. Permeable inter-floor constructions are uniformly, with 60-mm pitch, distributed throughout the height of the whole structure, from zero to upper mark.

During making of mock-up, a six-component force-torque transducer was built in to measure resultant forces and torques. The transducer is attached to a rigid metal plate at the bottom of the mock-up (Fig. 3). Assembled mock-up is then put on the automated rotating table.

Mock-up structure stands in the area of the wind tunnel where plane stationary air stream can be reduced to the one-dimensional irrotational stationary one. The wind stream the structure is exposed to and the angle of attack of the wind stream are considered to be constant along height, length and width of the external surface of the structure. The auto-similarity range under the Reynolds number for the mock-up under research was found by supporting researches that increased stream speedin the tunnel successively, $V\infty=$ from 5 to 30 m/s. Dependencies obtained define that, for

Fig. 3 Six-component force-torque transducer at the bottom of the mock-up

this model, auto-similarity range under the Reynolds number takes place at $V_\infty \geq 7$ m/s.

At the beginning of the experiment, the lattice structure is positioned at the angle of attack $\beta = 0^0$ and 'zero' (no-flow) indications of the strain-gage weigher are read; they are necessary for considering an initial displacement during loading of the transducer with a weight of the mock-up structure. Rotation of the rotary table with mock-up on it varies the angle of attack from 0^0 to 360^0, with 45^0 pitch, readings of the strain-gage weigher are recorded for each angle from which the wind attacks the model.

3 Zhucowski's Theorem Conditions for Periodic Profiles

A rectilinear lattice of plane profiles (Fig. 4) is treated as an infinite periodic structure of identical profiles [19], that are placed in line (lattice front) and have the same lattice

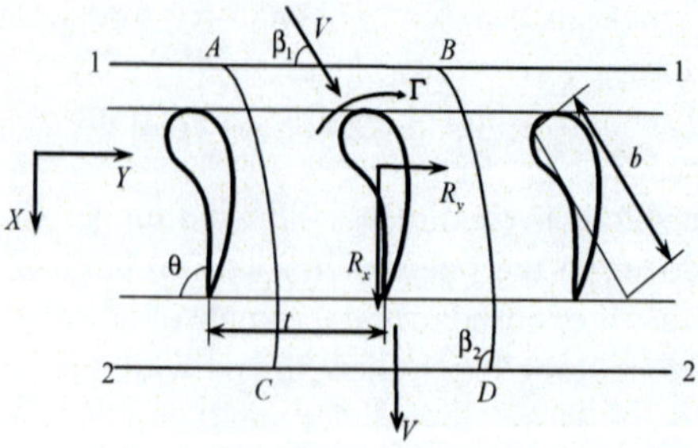

Fig. 4 Profiles lattice

pitch, t. An angle between a profile chord, b, and a lattice front, θ is a setting angle, constant for lattice profiles. For an insulated profile, an angle of attack is understood as an angle between speedof non-disturbed stream in front of the profile and a chord of the profile. Generally, the angle between direction of the approach stream speedand tangent to the arch of the lattice profile in the front point is called an angle of attack. A line perpendicular to the front is called a lattice axis.

Under consideration is a plane lattice [19], that consists of in infinite cylindrical bodies having the same sections in any planes parallel to the plane of the drawing. Curvilinear quadrangle *ABCD* is separated, periodicity condition is used at *AD* and *BC* streamlines. *AB* and *CD* sections are selected at s distance from the profiles lattice, so that pressure and velocities there may be considered as constant. Frontal axis u coincides with the axis *OY*, axial axis a, with the axis *OX*, all axes are shown in Fig. 4. In the direction perpendicular to the drawing, geometrical dimension makes 1.

Momentum projections at frontal axis u and axial axis a are recorded. With regard to rate circulation in the circuit *ABCD:* $\Gamma_{ABCD} = t(v_1 \cos \beta_1 - v_2 \cos \beta_2) = t(v_{1u} - v_{2u})$, we obtain a frontal component of the force that acts on the profile blade: $R_u = \rho v_{1a}\, \Gamma_{ABCD}$.

According to the Stokes theorem, in an irrotational stream rate circulation in the circuit *ABCD* is the same as circulation in the profile circuit. If a working fluid is incompressible, then, according to the consumption conservation equation: $v_{1a} = v_{2a} = v_a$.With regard to the Bernoulli equation for an incompressible fluid, an axial component of the force makes: $R_a = -\rho v_{u\,aver.}\,\Gamma$, where $v_{u\ aver.} = (v_{u\,1} + v_{u\,2})/2$ means an average frontal component of the velocity. As $v_a = v_{a\ aver.}$, the resultant force makes $|R| = \sqrt{R_a^2 + R_u^2} = \rho v_{aver.}\Gamma$, where $v_{aver.} = \sqrt{v_{a\,aver.} + v_{u\,aver.}}$. The expression for component forces:

$$\frac{R_u}{R_a} = -\frac{v_{a\ aver.}}{v_{u\ aver.}}.$$

Finding: force vector, $\overline{R}$, and speed vector, $\overline{v}_{\text{aver.}}$ are mutually perpendicular, as their scalar product makes zero.The direction of the resultant force $\overline{R}$ vector may be obtained by turning $\overline{R}$ vector 90° from the speeddirection in the direction opposite to circulation.

Let an insulated plane profile be treated for incompressible fluid at $t \to \infty$. As the insulated profile cannot vary the stream in front or behind itself, at $t \to \infty$ $v_{2u} \to v_{1u}$, $\beta_2 \to \beta_1$, $v_{\text{aver.}} \to v_\infty$(approach stream velocity), we obtain $t(v_{1u} - v_{2u}) \to \Gamma$. Final resultant force makes: $|R| = \rho\, v_\infty \Gamma$.

Finding: for insulated plane profile, the direction of the force vector $\overline{R}$, as is the case with flowing past the profile lattice, is perpendicular to the direction of speedvector $\overline{v}_\infty$ and is turned 90^0 opposite to the circulation direction.

4 Checking of Zhucowski's Theorem Conditions for Lattice Structures at Experimental Stationary Wind Exposure

The object of research is a lattice structure exposed to the wind. Plane section of the structure (Figs. 1 and 2) is treated as a plane periodic lattice. Figure 5 shows a rectilinear lattice of the structure profiles, treated as an infinite periodic structure of identical prismatic profiles spaced along the lattice front at constant intervals. The elements of lattice have identical sections of small square in any planes parallel to the bottom of the lattice. Under consideration is the ground floor of the lattice, parallel to its bottom and positioned close to the strain gauges for finding correlation between theoretical and experimental data. The ground floor is slightly raised over the bottom surface to reduce the turbulance of the approach stream (Fig. 3).

For an insulates lattice profile, angle of the wind attack is understood as an angle between speedof the undistrubed stream that approaches the structure, and a chord of the profile directed along its average cross-section (Fig. 5). The air stream that approaches the frontal section of the structure at the rate of $\mathrm{v}_{\infty} = 12\,\mathrm{m}^2/\mathrm{c}$ and angle of attack β, slightly changes direction and value of its velocity, and later, at some distance from the structure, restores its initial parameters.

Curvilinear quadrangle *ABCD* is separated, periodicity condition is used at *AD* and *BC* streamlines. *AB* and *CD* sections are selected to be such that pressure and velocities there may be treated as constant. Axes *OY* (frontal) and OX (axial), origin of coordinates are shown in Fig. 4. In addition, it demonstrates axes X_0, Y_0, fixed to coordinates of strain gages installation, i.e. positioned in the centre of the lattice profile plane section.

Pattern of the lattice structure plane section (Fig. 5) is similar to the plane profile lattice pattern (Fig. 4). With the help of experimentally obtained values of forces in the centre of the plane section of the structure, positioned in the area where force-torque sensors are installed, we can check the fulfilment of the Zhucowski's theorem conditions.

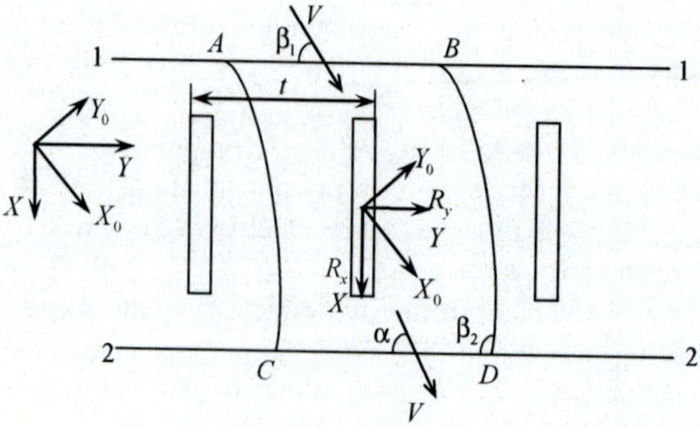

Fig. 5 Section of the lattice structure

Under consideration is a provision for perpendicularity of direction for the speed-vector of flow $\overline{v}_{\infty}$, that approaches the structure, and direction of the resultant force $\overline{R}$ in the sectional plane, as well as a condition for opposite directivity between the vector $\overline{R}$ and the circulation of the speedvector.

For experimentally obtained forces F_x, F_y the experiment finds a resultant force $R = \sqrt{F_x^2 + F_y^2}$ in plane XOY and its direction $tg\varphi = \frac{|F_x|}{|F_y|}$, angle φ. Then, disagreement Δ is found for angles of incoming and outcoming streams, which is necesaary for checking of theoreme conditions.

Figures 6 and 7 show examples of circuits for direction of forces in the structure section for 0, 135° angles of the wind attack.

Table 1 contains forces F_x, F_y for various angles of the wind attack β to the lattice, gives relation between axial and frontal components of the force: $tg\varphi$, angle φ, error Δ (in degrees) or deviation from the direction perpendicular to the stream speeddirection in the plane of the section.

Correlation between the resultant forces obtained from experiment and from conditions of the Zhucowski's theorem demonstrates that, with the measurement error and experiment performance without regard for the space component, they agree quite well, that confirms the Zhucowski's theorem can assess internal forces of the open lattice structure. The direction of the resultant force vector in plane $\overline{R}$ is perpendicular to the direction of the speedvector v_{∞} of the approach flow and is turned 90° opposite to the circulation direction. The direction of the resultan force

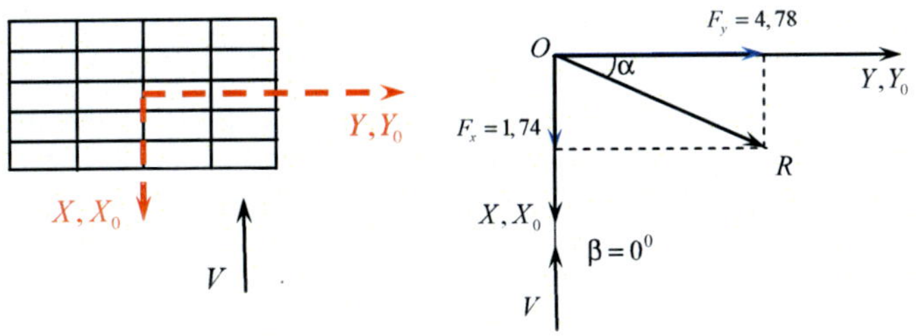

Fig. 6 Forces direction in the section for $\beta = 0°$ angle of attack

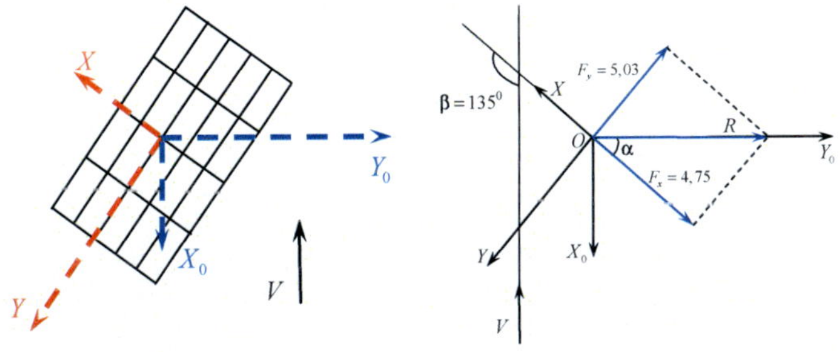

Fig. 7 Forces direction in the section for $\beta = 135°$ angle of attack

Table 1 Experimental data: angle of attack β (degrees), forces F_x, F_y(H), forces relation (tgφ), error Δ (degrees)

β	F_x	F_y	tgφ	φ	Δ
0	1.74	4.78	0.36	20	20
45	−1.88	4.25	0.44	23.5	21.5
90	−4.21	−0,74	0.17	10	10
135	−4.75	−5.03	1	45	0
180	−2.92	−5.58	1.93	63	27
225	2.79	−4.34	0.64	32	13
270	4.41	0.317	0.07	0	0
315	5.09	4.71	1.08	47	2

Table 2 Distribution of forces F_x, F_y, F_z (H), resultant of forces R in plane XOY, resultant of forces Rz (H) in space, tangent of the tilt angle $tg\gamma$ and angle γ (degree)

β	F_x	F_y	R	F_z	Rz	$tg\gamma$	γ
0	1,74	4,78	5.08	2,03	5,4765	0,39	21°
45	−1,876	4,250	4.64	2,65	5.34	0,57	29°
90	−4,210	−0,74	5.08	0,44	5.67	0,09	5°
135	−4,7564	−5,039	6.92	1,62	7.15	0,23	12°
180°	−2,927	−5,585	6.29	0,92	6.35	0,14	7°
225°	2,788	−4,342	5.15	0,27	5.15	0,05	2°
270°	4,419	0,317	4.43	0,12	4.43	0,02	1°
315	5,097	4,719	6.94	1,81	7.18	0,26	14

vector $\overline{R}$ can be obtained by turning vector $\overline{R}$ from the speeddirection, 90° opposite to the circulation direction.

Consider effect the space component F_z has on direction and value of the resultant force $\overline{R}_z$ On vectors $\overline{R}$ (resultant force in plane) and vector $\overline{F}_z$ a resultant force $\overline{R}_z$ will be plotted. This requires plotting of the space resultant of force vectors F_x, F_y and F_z. An angle between the plane XOY and the resultant vector R_z will be defined as γ. The values of resultant forces F_x, F_y and F_z and resultant force R_z are given in the Table 2.

The largest resultant force in the plane of the ground floor occurs at 135 and 315° angles of attack of the approach stream and makes the force of 7.0 N, slightly different of the space resultant that makes 7.2 N. The effect of the space force component F_z on the resultant force R_z is slight in the forces measurement point, which is why this component may be neglected with the error not greater than the error of the experiment performance.

5 Results

Aerodynamic experimental research of the open lattice structure rectangular in plan for various angles of wind attack were performed and obtained internal forces in the bottom of the structure.

The procedure of the experiment, finding the area of the stream stationary action to the lattice structure agrees with the solution of the aeroelasticity problem under the Zhucowski's theorem.

In the area of strain gages installation, the direction of the force resultant vector in the bottom plane $\overline{R}$ is perpendicular to the direction of the speedvector v_∞ of the approach flow and is turned 90° opposite to the circultion direction. The direction of the resultant force vector $\overline{R}$ can be obtained from turning vector $\overline{R}$ from the veloicity direction, 90° opposite to the circulation direction. The effect of the space force component F_z on the resultant force R_z is slight in the gauges installation area, which is why this component may be neglected with the error not greater than the error of the experiment performance.

Correlation between the resultant forces obtained from experiment and from conditions of the Zhucowski's theorem demonstrates that, with the measurement error and experiment performance, they agree quite well.

Applicability of the Zhucowski's theorem for assessment of the internal forces of the open lattice structure proves the applicability of the simplified model of the aeroelasticity problem for assessment of the experimental procedure for the exposure of the lattice structure to the wind.

6 Discussion

Under consideration is open lattice structure with axes of symmetry. Simplified procedure of experimental data analysis employs a plane lattice of prismatic bodies, similar to the system of plane profiles exposed to air flow. Such idealized construction presented as a regular lattice of plane profiles allows checking of the Zhucowski's theorem conditions.

Values of forces obtained from strain gage sensors are compared with the solution of the aeroelasticity problem under the Zhucowski's theorem, that affords the use of the mathematical model of the stream exposure to the lattice structure, that agrees with experimental data. The accepted mathematical model that agrees with the experimental data has certain assumptions.

Experimentally obtained forces and their directions agree with conditions of the aeroelasticity theorem, with an error not greater than that of the experiment. Certain angles of the wind attack to the structure practically agree with the data of the theorem conditions. Correlation between experimental and theoretical data substantiate the procedure for creating the conditions of the wind stationary exposure to the structure and finding the areas where these conditions are violated.

7 Conclusions

Experimental research of the exposure of the lattice structure to the wind requires the development of the procedure for the experiment and analysis of the data obtained, that can be justified by simplified mathematical models. This is how checking of the Zhucowski's theorem conditions for regular lattice of prismatic bodies helps in finding the area of stationary action of the wind to the lattice structure. Simplified mathematical model based on the experimental data is treated as a test model. It affords both improvement of the experimental procedure and correction of experimental conditions through close inspection look at the lattice structure exposed to the wind.

References

1. Krasnov NF (2010) Aerodynamics. Part 1, Moscow, pp 16–27
2. Simiu E, Skanlan R (2012) Wind Exposure to Buildings and Structures (actuated version. Stroyizdat, Moscow, pp 45–47
3. Magai AA (2015) Architectural and composition features of high-rise buildings. Acad J UralNIIproekt RAASN
4. Hill PG, Peterson C (2018) Mechanics and Thermodynamics of Propulsion. Prentice Hall (formerly by Addison-Wesley Publishing), Hoboken, pp 34–38
5. Bowen AJ (2019) The prediction of mean wind speeds above simple 2D hill shape. J Wind Eng Ind Aerodyn 15:220–259
6. Kent CW, Grimmond CSB, Gatey D, Barlow JF (2018) Assessing methods to extrapolate the vertical wind-speed profile from surface observations in a city centre during strong winds. J Wind Eng Ind Aerodyn 173:100–111
7. Gnyrya AI, Korobkov SV, et al (2011) Comprehensive experimental studies of the aerodynamics and the heat exchange of model buildings and facilities. Bull TGASU 4
8. Zhemochkin BN (2015) Calculation of frames (actuated versoin). Izdatelstvo MOSKVA, Moscow, pp 402–407
9. Egorychev OO, Churin PS, Poddaeva OI (2014) Experimental study of force and moments wind loads on high-rise buildings. Ind Civil Eng 9:28–30
10. Egorychev OO, Guvernyuk SV, Isaev SA, Poddaeva OI, Kornev NV, Usachov AE (2011) Numerical and physical modeling of wind effects on a group of high-rise buildings. Vestnik MGSU 3(1):185–191
11. Poddaeva OI, Kubenin AS, Churin PS (2015) Aerodynamics in architecture and construction. NIU MGSU, Moscow
12. Poddaeva OI, Churin PS, Pomelov VJ (2016) The research of wind loads on buildings and structures with increased level of responsibility, pp. 550–555
13. Frishter L, Lukin M (2020) Experimental study of wind impact on multilevel industrial scaffolding. In: E3S web conference volume, vol 157. https://doi.org/10.1051/e3sconf/202015706009
14. Babich YM, Rewinkel J-P, Filipchyk SV (2014) An experimental determination of the wind load acting on the roof structure of the magdeburg stadium. Bull Belorussian Russ Univ 1(42):126–135
15. Pomelov VY, Gribach DS, Churin PS (2016) Methods for making mock-ups of high-rise and uniqie buildings for aerodynamic testing 3(65):56–60

16. Requirements SP 20.13330.2016 (2010) Load and Effects. Updated edition of SNiP 2.01.07–85 Minregion of Russia: December 27, 2010, N787, valid since May 20, 2011 Minregion Rossii Publication, Moscow
17. American Society of Civil Engineers (2011) Minimum Design Loads for Buildings and other Structures. ANSI/ASCE 7-11. ASCE, New York, pp 9–21
18. Poddaeva OI, Churin PS, Fedosova AN (2018) Analysis of domestic and foreign regulatory and scientific and technical documents in the field of wind influence on buildings and structures that are part of hazardous production facilities. In: IOP conference series: materials science and engineering, MGSU, Moscow, vol 365(5)052025, pp 9–11
19. Sternin LE (2016) Basics of gas dynamics. Vuzovskaya kniga, Moscow
20. Samsonov VT (2019) Aerodynamics of industrial buildings: aerodynamic calculation during design of ventilation exhausts. INFRA-M, Moscow

Numerical Simulation of Accelerations of the Upper Floors of a High-Rise Building Under Wind Influence

Sergey Saiyan, Vladimir Andreev, and Alexander Paushkin

Abstract A new two-component method for calculating the dynamic response of high-rise buildings and structures to the action of wind is proposed, combining two numerical methods for solving two different physical problems. Computational fluid dynamics (CFD) tools are used to simulate wind flows and impacts on buildings and structures, after which finite element analysis (FEA) is carried out on the basis of superelement technology to determine the dynamic response of calculated objects to wind impacts. With this approach, it becomes possible to take into account the influence of real development, terrain, vortex excitation, the effect of disruption of vortices from neighboring buildings and structures, as well as the pulsation component of wind impact, which entails the need to take into account the inertial component of a high-rise building. To determine the characteristics of the impact of wind load on construction objects, physical modeling in wind tunnels is often used, however, it is known that this approach does not meet the full criteria of aerodynamic similarity, and to determine the dynamic response under wind influence, it is not able to fully meet the criteria of mechanical similarity. The methodology based on CFD and FEA means is devoid of such disadvantages. The proposed two-component numerical analysis was tested on the example of a high-rise building.

Keywords Numerical Simulation · Accelerations of the upper floors · High-rise building under wind influence · High-rise building vibrations

S. Saiyan (✉) · V. Andreev · A. Paushkin
Moscow State University of Civil Engineering, Moscow 129337, Russian Federation
e-mail: Berformert@gmail.com

V. Andreev
e-mail: asv@mgsu.ru

A. Paushkin
e-mail: alexglebov@bk.ru

P. Akimov et al. (eds.), *Proceedings of FORM 2022*, Lecture Notes in Civil Engineering 282, https://doi.org/10.1007/978-3-031-10853-2_25

1 Introduction

With an increase in the height of buildings and structures, wind exposure makes a more significant contribution to the stress–strain state of load-bearing structures. Dynamic loading caused by the action of the pulsation component of wind action entails the need to take into account the inertial component of a high-rise building. It affects both the stress–strain state of structures and the comfort of people staying in the rooms of the upper floors [1–6]. The main value characterizing the inertial effect of wind pulsations on buildings and structures is acceleration.

In this regard, it should be noted that the theoretical and experimental foundations of modern ideas about wind loads and their interaction with buildings and structures in Russian regulatory documents were developed in the late 50 s - early 60 s and have not changed fundamentally since then [7]. However, the recommendations on the precise dynamic analysis of N.A. Popov [8] does not exhaust the answers to questions about determining the effect of aerodynamic forces on structures, which are studied on the basis of solutions to problems of theoretical aerodynamics [9], numerical modeling [10, 11] and experimental research data [12], as well as determining the reaction of buildings and structures to wind effects, usually associated with problems of dynamics and statics.

The approaches adopted during the implementation in [7, 8] cannot be applied:

- when determining the reaction of buildings and structures caused by their resonant vortex excitation;
- to determine the reaction of coatings in the vertical direction;
- in all cases when the main contribution to the dynamic response of buildings and structures is made by the pulsation component in areas with negative average wind pressure (in "suction" zones);
- when determining torsional and in directions perpendicular to the wind flow of vibrations.

In addition to the approximation of the approaches used for the dynamic response to the pulsation of wind load used in [7, 8], the options for the location of a high-rise building and the interference of buildings are not considered. Additional wind impacts are not taken into account when the flow is disrupted from nearby buildings, as well as the terrain. The aerodynamic loads themselves, determined by the Building Codes, require clarification. Engineering approaches define the aerodynamic coefficient on the windward surface of a detached high-rise building as a constant value in height, independent of the height of the building. The velocity spectrum proposed by Davenport A.G. [13, 14] describes well the loading of only the windward side of the building. The existing approaches were suitable for buildings of small and medium height and sparse buildings. For high-rise buildings and structures (and, in particular, for high-rise residential complexes located in conditions of dense development), more accurate (in relation to engineering regulatory and analytical approaches) methods for determining wind impact are needed.

According to [7], the normative value of the main wind load w acting on the building in question is defined as the sum of the average w_m and pulsation w_g components:

$$w = w_m + w_g. \tag{1}$$

The normative value of the average component w_m of the wind load is determined by the formula:

$$w_m = w_0 k(z_e) c, \tag{2}$$

where: w_0 – standard value of wind pressure, z_e (m) – equivalent height, $k(z_e)$ – coefficient that takes into account the change in the average component of wind pressure at the height of z_e for the corresponding type of terrain, c – aerodynamic coefficient.

The normative value of the pulsation component of the main wind load w_g at an equivalent height z_e should be determined as follows:

a) for structures (and their structural elements) whose first natural oscillation frequency (Hz) is greater than the limit value of the natural frequency f_{lim},, determined by the formula:

$$w_g = w_m \zeta(z_e) v, \tag{3}$$

where w_m – average wind load, $\zeta(z_e)$ – coefficient of wind pressure pulsation for the equivalent height z_e, v – coefficient of spatial correlation of wind pressure pulsations.

b) for all buildings and structures that have ratios for the first two natural frequencies of vibrations $f_1 < f_{lim} < f_2$, the pulsation load should be determined by the formula:

$$w_g = w_m \varsigma \zeta(z_e) v, \tag{4}$$

The dynamic reaction is defined as for a one-mass system, by the methods of decomposition according to natural frequencies and forms of oscillations (linear spectral theory). The scheme for determining the dynamic reactions of the system to the action of wind pulsations is shown in Fig. 1.

Pulsating wind action makes a significant contribution to the stress–strain state with an increase in the height of buildings and structures. The dynamic contribution to the stress–strain state becomes so significant that its intensity can exceed even seismic impacts. Figure 2 shows the ratio of the relative spectral power density and frequencies (periods) of oscillations to the approximate frequency intervals of natural oscillations of some types of building structures. It can be seen from the figure that, although the seismic effect is more "powerful" than the wind effect, however, it is realized at higher oscillation frequencies, which covers the interval of the most

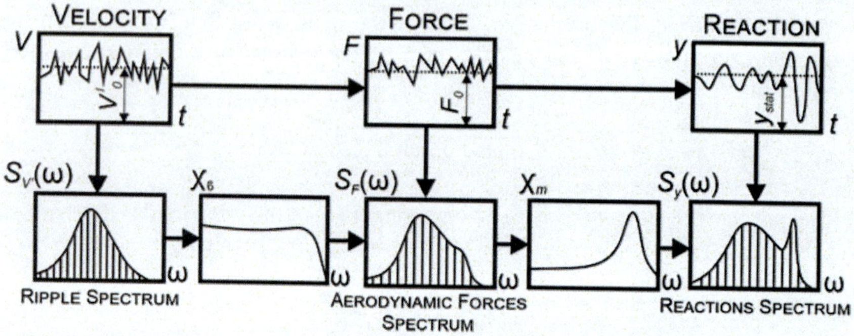

Fig. 1 Scheme for determining the dynamic reactions of the system to the action of wind gusts [15]

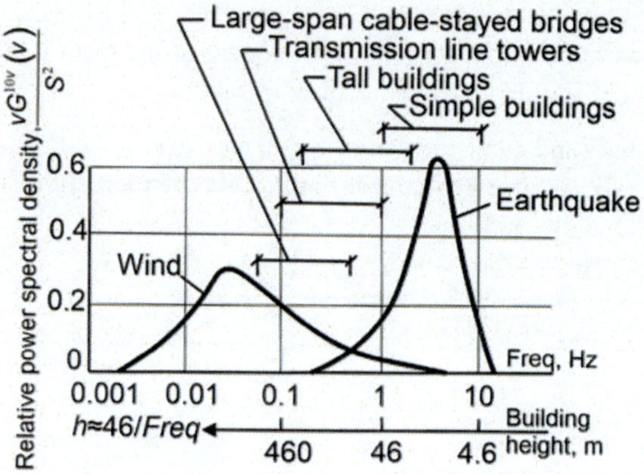

Fig. 2 Comparison of spectral power densities of wind speed and ground acceleration with natural vibration frequencies of the main types of structures [16, 17]

energy-intensive spectrum at frequencies close to the natural oscillation frequencies for normal-height buildings.

In Fig. 2, the values are presented in dimensionless form; the wind spectrum is calculated for $\mu_{V10} = 20$ m/s and $S_{V10}^2 = = 2400$ m/s^2 (where γ is a coefficient that takes into account the roughness of the ground surface, and is usually taken in the range from 0.005 to 0.015); the earthquake spectrum corresponds to the values $\mu_a = 0$ and $S_{V10}^2 = 0.38$ m^2/s^4.

Considering the physical essence of the wind, it is impossible not to consider its spectral characteristics in more detail. The description of the random wind speed using energy spectral functions characterizes the changes in wind speeds over time and in height, which are the distribution of the energy of the turbulent flow over frequencies. The energy spectrum of longitudinal pulsations of wind speeds can be described as consisting of four conditional base ranges of variation:

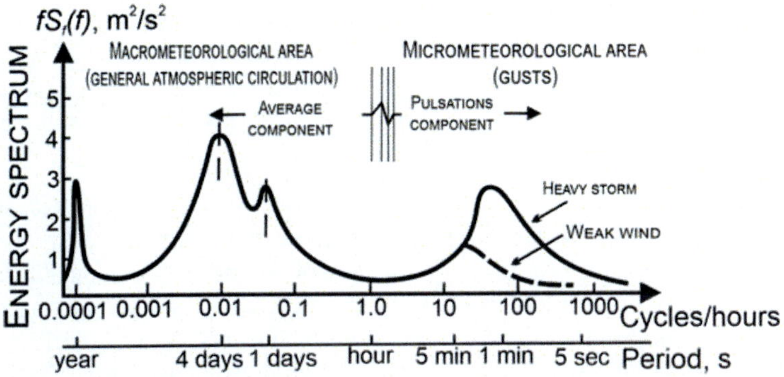

Fig. 3 Van der Hoeven energy spectrum [18]

1. The range of ultra-low frequencies has the largest vortices, comparable in size to the characteristic scale of the flow;
2. The low frequency region contains the main turbulent energy.
3. of the flow, this interval accounts for its maximum;
4. The inertial interval, the connection with large-scale vortices is lost, the spectrum is characterized only by the parameters of turbulent motion;
5. Viscous interval, the region of the highest frequencies in which the main scattering of turbulent energy occurs.

Figure 3 shows the frequency distribution of the power of the wind flow obtained by Van der Hoeven [18]. On the left, the spectrum has three peaks corresponding to three insolation cycles (annual, meteorological (4 days) and daily). The fourth peak, called micrometeorological in contrast to the first three macrometeorological peaks, is explained by turbulence arising in the air flow when it rubs against the earth's surface [15].

Between the third and fourth maxima, the spectrum has a large minimum lasting from several minutes to several hours.

The minimum available in the spectrum makes it possible to represent the wind load on the structure as the sum of two values. The first of them (the first three peaks on the left) corresponds to the steady wind speed and acts statically (the average component of the wind load). The second (the right peak, which has a larger (in a strong storm) or smaller (in a weak wind) value) determines the pulsation part of the longitudinal component of the velocity and acts dynamically (the pulsation component of the wind load) [19].

In the Russian Building Codes [7] it is also said that in all cases not provided for by internal rules, which include: other forms of structures, taking into account, with proper justification, other directions of wind flow or components of the total resistance of the body in other directions, the need to take into account the influence of nearby buildings and structures, terrain and similar cases, aerodynamic coefficients are established in the recommendations developed within the framework of scientific and technical support based on the results:

1. Physical (experimental) simulation – testing in wind tunnels;
2. mathematical (numerical) modeling of wind aerodynamics based on numerical schemes for solving three-dimensional equations of fluid and gas motion with adequate turbulence models implemented in modern verified licensed software complexes of computational fluid dynamics [20].

Physical (experimental) modeling in wind tunnels, with all its certain advantages, has a number of problems that cause difficulties in implementation, in processing and in understanding the processes of formation of the aerodynamic situation in the conditions of development. The main issue of discussion among specialists in the field of structural aerodynamics is the problem of applying experimental modeling under existing constraints when meeting the criteria of aerodynamic similarity. Taking into account the impossibility of implementing complete aerodynamic similarity, in order to avoid situations in which it will be impossible to conduct reliable experimental aerodynamic studies, they resort to weaker requirements for self-similarity of the flow, i.e. the independence of the values of aerodynamic coefficients from the size of the body and flow velocities (Reynolds numbers) is justified, which is not realized for all cases of body flow.

In addition, in the existing practice, during experimental modeling, special meteorological-type wind tunnels with a long working part are usually used, in which the flow structure corresponds to the so-called wall turbulence and is formed similarly to nature. For example, modeling the entire Moscow-City complex requires a wind tunnel of many large sizes to account for the interference of buildings. But for buildings with a height of 300–600 m, an experiment on models of 1:600, and, moreover, 1:1000 (which corresponds to a height of 20–60 cm) is unlikely to correctly assess the impact on real structures in close construction. The stage of estimating wind speeds in pedestrian zones is even more ambiguous, since the size of typical sensors is comparable to the height of a "pedestrian". With all its undoubted advantages, tests have an undoubted problem – the complexity and complexity of organization, processing and analysis. We note the problems associated with the construction of envelopes and the selection of the most unfavorable wind directions from the point of view of strength [10]. As for determining the dynamic response of experimental models under wind influences, the problem is compounded by the difficulty of observing the mechanical similarity of the object in order to adequately describe the stiffness and mass characteristics.

The problems of numerical simulation of wind flows are errors in mesh sampling and hypotheses introduced into mathematical models, as well as the inability, at the moment, to model three-dimensional unsteady flows due to the limited computational resources of such a task. Modern approaches to numerical modeling still need refinement, testing and verification of computational methods that could be used by researchers in scientific and project activities.

As for the problems associated with determining the dynamic response of high-rise buildings and structures, the main problem of regulatory approaches of various countries is not taking into account transverse vibrations in the direction perpendicular to the flow direction, as well as the absence of torsional vibrations, which have the most negative impact on human well-being (comfort of the upper floors). The main approach to determining the dynamic reaction consists in the linear-spectral theory corresponding to the solution of the linear dynamic problem of forced vibrations of structures under the action of pulsating wind loads by the decomposition of forces and displacements in rows according to orthonormal proper forms of the structure.

Based on the above, the problem associated with the need to create an alternative and modern calculation methodology for the study of the dynamic response of high-rise buildings and structures under wind influence, using advanced numerical modeling methods, was highlighted. The proposed method consists in a comprehensive study of the problem from the perspective of solving two different physical problems: modeling of unsteady wind flows and impacts on high-rise buildings and structures, as well as a refined definition of the mechanical response under dynamic loading under wind action.

2 Materials and Methods

We will describe successively the basic theory of numerical modeling methods for solving two problems within the framework of the developed methodology, as well as a specific formulation of the problem for testing.

Modeling of wind flows and impacts on high-rise buildings and structures is reduced to the numerical solution of three-dimensional non-stationary nonlinear equations of hydro-gas dynamics based on the Navier–Stokes equations describing the motion of a liquid or gas (3 equations for the components of velocity vectors), implemented in the ANSYS CFX software package [21]:

$$\begin{aligned}
&\rho\frac{\partial u}{\partial t}+\rho u\frac{\partial u}{\partial x}+\rho v\frac{\partial u}{\partial y}+\rho w\frac{\partial u}{\partial z}=-\frac{\partial p}{\partial x}+\mu\left[\frac{\partial^2 u}{\partial x^2}+\frac{\partial^2 u}{\partial y^2}+\frac{\partial^2 u}{\partial z^2}\right];\\
&\rho\frac{\partial v}{\partial t}+\rho u\frac{\partial v}{\partial x}+\rho v\frac{\partial v}{\partial y}+\rho w\frac{\partial v}{\partial z}=-\frac{\partial p}{\partial y}+\mu\left[\frac{\partial^2 v}{\partial x^2}+\frac{\partial^2 v}{\partial y^2}+\frac{\partial^2 v}{\partial z^2}\right];\\
&\rho\frac{\partial w}{\partial t}+\rho u\frac{\partial w}{\partial x}+\rho v\frac{\partial w}{\partial y}+\rho w\frac{\partial w}{\partial z}=-\frac{\partial p}{\partial z}+\mu\left[\frac{\partial^2 w}{\partial x^2}+\frac{\partial^2 w}{\partial y^2}+\frac{\partial^2 w}{\partial z^2}\right].
\end{aligned}\tag{5}$$

In addition, the equations of continuity of the flow (conservation of mass) and the state must be satisfied:

$$\frac{\partial\rho}{\partial t}+\frac{\partial(\rho u)}{\partial x}+\frac{\partial(\rho v)}{\partial y}+\frac{\partial(\rho w)}{\partial z}=0,\tag{6}$$

$$\rho=const,\tag{7}$$

where: u, v, w are the desired components of the velocity vector (along the x, y, z axes), p is pressure, t is physical time, μ is the dynamic viscosity coefficient for air, ρ is density.

The direct solution of Eqs. (5)–(7) taking into account vortices of all scales (DNS, Direct Numerical Simulation) with modern computing power is practically feasible only for very low flow rates and only for research tasks.

Existing approaches to turbulence modeling use the following approaches to turbulence modeling [10, 20, 22–25]:

- Large Eddy Simulation (LES): This approach is the second most labor-intensive of the existing ones after DNS. The idea of the approach is to "filter" the characteristics of the turbulent flow from short-wave turbulence, which is described by subgrid-scale models for vortices of the order of the filter, and for vortex turbulence exceeding the filter size, are solved exactly;
- Unsteady Reynolds Averaged Navier–Stokes – (RANS / URANS): the approach is based on the decomposition of velocity into time-averaged and pulsation components ($u_i(t) = \overline{u}_i + u_i'(t)$). As a result of the corresponding transformations of the Navier–Stokes equations, additional unknowns (shear "Reynolds" stresses) appear;
- Detached Eddy Simulation – (DES): This approach is a combination of LES and RANS. It consists in the fact that disconnected energy-carrying vortices in the separation zones are modeled by the LES approach, and the areas of the attached boundary layers are modeled by the RANS approach.

To solve practical knowledge-intensive engineering problems in modern calculation practice, the use of the RANS/URANS approach prevails, which has sufficient accuracy and simplicity for use.

Numerical simulation of the dynamic response of high-rise buildings and structures under wind influence is based on the finite element method (FEM) in the matrix form of displacements implemented in the ANSYS Mechanical APDL software package. To solve the basic system of equations of motion, global stiffness matrices [K], damping matrices [C], mass matrices [M], as well as time-dependent external nodal load vectors $\{F(t)$ are formed$\}$. The vector of nodal displacements $\{u(t)\}$ is taken as the desired functions.

The matrix equation of motion looks like this:

$$[M]\{\ddot{u}(t)\} + [C]\{\dot{u}(t)\} + [K]\{u(t)\} = \{F(t)\}. \tag{8}$$

Since direct integration of the equations of motion is computationally very expensive for large-scale systems, the technology of superelement modeling was used. Superelement modeling is a "convolution" of a large-scale system into a static and dynamic equivalent system (superelement), reflecting the most important integral parameters of the system, for example, frequencies and waveforms (basically, they show a good match only for the lowest, most energy-intensive oscillation frequencies), stiffness, mass, etc.

3 Results

To study the dynamic effect of wind load on buildings and structures, a structural scheme was developed in the form of a monolithic reinforced concrete frame measuring 42 × 42 × 201 m (Fig. 4), located in the first wind district.

The aerodynamic studies carried out in a non-stationary formulation made it possible to determine the total integral loads on the floors of a high-rise building, which were applied to an equivalent rod finite element system obtained from a full-size design scheme (Fig. 5).

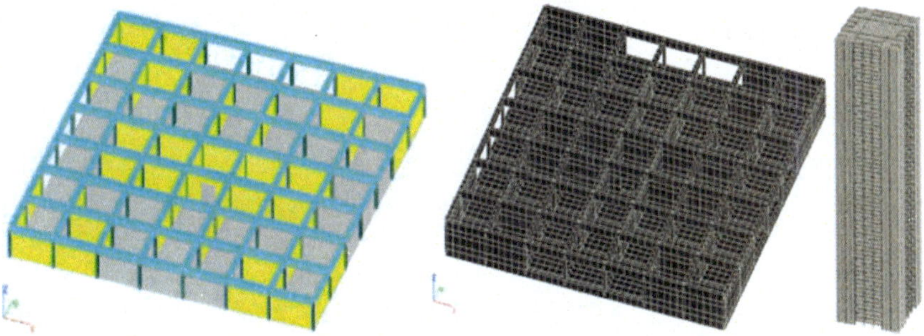

Fig. 4 Model of a typical floor (left) and finite element model of a building (right)

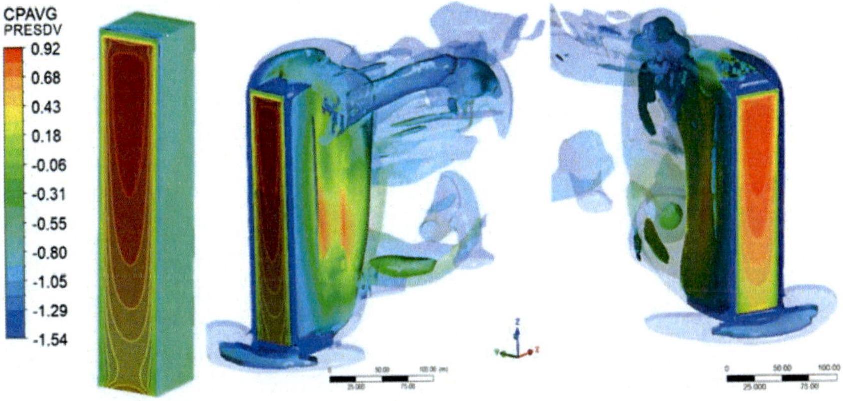

Fig. 5 Aerodynamic coefficients (left) and three-dimensional vorticity behind a high-rise building (right)

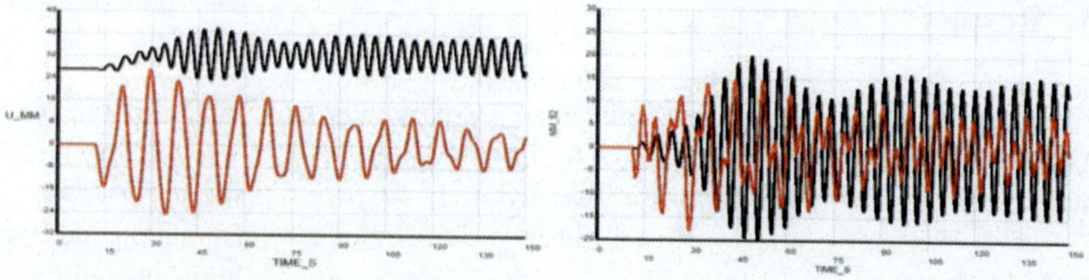

Fig. 6 Change in total displacements (*X* and *Y*) (left) and accelerations (*X* and *Y*) (right) for the upper floor of a high-rise building

The aerodynamic forces generated in this way were used as an external dynamic loading for strength finite element analysis of a high-rise building. The calculation results were the displacement and acceleration of the upper floor of the building (Fig. 6).

4 Conclusion

During the conducted research, it was done:

1. The existing approaches and problems to determining the dynamic response of high-rise buildings and structures under wind influence are described, which include: engineering normative and analytical methods, physical (experimental) modeling, as well as numerical modeling techniques.
2. The main spectral characteristics of wind flows that are necessary for the correct analysis of the pulsation component of wind flows are described.
3. A numerical modeling technique is proposed to determine the dynamic response of high-rise buildings and structures under wind influence, based on the combination of two numerical methods to solve two different physical problems: CFD (computational fluid dynamics for modeling wind flows and impacts on buildings and structures) + FEA (finite element analysis to determine the dynamic response of buildings and structures to wind impact). The basic mathematical relations and theory for the proposed methodology are described.
4. The methodology was tested on the example of a high-rise building. The total wind loads (average and pulsation) were obtained, which were applied to a superelement equivalent rod finite element system obtained from a full-size design scheme. Vibrations of a high-rise building (total displacements and accelerations) in the directions are determined: along the wind flow, as well as across the wind flow.

Acknowledgements This work was financially supported by the Ministry of Science and Higher Education of the Russian Federation (grant # 075-15-2021-686). All tests were carried out using research equipment of The Head Regional Shared Research Facilities of the National Research Moscow State University of Civil Engineering.

References

1. Kwok K, Burton M, Abdelrazaq A (2015) Wind-induced motion of tall buildings: designing for habitability. American Society of Civil Engineers, Reston, p 77
2. Kwon DK, Kareem A (2020) Hybrid simulation of a tall building with a double-decker tuned sloshing damper system under wind loads. Struct Design Tall Spec Build 29:1790. https://doi.org/10.1002/tal.1790
3. Ilyuhina EA, Lahman SI, Miller AB, Travush VI (2019) Konstruktivnye resheniya vysotnogo zdaniya «Lahta centr» v Sankt-Peterburge. Academia Arhitektura i stroitel'stvo (3). https://doi.org/10.22337/2077-9038-2019-3-110-121
4. Fei D, Kareem A (2020) Tall buildings with dynamic facade under winds. Engineering 6(12):1443–1453. https://doi.org/10.1016/j.eng.2020.07.020
5. Howarth H (2015) Human exposure to wind-induced motion in tall buildings: and assessment of guidance in ISO 6897 and ISO 10137. In: Conference: 50th United Kingdom conference on human responses to vibration
6. Landel P, Johansson M, Linderholt A (2021) Comparative study of wind-induced accelerations in tall timber buildings according to four methods. In: Conference: WCTE 2021 - world conference on timber engineering, Santiago, Chile
7. SP 20.13330 (2016) Nagruzki i vozdejstviya
8. Kucherenko VA (2000) Rekomendacii po utochnennomu dinamicheskomu raschetu zdanij i sooruzhenij na dejstvie pul'sacionnoj sostavlyayushchej vetrovoj nagruzki. Gosstroj Rossii, sost. Popov N.A. M.: GUP CNIISK, p 45
9. Landau LD, Lifshic EM (1986) Teoreticheskaya fizika: Uchebnoe posobie. V 10 t. T. VI. Gidrodinamika. Fizmatlit, Nauka, p 736
10. Dubinskij SI (2010) Chislennoe modelirovanie vetrovyh vozdejstvij na vysotnye zdaniya. Ph.D. thesis, MGSU, p 199
11. Belostockij AM, Akimov PA, Afanas'eva IN (2017) Vychislitel'naya aerodinamika v zadachah stroitel'stva. ASV, Izdatel'stvo, p 720
12. Kazakevich MI (2019) Osnovy raschetov sooruzhenij na vetrovye vozdejstviya. Monografiya. MGSU, p 180
13. Davenport AG (1961) The application of statistical concepts to the wind loading of structures. Proc Civil Eng 19:449–472
14. Davenport AG (1967) Gust loading factors. Proc Am Soc Civil Eng 93(1):No. ST3
15. Gordeev VN, Lantuh-Lyashchenko AI, Pavshinskij VA, Perel'muter AV, Pichugin SF (2009) Nagruzki i vozdejstviya na zdaniya i sooruzheniya. Izdatel'stvo S, p 528
16. Augusti G, Baratta A, Kashiati F (1984) Probabilistic method in structural engineering. Chapman and Hall, London, p 584. ISBN 5-274-00212-9
17. Ferry-Borges J, Castanheta M (1971) Structural safety. In: LNEC, Lisbon, p 326
18. Van der Hoven J (1968) Load fluctuations in turbulent flow. J Eng Mech Div Proc ASCE
19. Ostroumov BV (1985) Razrabotka, issledovanie i vnedrenie novyh konstruktivnyh form vysotnyh sooruzhenij na osnove eksperimental'no-teoreticheskih issledovanij ih vzaimodejstviya s vetrovym potokom. Ph.D. thesis, CNIIPSK im. Mel'nikova
20. Metodicheskoe posobie (2020) «Matematicheskoe (chislennoe) modelirovanie vetrovyh nagruzok i vozdejstvij» (k SP 20.13330.2016), M., FAU FCS Minstroya Rossii. (razrabotano ZAO NIC StaDiO)
21. ANSYS CFX-Solver Theory Guide (2021) Release 2021R2 ANSYS Inc., Canonsburg
22. Godunov SK (1976) Chislennoe reshenie mnogomernyh zadach gazovoj dinamiki. Nauka, p 400
23. Rouch P (1980) Vychislitel'naya gidrodinamika. Mir, p 618
24. Zheng X, Montazeri H, Blocken B (2020) CFD simulations of wind flow and mean surface pressure for buildings with balconies: comparison of RANS and LES. Build Environ 173:106747
25. Rezaeiha A, Montazeri H, Blocken B (2019) On the accuracy of turbulence models for CFD simulations of vertical axis wind turbines. Energy 180:838–857

Assessment of Thermal Aging of Bitumen-Mineral Materials in Various Ways

Muhammet Salihov and Evgeniy Veyukov

Abstract Bitumen-mineral materials prepared from mixtures of mineral materials with petroleum bitumen undergo aging during preparation, installation in engineering structures and during operation. For the accelerated study of the aging processes of bitumen in the composition of mineral materials, methods for warming up the samples of mixtures at high temperatures, methods of preheating and further freezing–thawing of samples, and others were obtained. The aging of bitumen is also judged by the change after heating in the furnace at + 163 °C for 5 h (RTFOT method). The main disadvantage of the latter method is that it does not take into account the influence of the interaction of the bitumen film with the surface of mineral components on the aging processes of bitumen-mineral materials. Since the thickness, structure and, accordingly, the stability of bitumen films on the surface of mineral components depend on their type and size, the aging processes of bitumen-mineral materials as a whole occur in different ways and this must be taken into account. The **purpose** of the work is to evaluate the thermal aging of bitumen-mineral materials by various methods. **Method.** The authors have proposed a special technique that allows us to study the aging processes of bitumen-mineral materials of both single-component and multicomponent structures at any high temperatures. The degree of aging of bitumen-mineral materials was carried out by establishing the values of the compressive strength of the materials under study, which were previously maintained respectively at +150 °C for 0, 1, 3, 5 and 7 h. **Results.** The analysis of the aging processes of bitumen-mineral materials was carried out by comparing the values of the aging coefficient at each stage of heating. The establishment of the values of the aging coefficient according to the new method makes it possible to accelerate the evaluation of comparative resistance against aging. Mathematical models have been developed that reflect the aging processes of bitumen-mineral materials depending on the preheating time at high temperatures. Models can be used to evaluate aging processes.

M. Salihov · E. Veyukov (✉)
Volga State University of Technology, Lenin Sq., 3, Yoshkar-Ola 424000, Russia
e-mail: VeukovEV@volgatech.net

M. Salihov
e-mail: SalihovMG@volgatech.net

P. Akimov et al. (eds.), *Proceedings of FORM 2022*, Lecture Notes in Civil Engineering 282, https://doi.org/10.1007/978-3-031-10853-2_26

Keywords Asphalt mixtures · Durability · Elevated temperature · AR grading · Paving asphalt

1 Introduction

As is known, the aging of bitumen in the composition of bitumen-mineral materials is the result of an irrevocable decrease in the value of properties under the influence of elevated and lowered temperatures, the action of aggressive media – water, solutions of salts and acids, radiation, biological media, mechanical forces, as well as chemical transformations as a result of evaporation of light fractions and spontaneous transformations in the astringent itself [1–8]. The study of the aging process of bitumen is an important condition for forecasting and evaluating the longevity and regulation of the operability of bitumen-mineral materials in expensive structures. The dynamics of the aging of BMM over time is studied by various methods [9–16]. One of the main methods is to study the changes in various physical and mechanical properties of BMM after holding their samples at high temperatures (preheating) [17, 18]. The main parameter in this case is the aging coefficient, calculated by dividing the values of the physico-mechanical property of BMM after preheating to the value of preheated samples for the same time intervals [19].

The values of the aging coefficient of complex BMMs can be obtained by calculations directly from the results of sample tests, using mathematical models and in the form of weighted averages from the values of single-component mixtures of their individual mineral components with bitumen. The latter is preliminarily established by special experiments. A unified methodology is used to perform laboratory experiments and calculate the values of the aging coefficient of BMM. It should be emphasized that the study of the aging processes of BMM is an important condition for the reasoned assignment of estimated service life and feasibility study of engineering structures with their use.

2 Materials and Methods

The following compositions were investigated:

1. Hot crushed fine-grained asphalt concrete (type B) of the following composition: granite crushed stone fr. 5…20 mm – 45.0% (by weight); crushed sand fr. 0.16…5 mm (M 1200) – 45.0%; mineral filler MF-1 – 10.0%; bitumen viscous BND 70/100 – 5.8% (over 100%);
2. Mixtures of granite crushed stone fr. 5… 20 mm with viscous bitumen in a ratio of 25:1; crushed sand with bitumen in a ratio of 10:1 and mineral powder with bitumen in a ratio of 5.67:1.

Some characteristics of the studied materials are presented in Table 1, 2, 3, 4 and 5.

Table 1 Some characteristics of bitumen 70/100

Indicator	Standard requirement	Actual value of the indicator
1. Needle penetration depth at temperature +25 °C, 0.1 mm	71–100	78
2. Softening point, °C, not lower	47	48
3. Tensile at °C, cm, not lower	3.7	3.7
4. Fragility temperature, °C, not lower	−18	−19
5. Flash point, °C, not lower	230	308
6. Change in sample mass after aging,%, not more than	0.6	0.3

Table 2 Some characteristics of bitumen 90/130

Indicator	Standard requirement	Actual value of the indicator
1. Needle penetration depth at temperature +25 °C, 0.1 mm	90–130	102
2. Softening point, °C, not lower	43	46
3. Tensile at °C, cm, not lower	4.0	4.3
4. Fragility temperature, °C, not lower	−17	−19
5. Flash point, °C, not lower	230	310
6. Change in sample mass after aging,%, not more than	0.6	0.4

Table 3 Granulometric composition of crushed stone M400 grade fr. 0.16…5 mm

Size of sieves, mm	Sieve fraction values								
	10	5	2.5	1.25	0.63	0.315	0.16	0.071	<0.071
Private balances, %	0	1.55	34.83	19.68	19.31	10.78	6.17	6.80	0.89
Total balances, %	0	1.55	36.38	56.06	75.37	86.15	92.31	99.11	100
Full passing, %	100	98.45	63.62	43.94	24.63	13.85	7.69	0.89	0

Table 4 Granulometric composition of crushed stone M400 grade fr. 5…20 mm

Size of sieves, mm	Sieve fraction values										
	20	15	10	5	2,5	1,25	0,63	0,315	0,16	0,07	<0,07
Private balances, %	0	3.3	51.9	44.99	1.7	0.14	0.05	0.1	0.24	0.83	0.07
Total balances, %	0	3.3	51.9	96.87	98.58	98.71	98.76	98.86	99.1	99.93	100
Full passing, %	100	96.7	48.1	3.13	1.42	1.29	1.24	1.14	0.9	0.07	0

Table 5 Granulometric composition of mineral filler MP-1 grade, fr. 0.05…0.071 mm

Size of sieves, mm	Sieve fraction values					
	1.25	0.63	0.315	0.16	0.07	<0.07
Private balances, %	0	0.05	0.78	21.12	6.16	71.89
Total balances, %	0	0.05	0.83	21.95	28.11	100
Full passing, %	100	99.95	99.17	78.05	71.89	0

The experiments used: standard sets of sieves with round hole cells for crushed stone, sand and mineral powder, electric technical scales, metal trays, bayonets, trowels, owls, steel knife, remote thermometer, cylindrical steel molds with punches with a diameter and height of 71.4 and 50.5 mm, stopwatch, caliper, ventilated electric oven with automatic temperature control within 20…350°, the electric furnace for heating mixtures before preparing standard samples is three-position with a heating temperature limit of 20…200°, hydraulic presses PSU-10, PS-120, IP-125.

Bitumen-mineral mixtures consist of a mixture of petroleum bitumen with mineral materials of crushed granite and limestone, crushed sand, limestone crushing waste and mineral filler of various grain sizes. In the process of preparation, laying in engineering structures and when in service, bitumen-mineral mixtures are subject to the aging process. The following methods were obtained to optimize the study process of the bitumen aging in the composition of mineral materials: samples heating at high temperatures, preliminary sample heating and further thawing and freezing test, studying the changes in the sample properties after prolonged water saturation.

The accuracy of the research results obtained for the aging process of bitumen-mineral materials largely depends on the chosen research methods. To date, there is a standard method for studying bitumen aging, the point of which is to measure the weight loss of bitumen during heating at +163 °C for 5 h according to GOST 33,140–2014 with free air supply. However, this method does not take into account the effect of base material on the condition of the films. In 2018, a new method for studying the aging processes of bitumen films on the surfaces of individual mineral grains of bitumen-mineral mixtures was developed at the Department of Building Technologies and Highways (Volga State University of Technology) whole [20, 21]. This method has the following significant advantages:

1. allows to study the aging process of bitumen films on the surfaces of any mineral components of bitumen-mineral mixtures and assess their role separately in the aging process;
2. allows to study the dynamic pattern of properties during aging, i.e. at any stages of structure formation, starting from the combining the components in a loose state until the complete loss of mechanical strength in a compacted state;
3. the study of the aging of bitumen-mineral mixtures is carried out under laboratory conditions, which makes it possible to adjust the mixture composition during the technical project development at the design stage;

4. allows to study the bitumen films aging by choosing, as the main, one of the most sensitive to temperature changes indicators according to GOST 9128–2013 or GOST R58406.2–2020.

The essence of the technique is to establish the values of some property indicators (for example, the compressive strength at +50 °C, average density, etc.) of samples of the materials under study according to standard GOST 12,801–97 methods, made of mixtures not heated and preheated at a temperature of +150 °C and the calculation of the values of the aging coefficient according to the formula:

$$K_{\text{aging}}^{R_{\text{C}}^{50\,^{\circ}\text{C},\,t_{\text{h}}}} = \frac{R_{\text{C}}^{50\,^{\circ}\text{C},\,t_{\text{h}}}}{R_{\text{C}}^{50\,^{\circ}\text{C},\,t_{\text{h}}=0}} \tag{1}$$

where $K_{aging}^{R_{\text{C}}^{50\,^{\circ}\text{C},\,t_h}}$ – the aging coefficient of the studied BMM according to the indicator $R_{\text{C}}^{50\,^{\circ}\text{C},\,t_h}$ at t_{h}; $R_{\text{C}}^{50\,^{\circ}\text{C},\,t_h}$ – the value of the compressive strength at +50 °C of the BMM sample from mixtures after warming up at high temperature during the time t_h; $R_{\text{C}}^{50\,^{\circ}\text{C},\,t_h=0}$ – the same is true for samples not preheated at high temperature $t_h = 0$.

The theoretical part of the work involves finding the values of the aging coefficient BMM in two ways:

1. According to established mathematical models;
2. As weighted averages of their values for mixtures consisting of their mineral components with bitumen according to the formula:

$$K_{aging(BMM)}^{R_{\text{C}}^{50\,^{\circ}\text{C}}} = \frac{\sum_{i=1}^{n} K_{aging(i)}^{R_{\text{C}}^{50\,^{\circ}\text{C}}} \cdot a_i}{\sum_{i=1}^{n} a_i} \tag{2}$$

where $K_{aging(BMM)}^{R_{\text{C}}^{50\,^{\circ}\text{C}}}$ – the weighted average values of the aging coefficient of BMM; $K_{aging(i)}^{R_{\text{C}}^{50\,^{\circ}\text{C}}}$ – the aging coefficient of samples from a mixture of each *i*-th mineral component of BMM separately with bitumen in terms of compressive strength at +50 °C; n – the number of mineral components in the BMM; a_i – the percentage (fraction) of each mineral component in the BMM.

To establish the reliability of various methods of finding the values of the aging coefficient, it is proposed to perform appropriate experiments and calculations and make a comparative analysis of their convergence.

3 Performing Experiments and Analyzing Results

The experimental part of the work was performed according to the above-mentioned method. The production and testing of standard cylindrical samples made from preheated at a temperature of +150 °C for 0, 1, 3, 5 and 7 h was carried out in accordance with the GOST 12,801–97 methods (Fig. 1).

According to the test results, the values of the aging coefficient were calculated according to the index of compressive strength at +50 °C. Their values are given in Table 6.

As can be seen from Table 6, preheating at a temperature of + 150 °C leads to a decrease in the values of the aging coefficient of BMM. At the same time, asphalt concrete of type B and a sample from a mixture of granite crushed stone with bitumen, aging occurs in the direction of a constant decrease, and mixtures of fine-grained components (crushed sand and mineral powder) with bitumen in the initial periods of warming up (for up to 5 h), there is some increase in the values of the aging coefficient, and with further warming, the reverse process begins - its values begin to decrease. Such dynamics of aging processes can be explained as follows:

1. crushed fine-grained asphalt concrete and mixtures of acidic large aggregates (granite) with bitumen occur, firstly, due to the relatively weak relationships between the acidic surfaces of stone particles with films of petroleum bitumen containing mainly acidic surfactants – asphaltogenic acids; secondly, thermal

(a) (b)

Fig. 1 **a** – The standard cylindrical specimens size of 71.4 mm. **b** – The Bitumen-limestone crushed stone on metal trays in ventilated thermal furnace

Table 6 Experimental values of the aging coefficient in terms of compressive strength at +50 °C

t_h, hour	$K_{aging}^{R_C^{50\,°C}}$			
	Asphalt concrete of type B	Mixture of granite rubble with bitumen	Mixture of crushed sand with bitumen	Mixture of mineral filler with bitumen
0	1.00	1.00	1.00	1.00
1	0.87	0.32	1.02	1.00
3	0.78	0.15	1.43	1.13
5	0.65	0.01	2.00	1.63
7	0.57	0.01	1.07	1.38

transformations in the structure of bitumen on contact surfaces with mineral particles, they develop faster as a result of less stratification of bitumen films and a higher content of the non-oriented part of hydrocarbons in them; thirdly, the free and loosely bound zones in bitumen films on the surfaces of particles polymerize faster during heating due to their less binding to the surfaces of aggregates; at the same time, the proportion of saturated hydrocarbons in the composition of bitumen films increases, which leads to a decrease in their cohesive and adhesive strength and, consequently, to a decrease in adhesive ability;

2. in mixtures with dispersed aggregates in the initial heating period, an increase in the values of the aging coefficient is observed due to an increase in the strength of bitumen films in free and loosely bound (weakly oriented) zones as a result of the embrittlement phenomenon developing in them. At the same time, it should be expected that the greater the content of more dispersed mineral particles in the mixture, the greater the process of initial growth of the strength of the system will be. The latter, apparently, is explained by the greater connectivity of bitumen films with the surfaces of smaller mineral particles. With further heating, the reverse process begins due to the achievement of internal stresses in bitumen films of its ultimate cohesive strength.

Analysis of the dynamics of changes in the values of the aging coefficient depending on the duration of preheating of mixtures at a temperature of + 150 °C allowed us to obtain a number of mathematical models of an exponential nature (3…6). To do this, we turned to the CurveExpert 1.4 software product:

1. For asphalt concrete of type B:

$$K_{aging}^{R_C^{50\,^\circ C}} = 0.9769 * e^{-0.0791 * t_h} \tag{3}$$

the approximation rank: *r=0.9769*; the average deviation: *s=0.046*.

2. For a mixture of granite rubble with bitumen:

$$K_{aging}^{R_C^{50\,^\circ C}} = 0.9900 * e^{-1.0080 * t_h} \tag{4}$$

$r = 0.9909;\ s = 0.06398.$

3. For a mixture of crushed sand with bitumen:

$$K_{aging}^{R_C^{50\,^\circ C}} = 1.7570 * e^{-\frac{(t_{h-4.1683})^2}{22.1738}} \tag{5}$$

$r = 0.8439;\ s = 0.3238.$

4. For a mixture of limestone mineral filler MP-1 with bitumen:

$$K_{aging}^{R_C^{50\,^\circ C}} = 1.2810 * e^{-\frac{(t_{h-4.21})^2}{84.6820}} \tag{6}$$

$r = 0.8594;\ s = 0.1975.$

Assuming the presence of the influence of the type, size and content of mineral components on the aging process of BMM, a hypothesis is further put forward about the effect on the values of the aging coefficient of the content of each of them in proportion to their amount in the mixture. To test this hypothesis, weighted average values of the aging coefficient were calculated according to the formula (2). Comparative results of calculations are given in Table 7 and Fig. 2.

The analysis given in Table 7 data show that the obtained mathematical models (2–6) adequately describe the aging processes of the studied bitumen-mineral materials. Consequently, these models in general can be recommended for use to predict the aging processes of hot bitumen-mineral materials designed for use in road structures.

Table 7 Comparative values of the aging coefficient of BMM in various ways

t_h, hour	$K_{aging}^{R_{C}^{50\,°C}}$								
	Asphalt concrete of type B			Mixture of granite rubble with bitumen		Mixture of crushed sand with bitumen		Mixture of mineral filler with bitumen	
	By experiment	Using a formula (3)	Using a formula (2)	By experiment	Using a formula (4)	By experiment	Using a formula (5)	By experiment t	Using a formula (6)
0	1.00	0.98	1.00	1.00	0.99	1.00	0.80	1.00	1.04
1	0.87	0.90	0.70	0.32	0.36	1.02	1.14	1.00	1.13
3	0.78	0.78	0.82	0.15	0.05	1.43	1.65	1.13	1.26
5	0.65	0.66	1.17	0.01	0.006	2.00	1.70	1.63	1.27
7	0.57	0.56	0.62	0.01	0.0008	1.07	1.29	1.38	1.24

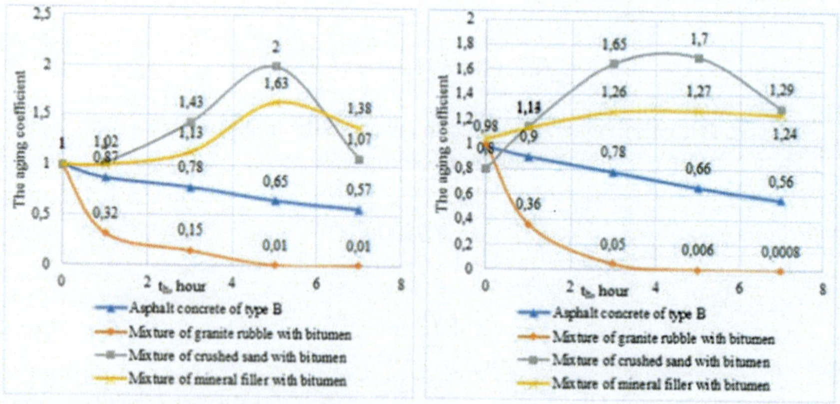

(a) (b)

Fig. 2 **a** – The values of the aging coefficient by experiment. **b** – The values of the aging coefficient by using a formula

4 Conclusions

1. Finding the values of the aging coefficient by conducting laboratory experiments using a new technique allows us to rapidly assess the comparative resistance against aging of bitumen mineral materials of various compositions.
2. Based on the performed experiments and theoretical analysis, mathematical models have been established that reflect the aging processes of bitumen-mineral materials depending on the time of preliminary heating at high temperatures. Models (3–6) in this case can be used to evaluate the ability of various bitumen-mineral materials to resist thermal aging in road structures.
3. Model (2) can be used to obtain the values of the aging coefficient of bitumen-mineral materials at pre-determined values of the aging coefficient for mixtures of their individual mineral components with bitumen. The advantage of its use over the previous methods is the possibility of obtaining estimated data on the thermal aging of bitumen-mineral materials of new compounds without further additional experiments.

Acknowledgements The authors thank the staff of the Department of construction technologies and highways of the Volga state technological University (Yoshkar-Ola, Russian Federation).

References

1. Saedi S, Oruc S (2020) Fibers
2. Porto M, Caputo P, Loise V, Teltayev BB, Angelico R, Calandra P, Rossi RC (2019) News Natl Acad Sci Repub Kazakhstan, Ser Geol Tech Sci
3. Wei H, Bai X, Qian G, Wang F, Li Z, Jin J, Zhang Y (2019) Materials (Basel)
4. Loise V, Vuono D, Policicchio A, Teltayev B, Gnisci A, Messina G, Rossi GO (2019) Colloids surfaces a physicochem. Eng Asp
5. Manzone M, Ruffinengo B (2019) J Agric Eng
6. Liu T, Zankavich VN, Aliakseyeu YN, Khroustalev BM (2019) Sci Tech
7. Molenaar AAA, Hagos ET, van de Ven MFC (2010) J Mater Civ Eng 22:779
8. Der Van Poel C (2007) J Appl Chem 4:221
9. Zhurinov MZ, Teltayev BB, Kalybai AA (2019) News Natl Acad Sci Repub Kazakhstan, Ser Geol Tech Sci
10. Teltayev B, Radovskiy B (2018) Road Mater Pavement Des 19:1832
11. Teltayev BB, Seilkhanov TM (2018) Eurasian Chem J
12. Teltayev B, Radovskiy B (2018) Road Mater Pavement Des
13. Zakrevskaya L, Handelsman I, Provatorova G (2018) MATEC Web Conf
14. Baek C, Underwood B, Kim Y (2012) Transp Res Rec
15. Saoula S, Soudani K, Haddadi S, Munoz ME, Santamaria A (2013) Mater Sci Appl 04:312
16. Pasandín AR, Pérez I, Oliveira JRM, Silva HMRD, Pereira PAA (2015) J Clean Prod 101:165
17. Salikhov M, Veyukov E, Conf IOP (2020) Ser Mater Sci Eng 896:012073
18. Salihov M, Veyukov E, Lomakina E, Malyanova L (2021) E3S Web Conf 263: 01005
19. Salikhov MGMLI, Veyukov EV, Sabirov LR (2018) A method for determining the speed and intensity of aging of asphalt concrete. Patent for Invention No. 2654954. Accessed 13 Feb 2017

20. Salikhov MG, Veyukov EV, Kalugin RY, Yendeletov AO, Conf IOP (2020) Ser Mater Sci Eng 832:012031
21. Salikhov MG, Veyukov EV, Vainshtein VM, Malianova LI, Conf IOP (2021) Ser Mater Sci Eng 1083:012032

Prospects for the Use of Zeolite in Multicomponent Fine-Grained Concretes

Anastasiia Puzatova, Maria Dmitrieva, Vladimir Leitsin, Alina Panfilova, and Maria Shinyaeva

Abstract The article provides a review of scientific literature and experimental data on the use of zeolites in construction, namely on the use of this mineral as an active mineral additive in concrete. Based on the data obtained by experimental methods, a conclusion was made about the effectiveness of the use of zeolite for multicomponent fine-grained concretes used in concrete 3D printing. The increase in the strength of samples with different content of zeolite, which replaces part of the cement, at the age of 7 and 28 days was determined. The best performance was determined for samples with 5–15% zeolite content. The data on the setting time of the concrete mixture are given. For compositions with 5–15% zeolite, the setting time is reduced by 1–1 h 20 min compared to the control composition, and the time interval until the end of setting can be optimized by choosing the zeolite concentration for a certain portion of the 3D printer "ink". The conclusion is made about the effectiveness of the use of zeolite in building 3D printing.

Keywords Zeolite · Mineral additive · Fine-grained concrete · Strength · Setting time · Additive manufacturing technologies · Concrete 3D-printing

A. Puzatova (✉) · M. Dmitrieva · V. Leitsin · A. Panfilova · M. Shinyaeva
Immanuel Kant Baltic Federal University, 14 Nevskogo street, Kaliningrad 236016, Russian Federation
e-mail: asharanova@kantiana.ru

M. Dmitrieva
e-mail: admitrieva@kantiana.ru

V. Leitsin
e-mail: vleitsin@kantiana.ru

A. Panfilova
e-mail: adpanfilova@stud.kantiana.ru

M. Shinyaeva
e-mail: mshinyaeva@stud.kantiana.ru

P. Akimov et al. (eds.), *Proceedings of FORM 2022*, Lecture Notes in Civil Engineering 282, https://doi.org/10.1007/978-3-031-10853-2_27

1 Introduction

Zeolites are a large group of similar in composition and properties minerals of volcanic-sedimentary origin, aqueous calcium and sodium aluminosilicates from a subclass of framework silicates with a glassy or pearly luster. Zeolites have a special "porous" structure, the occurrence of which is explained by the substitution of Si_4^+ for Al_3^+. They have unique properties, such as: molecular sieve effect, high ion-exchange, sorption and catalytic properties, which appear due to the presence of cavities and channels in the crystal lattice. These properties of minerals determine a wide range of applications of zeolites and wide distribution in the world.

Although zeolites have been known for more than 200 years, for a long time they were considered as rare minerals that do not form industrial accumulations and have no practical application. However, later it became known that zeolites are the most valuable minerals in industry, having an open frame-cavity structure of the $[(Si,Al)O_4]$ type, which has a negative charge. The latter is compensated by neutralizing, positively charged cations (Na, K, Mg, Ca, Sr, Ba, Ca) contained in the zeolite [1].

Zeolites are subdivided into natural and synthetic and are used in many economic fields of activity - in industry, agriculture and ecology [2].

Natural zeolite is a fairly effective cheaper substitute for synthetic zeolite and some natural mineral compounds such as chalk, kaolin, diatomite. In large-tonnage technological processes, where the use of synthetic zeolites is not profitable or limited, the use of natural zeolites is of great importance. Their practical application is determined by their ion-exchange and adsorption properties, thermal and chemical resistance, mechanical strength, structure of the crystal lattice, cationic composition [1].

The use of zeolites in the construction industry is the most popular. Back in the 70 s. of the last century, most of the zeolites mined in Western Europe were used as building stone, low-density insulating materials and in the production of pozzolanic cement. Zeolites are especially effective in the production of hydraulic cements that are resistant to the chemical attack of the environment, such as sea water [2].

Zeolites have pozzolanic properties and are among the natural pozzolans due to their reactive silica (SiO2) and alumina (Al2O3). These compounds, reacting with Portlandite during Portland cement hydration, convert it into C-S–H gels and aluminates, while the microstructure of the hardened cement concrete improves, and the water resistance of concrete increases [3]. Due to the introduction of natural zeolite into the concrete mixture, it is possible to significantly reduce the content of expensive clinker without deteriorating the properties of cement [4–10]. Also, zeolites are used as a drying filler for building materials, they have the ability to absorb excess moisture in the cement mortar and gradually release it as needed, thereby contributing to the complete hydration of the cement. The addition of zeolite improves the setting of building mixtures and increases their strength [1]. It has been proven that the introduction of 5% natural zeolite into the composition of fine-grained concrete leads to an increase in compressive strength by 37%, the introduction of 10%

zeolite increases the tensile strength by 19%, without increasing the water demand of the mixture [3].

Zeolite is intended, also as an active mineral additive, for the manufacture of cements, concrete mortars, foam and aerated concrete. Zeolite can be used: to introduce into the composition of concrete in order to save cement and reduce concrete cracking; as fillers of concrete solutions, as it is not subject to caking, which makes it possible to exclude the use of anti-caking agents; for the production of high-strength concrete; as a lime-zeolite binder for autoclaved and non-autoclaved silicate concretes; as a gypsum cement pozzolanic binder component and concretes based on it; in the production of silicate bricks; for the production of dry mixes, including for the production of drying plaster [1]. In addition, the introduction of natural zeolite reduces corrosion processes in heavy concrete [11]. The optimal degree of concrete filling with zeolite is noted in the range of 20–25% wt. [12].

Zeolites, acting as active mineral additives, intensively bind calcium hydroxide and low-basic hydrosilicates and calcium hydroaluminates formed during the hardening of Portland cement. When zeolite is introduced into the system, free calcium hydroxide will be taken up by the zeolite. This leads to an acceleration of structure formation in the system. Therefore, the introduction of natural zeolite-containing rocks increases the durability and frost resistance of concrete by 2–3 times. This does not require the use of various types of chemical concrete modifiers, which is undoubtedly a positive fact [13].

The development of 3D printing is entering the leading positions in all areas of science and industry. Specifically, in construction, compositions for three-dimensional printing, their properties and characteristics are mainly studied. As in ordinary concrete, a binder is taken as the basis (cement, fly ash, microsilica, kaolin clay, etc.). But, if in ordinary concrete sand, crushed stone or gravel are used as aggregates, then in concrete for three-dimensional printing, due to the small diameter of the extruder nozzle, only sand of various fractions, reinforcement with glass or polypropylene fiber, a small amount of water for setting are used. An integral part is chemical additives that give concrete the properties necessary for printing (plasticizers, hardening accelerators or retarders, water-reducing additives, etc.). To achieve the full benefits of concrete printing capable of creating complex shapes, it is necessary to study the strength properties of the samples, as well as conduct in-depth studies to improve the structural, mechanical and durability properties.

One of the main tasks in the development of additive 3D printing technology using concrete mixtures is to substantiate the requirements for it and for hardened concrete, based on the characteristics of technological equipment and molding modes. An analysis of the literature data [14] showed that in order to obtain concrete with the required characteristics, the concrete mix must have certain properties corresponding to the additive technology. At the stage of molding (extrusion) and hardening of the mixture in the field, these are: - dispersion; - rheological properties (formability, possibility of transportation through pipes, plastic strength); - high adhesion between the layers of the mixture, tight fit of the layers; - absence of breaks in the mixture; - no cracking, low shrinkage; - uniformity of hardening (setting); - high setting speed after extrusion; For the finished product, these are: - the required strength; - high uniformity and stability of properties; - low density and thermal conductivity; - high adhesion strength; - frost resistance. It is important that the required characteristics are formed already at the stage of substantiating the composition of concrete mixtures and developing their compositions [15]. The use of finely dispersed additives, such as zeolites [16, 17], in concrete mixtures for additive technologies increases the strength of cement stone in the early stages of hardening, which is a determining factor in the formation of construction and technical characteristics for 3D printing technology [15].

2 Materials and Methods

To assess the prospects for the development of zeolite-bearing deposits and the efficiency of using this material as an additive in concrete, the available volcanic zeolite according to TU 08.11.12–003-00,327,325,428–2019, manufacturer "Transcarpathian zeolite plant", Sokirnitskoe deposit, Ukraine was used. Properties are presented in Table 1.

The influence of zeolite on the characteristics of concrete both in the freshly mixed and in the hardened state was studied. Experimental samples were made on the basis of a multicomponent composition [18] containing microsilica and highly active metakaolin, based on different degrees of cement substitution by zeolite. The following compositions were studied:

- 0 - control;
- 1 - 5% zeolite;
- 2 - 10% zeolite;
- 3 - 15% zeolite;
- 4 - 20% zeolite;
- 5 - 25% zeolite;
- 6 - 30% zeolite.

The consumption of materials per 1 m^3 of concrete mixture with different percentages of zeolite is presented in Table 2.

The compressive and flexural strengths in bending at the age of 7 and 28 days were determined on beam samples 40 × 40 × 160 mm. For the manufacture of samples, a standard technique was used according to GOST 310.4–81 "Cements. Methods of bending and compression strength determination" [19].

In addition to the strength characteristics, density, consistency using Vicat apparatus and setting time were measured for each composition. The consistency of the obtained solutions, as well as the beginning and end of setting, were determined according to the standard methods of GOST 310.3–76 "Cements. Methods for determination of standard consistency, times of setting and soundness" [20].

Table 1 Properties of zeolite

Characteristic	Meaning
Mineral composition, [%]	
– clinoptilolite	75
– X-ray amorphous phase	8
– quartz	9
– plagioclase	3
– dolomite	1
– hydromica	4
Chemical composition, [%]	
– SiO_2	71.4
– Al_2O_3	11.9
– CaO	1.26
– K_2O	3.02
– Fe_2O_3	1.14
– MgO	0.45
– Na_2O	2.086
– TiO_2	0.16
– MnO	0.034
– P_2O_5	0.02
Granulometry, [mm]	0–0.5
Appearance	Granules of light gray color
Porosity, [%]	38–48
Density, [g/cm^3]	2,2–2,3
Mohs hardness	3.5–4
Effective pore diameter, [nm]	0.4
pH	6,8–7,2
Water adsorption capacity, [%]	34–38

Table 2 Material consumption per 1 m^3

Composition No	Cement, [kg/m^3]	Sand, [kg/m^3]	Microsilica (MS), [kg/m^3]	Metakaolin (MK), [kg/m^3]	Zeolite, [kg/m^3]	Hyperplast., [l/m^3]	Water, [l/m^3]
0	545.3	1168	156	78	–	11.7	399.4
1	518.03	1168	156	78	27.27	11.7	368.68
2	490.77	1168	156	78	54.53	11.7	368.68
3	463.5	1168	156	78	81.8	11.7	399.4
4	436.24	1168	156	78	109.06	11.7	399.4
5	408.97	1168	156	78	136.33	11.7	399.4
6	381.71	1168	156	78	163.59	11.7	399.4

3 Results

The results of tests for compressive and flexural strength are shown in Figs. 1 and 2.

The results of experimental studies allow us to note an increase in the strength of samples containing zeolite in comparison with the control composition without zeolite. The best performance in terms of strength increase is observed in compositions containing 5, 10 and 15% zeolite. The compressive and flexural strength at the age of 7 days for samples of these compositions increases by 24.6–49.4 and 34.4–49.9%, respectively; compressive and flexural strength at the age of 28 days -

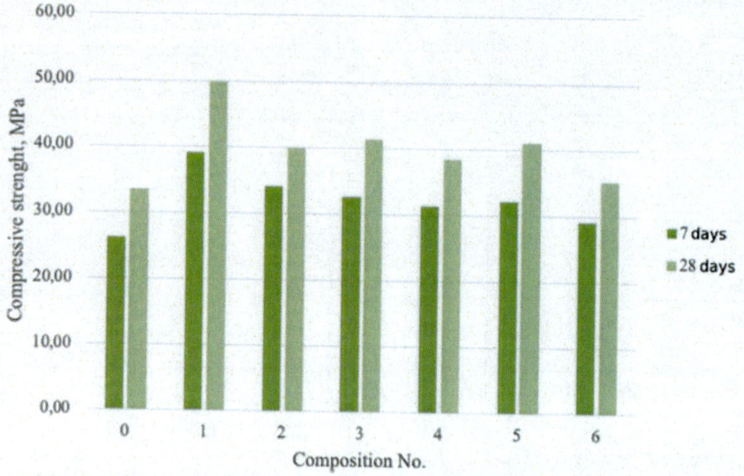

Fig. 1 Graph of compressive strength

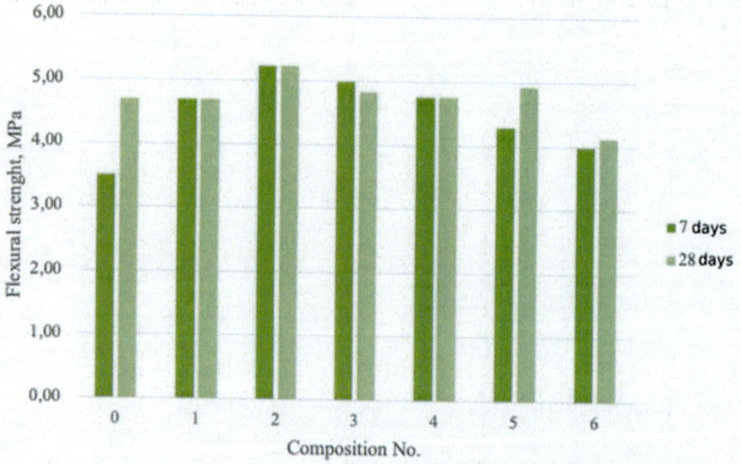

Fig. 2 Graph of flexural strength

Table 3 Characteristics of fresh concrete mix

Composition no.	Characteristic				
	Average sample weight, [g]	Average sample density, [g/cm^3]	Vicat apparatus consistency, [mm]	Initial setting time	Final setting time
0	505.6	1.975	38.5	2 h 45 min	3 h 15 min
1	513.9	2.078	38.5	1 h 45 min	2 h 50 min
2	512.5	2.001	38	1 h 35 min	2 h 40 min
3	520.3	2.032	38.5	1 h 25 min	3 h 10 min
4	512.63	2.002	38	2 h 00 min	3 h 30 min
5	516.27	2.017	38	1 h 55 min	3 h 05 min
6	501.22	1.958	38.5	1 h 20 min	2 h 35 min

by 19.1–48.7 and 0–11.5%, respectively. Further, there is a tendency to reduce the effectiveness of the zeolite additive with its percentage increase in the composition. However, the strength of the samples containing zeolite was higher than the strength of the control samples, with the exception of the flexural strength of samples with 30% zeolite at the age of 28 days. The latter indicates the existence of a maximum allowable dose of the introduced zeolite. Thus, it can be concluded that the use of zeolite is effective in controlling the properties of multicomponent fine-grained concretes. Moreover, the achieved active increase in strength in all compositions is noted at the age of 7 days, and by 28 days the increase in strength decreases, but remains at a fairly high level relative to control samples. From this we can conclude that the use of zeolite is exceptionally effective for modifying concrete compositions for 3D printing, providing an early set of strength.

The results for determining the density, consistency according to the Vicat apparatus, as well as the setting time are presented in Table 3.

The presented experimental results make it possible to note that the density and density of mixtures practically do not change and remain stable regardless of the percentage of replacement of cement by zeolite. At the same time, with an increase in the concentration of zeolite, a significant decrease in the setting time is observed. Moreover, the dependence of the reduction in the setting time on the concentration of the zeolite is nonmonotonic, which indicates the presence of the optimal concentration of the additive. In the considered case, the optimal zeolite concentration for 3D printing can be considered as 15% of the mass of cement (composition 3). Despite the fact that the observed increase in strength characteristics is maximum for a zeolite concentration of 10% (composition 2), the preferred choice of zeolite additive in the amount of 15% (composition 3) is explained by a significant increase in the time interval until the end of setting, which entails an increase in the life of the 3D printer with one batch of mixture.

4 Conclusions

The effectiveness of the use of zeolite in the production of multicomponent fine-grained concretes for construction 3D printing has been proven. The greatest increase in strength both in compression tests and in flexural is observed in samples with a content of 5–15% zeolite, the setting time for these samples is also significantly reduced. The time interval until the end of setting can be optimized by choosing the concentration of zeolite for a certain portion of the "ink" of the 3D printer.

Acknowledgements The reported study was funded by RFBR and Kaliningrad region, project number 19-48-390001 *Multicomponent nanomodifiers of concrete mixes.*

References

1. Toturbiev BD (2012) Proc Inst Geol Dagestan Sci Center Russ Acad Sci 58:47–51. (in Russian)
2. Vasilyanov LS, Lazareva EA (2016) News Sci Kazakhstan 1(127): 61–85. (in Russian)
3. Morozova NN, Kais HA (2016) Bull Kazan Technol Univ 19(10):64–68. (in Russian)
4. Rahul P, Ravella DP, Rao PVCS (2022) Mater Today: Proc. in press
5. Zheng X, Zhang J, Ding X, Chu H, Zhang J (2021) Constr Build Mater 288:123062
6. Dabbaghi F, Sadeghi-Nik A, Libre NA, Nasrollahpour S (2021) Structures 34:2617–2627
7. Das M, Adhikary SK, Rudzionis Z (2021) Mater Today: Proc. in press
8. Teja GS, Ravella DP, Rao PVCS (2021) Mater Today: Proc 43(2):2355–2360
9. Madhuri PV, Rao BK, Chaitanya A (2021) Mater Today: Proc 47(15):5369–5378
10. Erfanimanesh A, Sharbatdar MK (2020) J Build Eng 32:101781
11. Potapova LI, Kais HA (2016) Innov Sci 6(2):132–134. (in Russian)
12. Lankin SV (2014) Prob Ecol Upper Amur Reg 16:10–17. (in Russian)
13. Morgun AN (2015) Sci Technol Educ 7(13):101–105. (in Russian)
14. Canessa E, Fonda C, Zennaro M, Deadline N (2013) Low-cost 3D printing for science, education and sustainable development. In: The Abdus Salam international centre for theoretical physics, p 192
15. Demyanenko OV, Kopanitsa NO, Sorokina EA (2018) Bull Tomsk State Univ Arch Build 20(4):122–134. (in Russian)
16. Panda B, Singh GVPB, Unluer C, Tan MJ (2019) J Clean Prod 220:610–619
17. Khan SA, Koç M, Al-Ghamdi SG (2021) J Clean Prod 303:127027
18. Sharanova A, Dmitrieva M (2019) E3S Web Conf 97:06018
19. GOST 310.4-81 (198) Cements. Methods of bending and compression strength determination, Izdatelstvo standartov, Moscow, p 11
20. GOST 310.3-76 (1976) Cements. Methods for determination of standard consistency, times of setting and soundness, Izdatelstvo standartov, Moscow, p 6

Influence of the Carnegie Ridge on the Development of Seismogenic Sources and Seismicity of Ecuador

David Cajamarca-Zuniga and Oleg Kabantsev

Abstract The subducting process of Nazca plate beneath the South American tectonic plate governs the seismicity at the Andes region, making it one of the world's most active seismic regions. The subduction and collision of the aseismic Carnegie Ridge (CR), and the segmentation and drift of the continental North-Andean Block (NAB) with an intricate system of geological faults, are the main seismogenic sources in Ecuador. The oblique convergence of the CR is causing a ploughing effect and shallow-focus earthquakes on the Ecuador shoreline and deep seismic activity in the Pastaza-Napo region at the Amazon basin. The North Andean Block ejection by the Carnegie Ridge collision develops a complex system of active faults along the Andes and shallow-focus earthquakes along Chingual-Cosanga-Pallatanga-Puná fault system. Geodynamic and seismic activity suggests that subduction of the prolongation of CR beneath Ecuador requires about 400 km from the Colombia-Ecuador-Peru trench and is the most influential factor in the development of Ecuador's seismicity.

Keywords Ecuador · Carnegie ridge · North andean block · Nazca plate · Seismic hazard · Subduction · Seismicity

1 Introduction

The greatest earthquakes are known to occur at the tectonic plate interface of subduction zones. Ecuador is located in one segment of the Rim of Fire, at the northwest interface of the South American plate with the Nazca Plate. The Rim of Fire is the world's most seismically active region, where about 90% of all earthquakes and about 80% of the strongest earthquakes have occurred [1–3]. The Andes are part of the Rim of Fire, and are one of the highest seismic activity regions in the world, accounting

D. Cajamarca-Zuniga (✉)
Catholic University of Cuenca, Av. de las Americas and Humboldt, Cuenca 010101, Ecuador
e-mail: cajamarca.zuniga@gmail.com

D. Cajamarca-Zuniga · O. Kabantsev
Moscow State University of Civil Engineering, Moscow 129337, Russian Federation

P. Akimov et al. (eds.), *Proceedings of FORM 2022*, Lecture Notes in Civil Engineering 282, https://doi.org/10.1007/978-3-031-10853-2_28

for around 20% of the total seismic energy released by our planet [4]. The subduction of the Nazca tectonic plate and the complex system of active geological faults generates mostly shallow earthquakes in Ecuador. About 90% of continental territory of Ecuador presents high seismic hazard. Here, the 7th world's largest registered earthquake (Mw = 8.8) occurred in 1906 [5–9].

In this article we present a brief but comprehensive review of the tectonic setting and geodynamics of the North-Andean region at the South American plate, as well as the influence of the Carnegie Ridge (CR) and the migration of the North-Andean block (NAB) in the development of seismogenic sources and seismicity of Ecuador. Finally, we present some maps of the inferred prolongation of the CR beneath the territory of Ecuador.

2 Methods

A comprehensive study of scientific works on the tectonic settings, cortical movement and geodynamics of the North-Andean region of the South American plate, as well as the influence of the Carnegie Ridge (CR) and the segmentation of the North-Andean block (NAB) have been performed in order to determine main seismogenic sources and their influence on the seismicity of Ecuador. We have correlated the results of researches on geodynamics with epicentral location, magnitude and depth of historical high intensity earthquakes. To obtain the information about earthquakes we reviewed and combined pre-instrumental and instrumental information from several local, regional, and global sources such as IG-EPN, CERESIS, EHB, ISS, ISC, CENTENNIAL, NEIC, GCMT, USGS, and from other published scientific researches. The results were mapped GIS software.

3 Results

The current seismogenic structures give evidence of progressively evolving tectonic forces across geological time, but remain unchanged in terms of the relatively small return periods used for seismic hazard assessment [10]. Three lithospheric plates dominate the current tectonic mosaic of north-western region of South America: the Caribbean plate by the north, the Nazca plate by the west and the South American plate by the east, cored in this region by the Amazonian Craton (Guiana shield) Fig. 1(a) [11]. The studies analysed show that in the last 30 years several investigations have been carried out on active tectonic deformation in the Andes region of Ecuador. It is established that in the Equatorial latitudes the subducting process of the Nazca plate (N83; 55–75 mm/year) beneath the South American tectonic plate is the most evident and main geodynamic process in the northwestern region of South America and in the territory of Ecuador Fig. 1(b) [5, 10, 12–17].

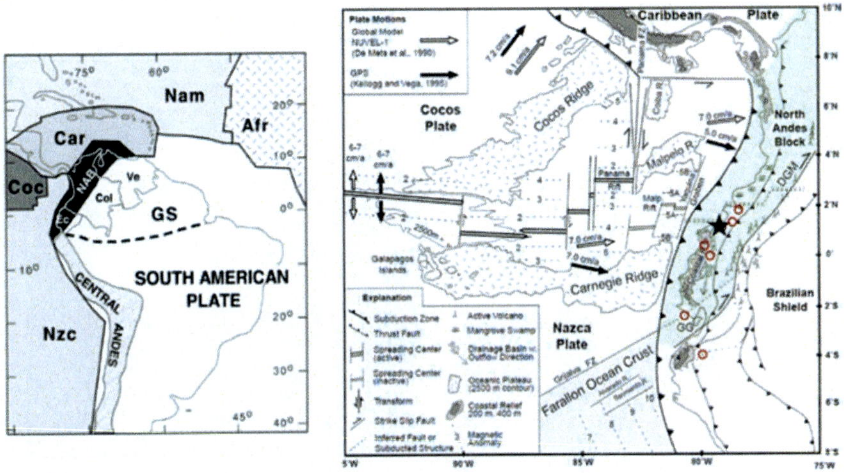

(a) (b)

Fig. 1 **a** Tectonic setting of the Northwest of South America (from Cediel, 2005 [11]). North American Plate (Nam); Caribbean Plate (Car); Cocos Plate (Coc); Nazca Plate (Nzc); North Andean Block (NAB); Guiana Shield (GS); African Plate (Afr); Venezuela (Ve), Colombia (Col), Ecuador (Ec). **b** Tectonic setting of Ecuador, major faults and relative plate motions. Locations of the 1906 earthquake (Mw = 8.8, black star); from north to south the 1979, 1958, 2016, 1942, 1901, 1953 earthquakes ($M \geq 7.8$, red circles). Edited after Gutscher et al., 1999 [15]

The subduction of the Nazca oceanic plate beneath the South America continental plate takes place along the Colombia-Ecuador-Peru trench which extends for more than 9000 km [4, 18]. In addition to activity associated to plate interfaces, the convergent tectonics causes active internal deformation in the Andean block, as demonstrated by neotectonics and seismicity concentrations along the coastal and interandean valleys fault systems [18]. The segment of the Northern Andes where both thrust faulting and crustal shortening are observed coincides with the region where the Carnegie Ridge is being subducted. Along the subandean zone and the eastern cordillera, there is a large system of thrusts, as well as transpressive and strike-slip faults, that suggest that "this is a (growing) uplifting mountain belt linked with an important component of NNE-directed transcurrent motions" [19, 20].

The Ecuadorian Andes mainly defines a compression zone featured by reverse faults in the foothills essentially orthogonal to the plate convergence vectors and slip faults as the Dolores-Guayaquil Megashear (DGM) [21, 22], and the Chingual-Cosanga-Pallatanga-Puná (CCPP) fault system [17, 18], which are segments of the Guayaquil-Caracas Continental Megashear (GCM). In the north of Ecuador the dextral slip occurs 300–400 km from the trench (Fig. 2), and "appears to be caused by increased coupling above the Carnegie Ridge flat slab segment" [15, 21]. The subduction process of the Carnegie Ridge causes a shortening effect of the North Andean Block in the collision zone and the consequent ejection to the Northeast along the GCM at a rate of ~ 6–10 mm/year [23, 24], as well as the uplifting of

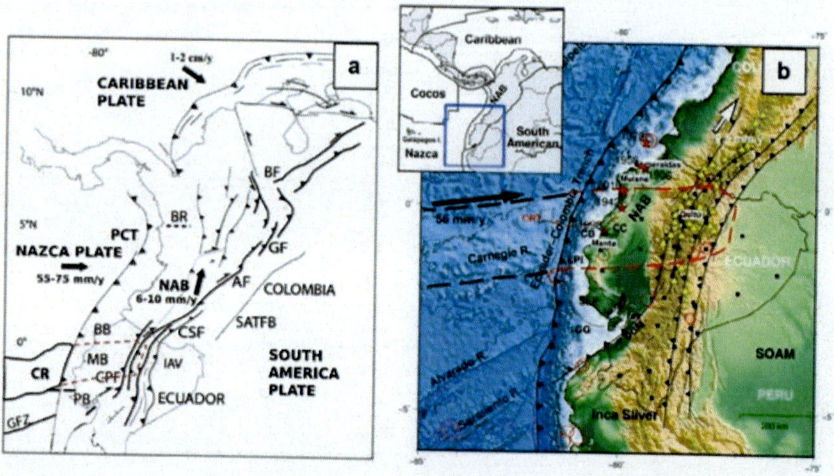

Fig. 2 Inferred continuation of Carnegie Ridge (red dashed line). Structural arrangement of the north-western region of South America. Here: NAB = North Andean Block, GFZ = Grijalva fracture zone, CR = Carnegie Ridge, PB = Progreso basin, MB = Manabí basin, BB = Borbón basin, CPF = Calacalí–Pallatanga fault, IAV = Interandean valley, CSF = Chingual–La Sofía fault, AF = Algeciras fault, GF = Guaicáramo fault, BF = Bocono fault, BR = Baudo range, PCT = Colombia-Ecuador-Perú trench, SAFTB = Sub-Andean fault and thrust belt. Figure 2a modified after Witt et al., 2006 [29]. Figure 2b modified after Staller et al., 2018 [30]

the coastal region at a rate of 0.1–0.5 mm/year and the Pastaza-Napo region at the Amazon basin [25–28]. The uplift is associated to subducting Carnegie Ridge (CR) as a part of the Nazca plate. Thus, geodynamics of Ecuador departs from the classical and conceptual subduction type. There are 3 main seismogenic sources that affect the seismicity of Ecuador. The subducting oceanic Nazca plate beneath the South American plate is the major process which controls the tectonic setting of this region and originates two other seismogenic factors: the subduction and collision of the CR, and the segmentation and "escape" of the North-Andean Block (NAB) with an intricate strike-slip fault system Fig. 2 [29].

3.1 The Influence of the Carnegie Ridge

To understand the seismicity of Ecuador caused by the west–east subducting Nazca tectonic plate we should consider the influence of the Carnegie Ridge. The Carnegie Ridge is a major 2-km-high by 200-km-width seamounts of the Nazca oceanic plate subducting beneath South American continental plate at the centre of the coastal line of Ecuador between 1°N and 2°S latitude, and moving beneath the territory of Ecuador (Fig. 2, Fig. 3) [15, 29–31].

Gutscher et al., 1999 [15] proposed a model of "two lithospheric tears bound the Carnegie Ridge collision zone" (Fig. 3b), where the buoyant prolongation of the

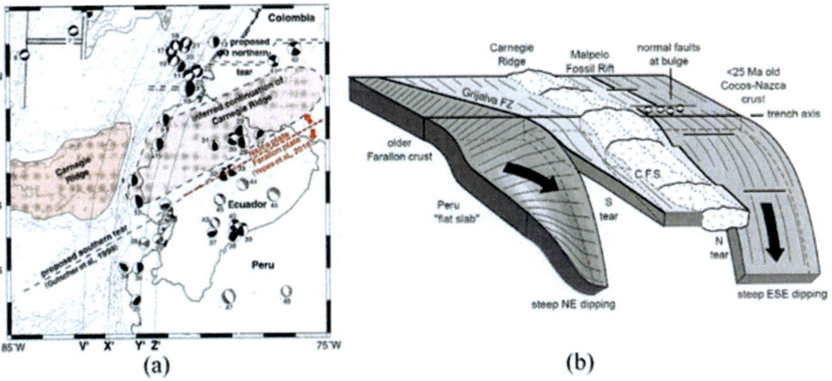

Fig. 3 **a** Inland prolongation of Grijalva rifted margin and the interface between Nazca and Farallon plates (red dashed line) proposed by Yepes et al., 2016 [10]. **b** Three-dimensional view of a two-level model of the CR collision with the South American plate. Figure 3 edited after Gutscher et all 1999 [15]

CR was estimated "to continue at least 110 km and probably up to 500 km from the trench". Between the two tears, the buoyant flat slab segment must have ~17 km crustal thickness to resist subduction. The northern flank is more buoyant and thicker (19 km) compared to the southern flank (14 km) [31]. In the subducting process we should take into attention that the ~2-km-high Carnegie Ridge at the collision zone causes permanent distortions as it is bulldozing the front of the fore-arc block [32]. Although Michaud, 2009 [33] proposes that the geological expression of Carnegie Ridge subduction requires about 300 km of CR penetration beneath Ecuador (Fig. 4).

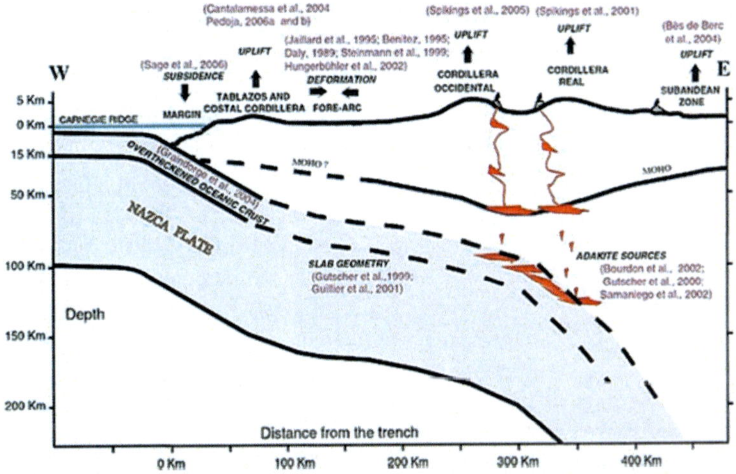

Fig. 4 West–East cross-section of Ecuador at 1°11' S, showing geological events related to Carnegie Ridge subduction (source: Michaud et al., 2009) [33]

The southern tear occurs along the Grijalva fracture zone (Grijalva FZ), while the northern tear is estimated to be approximately to the north of the CR (Fig. 1b, Fig. 3) [15].

The subduction of the Carnegie Ridge is a process that controls the locations and clusterisation of large earthquake activity along the northwest coastline of Ecuador, as well as the evolution of the foothill basins of Borbon (BB) and Manabi (MB), and the uplifting of both the coast region and the Pastaza-Napo region at the Amazon basin [28, 29, 34–36]. The schematic geological cross-section of the subduction process of Carnegie Ridge at the collision zone between latitudes 1°N-2°S suggested in [33] shows some geological events related to this subduction process (Fig. 4).

In the collision zone of Carnegie Ridge, the orientation of crustal faults changes from north–south to northeast-southwest along the subduction margin [34, 37]. The morphology of the Ecuadorian subandean zone and its strong tectonic activity can be attributed to a major geodynamic event, such as the Carnegie Ridge's subduction. According to [33] "The oblique convergence of the Carnegie Ridge would have caused the northward drift of the North Andean block and the opening of the Gulf of Guayaquil".

The bulldozing effect of the CR subduction contributed to uplift of the coastal ranges at rates of 0.30–0.50 mm/year for the Manta Peninsula [25], generate the north-eastward motion of the North Andean crustal block along the Guayaquil-Caracas continental Megashear [24, 30].

3.2 *The North Andean Block (NAB) and the Right-Lateral Strike-Slip Shear*

In order to understand the geodynamic scheme of Ecuador, it is also necessary to consider the movement of North Andean Block as a detaching "mini-plate", sliding through the northwest corner of South America on a major right-lateral strike-slip fault. Northwest South America is a distinct geological segment of the Andes [11] and a broad convergent plate boundary zone characterized by subduction, ongoing plate collision, and active seismicity. The North Andean Block is bordered on the west by the Colombia-Ecuador trench (Ecuador subduction zone) and Panama Block, and on the north by the South Caribbean deformed belt, and on the east by the Guayaquil-Caracas continental Megashear (GCM) (Fig. 5). The oblique subduction in this region induces deformation of the overriding continental plate. As a result of this process the NAB is being ejected to the northeast following the front of the Eastern Cordillera along a transpressive system of faults [10, 12, 35, 38]. According to recent studies, the NAB is migrating relatively fast, just as the Nazca Plate is moving to the east relative to the Amazonian Craton (~56 mm/year) [16, 17], the NAB is moving to the north-east relative to the South American plate at 6–10 mm/yr [10, 13, 17, 24, 38]. The Guayaquil Gulf opens at the southern junction between two fracture zones that isolate the NAB, the Colombia-Ecuador Trench and the GCM (Fig. 5) [35]. The east

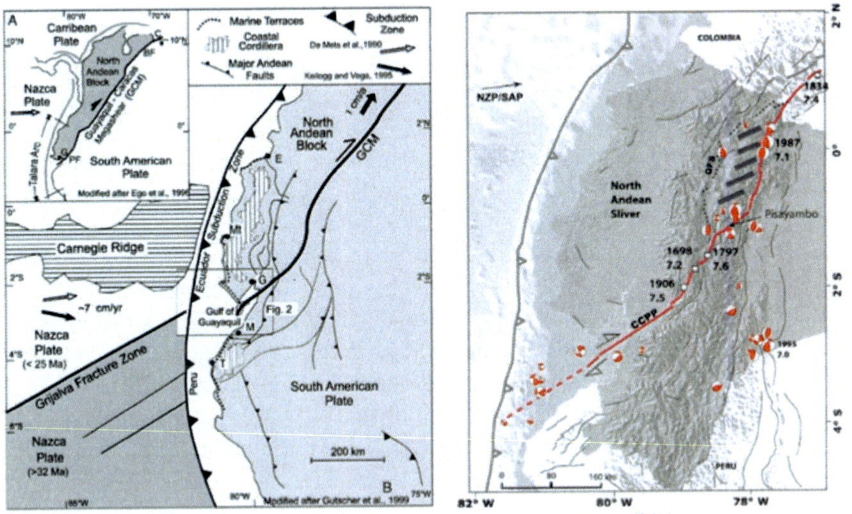

(a) (b)

Fig. 5 **a** Structural arrangement of the north-western region of South America, showing the NAB, limited by Colombia-Ecuador-Peru Trench, Panama Block and GCM. Source: Dumont et al., 2005 [35]. **b** Chingual-Cosanga-Pallatanga-Puna (CCPP) fault system [17]

Andean front fault zone starts east of the Gulf of Guayaquil as the dextral Pallatanga Fault [23, 39, 40]. Northern this fault continues as the Chingual-La Sofia Fault [10, 17, 23, 39, 41]. The dextral motion of NAB along the eastern border is documented by several studies [15, 17, 23, 40, 42].

The motion of the North Andean Block "produces a complex system of active faults that generate shallow-focus earthquakes on the eastern front of the Andes" [13, 30]. The Chingual-Cosanga-Pallatanga-Puna (CCPP) fault system (Fig. 5) is the most developed fault system in the territory of Ecuador and constitutes the eastern tectonic border of the NAB in Ecuador, where several shallow earthquakes have been determined [23]. The CCPP fault system extends from Guayaquil Gulf in Ecuador to the eastern Cordillera Real on the border between Ecuador and Colombia [17], then continues into Colombia as the Algeciras Fault [10]. It should be noted that multiple large earthquakes have occurred in the north–south trending segments of the CCPP fault and in the CR collision zone. The shear zone of CCPP accounts for high cortical seismic activity in the central-northern Andes Cordillera of Ecuador. Large earthquakes have been registered mainly along NE-SW fault systems governed by the Guayaquil-Caracas continental Megashear (GCM).

4 Discussion

The main seismogenic source in Ecuador is the subducting process of the Nazca tectonic plate beneath the South American continental plate. In the background of this process, is important to consider two factors: the influence of CR, which causes a "ploughing" effect (expression suggested in this study) on the shoreline and deep seismic activity in the Pastaza-Napo region at the Amazon basin, and the northward drift of North-Andean block, which produces a complex system of active strike-slip faults and generates shallow-focus earthquakes along the CCPP fault system.

The tectonic activity along the Ecuadorian coastline have stress direction alternately ~NS and ~OE in northern Ecuador [26, 35, 43], and ~NS direction in front of the Carnegie Ridge in the Manta Peninsula and in the Gulf of Guayaquil (Fig. 6) [25, 30], which demonstrates the "ploughing" effect of the inferred continuation of the Carnegie Ridge beneath the territory of Ecuador as a result of the subduction process of the Nazca plate.

The shoreline of Ecuador experienced great megathrust earthquakes along the northern flank of Carnegie Ridge collision zone: Mw 8.8 (1906), Mw 8.2 (1979), Mw 7.8 (1942 and 2016), Mw 7.7 (1958) [27, 31]. The central-north Andean region of Ecuador shows high crustal activity and registers large historical earthquakes along the CCPP fault system: Mw 8.3 (1797), Mw 7.7 (1698), Mw 7 (1645 and 1868), Mw 6.7 (1949), Mw 6.3 (1938), Mw 5.5 (1996). The sub-Andean zone at the Amazon basin shows an intermediate-depth seismicity in the Pastaza-Napo region

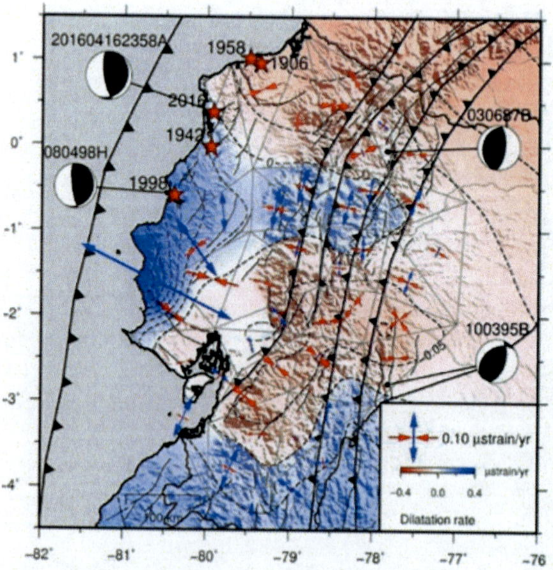

Fig. 6 Principal vectors of the crustal horizontal strain rate tensors in Ecuador. Compression is represented by inside pointing arrows, whereas extension is represented by outward pointing arrows. From: Staller et al., 2018 [30]

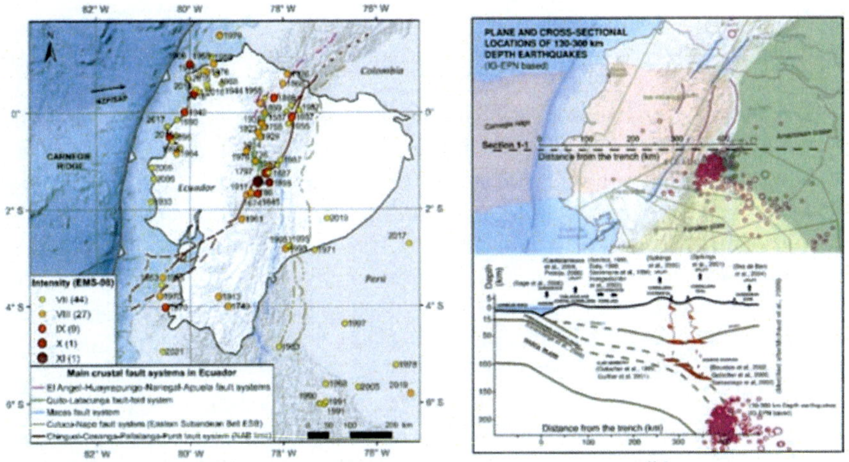

(a) (b)

Fig. 7 **a** Epicentres and years of 1900–2021 earthquakes with macroseismic intensities ≥ VII, where we can see the major seismic intensity at the collision zone, as well as along the CCPP fault system. **b** Plane and cross-sectional locations of 130–300 km depth earthquakes, that show the inferred 400 km prolongation of Carnegie Ridge beneath the territory of Ecuador

and a high shallow-focus activity to the south, between the Macas and Quito-Napo fault systems: Mw 7.5 (2019), Mw 7.4 (1971), Mw 6.6 (1961), Mw 6.5 (1995), Mw 6.4 (1987) [44–47] (Fig. 7a).

The geodynamic and seismic activity in the territory of Ecuador suggest that the prolongation of the Carnegie Ridge and its geological expression of subduction beneath Ecuador requires about 400 km from the Colombia-Ecuador-Peru trench (Fig. 7b), in this figure the locations of depth earthquakes epicentres are according to IG-EPN earthquake catalogue [48]. The southern tear of CR occurs along the Grijalva FZ. The eastern CR interface boundary may be a triple junction of the subducting Nazca tectonic plate with the Farallon plate and the Amazonian Craton, which probably controls the intermediate-depth seismicity in the Pastaza-Napo region at the Amazon basin.

5 Conclusion

The main seismogenic sources in Ecuador are linked to the subduction process of the Nazca tectonic plate beneath the South American continental plate. In this process, the convergence of the Carnegie Ridge, on one hand causes a ploughing effect resulting in shallow-focus earthquakes on the shoreline and deep seismic activity in the Pastaza-Napo region at the Amazon basin, and, on the other hand, the oblique collision of CR generates the northward drift or so called "escape" of the NAB, which produces a complex system of active faults (CCPP fault system as a part of GCM) and produces high seismic activity and shallow-focus earthquakes along the Central and Northern Andes.

The shoreline of Ecuador experienced large megathrust earthquakes along the northern flank of Carnegie Ridge collision zone. The central-north Andean region of Ecuador shows high crustal activity and registers large historical earthquakes along the CCPP fault system. The sub-Andean zone at the Amazon basin shows an intermediate-depth seismicity in the Pastaza-Napo region and a high shallow-focus activity to the south, between the Macas and Quito-Napo fault systems. The geodynamic and seismic activity in the territory of Ecuador suggest that the prolongation of the CR and its geological expression of subduction beneath Ecuador requires about 400 km from the Colombia-Ecuador-Peru trench. The eastern CR interface boundary may be a triple junction of the subducting Nazca Plate with the Farallon plate and the Amazonian Craton, which probably controls the intermediate-depth seismicity in the Pastaza-Napo region at the Amazon basin. The subduction of the prolongation of CR beneath Ecuador is the most influential factor in the development of Ecuador's seismicity.

Acknowledgements The authors gratefully acknowledge the Department of Research and Innovation at the Catholic University of Cuenca in funding this publication.

References

1. Kious WJ, Tilling RI (1996) This dynamic eafth: the story of plate tectonics. U.S. Geological Survey, Washington, D.C., U.S.A
2. United States Geological Survey. Ring of Fire. In: USGS Earthq Gloss. https://earthquake.usgs.gov/learn/glossary/?termID=150. Accessed 20 Feb 2018
3. National Ocean Survey's National Earthquake Information Center (1971) Seismology. About Earthquakes. United States National Earthquake Iinformation Center, Rockville, USA
4. Giesecke A, Gómez Capera AA, Leschiutta I et al (2004) The CERESIS earthquake catalogue and database of the andean region: background, characteristics and examples of use. Ann Geophys 47:421–435. https://doi.org/10.4401/ag-3310
5. Soto-Cordero L, Meltzer A, Bergman E, et al (2020) Structural control on megathrust rupture and slip behavior: insights from the 2016 Mw 7.8 Pedernales Ecuador earthquake
6. Nocquet JM, Jarrin P, Vallée M et al (2017) Supercycle at the Ecuadorian subduction zone revealed after the 2016 Pedernales earthquake. Nat Geosci 10:145–149. https://doi.org/10.1038/ngeo2864
7. Alvarado A, Audin L, Nocquet JM et al (2014) Active tectonics in Quito, Ecuador, assessed by geomorphological studies, GPS data, and crustal seismicity. Tectonics 33:67–83. https://doi.org/10.1002/2012TC003224
8. Mayorga EF, Sánchez JJ (2016) Modelling of Coulomb stress changes during the great (Mw = 8.8) 1906 Colombia-Ecuador earthquake. J South Am Earth Sci 70:268–278. https://doi.org/10.1016/j.jsames.2016.05.009
9. Beauval C, Yepes H, Palacios P et al (2013) An earthquake catalog for seismic hazard assessment in Ecuador. Bull Seismol Soc Am 103:773–786. https://doi.org/10.1785/0120120270

10. Yepes H, Audin L, Alvarado A et al (2016) A new view for the geodynamics of Ecuador: implication in seismogenic source definition and seismic hazard assessment. Tectonics 35:1249–1279. https://doi.org/10.1002/2015TC003941
11. Cediel F, Shaw RP, Cáceres C (2005) Tectonic assembly of the Northern Andean Block
12. Taboada A, Rivera LA, Fuenzalida A et al (2000) Geodynamics of the northern Andes: subductions and intracontinental deformation (Colombia). Tectonics 19:787–813
13. Nocquet JM, Villegas-Lanza JC, Chlieh M et al (2014) Motion of continental slivers and creeping subduction in the northern Andes. Nat Geosci 7:287–291. https://doi.org/10.1038/ngeo2099
14. Kendrick E, Bevis M, Smalley R et al (2003) The Nazca-South America euler vector and its rate of change. J South Am Earth Sci 16:125–131. https://doi.org/10.1016/S0895-9811(03)00028-2
15. Gutscher MA, Malavieille J, Lallemand S, Collot JY (1999) Tectonic segmentation of the North Andean margin: impact of the carnegie ridge collision. Earth Planet Sci Lett 168:255–270. https://doi.org/10.1016/S0012-821X(99)00060-6
16. Champenois J, Baize S, Vallee M et al (2017) Evidences of surface rupture associated with a low-magnitude (Mw5.0) shallow earthquake in the ecuadorian andes. J Geophys Res Solid Earth 122:8446–8458. https://doi.org/10.1002/2017JB013928
17. Alvarado A, Audin L, Nocquet JM et al (2016) Partitioning of oblique convergence in the Northern Andes subduction zone: migration history and the present-day boundary of the North Andean Sliver in Ecuador. Tectonics 35:1048–1065. https://doi.org/10.1002/2016TC004117
18. Dimate C, Drake L, Yepez H et al (1999) Seismic hazard assessment in the Northern Andes (PILOTO project). Ann di Geofis 42:1039–1055. https://doi.org/10.4401/ag-3787
19. Fiorini E, Tibaldi A (2012) Quaternary tectonics in the central interandean valley, Ecuador: fault-propagation folds, transfer faults and the cotopaxi volcano. Glob Planet Change 90–91:87–103. https://doi.org/10.1016/j.gloplacha.2011.06.002
20. Tibaldi A, Rovida A, Corazzato C (2007) Late quaternary kinematics, slip-rate and segmentation of a major cordillera-parallel transcurrent fault: the cayambe-afiladores-sibundoy system, NW South America. J Struct Geol 29:664–680. https://doi.org/10.1016/j.jsg.2006.11.008
21. Baudino R, Hermoza W (2014) Subduction consequences along the andean margin: thermal and topographic signature of an ancient ridge subduction in the Marañón Basin of Perú. Geol Acta 12:287–306. https://doi.org/10.1344/GeologicaActa2014.12.4.2
22. Kellogg JN, Vega V, Stailings TC, Aiken CLV (1995) Tectonic development of Panama, Costa Rica, and the Colombian Andes: constraints from global positioning system geodetic studies and gravity. Spec Pap Geol Soc Am 295:75–90. https://doi.org/10.1130/SPE295-p75
23. Ego F, Sébrier M, Lavenu A et al (1996) Quaternary state of stress in the Northern Andes and the restraining bend model for the Ecuadorian Andes. Tectonophysics 259:101–116. https://doi.org/10.1016/0040-1951(95)00075-5
24. Trenkamp R, Kellogg JN, Freymueller JT, Mora HP (2002) Wide plate margin deformation, southern Central America and northwestern South America, CASA GPS observations. J South Am Earth Sci 15:157–171. https://doi.org/10.1016/S0895-9811(02)00018-4
25. Pedoja K, Dumont JF, Lamothe M et al (2006) Plio-quaternary uplift of the Manta Peninsula and La Plata Island and the subduction of the Carnegie Ridge, central coast of Ecuador. J South Am Earth Sci 22:1–21. https://doi.org/10.1016/j.jsames.2006.08.003
26. Pedoja K, Ortlieb L, Dumont JF et al (2006) Quaternary coastal uplift along the Talara Arc (Ecuador, Northern Peru) from new marine terrace data. Mar Geol 228:73–91. https://doi.org/10.1016/j.margeo.2006.01.004
27. Salocchi AC, Minarelli L, Lugli S et al (2020) Liquefaction source layer for sand blows induced by the 2016 megathrust earthquake (Mw 7.8) in Ecuador (Boca de Briceño). J South Am Earth Sci 103:102737. https://doi.org/10.1016/j.jsames.2020.102737
28. de Berc SB, Soula JC, Baby P et al (2005) Geomorphic evidence of active deformation and uplift in a modern continental wedge-top - foredeep transition: example of the eastern Ecuadorian Andes. Tectonophysics 399:351–380. https://doi.org/10.1016/j.tecto.2004.12.030

29. Witt C, Bourgois J, Michaud F et al (2006) Development of the Gulf of Guayaquil (Ecuador) during the quaternary as an effect of the North Andean block tectonic escape. Tectonics 25:1–22. https://doi.org/10.1029/2004TC001723
30. Staller A, Álvarez-Gómez JA, Luna MP et al (2018) Crustal motion and deformation in Ecuador from cGNSS time series. J South Am Earth Sci 86:94–109. https://doi.org/10.1016/j.jsames.2018.05.014
31. Collot JY, Sanclemente E, Nocquet JM et al (2017) Subducted oceanic relief locks the shallow megathrust in central Ecuador. J Geophys Res Solid Earth 122:3286–3305. https://doi.org/10.1002/2016JB013849
32. Cloos M (1993) Lithospheric bouyancy and collisional orogenesis subduction of continental margins, island arcs and oceanic plateaus (abstract). Geol Soc Am Annu Bull 25:70–71. https://doi.org/10.1130/0016-7606(1993)105<0715
33. **Michaud F, Witt C, Royer JY (2009) Influence of the subduction of the Carnegie volcanic ridge on Ecuadorian geology: reality and fiction. Memoir Geol Soc Am 204:217–228**
34. Hoskins MC, Meltzer A, Font Y et al (2021) Triggered crustal earthquake swarm across subduction segment boundary after the 2016 Pedernales, Ecuador megathrust earthquake. Earth Planet Sci Lett 553:116620. https://doi.org/10.1016/j.epsl.2020.116620
35. Dumont JF, Santana E, Vilema W et al (2005) Morphological and microtectonic analysis of Quaternary deformation from Puná and Santa Clara Islands, Gulf of Guayaquil, Ecuador (South America). Tectonophysics 399:331–350. https://doi.org/10.1016/j.tecto.2004.12.029
36. Pedoja K (2003) Les terrasses marines de la marge Nord Andine (Equateur et Nord Pérou): relations avec le contexte géodynamique. Pierre and Marie Curie University
37. Manchuel K, Régnier M, Béthoux N et al (2011) New insights on the interseismic active deformation along the North Ecuadorian-South Colombian (NESC) margin. Tectonics 30:1–25. https://doi.org/10.1029/2010TC002757
38. Egbue O, Kellogg J (2010) Pleistocene to present north andean "escape." Tectonophysics 489:248–257. https://doi.org/10.1016/j.tecto.2010.04.021
39. Soulas JP, Eguez A, Yepes H, Perez H (1991) Active tectonics and seismic hazard in the Ecuadorian Andes and the extreme south of Colombia. Ecuadorian Geol Bull 2:3–11
40. Winter T, Avouac J-P, Lavenu A (1993) Late Quaternary kinematics of the Pallatanga strike-slip fault (Central Ecuador) from topographic measurements of displaced morphological features. Geophys J Int 115:905–920. https://doi.org/10.1111/j.1365-246X.1993.tb01500.x
41. Tibaldi A, Ferrari L (1992) Latest pleistocene-holocene tectonics of the ecuadorian andes. Tectonophysics 205:109–125. https://doi.org/10.1016/0040-1951(92)90421-2
42. Wong I, Dober M, Hemphill-Haley M, Terra F (2012) Seismic hazard along the southern coast of Ecuador. In: 15 WCEE, Lisboa, p 10
43. **Dumont JF, Navarrete E, Martillo C (2010) Observatory of the Ecuadorian coast. Scientific** Report. (in Spanish: Observatorio del Litoral Ecuatoriano). Guayaquil
44. Beauval C, Yepes H, Bakun WH et al (2010) Locations and magnitudes of historical earthquakes in the Sierra of Ecuador (1587–1996). Geophys J Int 181:1613–1633. https://doi.org/10.1111/j.1365-246X.2010.04569.x
45. Swenson JL, Beck SL (1996) Historical 1942 Ecuador and 1942 Peru subduction earthquakes, and earthquake cycles along Colombia-Ecuador and Peru subduction segments. Pure Appl Geophys 146:67–101. https://doi.org/10.1007/bf00876670
46. United States Geological Survey. Earthquake catalog. In: USGS Earthq Hazards program. https://earthquake.usgs.gov/earthquakes/search/. Accessed 30 July 2021
47. Instituto Geofísico de la Escuela Politécnica Nacional (IG-EPN) Catálogo Homogenizado 1587–2011. In: Sismicidad. https://igepn.edu.ec/mapas/sismicidad/mapa-catalogo-homogenizado. Accessed 30 July 2021
48. Instituto Geofísico de la Escuela Politécnica Nacional (IG-EPN) (2021) Catalogue of earthquakes in Ecuador. In: Mapa Digit. Interact. peligro sísmico para Ecuador. https://www.igepn.edu.ec/mapas/sismicidad/mapa-peligro-sismico. Accessed 23 Sept 2021

Consideration of External Reinforcement Based on Carbon Fibers for the Redistribution of Bending Moments in Slabs

Oleg Simakov

Abstract The use of external reinforcement systems to strengthen reinforced concrete structures has become widespread due to its high efficiency and manufacturability. At the same time, despite a sufficient number of regulatory documents on the main areas of calculations, there are a number of practical application issues that require detailed study and elaboration. One of the issues is the redistribution of bending moments in statically indeterminate systems. In this article, the question of the influence of external reinforcement on the redistribution of bending moments in a monolithic reinforced concrete floor is considered. Various options for the use of external reinforcement to strengthen the span and support zones are considered. The requirements of the norms and the results of real experiments are taken into account. Based on the results, a conclusion is made about the influence of external reinforcement on the redistribution of bending moments in the plate, as a result of the influence on the utilization factor of the load-bearing capacity of the reinforced section.

Keywords FRP · Moment redistribution · Slab

1 the Status of the Issue of the Application of External Reinforcement

External reinforcement systems are actively used to strengthen reinforced concrete structures of monolithic girderless floors (Fig. 1).

At the moment, on the basis of experimental studies, the nature of deformation of bent reinforced concrete elements with reinforcement by external reinforcement based on carbon fibers has been determined [1–21]. The influence of external reinforcement on the redistribution of forces in the bent elements, taking into account the physical nonlinearity of reinforced concrete, is also determined.

O. Simakov (✉)
Moscow State University of Civil Engineering, Moscow 129337, Russian Federation
e-mail: simakov.oa@mgsu.ru

P. Akimov et al. (eds.), *Proceedings of FORM 2022*, Lecture Notes in Civil Engineering 282, https://doi.org/10.1007/978-3-031-10853-2_29

(a) (b)

Fig. 1 Reinforcement of girderless ceilings **a** span zone; **b** support zone

In general, the dependence of the stiffness change is analytically described only in [2] and has the form (in relation to the moment of inertia of the section):

$$I_{red} = I_b + f \cdot I_s + f \cdot I'_s + f \cdot I_f \tag{1}$$

where f is the reduction coefficient.

In this case, the stiffness of the element, as the main desired value, is determined by the formula:

$$D = E_{b1} \cdot I_{red} \tag{2}$$

where Eb1 is the modulus of deformation of compressed concrete taking into account the duration of the load;

At the same time, for the case of crack formation, the moment of inertia is determined by the formula:

$$I_{red} = \frac{bx^3}{12} + bx\left(\frac{x}{2}\right)^2 + \alpha \cdot A_s\left(x - a'\right)^2 + \alpha \cdot A'_s(x - a)^2 + \alpha_f \cdot A_f(h - x)^2 \tag{3}$$

Taking into account the lowering coefficients:

$$\mu_s = \frac{A_s}{bh_0}; \mu'_s = \frac{A'_s}{bh_0}; \mu_s = \frac{A_f}{bh_0}$$

As can be seen from this dependence, the potential contribution of external reinforcement to the formation of the overall rigidity of the reinforced element is quite small (the cross-sectional area of external reinforcement is not commensurably small in relation to the area of concrete and steel reinforcement).

The experiments carried out (mainly single-span and multi-span beams) showed a significant deviation of the analytical stiffness from the experimental one. The generalized dependence is shown in Fig. 2.

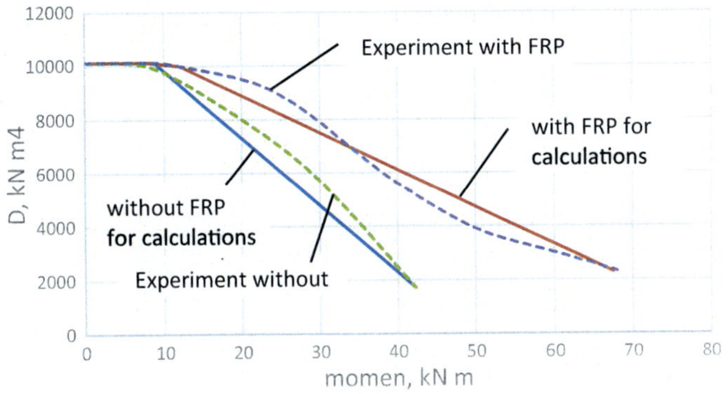

Fig. 2 The dependence of the rigidity of a reinforced concrete element with external reinforcement on the current moment (take into account the moment of crack formation and adjustment of the calculation formula)

2 Numerical Simulation

Based on the presented graph, the following modeling options for external reinforcement accounting were considered:

- analytical calculation of stiffness according to the method [2] and subsequent numerical calculation of bending moment isofields;
- determination of stiffness based on experimental data followed by numerical modeling.

The construction of the floor slab with a 3 × 3 cell cross section and a size of 7 × 7 m was chosen as the object of the study (Fig. 3).

The following schemes were selected as the studied ones (Fig. 4):

Calculation results.

1. Without taking into account the effect of changes in stiffness (method [2])

   This technique is characterized by almost no effect of external reinforcement on the final stiffness of the section.

   Based on the results of calculations (Fig. 5):

   - reference torque 10.6 t m
   - span torque 2.86 t m

2. Strengthening of span zones, determination of stiffness based on experimental data.

   Based on the results of calculations (Fig. 6):

   - reference moment 10.18 t m
   - span moment 3.2 t m

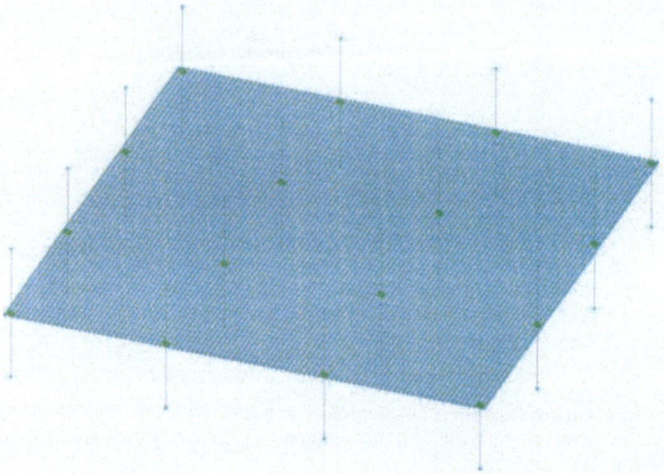

Fig. 3 Design scheme of overlap

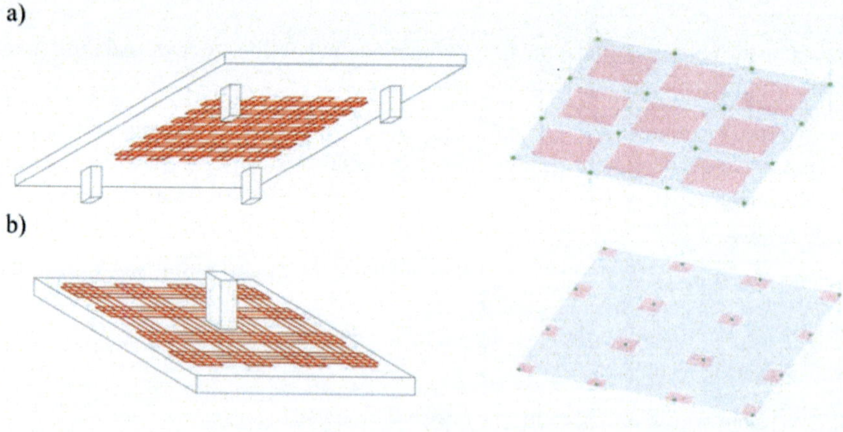

Fig. 4 Design schemes: **a** strengthening of the span zone, **b** strengthening of the reference zone

3. Reinforcement of support zones, determination of stiffness based on experimental data.

   Based on the results of calculations (Fig. 7):

   – reference moment 12.06 t m
   – span moment 2.61 t m

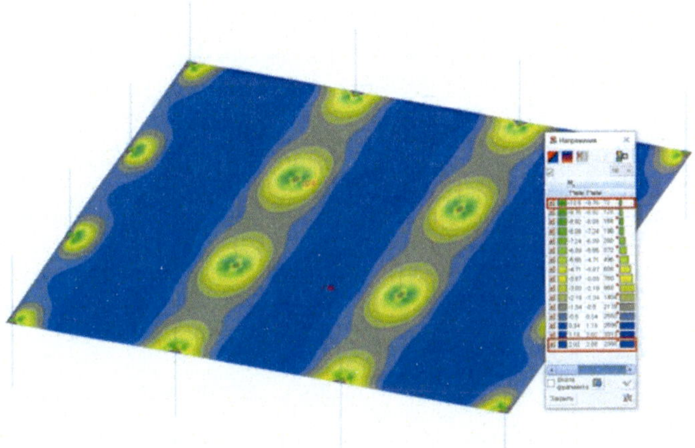

Fig. 5 Calculation results for the calculated modulus of elasticity

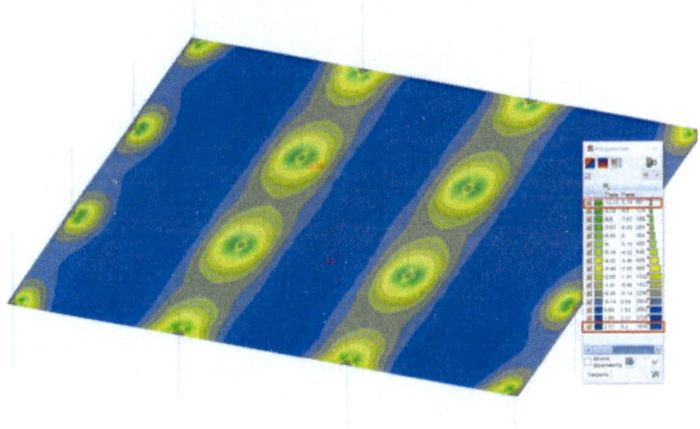

Fig. 6 The results of calculations when correcting the elastic modulus of the span zone

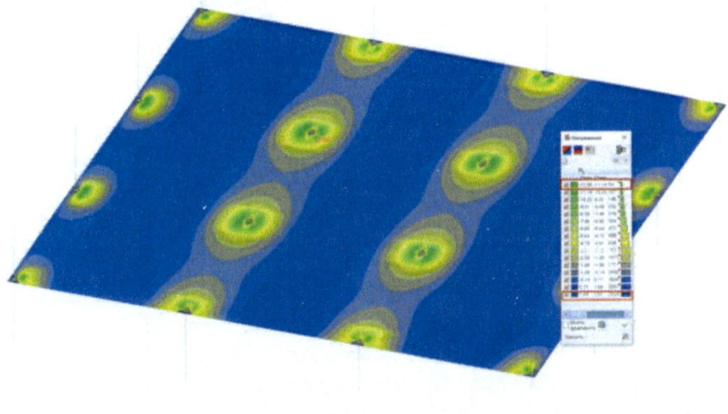

Fig. 7 The results of calculations when correcting the elastic modulus of the base zone

Table 1 The numerical results of the analysis

	Analytical calculation of stiffness	Strengthening of the span zone (stiffness according to experimental data)	Reinforcement of the reference zone (stiffness according to experimental data)
Span moment, t m	2,86/0	3,2/ + 12%	2,61/–8,7%
Base moment, t m	10,6/0	10,18/–4,0%	12,06/ + 13,8%

3 Conclusion

Based on the results of the numerical analysis, the following conclusions can be drawn:

1. The influence of external reinforcement on the amount of bending moment when reinforcing the floor slabs is in the range of up to 14% (within the framework of the considered scheme). The summarized data are given in Table 1.
2. It is advisable to evaluate stiffness based on experimental data, since the existing model for determining stiffness does not show good convergence of results. To do this, basic nomograms can be developed at the initial stages.
3. In general, taking into account the previously determined load-bearing capacity coefficient (at least 20%), in most cases, failure to take into account the redistribution of moments during the reinforcement of floor slabs will not lead to critical cases (failure).

Acknowledgements This work was financially supported by the Ministry of Science and Higher Education of the Russian Federation (grant # 075-15-2021-686). All tests were carried out using research equipment of The Head Regional Shared Research Facilities of the National Research Moscow State University of Civil Engineering.

References

1. Tamrazyan AG, Fedorova NV (2016) Ocenka nadezhnosti zhelezobetonnyh konstrukcij, usilennyh ugleplastikovym vneshnim armirovaniem. Izvestiya vysshih uchebnyh zavedenij Tekhnologiya tekstil'noj promyshlennosti 6(366):226–231
2. Mukhamediev TA, Kuzevanov DV (2013) Calculation of the strength of normal sections of reinforced concrete structures reinforced with composite materials. Conc Reinforced Conc 6:20–24
3. Cohn MZ, Lounis Z (1991) Moment redistribution in structural concrete codes. Can J Civ Eng 18(1):97–108
4. Gribniak V, Arnautov A, Kaklauskas G, Jakstaite R, Tamulėnas V, Gudonis E (2014) Deformation analysis of RC ties externally strengthened with FRP sheets. Mech Comp Mater 50:669–676. https://doi.org/10.1007/s11029-014-9454-7

5. Yeghnem R, Meftah SA, Benyoucef S, Tounsi A, Bedia EAA (2013) A unite-element model for the lateral stiffness and vibration characteristics of RC shear walls strengthened with composite sheets: creep and the shrinkage effect. Mech Compos Mater 49(2):181–192
6. Casadei P, Nanni A, Galati N, Ibell T, Denton S (2103) Moment redistribution in continuous CFRP strengthened concrete members: experimental results. In: International conference composites in construction - CCC2003, Cosenza, Italy, 16–19 September 2003, pp 307–312
7. Kesminas D, Tamulenas V (2014) Investigation of deformation and cracking behavior of RC ties strengthened with FRP sheets. Sci Fut Lithuania Civ Eng (2014)
8. Gribniak V, Cervenka V, Kaklauskas G (2013) Deflection prediction of reinforced concrete beams by design codes and computer simulation. Eng Struct 56:2175–2186
9. Gribniak V, Kaklauskas G, Kliukas R, Jakubovskis R (2013) Shrinkage effect on short-term deformation behavior of reinforced concrete: when it should not be neglected. Mater Design 51:1060–1070
10. Ashour AF, El-Refaie SA, Garrity SW (2004) Flexural strengthening of RC continuous beam using CFRP laminates. Cement Conc Compos 26:765–777
11. Tajaddini A (2015) Investigation of moment redistribution in FRP-strengthened slabs and T-beams. Ph.D. thesis. University of Bath, UK
12. Tajeddini A, et al (2016) Quantifying moment redistributi on in FRP-strengthened RC beams. Struct Build
13. Garden HN, Hollaway LC (1998) An experimental study of the influence of plate end an chorage of carbon fibre composite plates used to strengthen reinforced concrete beams. Compos Struct 42:175–188
14. Breveglieri M, Barros JA, Dalfré GM, Aprile A (2012) A parametric study on the effectiveness of the NSM technique for the flexural strengthening of continuous RC slabs. Compos B Eng 43(4):1970–1987
15. Lou T, Lopes SM, Lopes AV (2015) Neutral axis depth and moment redistribution in FRP and steel reinforced concrete continuous beams. Compos B Eng 70:44–52
16. Biscaia HC, Chastre C, Silva MA (2013) Linear and nonlinear analysis of bond-slip models for interfaces between FRP composites and concrete. Compos Pt B: Eng 45(1):1554–1568
17. Lushnikova VY, Tamrazyan AG (2018) The effect of reinforcement corrosion on the adhesion between reinforcement and concrete. Mag Civ Eng 80(4):128–137
18. Silva PF, Ibell TJ (2008) Evaluation of moment distribution in continuous fibre-reinforced polymerstrengthened concrete beams. ACI Struct J 105(6):729–739
19. Tamrazyan AG (2018) Calculation of reinforced concrete plates with hole at long-term loading. IOP Conf Ser Mater Sci Eng 365(5):052021
20. Tamrazyan AG, Fedorova NK (2016) Reliability assessment of reinforced concrete structures, strengthening by external reinforcement with carbon fiber. Izvestiya Vysshikh Uchebnykh Zavedenii Seriya Teknologiya Tekstil'noi Promyshlennosti 366(6):226–230
21. Tamrazyan AG, Alekseytsev AV (2019) Evolutionary optimization of reinforced concrete beams, taking into account design reliability, safety and risks during the emergency loss of supports. In: E3S web of conferences, vol 97, p 04005

Hardening Kinetics and Strength of Nanomodified Cement Composites

Anna Grishina and Evgenij Korolev

Abstract Nanomodification of building materials requires the involvement of theoretical knowledge to justify the choice of the type of nanomodifier, as well as experimental studies to establish the method of its introduction, concentration and distribution in the volume of the composite. Failure to comply with this leads to the occurrence of cases in which the low efficiency of nanomodification is demonstrated. One of the effective ways to reduce the complexity in controlling the properties of a composite using a nanomodifier is to establish the kinetics of structure formation of the modified composite. Therefore, the aim of the work is to establish the effect of the composition and concentration of nano- and micromodifiers on the strength and kinetics of strength development of cement composites. For research, 2 types of composite cement binders and nanomodifiers based on zinc hydrosilicates were used. An analysis was made of the kinetics of hardening and compressive strength of nanomodified composite stones. It has been established that the effectiveness of nanomodification is determined by the microstructure of the modified binder and the composition of the nanomodifier. The highest strength of the composites is observed when using nanosized zinc hydrosilicates manufactured at $\alpha = 1.0$. The strength of the composite binder depends on the rate of strength gain by the composite; at a high rate of strength gain, its value is lower. This makes it possible to control the strength of the obtained materials by adjusting the rate of curing and optimizing the microstructure of the nanomodified binder.

Keywords Strength · Composite binder · Nanomodification · Nanosized zinc hydrosilicates · Hardening kinetics

A. Grishina (✉)
Moscow State University of Civil Engineering, Yaroslavskoye Shosse, 26, Moscow 129337, Russian Federation
e-mail: GrishinaAN@mgsu.ru

E. Korolev
Saint Petersburg State University of Architecture and Civil Engineering, Vtoraya Krasnoarmeiskaya street, 4, Saint Petersburg 190005, Russian Federation
e-mail: prorector_nr@spbgasu.ru

P. Akimov et al. (eds.), *Proceedings of FORM 2022*, Lecture Notes in Civil Engineering 282, https://doi.org/10.1007/978-3-031-10853-2_30

1 Introduction

One of the effective ways to control the properties of cement stones is the use of nano-modifying additives [1]. However, the results obtained are sometimes contradictory [2, 3], which raises questions regarding the advisability of using nanomodifiers [4]. One of the reasons for the difficulties in controlling the properties of a composite with the help of nanomodifiers may be the features of the formation of the microstructure of the composite in the presence of a nanomodifier [5, 6]. At the same time, the modification of composite binders pre-optimized with microsized additives is of interest. This approach provides comprehensive control of the structure and properties of the material at the micro- and nanolevels. Therefore, nanomodifiers of the same composition and concentrations will probably make it possible to obtain composites of various strengths while optimizing the microstructure of the material. Also of interest is the kinetics of strength development of such composites. For cement systems, it is advisable to use modifiers of a silicate nature as modifiers, for example, metal hydrosilicates [6–10]. Therefore, the aim of the research is to establish the effect of the composition and concentration of nano- and micromodifier of silicate nature on the strength and kinetics of strength development of composites.

2 Methods and Materials

Two types of composite binders were used for research: 1) a binder containing cement and micro-sized barium hydrosilicates in an amount of 10% by weight of cement, and 2) a binder containing cement, micro-sized barium hydrosilicates in an amount of 10% and zinc hydrosilicates in an amount of 2.0% by weight of cement. The synthesis of micro-sized metal hydrosilicates was carried out using the technology of hydrochemical synthesis. The average particle diameter of the obtained modifiers was 5…7 microns. The chemical composition of the modifying additive based on barium hydrosilicates additionally contains a small amount of barium carbonates formed during carbonation of barium salt as a result of mixing during synthesis. A modifying additive based on zinc hydrosilicates additionally contains a small amount of $Zn_4SO_4(OH)_6 \cdot nH_2O$. Nanoscale zinc hydrosilicates manufactured according to the technology presented in [11] and possessing sedimentation stability [12] were used for the nanomodification of the binder. The gross formulas of the nanomodifiers used are given in Table 1. The concentrations of the nanomodifiers used are shown in Table 2. The average particle size is shown in Table 3.

Composite binders are made by mixing cement with metal hydrosilicates until homogeneous. The resulting binders were mixed with colloidal solutions of zinc hydrosilicates until a dough of normal density (water-binding ratio was 0.32) was obtained, from which samples were molded and stored in a normal hardening chamber. The compressive strength was determined through 1; 3; 7; 14 and 28 days curing on an Advantest 9 hydraulic press.

Table 1 Gross formulas of colloidal solutions

α	Concentration of $Fe(OH)_3$ ($C(Fe(OH)_3)$), %		
	0.3	0.5	0.7
0.5	$ZnO{\cdot}2.35SiO_2{\cdot}nH_2O$	$ZnO{\cdot}3.92SiO_2{\cdot}nH_2O$	$ZnO{\cdot}5.49SiO_2{\cdot}nH_2O$
1.0	$ZnO{\cdot}4.70SiO_2{\cdot}nH_2O$	$ZnO{\cdot}7.84SiO_2{\cdot}nH_2O$	$ZnO{\cdot}10.97SiO_2{\cdot}nH_2O$
1.5	$ZnO{\cdot}7.05SiO_2{\cdot}nH_2O$	$ZnO{\cdot}11.76SiO_2{\cdot}nH_2O$	$ZnO{\cdot}16.46SiO_2{\cdot}nH_2O$

Table 2 Concentrations of solutions

α	Concentration of $Fe(OH)_3$ ($C(Fe(OH)_3)$), %		
	0.3	0.5	0.7
0.5	0.0282	0.0410	0.0539
1.0	0.0474	0.0732	0.0989
1.5	0.0667	0.1053	0.1438

Table 3 The average particle diameter of metal hydrosilicates, nm

α	Concentration of $Fe(OH)_3$ ($C(Fe(OH)_3)$), %		
	0.3	0.5	0.7
0.5	13.3	12.9	14.7
1.0	11.3	15.8	21.0
1.5	13.1	20.8	29.7

3 Results

The general appearance of the strength gain curves is identical and has a classical shape. Examples of the kinetics of increasing the strength of the obtained nanomodified artificial composite cement stones are shown in Figs. 1, 2, 3, 4 and 5.

The obtained dependences of strength development (Figs. 1, 2, 3, 4 and 5) are described by a function of the form (determination coefficient $R^2 > 0.9$):

$$R_{st} = R_{max}(1 - e^{-bt}) \quad (1)$$

where R_{st} – the compressive strength of the cement stone, MPa; R_{max} – maximum strength of cement stone, MPa; b – empirical coefficient characterizing the rate of strength increase; t – time, day.

The values of the empirical coefficients for the composition containing cement and microsized barium hydrosilicates are presented in Table 4. The values of empirical coefficients for a composition containing cement and microsized barium and zinc hydrosilicates are presented in Table 5.

The grade strength values of nanomodified composite cement stone containing barium hydrosilicates are shown in Fig. 6. Grade strength values of nanomodified

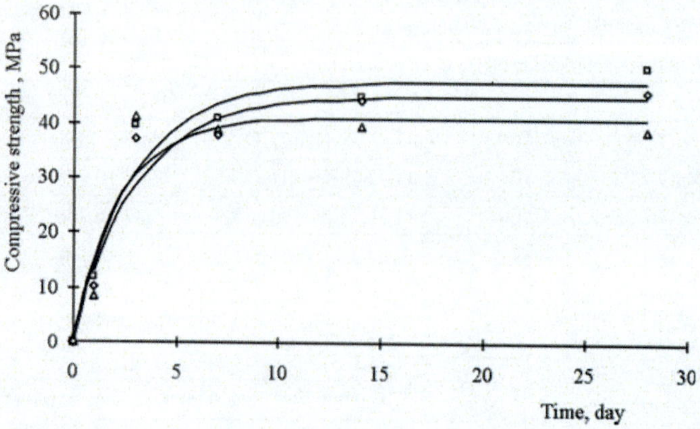

Fig. 1 Kinetics of strength increase by nanomodified ($C(Fe(OH)_3) = 0.5\%$) composite cement stone containing microsized barium hydrosilicates, where $\Delta - \alpha = 0.5$; $\square - \alpha = 1.0$; $\lozenge - \alpha = 1.5$

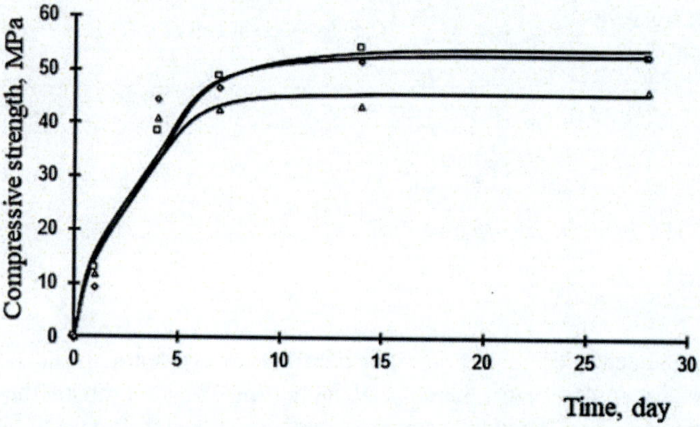

Fig. 2 Kinetics of strength increase by nanomodified ($C(Fe(OH)_3) = 0.7\%$) composite cement stone containing microsized barium hydrosilicates, where $\Delta - \alpha = 0{,}5$; $\square - \alpha = 1{,}0$; $\lozenge - \alpha = 1{,}5$

composite cement stone containing barium and zinc hydrosilicates are shown in Fig. 7.

4 Discussion

Analysis of the R_{max} values in Tables 4 and 5 and in Figs. 6, 7 shows that the strength of nanomodified cement stones differs significantly depending on the composition of the nanomodifier, as well as depending on the microstructure of the composites. It should be noted that the presence of microsized zinc hydrosilicates in the binder has

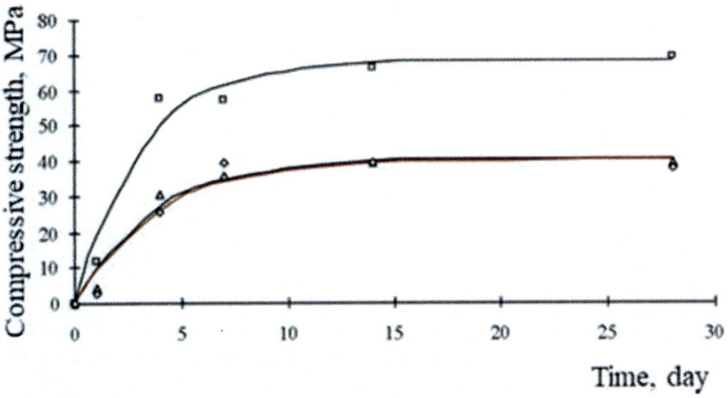

Fig. 3 Kinetics of strength increase by nanomodified ($C(Fe(OH)_3) = 0.3\%$) composite cement stone containing microsized hydrosilicates of barium and zinc, where Δ – α = 0,5; □ – α = 1,0; ◊ – α = 1,5

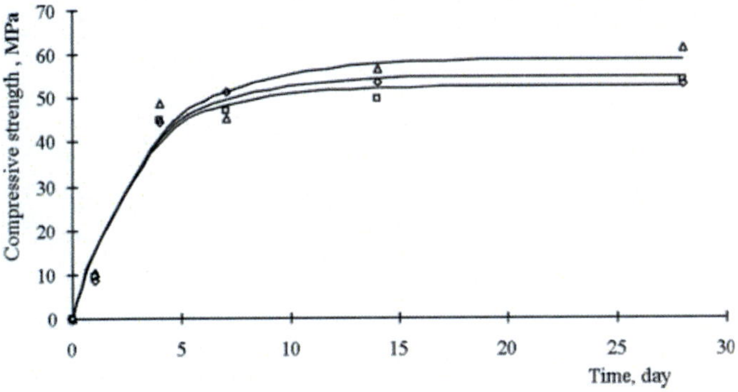

Fig. 4 Kinetics of strength increase by nanomodified ($C(Fe(OH)_3) = 0.5\%$) composite cement stone containing microsized hydrosilicates of barium and zinc, where Δ – α = 0.5; □ – α = 1.0; ◊ – α = 1.5

a positive effect on the strength of the obtained nanomodified stones. Thus, the structure of the composite at the microlevel affects the efficiency of nanomodification of the composite. Therefore, the efficiency of using a nanomodifier should be evaluated on composites with different microstructures. As microsize modifiers for controlling the microstructure of a composite, it is expedient to use additives that give the composites specified special properties. In this paper, microsized zinc hydrosilicates were additionally used to control the structure of the material, which can increase the biostability of the composite [13–15]. In addition, it is known from articles [16, 17] that a modifier based on zinc hydrosilicates can affect the rate of cement hydration. So, with the content of zinc hydrosilicates in the amount of 2%, the rate of cement hydration increases. However, a change in the concentration of the silicate phase in can significantly change the rate of hydration, both increasing and decreasing its

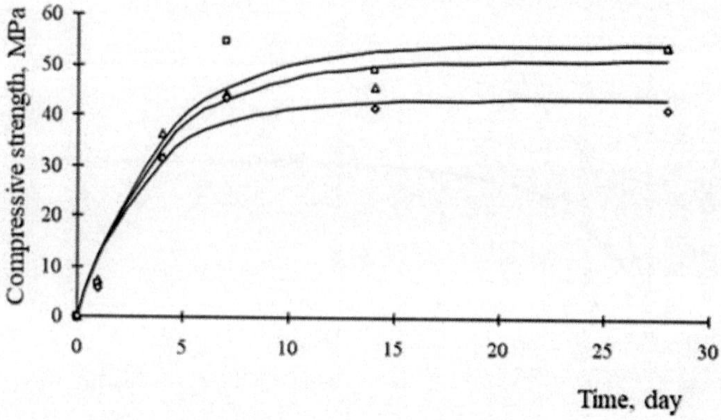

Fig. 5 Kinetics of strength increase by nanomodified ($C(Fe(OH)_3) = 0.7\%$) composite cement stone containing microsized hydrosilicates of barium and zinc, where Δ – α = 0,5; □ – α = 1,0; ◊ – α = 1,5

Table 4 Values of empirical coefficients

Concentration of $Fe(OH)_3$ ($C(Fe(OH)_3)$), %	α	R_{max}	*b*
0.3	0.5	40.82	0.463
	1.0	47.71	0.351
	1.5	44.94	0.340
0.5	0.5	42.16	0.476
	1.0	49.24	0.323
	1.5	44.14	0.425
0.7	0.5	45.42	0.422
	1.0	53.89	0.308
	1.5	52.75	0.340

Table 5 Values of empirical coefficients

Concentration of $Fe(OH)_3$ ($C(Fe(OH)_3)$), %	α	R_{max}	*b*
0.3	0.5	40.60	0.281
	1.0	68.20	0.330
	1.5	40.60	0.262
0.5	0.5	58.67	0.302
	1.0	52.68	0.352
	1.5	54.76	0.337
0.7	0.5	51.26	0.263
	1.0	54.42	0.255
	1.5	43.36	0.309

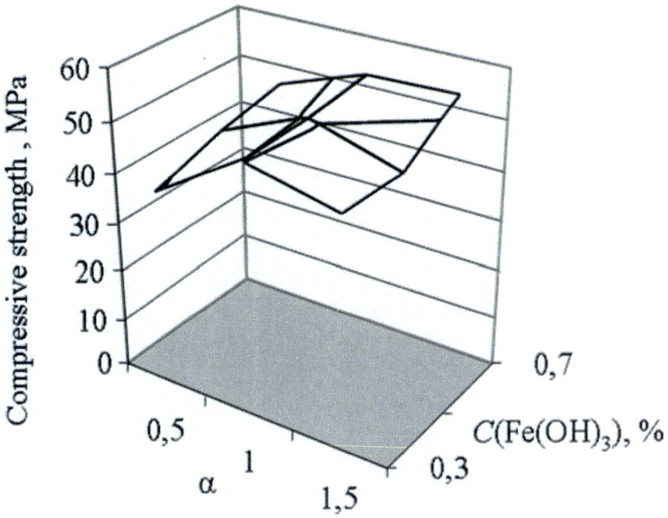

Fig. 6 Strength of nanomodified composite cement stone containing barium hydrosilicates

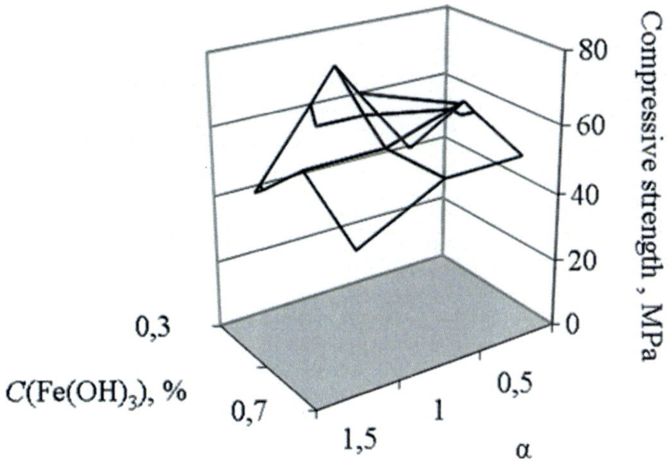

Fig. 7 Strength of nanomodified composite cement stone containing barium and zinc hydrosilicates

value [18–20]. In this case, the introduction of a nanomodifier additionally changes the concentration of silicic acid in the hydrating system and, accordingly, the rate of formation of reaction products. Analyzing the change in the strength of composites containing cement and barium hydrosilicates, it should be noted that the highest strength is achieved when using nanosized zinc hydrosilicates manufactured at $\alpha = 1.0$, regardless of the concentration of $Fe(OH)_3$. It is also important to note that for these compositions, a relatively low rate of strength development is observed, which decreases with increasing $Fe(OH)_3$ concentration. At the same time, when comparing concentrations, silicate modules, and particle sizes of zinc hydrosilicates (Tables 1, 2, and 3), it should be noted that there are intersections with the indicated indicators for other nanomodifiers, however, slowing down the rate of increase in strength when using them. The strength of nanomodified composite cement stones containing barium and zinc hydrosilicates is also higher at $\alpha = 1.0$, the deviation is observed only at $C(Fe(OH)_3 = 0.5\%$ and $\alpha = 1.0)$.

The activity of the silicate nanomodifier is significantly affected by the ratio of ions in the nanomodifier and the pH of the solution [21, 22]. Analyzing the values of R_{max} and b, it should be noted that at a high value of the rate of strength development (the value of the coefficient $b = 0.4$), the formed grade strength of the composite is lower. With a decrease in the hardening rate, the strength of the composite may increase. It is obvious that the use of a microsized additive based on zinc hydrosilicates makes it possible to reduce the rate of strength development, however, the graded strength of the obtained composites increases.

5 Conclusion

The strength of nanomodified composite binders is determined both by the composition and state of the nanomodifier (ion ratio, pH) and by the microstructure of the modified composite. Preliminary optimization of the microstructure of the composite will improve the efficiency of the use of nanomodifiers. Also, the value of the grade strength is determined by the kinetics of the strength gain of the composites. Thus, at a high rate of strength development, which is characterized by an empirical coefficient b (at $b = 0.4$), the grade strength of the forming composite is lower.

Acknowledgements This work was financially supported by the Ministry of Science and Higher Education of the Russian Federation (Project: Theoretical and experimental design of new composite materials to ensure safety during the operation of buildings and structures under conditions of technogenic and biogenic threats #FSWG-2020-0007).

References

1. Ahmed H, Bogas JA, Guedes M, Pereira MFC (2019) Dispersion and reinforcement efficiency of carbon nanotubes in cementitious composites. Mag Conc Res 71(8):408–423
2. Ramezan M, Kim YH, Sun Z (2019) Mag Conc Res 72(20):1047–1071
3. Korolev EV (2014) Build Mater (In Russian – Stroit Mater) 11:47–79
4. Korolev EV, Inozemtcev AS (2012) Build Mater (In Russian– Stroit Mater)4:76–79
5. Korolev EV (2013) Build Mater (In Russian – Stroit Mater) 6:60–64
6. Loganina V, Zhegera K, Fediuk R, Timokhin R, Zayakhanov M, Liseitsev Y (2020) J Mat Civ Eng 32(5):06020004
7. Loganina VI, Makarova LV, Tarasov RV, Zhegera KV (2014) Adv Mat Res 1022:3–6
8. Kadri EH, Kenai S, Ezziane K, Siddique R, De Schutter G (2011) Appl Clay Sci 53:704–708
9. Shah V, Scott A (2021) Cem Concr Compos 121:104068

10. Maksimova IN, Makridin NI, Surov IA (2014) Reg Archit Constr (In Russian – Reg Archit Constr) 2:63–68
11. Grishina A, Korolev E (2018) E3S web conf 33:02035
12. Grishina AN, Korolev EV, Satukov AB (2013) Build Mater (In Russian– Stroit Mater) 9:91–93
13. Riduan SN, Zhang Y (2021) Chemistry 16:2588–2595
14. Erofeev V, Rodin A, Rodina N, Kalashnikov V, Erofeeva I (2016) Procedia Eng 165:1448–1454
15. Janczak K, Kosmalska D, Kaczor D, Wedderburn L, Malinowski R (2021) Materials 14:4228
16. Grishina AN, Korolev EV, Gladkikh VA (2022) Materials 15:292
17. Ramachandran SV, Feldman RF, Kollepardi M, Malhotra VM, Dolch VL, Mehta PK, Ohama I, Ratinov VB, Rosenberg TI, Mailwagans NP (1988) Concrete admixtures ref. Stroyizdat, Moscow, p 142
18. Langan BW, Weng K, Ward MA (2002) Cem Conc Res 32(7):1045–1051
19. Wang F, Chen P, Li X, Zhu B (2018) Materials 11(10):1849
20. Zelić J, Rušić D, Veža D, Krstulović R (2000) Cem Conc Res 30(10):1655–1662
21. Shabanova NA, Belova IA, Markelova MN (2020) Glass Phys Chem 46:84–89
22. Shabanova NA (1998) J Coll (In Russian – Kolloid J) 60:651–654

End-Plate Beam-to-Column Joints in the Steel Framework

Alexander Tusnin and Valeria Platonova

Abstract An integral element in the steel frame of a multistoried building is the joints of beams with columns. Accounting for their actual response allows more accurately and reliably assess the stress-strain state of the entire frame and correctly designing the building. End-plate beams-to-columns joints are widespread and commonly used due to reliability, simplicity of design and installation. By the response, end-plate joints the most often occupy an intermediate position between rigid joints and flexible joints. The flexibility of joints is affected by their design, bolts pretension, and ratio of column and beam stiffness. Initial imperfections have a significant influence on the response of the end-plate joints. It can be caused by tolerances and offset in rolling-steel, deviations from the design geometry during building erection. This article analyzes Russian and international experience in structural calculating and designing end-plate joints in steel constructions. The issues of the importance of taking into account the response of the elements of joint and the necessity to develop a methodology for calculating this type of connections, including initial gaps between adjacent surfaces. A program for investigating the influence of initial gaps with the choice of mathematical methods for analyzing numerical data is presented. A method for estimating the confidence interval of the change in the characteristics of the joint, considering by the rolling-steel imperfections is proposed.

Keywords Beam-to-column connection · End-plate bolted steel joints · Initial gaps · Semi-rigid joint

1 Introduction

The connection of beams with columns in steel frames must be reliable, durable, technologically advanced and easy to build erection. The actual response of joints must be in accordance with the design assumptions and regulatory requirements for

A. Tusnin (✉) · V. Platonova
Moscow State University of Civil Engineering, Moscow 129337, Russian Federation
e-mail: TusninAR@mgsu.ru

P. Akimov et al. (eds.), *Proceedings of FORM 2022*, Lecture Notes in Civil Engineering 282, https://doi.org/10.1007/978-3-031-10853-2_31

ensuring strength, stability and deformability. It is necessary to take into account the influence of the flexibility of joints on the response of the steel framework.

It is convenient to assemble the frame elements at the installation using bolted connections. With the use of welding, it is possible to perform quite complex assemblies; however, for high-quality welding work, compliance with certain conditions is required, which cannot always be provided on the building yard. One of the most convenient solutions from the point of view of simplicity of design and speed of building installation are end-plate connections of steel beams and columns. These joints, with proper manufacturing quality and a minimum number of added elements, are able to ensure reliable coupling of the elements being joined. To increase the rigidity of the joints, transverse stiffener connections are arranged on the column, and the thickness of the end plate of the beam increases [1–4]. A rise in the rigidity of the connections leads to an excessive margin of safety, increases its material consumption and complicates the design. Joints assemblies without transverse ribs on the column and flanges, the thickness of which is determined by strength, are more rational. These joints are more compact, have less material consumption, but have a higher yielding (Fig. 1a, b).

Bolts are an important element that ensures the operation capacity of connections. The effective inclusion of bolts in the operation of the joint is ensured by their pretension. The bolts in the end-plate connection are in a complex stressed state and work in tension with bending. The reason of bolt's flexure is the presence of gaps due to rolling-steel tolerances and building installation inaccuracy. These gaps increase under the action of the load. The presence of gaps leads to the fact that previously tightened high-strength bolts lose up to 20–30% of the tension forces when installing subsequent rows of bolts. The greatest losses occur in swamps located in the area of the beam flange [5]. According to paper [6], deformations in the tension area of the column web grow faster when using bolts of class 10.9. The use of bolts of class 8.8 allows for a uniform character of column web deformation due to the greater plasticity of the bolts. Before the development of plastic flow in bolts, the strength class does not significantly affect deflections, and with an increase in load, higher bolt strength can reduce the bearing capacity of the joint. It is noted in [7] that the presence of pre-stressing bolts does not affect the load-bearing capacity of a joint operating under static loads, but significantly affects the strain capacity of the joint.

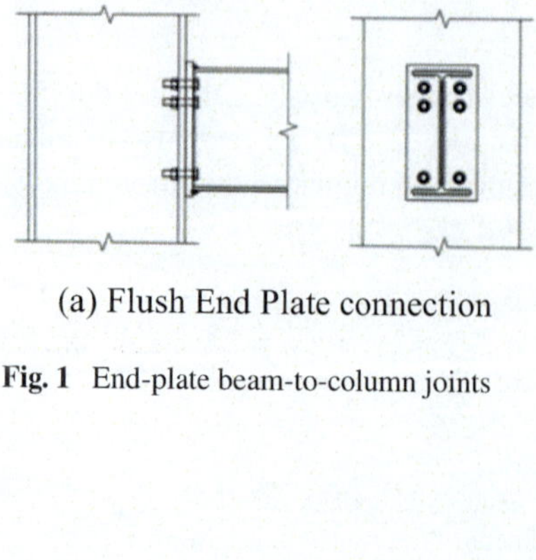

(a) Flush End Plate connection

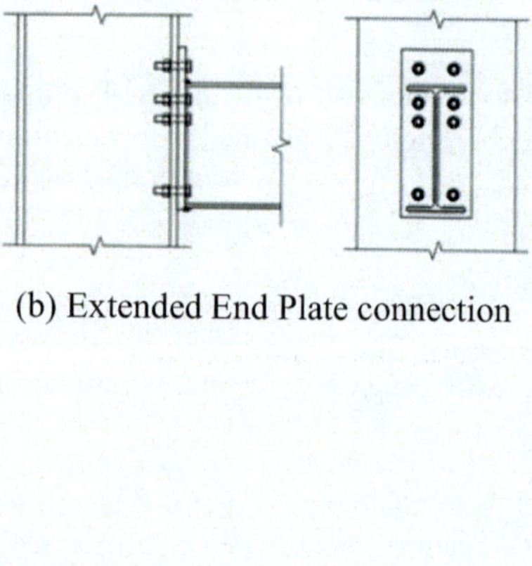

(b) Extended End Plate connection

Fig. 1 End-plate beam-to-column joints

The main feature of the selection of the end-plate thickness is the exclusion of the possibility of sharp-crack fracture of the joint. Therefore, with a large thickness of the end-plate relative to the diameter of the bolts, destruction occurs in the bolts of the upper rows from tensile forces [8].

The issue of studying the flexibility of the compound is also essential [9, 10]. Its classification depends on the bending stiffness of the joint. In engineering practice, two models are widely used—flexible and rigid joints. Ideal hinges do not perceive the moment. An absolutely rigid connection ensures rotation at the same angle of the column and the beam at the joint transmits the bending moment from the beam to the column without loss. In addition to these connections, joints with a certain bending flexibility are often used. European standards [11] classify joints as flexible, semi-rigid and rigid.

In general, foreign research is mostly focused on the development of methods for determining the characteristics of various configurations of connection solutions based on numerical and experimental data [12–18, 23–35]. Most of them have a common structure: full-scale tests of samples are carried out, according to which the corresponding finite element model is verified. A numerical method is used to collect a sample from models with varying values of the target parameters of the compound. Further, based on the calculation data, analytical formulas and graphs are derived in to software complexes to determine the desired values, such as the rotational stiffness of the joint and the stresses in the elements of the joint. The question of the influence of initial gaps is not considered in depth. More attention is paid to the operation of connections under the action of special loads.

The classic approach to the solution documented in Eurocode is the component method. According to it, the response of each of the elements of the compound is replaced by an idealized spring with a stated stiffness. The stiffness of the elements is determined by the tables, graphs and analytical formulas given in the norms. This approach is considered time-consuming; therefore, foreign practice is focused not on developing a general calculation method, but on clarifying the operation of individual joint configurations [12].

Investigations are also being conducted on stainless steel compounds, which have better plastic characteristics compared to carbon steel. Generalized methods in this case do not consider the peculiarities of the material. The issue of corrosion resistance is investigated, the analysis of technological operations leading to deterioration of the properties of the material and its mechanical damage, for example, drilling holes for bolts and the use of combined materials in the joint [13].

The plastic properties of the compound are considered in strength, including in the case of progressive collapse [14]. For this purpose, experiments and studies of the end-plate joints under the action of dynamic loads [15], seismic loads [16, 17], cyclic affects [18] are carried out, the effect of taking into account the stages of installation of structures is considered.

In foreign practice, connections without additional stiffeners are widely used (Fig. 1). The absence of stiffeners leads to the fact that due to the deformation of the end-plate, the bolts are affected by off-center tension and collapse earlier than in a similar joint with stiffeners. The shape of the curvature of the end-plate, depending on its thickness, in turn affects the development of lever forces [19].

Experimental studies show that end-plate joints without additional structural elements are able to withstand loads greater than those recommended by existing standards. According to [19], the load-bearing capacity of the full-scale model is 2.5 times higher than the value determined by the norms. It is also established that the actual destructive loads exceed the calculated values for SP 16.13330.2017 by at least 26%. The margin of safety according to the calculated load values is also valid for connections with unfilled plates with gaps up to 1.5 mm in the bolt area. According to the test results, it was found out that the shrinkage distortion of the end-plate has an insignificant effect on the work of themselves on tension. The maximum effect (less than 2.6%) was found for groups of samples without additional stiffeners. It was found that gaps up to 1.5 mm in the bolt area and the shrinkage distortion of the end-plate reduce the bearing capacity by only 3% in comparison with samples with milled contact surfaces [19].

Recommendations for the design of end-plate connections are included in Amendment No. 2 to SP 16.13330.2017, where the requirements for pre-tensioning of bolts and the method of transmission of transverse forces in the joint are planned, and the necessary checks are indicated. The initial imperfections are proposed to be taken into account by entering a reducing coefficient to the calculated resistance of a single-bolt joint for tension and normalizing the permissible gaps in a wider range.

The problem of the influence of the joint design on the response of the end-plate connection requires additional experiments, as well as the development and verification of numerical models to establish the influence of the variability of the components of the connection of various configurations on its strength and stiffness properties.

The literature review has shown that taking into account the influence of initial imperfections is the actual problem of designing end-plate joints working on tension.

2 Methods

When using columns made of rolled profiles in steel frames of multi-storey buildings, it is necessary to take into account possible deviations in cross-section sizes due to rolling features. The analysis of possible deviations in the geometrical sizes rolled steel profiles according to GOST R 57,837-2017 allowed us to establish that the most unfavorable for the formation of gaps is the flange skew, determined by the formula $\Delta = 0.015b$, where Δ is the skew, i.e. the difference in the height of the profile between the extreme points of the flange, b is the width of the flange.

At the same time, the skew does not exceed for profiles up to 120—1 mm high, at a height of 120–290—3 mm, at a height above 290—4 mm.

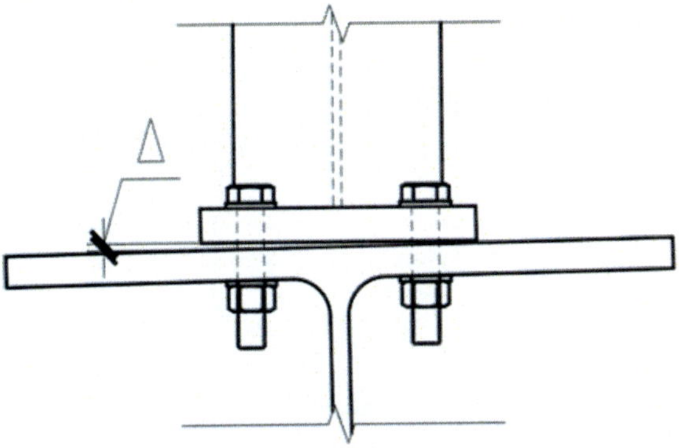

Fig. 2 Gap in the end-plate connection due to the skewed of column flange

The article considers the influence of the rolling steel profile misalignment, which leads to the non-parallelism of the column flange and the beam flange. The consequence of this is the appearance of a gap in the horizontal plane of the joint with the opening of the gap in the horizontal plane of the joint (Fig. 2).

The size of possible gaps in the area of the location of the bolts end-plate connection is a combination of various random factors, such as misalignment profile rolling, manufacturing of elements and inaccuracies of building installation. It is necessary to study the effect of the gap value Δ on the operation of the joint's response. Further calculation of the stress-strain state of the end-plate connection is carried out according to the following method:

1. In accordance with the design requirements of SP 16.13330.2017 and [3, 5] for flange connections, we will form a sample of 50 samples of numerical models with variability in the parameters of the column profile, beam, end-plate thickness and bolt diameter.
2. Partitions the sample into two groups in the proportion of 80/20—40 samples are used to restore analytical dependencies, and on the remaining 10 samples we will check the accuracy of the equations obtained.
3. To determine the values of stresses in the elements of the node and the overall compliance of the connection by regression analysis, we restore the following dependencies:

$$f\left(W_x^b, I_x^b, W_x^c, I_x^c, t, d_b, M, Q\right) = \varphi \tag{1}$$

$$f\left(W_x^b, I_x^b, W_x^c, I_x^c, t, d_b, M, Q\right) = \sigma_f \tag{2}$$

$$f\left(W_x^b, I_x^b, W_x^c, I_x^c, t, d_b, M, Q\right) = N_b \tag{3}$$

$$f\left(W_x^b, I_x^b, W_x^c, I_x^c, t, d_b, M, Q\right) = M_b \tag{4}$$

where:

$W_x{}^b$—the moment of resistance of the beam section relative to the axis x-x, cm^3;

$I_x{}^b$—the moment of inertia of the beam section relative to the x-x axis, cm^4;

$W_x{}^c$—the moment of resistance of the column section relative to the axis x-x, cm^3;

$I_x{}^c$—the moment of inertia of the column section relative to the x-x axis, cm^4;

t—thickness of the beam end-plate, cm;

d_b—diameter of bolts, cm.

φ—rotation capacity of joint, rad/kN·m;

σ_f—maximum von-Mises stress in the beam end-plate, Pa;

N_b—maximum axial force in bolts, kN;

M_b—maximum bending moment in bolts, kN·м.

4. The maximum size of the gap is determined including the width of the flange b_f by the formula $\Delta_{max} = 0.015b_f$. The minimum gap value is $\Delta_{min} = 0$, and the average value is $\Delta = \Delta_{max}/2$.
5. Plotting graphs of the distribution of the density of the voltage differences $\Delta\sigma$ in the connection elements with and without the gap (Fig. 3). Separate graphs are constructed for bolts and end-plate.
6. Setting a confidence interval of 0.99 for the sample under study and determine the maximum value of the voltage difference in a given interval. Depending on the distribution of the graph (Fig. 3), we will use the standard 3-sigma method, or a nonparametric analogue.

The data for regression analysis are determined as a result of numerical calculations. To carry out numerical calculations, a finite element model of the flange connection was developed. Connections are modeled in the Femap with Nastran software package by axial element, plate and volumetric finite elements. In order to optimize the calculation, volumetric elements were used only in the end-plate and

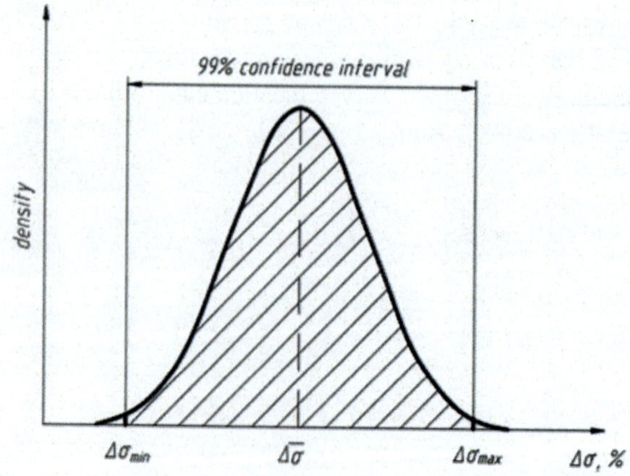

Fig. 3 Density distribution of voltage differences

Fig. 4 Finite-element model of the connection

part of the column flange within the end-plate. The remaining parts of the column and beams are modeled by plate elements of the SURFACE type. Bolts are set by BEAM type elements with nut and washer modeling using special RBE2 type elements. All linear and nodal displacements are connected to the ends of the column. A grid of finite elements is adopted for plate elements equal to 20 mm, for bulk elements—10–15 mm with a thickness division into layers. A special element of the RBE2 type was used to distribute the torque component along the beam section in the end face (Fig. 4).

To assess the reliability of numerical results, it is planned to conduct experimental studies on an experimental structure consisting of a column and a beam with end-plate connections. The experimental design model has the ability to vary the size of the gap between the end-plate and the column flange. In addition to experimental studies, an analytical calculation of the flange is performed according to the methodology described in [5].

Numerical and analytical calculations are performed for the action of a bending moment of 100 kN·m. During the calculation, the pre-tension of the bolts was not taken into account. The calculation is performed for the node according to Fig. 1, and with the following parameters:

- Beam—30B2 GOST R 57,837-2017;
- Column—35K2 GOST R 57,837-2017;
- End-plate—thickness 36 mm; width 170 mm, height 329 mm;
- Bolts—M18 GOST R 52,643-2006.

According to [5], the axial forces in the bolts are determined by the following formula:

$$N_{b\Sigma} = N_b + N_{bV} \tag{5}$$

where:

$N_b = \sigma_f \frac{a \cdot t_f}{2} + \sigma_w \frac{b \cdot t_w}{2}$—the main part of axial force;

$N_{bV} = b \cdot K_N \cdot R_{yfl} \cdot c_\varepsilon$—additional axial force from the lever forces;

σ_f, σ_w—the tension in the column flange and the average tension the adjacent section of the column web;

t_f, t_w—thickness of the stretched flange and web of column;

a, b—distances between the bolts along the shelf and wall, respectively;

K_N—a tabular coefficient depending on the thickness of the end-plate;

R_{yfl}—resistance of end-plate steel;

$c_\varepsilon = 1$—coefficient depending on fiber stresses.

The stresses in the beam will be determined taking into account the bias of the neutral axis of the connection. Its position is determined by the following formula:

$$a \cdot \xi^2 + b \cdot \xi + c = 0 \quad (6)$$

where:

$a = -0,5(\alpha + \lambda + \lambda \cdot \alpha)$; $b = \alpha + 0,5\lambda(2\Psi + 1)$; $c = \alpha^2 + 0,5\lambda \cdot \alpha - \Psi(\alpha + \lambda)$;

$\alpha = \frac{A_f}{t_w \cdot h_{ef}}$;

$\lambda \approx \frac{t_f}{t_w}$; $\Psi = \frac{M}{R_y \cdot t_w \cdot h_{ef}^2}$

3 Results and Discussion

According to the results of numerical calculation, the forces in the bolts were (Fig. 5):

$N_1 = 162.7$ kN; $N_2 = 59.9$ kN; $N_3 = -0.69$ kN

To determine the forces in the bolts, we present a diagram of the normal stresses in the joint (Fig. 6):

Axial forces in the bolts along the rows will matter:

$$N_1 = N_b = \sigma_f \frac{a \cdot t_f}{2} + \sigma_w \frac{b \cdot t_w}{2} = \frac{28,7\frac{kN}{cm^2} \cdot 8,95 \cdot 0,9\text{cm}}{2} + \frac{0,5 \cdot (28,7 + 18,2)\frac{kN}{cm^2} \cdot 7,05 \cdot 0,65\text{cm}}{2} = 169,3\text{kN}$$

$$N_2 = N_b + N_{bV} = \frac{0,5 \cdot (4,5 + 18,2)\frac{kN}{cm^2} \cdot 10 \cdot 0,65\text{ cm}}{2} + 10\text{ cm} \cdot \frac{24kN}{cm^2} \cdot 0,163$$
$$1 = 36,9 + 39,1\text{ kN} = 76\text{ kN}$$

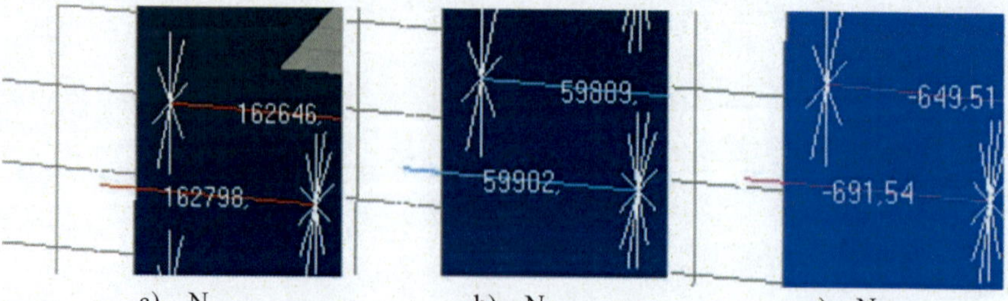

a) N_1 b) N_2 c) N_3

Fig. 5 Axial forces in bolts, N

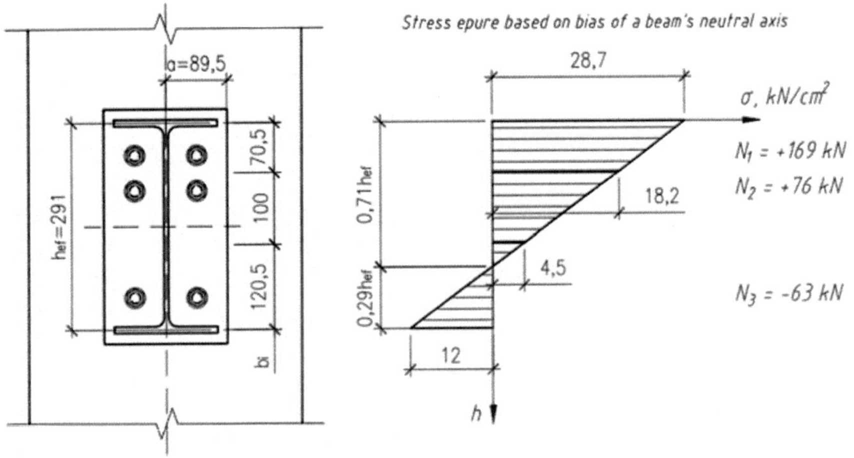

Fig. 6 Additional parameters for determining axial forces in bolts

Table 1 Numerical and analytical axial forces in bolts

Row of bolts	Analytical method, kN/%	Numerical method, kN/%
N_1	169.3/100	162.7/96
N_2	76.0/100	59.9/79
N_3	– 63/100	– 0.69/1

$$N_3 = N_b = \sigma_f \frac{a \cdot t_f}{2} + \sigma_w \frac{b \cdot t_w}{2} = \frac{-12 \frac{kN}{cm^2} \cdot 8{,}95 \cdot 0{,}9\,\text{cm}}{2} + \frac{0{,}5 \cdot (4{,}5 - 12) \frac{kN}{cm^2} \cdot 12{,}05 \cdot 0{,}65\,\text{cm}}{2} = -63\,\text{kN}$$

Table 1 compares the numerical and analytical results.

Based on the calculations carried out, a good qualitative correspondence of the results obtained in different ways was established. At the same time, the maximum forces in the bolts, which determine the strength of the joint, almost coincided. The forces in the remaining rows of bolts differ more significantly, which is due to the disadvantages of the analytical calculation method.

4 Conclusion

Based on the analysis, the following conclusion can be drawn:

1. The issue concerning influence of initial gaps in flange connections has not been studied sufficiently.
2. Determining the size of the gap between the end-plate and the column flange is the misalignment of the rolling profile of the column.

3. Due to the initial gaps and gaps that open under the action of the load, in addition to axial forces, bending occurs in the bolts. This factors mentioned above lead to an earlier exhaustion of strength.
4. A sequence of investigation of the operation of the end-plate connection is proposed, which allows considering the random nature of the misalignment of the flange of the rolling profile of the column.
5. A finite element model of a flange connection has been worked out. Its verification is carried out using a well-known analytical method. Further verification of the reliability of numerical calculations will be implemented according to experimental studies.

References

1. Series 2.440-2 (1994) Joints of steel structures of industrial buildings of industrial enterprises. Issue 7. Bolted flanged frame connections of beams with columns of steel frames of buildings and structures, NIPIPromstalkonstruktsiya, Moscow
2. Recommendations for the calculation, design, manufacture and installation of flanged joints of steel building structures (1988) CO "Stalmontazh", VNIPI "Promstalkonstruktsiya", TsNIIPSK Melnikov, Moscow
3. Guidelines for the design, manufacture and assembly of mounting flange joints of roof trusses with belts from wide-shelf I-beams (1981) TsNIIPSK Melnikov, Moscow
4. Son MP (2017) Int Res J 6-3(60):164–167
5. Katyushin VV (2005) Buildings with frames made of steel frames of variable section (calculation, design, construction). OAO Stroyizdat Publishing House, Moscow
6. Son MP (2019) Bull Perm Nat Res Polytechnic Univ Appl Ecol Urbanistics 1:142–152
7. Volodin MV et al (2020) Constr Mech Calculation Struct 1(288):62–73
8. Kashevarova GG, Son MP, Zemlyanukhin AD (2020) Modern technologies in construction. Theory Pract 1:44–49
9. Tusnina VM, Platonova VD (2020) Ind Civ Constr 9:28–33
10. Tusnina VM (2017) Eng Constr J 5(73):25–39
11. EN 1993-1-8 (2005) Eurocode 3. Design of Steel Structures. Part 1.8: Design of joints. CEN
12. Wald F, Švarc M (2001) Contributions to experimental investigation of engineering materials and structures. CTU Rep 2–3:1–13
13. Elflah MAH (2018) Structural behavior of stainless steel bolted beam to column joints, Ph.D. thesis, University of Birmingham
14. Baldassino N, Zandonini R (2009) Structures Congress 2009: Don't Mess with Structural Engineers: Expanding Our Role 1–10
15. Grimsmo EL et al (2015) Int J Impact Eng 85:132–145
16. Tartaglia R et al (2017) 03.10: Seismic design and performance of extended stiffened end-plate joints (ce/papers) 1(2–3):570–579
17. Cassiano D, D'Aniello M, Rebelo C (2018) Steel Compos Struct 26(5):621–634
18. Yoganata YS et al (2020) IOP Conf Ser Mater Sci Eng 930(1):012050
19. Son MP (2018) Constr Mech Eng Struct Struct 4:348–356
20. Tusnina VM, Kolyago AA (2018) Ind Civ Constr 2:28–34
21. Agafonkin VS (2015) Proceedings of the Kazan State University of Architecture and Civil Engineering, vol 2, pp 130–136
22. Kriksunov EE, Perelmuter AV, Yurchenko VV (2010) Bull Constr Equipment 56–59
23. Haapio J (2010) J Struct Mech 43(1):1–11

24. Wang Q, Wang L, Jlang B, Li H, Liu FQ (2011) Adv Mater Res 163–167:102–105
25. Hu XB, Yang YW, He GJ, Fan Y, Fan L, Zhou P (2013) Appl Mech Mater 256–259:821–825
26. Bzdawka K, Heinisuo M (2010) Rakenteiden Mekaniikka (J Struct Mech) 43(1):25–43
27. Hasan MJ (2017) J Constr Steel Res 149–161
28. Sokol Z, Wald F, Delabre V, Muzeau JP, Svarc M (2002) Eurosteel Coimbra 1219–1228
29. D'Alessandro E, Brando G, De Matteis G (2018) Proc Inst Civ Eng Struct Build 171(6):444–462
30. Luo L et al (2020) Materials 13(22):5133
31. Pisarek Z (2013) J Civ Eng Environ Archit 30(60):2
32. Costa R et al (2019) Eng Struct 188:277–289
33. Al-Rifaie AK (2018) Lateral impact responses of steel end plate beam-to-column connections. Ph.D. thesis, University of Liverpool
34. Zhao D et al (2020) Appl Sci 10(14):4812
35. Ostrowski K, Kozłowski A (2017) 01.25: The influence of end–plate joints stiffening on the rotation capacity (ce/papers), 1(2–3):381–388

Increasing the Bearing Capacity of the Joints of Wooden Elements on Metal Plates

Maksim Dezhin and Aleksandr Ibragimov

Abstract Today, the connections of wooden elements on metal plates are relevant. The purpose of the work is to evaluate the effect of the use of epoxy glue on the strength characteristics and deformation of the joints of wooden elements on metal plates using screwed screws. The article describes the developed design of a single-cut sample of the connection of a glued wood element with a metal overlay, which made it possible to simulate the actual operation of the nodal mates of the main and secondary wooden beams on metal overlays. The results of comparative tests of samples of connections of wooden elements from glued wood on metal linings using screwed and glued screws, 3 samples of each of 5 types of connections (one glueless and four types of glued screws) are presented. It was revealed that the use of epoxy adhesives changes the nature of the joint deformation on screwed rods. In the presence of glue, the connection works more elastically. It has been established that the use of epoxy adhesives increases the strength characteristics of a glueless joint by the criterion of breaking load by 1.46–1.78 times, by the criterion of elastic work of the joint by 1.3–1.82 times for screwed joints. The article defines the deformation of the samples within their bearing capacity. Deformation $D_{ult} = 2$ mm was achieved by individual samples of glued-in joints at loads 1.77, 1.40 times greater than the load in a non-adhesive joint. Total deformations D_t at the same load within the limits of the elastic operation of the joints are reduced by 7–10 times for glued-in joints. The conducted studies and the literature studied during the study made it possible to formulate the main proposals for increasing the bearing capacity of the connection of wooden elements based on metal linings for further study of this direction.

Keywords Glued wood · Metal linings · Testing of joint samples · Joint bearing capacity · Joint deformability · Upper limit of the elastic work region · Total deformation · Residual deformation · Elastic deformation

---

M. Dezhin (✉) · A. Ibragimov
Moscow State University of Civil Engineering, Moscow 129337, Russian Federation
e-mail: maksim1403@yandex.ru

P. Akimov et al. (eds.), *Proceedings of FORM 2022*, Lecture Notes in Civil Engineering 282, https://doi.org/10.1007/978-3-031-10853-2_32

1 Introduction

Currently, there are various options for connecting wooden elements of building structures. To date, connections on metal plates such as Sherpa are relevant. The fastener consists of two metal plates that are attached to wooden structural elements with screws and form a rigid dovetail joint. T. Bogensperger, F. Hude [1], M. Augustin, G. Flatscher [2, 3], H. Schinner [4], Xu Yun [5] were engaged in the study of the connections of elements of wooden structures on metal plates of the Sherpa type. The performed analysis of the state of the issue and the evaluation of earlier tests made it possible to formulate the main directions of research, including the program of experiments [6]. The purpose of the work is to evaluate the effect of the use of epoxy glue on the strength characteristics and deformation of the joints of wooden elements on metal plates using screwed screws.

2 Methods

By the nature of the destruction of the previously tested samples, it was clear that the determining factor in the destruction of the samples was the insufficient value of the pull-out force that the screws could withstand. Therefore, it has been proposed to increase the pull-out force of the screws by adding epoxy to the pre-drilled screw holes.

The design of a single-cut sample of the connection of a glued wood element with a metal overlay was developed, which made it possible to simulate the actual operation of the nodal mates of the main and secondary wooden beams on metal overlays. The developed design consists of a board 50(b) * 100(h) * 250(l) mm and a metal plate screwed to its end. The cross section of the board 100 * 50 mm was taken from the condition of placing on its end part a metal lining measuring 35(b) * 10(h) * 110(l) mm. Metal slips are made of steel 40X. Screws with a diameter of 4.5 mm and a length of 50 mm were used as fasteners. Compound samples were prepared according to the accepted scheme shown in Fig. 1.

The manufactured wooden elements with drilled holes are shown in Fig. 2. The fabricated metal plate is shown in Fig. 3.

One of the fabricated samples is shown in Fig. 4.

Samples were tested according to a compressed scheme in accordance with the requirements of GOST 33,082-2014 "Wooden structures. Methods for determining the bearing capacity of nodal joints "with bringing them to failure by loading steps with periodic unloading. For the tested samples, the load was applied to the upper end of the metal plate. During the study, 3 samples of each of 5 types were tested: type 1—connection of the lining and the board with screwed screws without glue—samples № 1–3; type 2—connection of the overlay and board with glued screws (100 parts by weight of epoxy resin ED-20 and 10 parts by weight of PEPA hardener)—samples № 4–6; type 3—connection of the overlay and board with glued screws (100 parts by

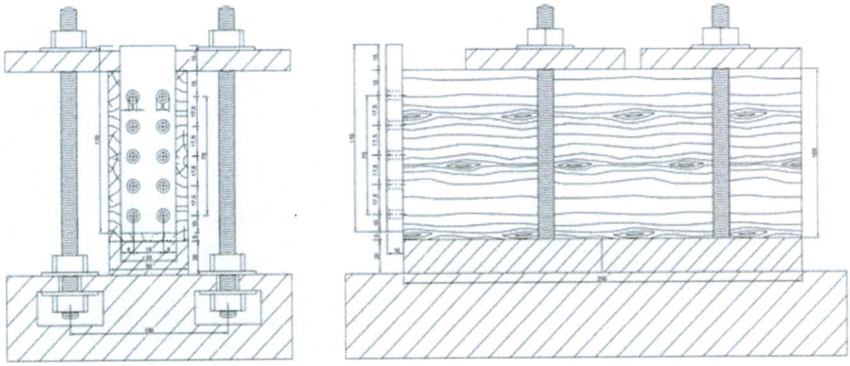

Fig. 1 Test sample schematic

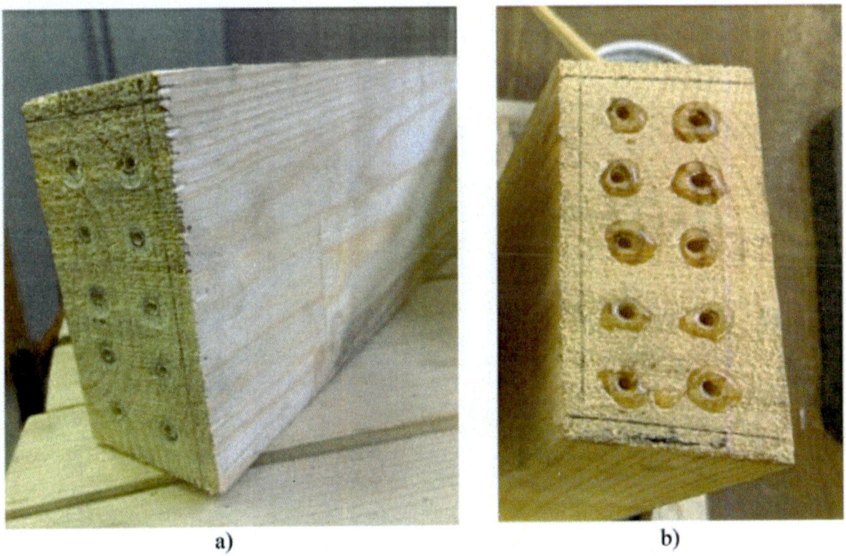

Fig. 2 **a** Manufactured wooden element with drilled holes (samples № 1–3); **b** Manufactured wooden element with drilled holes filled with glue (samples № 4–15)

weight of epoxy resin ED-20 and 10 parts by weight of TETA hardener)—samples № 7–9; type 4—connection of the overlay and board with glued screws (100 weight parts of DER-331 epoxy resin and 13 weight parts of PEPA hardener)—samples № 10–12; type 5—connection of the overlay and board with glued screws (100 weight parts of DER-331 epoxy resin and 13 weight parts of TETA hardener)—samples № 13–15.

a)

b)

Fig. 3 **a** Manufactured metal plate; **b** Manufactured metal plate. View from the back

a)

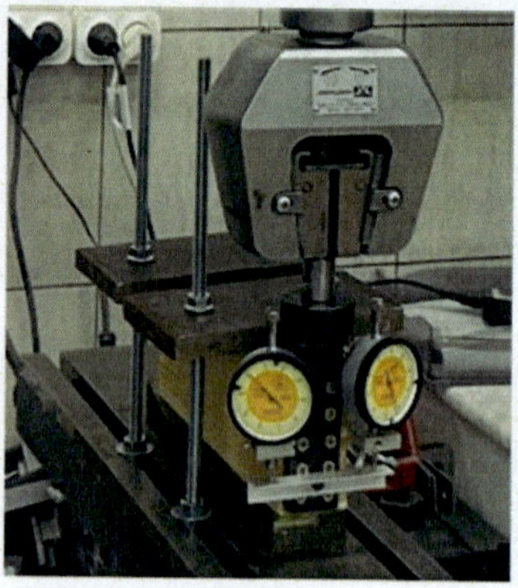

b)

Fig. 4 **a** General view of the manufactured sample; **b** General view of the manufactured sample with installed dial gauges in the testing machine

3 Results

When the samples were loaded, the metal plates were displaced relative to the wooden elements to which they were attached, up to destruction. During the tests, there was no delamination of metal linings from the surface of the boards, no cracks, distortions, or crushing of the connection elements were detected. After the tests, the opening of the samples showed that at the beginning of the application of the load, the screwed/glued screws were switched on, then the pulling out of the screws began, accompanied by deformations of the joint and crushing of the wood in the nests of the screws.

Comparative test results of joint samples on screwed and glued screws are presented in Table 1. Comparative graph "Load—total deformation D_t" of all types of tested joints is shown in Fig. 5.

Table 1 Comparative test results of joint samples on screwed and glued screws

			Load level			The average
			N_t	$N_{I\text{-}II}$	N_2	
Type 1 (driven screws)-№ 1,2,3	N screwed., kN		10.97	3	9.48	–
	D_t, mm		2.66	0.112	2	–
	Share of elastic deformations, D_e/D_t		0,19	0.40	0.20	0.26
Type 2 (adhesive screws—ED-20 + PEPA)-№ 4,5,6	N pasted., kN		15.99	10	–	–
	D_t, mm		1.40	0.08	2	–
	Share of elastic deformations, D_e/D_t		0.14	0.625	–	0.38
	Effect type2/type1	K from N	1.46	3.33	–	2.40
		K share D_e by D_t	0.74	1.56	–	1.15
Type 3 (glue screws—ED-20 + TETA)-№ 7,8,9	N pasted, kN		19.52	16	16.78	–
	D_t, mm		1.52	0.08	2	–
	Share of elastic deformations, D_e/D_t		0.15	0.625	0.13	0.30
	Effect type3/type1	K from N	1.78	5.33	1.77	2.96
		K share D_e by D_t	0.79	1.56	0.65	1.00
Type 4 (adhesive screws—DER-331 + PEPA)-№10,11,12	N pasted, kN		18.70	14	–	–
	D_t, mm		0.69	0.11	2	–
	Share of elastic deformations, D_e/D_t		0.26	0.727	–	0.49
	Effect type4/type1	K from N	1.70	4.67	–	3.19
		K share D_e by D_t	1.37	1.82	–	1.60
Type 5 (adhesive screws—DER-331 + TETA) -№13,14,15	N pasted, kN		17.41	13.5	13.26	–
	D_t, mm		1.91	0.25	2	–
	Share of elastic deformations, D_e/D_t		0.19	0.52	0.15	0.29
	Effect type5/type1	K from N	1,59	4.5	1.40	2.50
		K share D_e by D_t	1.00	1.3	0.75	1.02

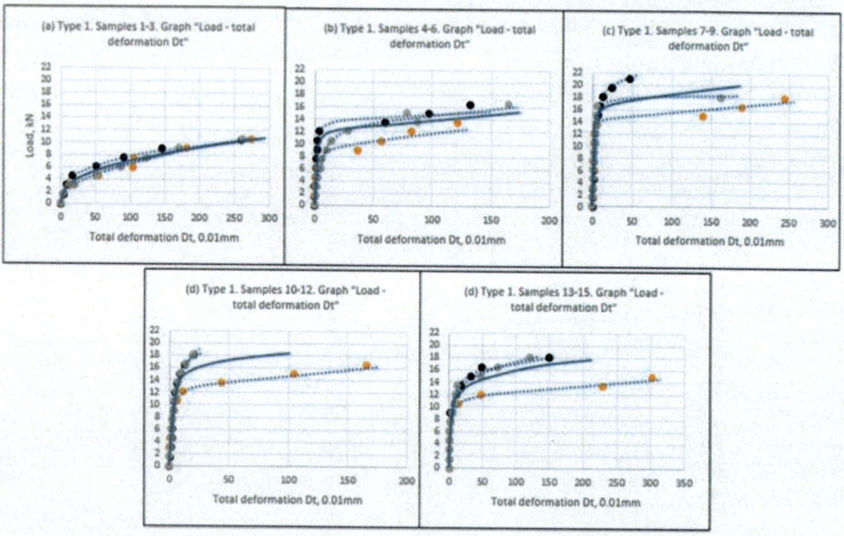

Fig. 5 Comparative graph "Load—total deformation D_t of connections: (a) type 1) on screwed screws; (b) type 2) on glued screws (ED-20 + PEPA); (c) type 3) on glued screws (ED-20 + TETA); (d) type 4) on glued screws (DER-331 + PEPA); (e) type 5) with glued screws (DER-331 + TETA)"

4 Discussion

1) It has been determined that the installation of driven screws with adhesive increases the load-bearing capacity of the connection. The bearing capacity of joints on glued screws using epoxy resin ED-20 and hardener PEPA is 1.46 times greater than that of joints without the use of glue: using epoxy resin ED-20 and hardener TETA—1.78 times; with the use of DER-331 epoxy resin and PEPA hardener—1.70 times; with the use of DER-331 epoxy resin and TETA hardener—1.59 times. The maximum bearing capacity of the N_t joints was 10.97, 15.99, 19.52, 18.70 and 17.41 kN, the load of $N_{I\text{-}II}$ was 3.0, 10.0, 16.0, 14.0 and 13.5 kN for glueless and glue-screwed joints (ED-20 + PEPA, ED-20 + TETA, DER-331 + PEPA, DER-331 + TETA), respectively;
2) The deformations of the samples were determined within their bearing capacity. At the level of the upper boundary of the elastic work area, the total deformations of the samples averaged $D_t = 0.11$ mm, $D_t = 0.08$ mm, $D_t = 0.08$ mm, $D_t = 0.11$ mm, $D_t = 0.25$ mm, which is less limit deformation established by the design standards SP 64.13330.2017 for dowel joints $[D_u] = 2$ mm in 18.18, 25.00, 25.00, 18.18, 8.00 times for glueless and glue-screwed joints (ED-20 + PEPA, ED-20 + TETA, DER- 331 + PEPA, DER-331 + TETA), respectively;
3) It was found that the use of epoxy adhesives changes the nature of the joint deformation on screwed rods. In the presence of glue, the connection works more elastically, the proportion of elastic deformations in the composition of

the total deformations of the joints is greater on average for three samples of each type of joints by 1.56, 1.56, 1.82, 1.30 times for glued rods (ED-20 + PEPA, ED-20 + TETA, DER-331 + PEPA, DER-331 + TETA), respectively, than in joints without the use of glue, while the absolute values of deformations D_t within the elastic work for glued joints (ED-20 + PEPA, ED-20 + TETA, DER-331 + PEPA, DER-331 + TETA) differ from glueless joints by 37, 37, 0, 127%, respectively;

4) The average design values of the bearing capacity of the joints of wooden elements on metal plates are established based on the loads $N_{I\text{-}II} = 3$ kN, $N_{I\text{-}II} = 10$ kN, $N_{I\text{-}II} = 16$ kN, $N_{I\text{-}II} = 14$ kN, $N_{I\text{-}II} = 13.5$ kN, corresponding to the upper boundary of the area of elastic operation of the joints and the required safety factor $K_{I\text{-}II} = 1.3$, which amounted to $N_t = 3/1.3 = 2.3$ kN, $N_t = 10/1.3 = 7.7$ kN, $N_t = 16/1.3 = 12.3$ kN, $N_t = 14/1.3 = 13.1$ kN, $N_t = 13.5/1.3 = 10.8$ kN for glueless and glue-screwed joints (ED-20 + PEPA, ED-20 + TETA, DER-331 + PEPA, DER-331 + TETA) respectively.

The goal was achieved in the work—an assessment was made of the effect of the use of epoxy glue on the strength characteristics and deformation of the joints of wooden elements on metal plates using screwed screws.

5 Conclusions

It was revealed that the use of epoxy adhesives changes the nature of the joint deformation on screwed rods. In the presence of glue, the connection works more elastically. It has been established that the use of epoxy adhesives increases the strength characteristics of a glueless joint by the criterion of breaking load by 1.46–1.78 times, by the criterion of elastic work of the joint by 1.3–1.82 times for screwed joints. The article defines the deformation of the samples within their bearing capacity. Deformation $D_{ult} = 2$ mm was achieved by individual samples of glued-in joints at loads 1.77, 1.40 times greater than the load in a non-adhesive joint. Total deformations D_t at the same load within the limits of the elastic operation of the joints are reduced by 7–10 times for glued-in joints. Conducted studies and scientific articles studied in the course of the study V. Linkov [7–13], A. Naichuk, M. Babaev [14], M. Dezhin [15, 16], G. Pirnbacher [17], T. Uibel [18], Y. Pluess, R. Brandner [19], U. Huebner [20] made it possible to formulate the main proposals for increasing the bearing capacity of the connection of wooden elements based on metal linings for further study of this direction.

Acknowledgements A huge contribution to the research was made by Professor, Doctor of Technical Sciences, Head of the Department of Metal and Wooden Structures of NRU MGSU Vladimir Ivanovich Linkov.

This work was carried out with the support of the Head Regional Center for the Collective Use of Scientific Equipment and Installations of the NRU MGSU, as well as with the direct participation of the head of the laboratory of the Department of Metal and Wooden Structures Bekker Dmitry Aleksandrovich.

References

1. Bogensperger T, Hude F (2007) Development of a heavy-duty connection for main-sub-beam connections. In: Burri S et al (eds) 6 international timber construction conference. Publisher Forum Holzbau, Graz, pp 209–224
2. Augustin M, Flatscher G (2010) Verification of SHERPA connections based on the SHERPA manual. In: Burri S et al (eds) 16 international timber construction conference. Publisher Forum Holzbau, Graz, pp 1–16
3. Augustin M (2009) Transfer of high loads with Sherpa system connectors. In: Burri S et al (eds) 15 international timber construction conference. Publisher Forum Holzbau, Graz, pp 1–21
4. Schinner H (2012) Fastening wooden components to reinforced concrete substrates using system connectors. In: Burri S et al (eds) 18 international timber construction conference. Publisher Forum Holzbau, Graz, pp 1–20
5. Xu Y (2015) Ph.D. thesis, Saint Petersburg State University of Architecture and Civil Engineering
6. Dezhin M (2021) Eng J Don 78:264–271
7. Linkov V (2020) Eng J Don 77:339–346
8. Linkov V (2019) Eng J Don 52:159–168
9. Linkov V (2021) Proceedings of higher education institutions. Text Ind Technol 394:179–185
10. Linkov V (2017) Proceedings of higher education institutions. Text Ind Technol 369:212–217
11. Linkov V (2019) Proceedings of higher education institutions. Text Ind Technol 380:153–158
12. Linkov V (2018) Proceedings of higher education institutions. Text Ind Technol 375:84–89
13. Linkov V Sci Rev 23:11–15
14. Najchuk A, Babaev M (2010) Ind Civ Eng 1:21–23
15. Dezhin M (2022) Eng J Don 85:470–478
16. Dezhin M (2022) Bull Constr Equipment 1050:28–30
17. Pirnbacher G (2009) Stress and optimization potential of self-drilling wood screws. In: Burri S et al (eds) 15 international timber construction conference. Publisher Forum Holzbau, Graz, pp 1–20
18. Uibel T (2009) Assessment of the risk of splitting when screwing self-drilling wood screws. In: Burri S et al (eds) 15 international timber construction conference. Publisher Forum Holzbau, Graz, pp 1–16
19. Pluess Y, Brandner R (2014) Investigations into the load-bearing behavior of axially loaded screw groups in the narrow side of cross-laminated timber (CLT). In: Burri S et al (eds) 20 international timber construction conference. Publisher Forum Holzbau, Garmisch-Partenkirchen, pp 1–14
20. Huebner U (2009) Withdrawal resistance of wood screws in ash glued laminated timber. In: Burri S et al (eds) 15 international timber construction conference. Publisher Forum Holzbau, Graz, pp 1–16

Semi-rigid Steel Beam-to-Column Connections

Valentina Tusnina and Liubov Chechulina

Abstract The article provides an analytical review of existing studies and design solutions of the beam-column joints in steel construction. The approaches to the study of the behavior of the beam-column joints of steel structures under static and cyclic dynamic loadings are reviewed. The existing methods for determining the stiffness are based mainly on regulatory engineering techniques, experimental and numerical studies. The methods are compared and the advantages and disadvantages of each are described. It is possible to achieve a reliable picture of the stress-strain state of joints in the elastic-plastic stage of work based on finite elements method. Because of large variety of joints, the further research remains relevant. The analysis of the applicability of the obtained characteristics, in particular moment-rotation diagrams, with the help of numerical calculation methods is carried out. Modern software based on FEM allows to use the obtained characteristics by different approaches of special zero-length rotational spring element. Computational power grows and so grows the skill of engineering modeling and calculation. However, the pinned and rigid joints concept is still remains basic in engineering practice; all the studies show that consideration of the flexibility of joints in the work of the frame is necessary and possible to process with the help of CAE programs.

Keywords Semi-rigid joint · Steel frame · Beam · Column · Stiffness · Rotational angle · Support moment

1 Introduction

The connections of beams with columns are one of the most stressed zones of multi-storey frame buildings. Bending and torques, longitudinal and transverse forces are concentrated there in very short sections. Due to the irregularities of the contact zones, manufacturing and installation errors in connections of prefabricated structures, it is almost impossible to achieve a uniformly distributed load transfer between

V. Tusnina (✉) · L. Chechulina
Moscow State University of Civil Engineering, Moscow 129337, Russian Federation
e-mail: tusninavm@mgsu.ru

P. Akimov et al. (eds.), *Proceedings of FORM 2022*, Lecture Notes in Civil Engineering 282, https://doi.org/10.1007/978-3-031-10853-2_33

the elements being joined. Therefore stress concentrators and uneven deformations occur. In addition, each of connection elements functions under different conditions and at different stages of the stress-strain state.

In high-rise and large-span construction, there is a trend to use composite connections that require careful analysis. To ensure mechanical safety, a more detailed study of the stress-strain state of such connections under various impacts is needed.

The construction solution of the beam-to-column connection affects not only the bearing capacity of the joint itself and the connected elements (column, beam), but also the behavior of the frame as a whole [1–5]. Due to the wide variety of such joints, the study of their actual behavior is still relevant.

The calculation and design of steel frames in practice is commonly based on a simplified division of these joints into 2 idealized models—rigid and pinned. In reality, most of the joints take an intermediate position, being more or less flexible, characterized by the dependence between the angle of rotation of the end of the beam in the connection and the reference moment.

The "moment—rotation" relationship is nonlinear (Fig. 1). This is mainly because of the fact that the joint is a combination of several elements that interact differently at different levels of load application and undergo large deformations. Heterogeneity of the connection material, local flexibility of the connecting elements, stress concentrations, local buckling, etc.—all contributes to non-linearity.

A semi-rigid model of joint behavior has been proposed for a long time. It is reflected in Eurocode 3 and the US (AISC LRFD) regulatory documentation (Fig. 2), which specify the stiffness criteria and provide recommendations for the calculation of frames with such joints [6, 7].

The purpose of this article is to review the generalizing experience of research on the behavior of steel structure connections and ways to take into account their influence in the structure of the frame. Such an analysis will help to determine the

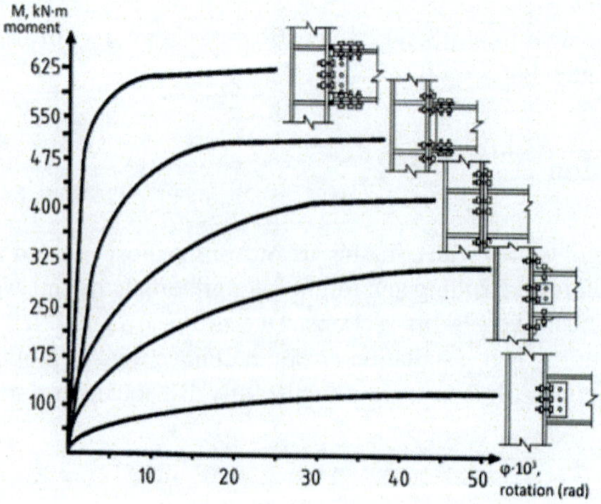

Fig. 1 Connection moment-rotation curves

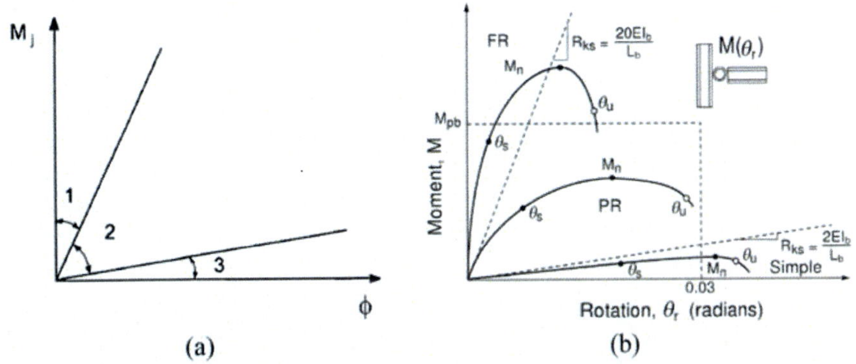

Fig. 2 Joint stiffness classification according to Eurocode 3 (**a**) [6] and AISC LRFD (**b**) [7]

current state of the mentioned topic, as well as the main trends in the development of numerical and constructive studies of composite beam-to-column steel frame connections.

2 Analysis Methods of Semi-rigid Joints

The complexity of the calculations is caused by the determination of the stiffness of the joint itself, since it depends on many parameters (type of connection, geometric characteristics of elements, etc.). To obtain a moment-rotation curve for each type of joint, a separate study is carried out. The basic methods are experimental, analytical (mechanical) and numerical. Other classifications are also possible, in which methods based on these are considered.

The most reliable results come from experimental studies. Large databases of tests, which indicate the geometric and mechanical characteristics of components, moment-rotation diagrams and angular stiffness of each joint, have been created. Among them are Goverdhan data bank [8], Nethercot data bank [9], Steel connection data bank [10], and SERICON data bank [11]. There are also many separate studies that are not included in these databases. However, experimental data are used primarily for validation of analytical and numerical models, but not in ordinary engineering practice.

Mechanical method is also known as component method adopted by Eurocode 3. Each component is being modelled as an elastic spring with a specific stiffness and strength (Fig. 3). The dependencies of the graphs for many of the typical joints have already been identified through the construction of mechanical models. The research [12] shows that some of such Eurocode 3-Annex J models are imperfect. Further improvement of the model gives more accurate result, which is in a good agreement with finite element analyses and with available experimental data.

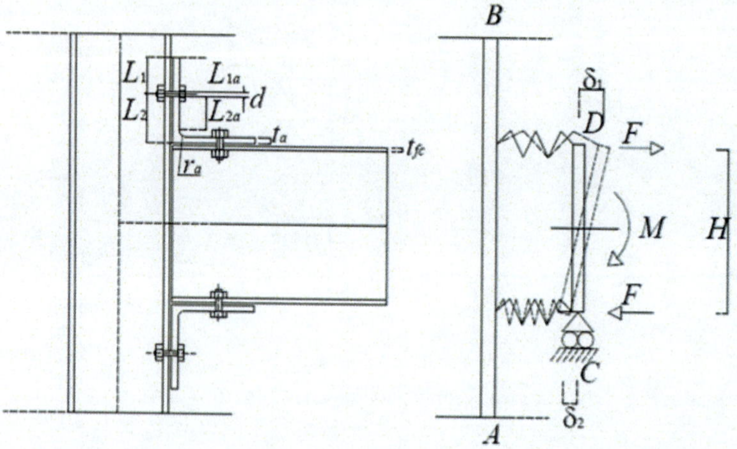

Fig. 3 Mechanical model of top and seat angle joint [12]

With the increase in computing power and the development of software systems, numerical simulation has become most widespread in the study of the joints behavior, as the most powerful and effective method for determining a reliable stress-strain state under various affects.

In article [13] on the basis of numerical calculation in the ABAQUS 6.13 computing complex, the rigidity of beam-to-column joint with double vertical angles was investigated (Fig. 4). It has been established that the main factors determining the deformability of the joint are the deformation of the angles, which accounts for up to 87% of the total joint deformation. The structure is able to support up to 8% of the bending span moment within the beam.

For Single Web Angle Connection, the moment-rotation curve is investigated in [14] also using the ABAQUS software. Several models have been developed to represent connection flexibility. The finite element model gives good curve fitting to

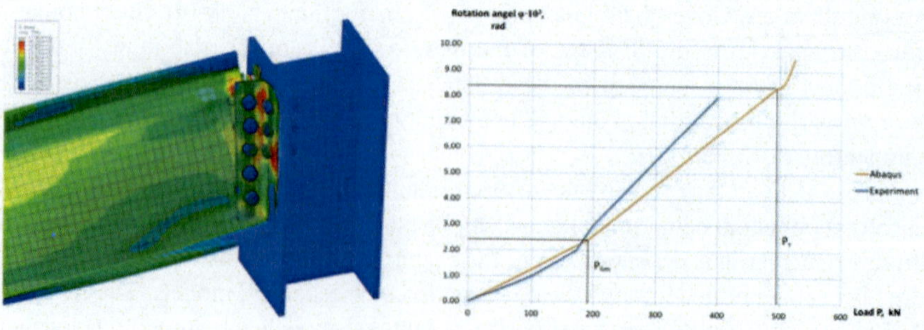

(a) (b)

Fig. 4 Double vertical angles joint finite-element model under load (**a**) and load-rotation diagram (**b**) [13] the test data. The popular Kishi and Chen's as well as Lee and Moon's models are as accurate as finite element models, but only for a limited mechanisms range as shown in Fig. 5.

Other examples of joint is the flush end plate one with pretension bolts (Fig. 6a) and the extended end plate one from [15]. The behavior depends on the plate thickness, column stiffeners, and bolts type. Which proves that joint stiffness relies on connection components. Again, FE model is close to the experimental one with average capacity ratio of 1.13.

The comparison between the joints with unstiffened flange cleats (Fig. 7a) and with stiffened flange cleats (Fig. 7c) is introduced in [16] by ANSYS CAE. The second one has a smaller rotation with much larger bending moments. Bending resistance and initial stiffness obtained from numerical analyses and experimental tests show a difference of less than 3%.

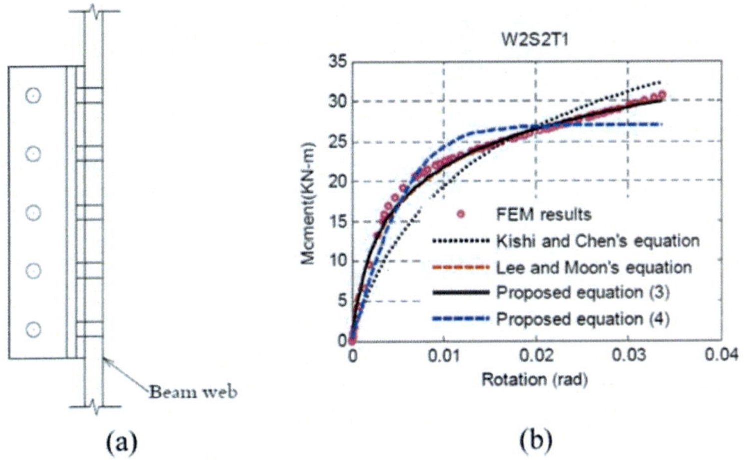

Fig. 5 Single web angle joint model (**a**) and moment-rotation diagram (**b**) [14]

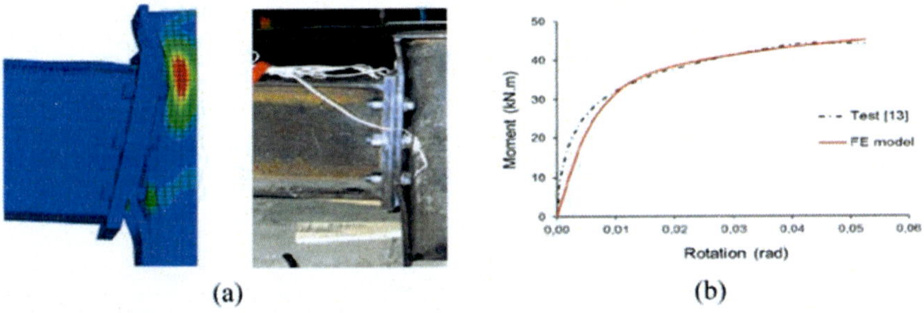

Fig. 6 Flush end plate failure mode results (**a**) and its moment-rotation curve [15]

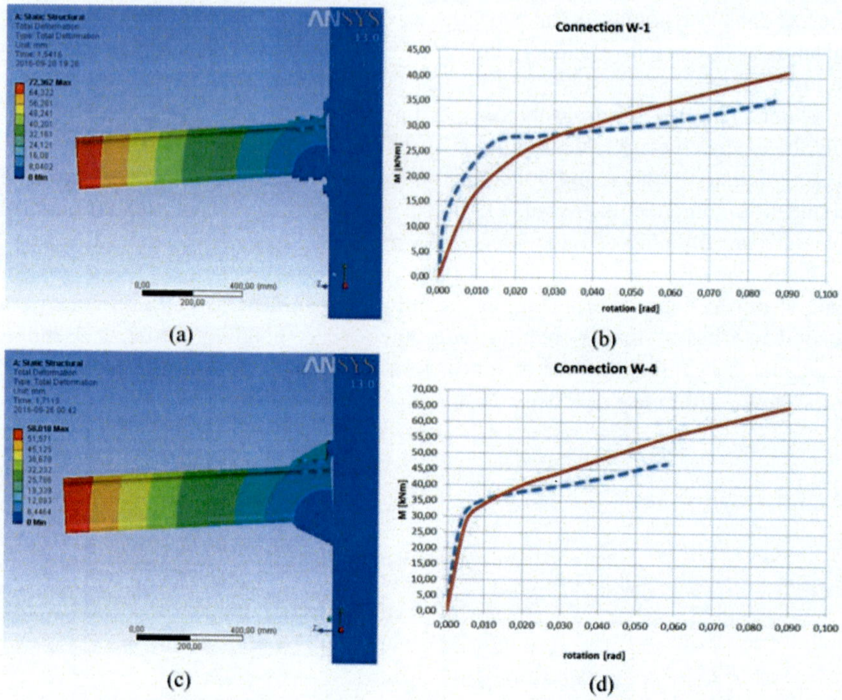

Fig. 7 FEM vertical displacements (**a, c**) and moment-rotation curves (**b, d**) [16]

As studies of the stress-strain state of the beam-column joints show, numerical calculations in software complexes based on the finite element method give the closest results to experimental data.

Special characteristics of the methods are given in Table 1.

Table 1 Advantages and disadvantages of methods for obtaining M-ϕ diagram

Method	Advantages	Disadvantages
Experimental	Gives the most reliable results	Expensive to carry out Requires special equipment
Analytical	Needs low computing power Ease of application Parametrization available	Requires verification with experimental or numerical data Applicable to simple joints
Numerical (FEM)	Applicable to complicated joints Most clearly shows the zones of stresses	Requires large computing power

3 Modeling Semi-rigid Steel Frames

Taking into account the flexibility of the connections significantly complicates the calculation of structures. However, modern software systems make it possible to obtain a reliable picture of the stress-strain state of frames without much effort. Several methods have been proposed to increase accuracy.

Article [17] provides a nonlinear calculation of two space frames. Semi-rigid connections are modeled using special zero-length spring elements. The basic matrix equation for the flexibility of a joint is a diagonal stiffness matrix that relates the force in the joints to its displacements as follows:

$$\{\Delta F_s\} = [K_s]\{\Delta U_s\} \tag{1}$$

$$[K_s] = \begin{bmatrix} k_x^{lin} & 0 & 0 & 0 & 0 & 0 \\ 0 & k_y^{lin} & 0 & 0 & 0 & 0 \\ 0 & 0 & k_z^{lin} & 0 & 0 & 0 \\ 0 & 0 & 0 & k_x^{non} & 0 & 0 \\ 0 & 0 & 0 & 0 & k_y^{non} & 0 \\ 0 & 0 & 0 & 0 & 0 & k_z^{non} \end{bmatrix} \tag{2}$$

where the tangent stiffness for the linear springs k_n^{lin} or the nonlinear springs k_n^{non} is:

$$k_n^{lin} = R_{k,n}^{lin} \tag{3}$$

$$k_n^{non} = R_{kt,n}^{non} \tag{4}$$

$R_{k,n}^{lin}$ is a constant scalar of a linear spring, and $R_{kt,n}^{non}$ is the tangent stiffness of a nonlinear spring with respect to the n axis.

The proposed spring connection element can be employed to model any connection types with known stiffness diagram. The nonlinear semi-rigid connections dampen the deflection due to energy dissipation and the consonance does not occur in semi-rigid frames. Thus, proper consideration of semi-rigid connections is essential in predicting the realistic dynamic behavior of steel structures.

A more complex connection finite element model is applied in the article [18]. It consists of several rod and spring elements as shown in Fig. 8 and is used to calculate the spatial frame for dynamic load.

Researchers, like in [17], use computer programs written by themselves. The most popular FEM software also allows such calculations, but some of them have limitations. They are shown in Table 2.

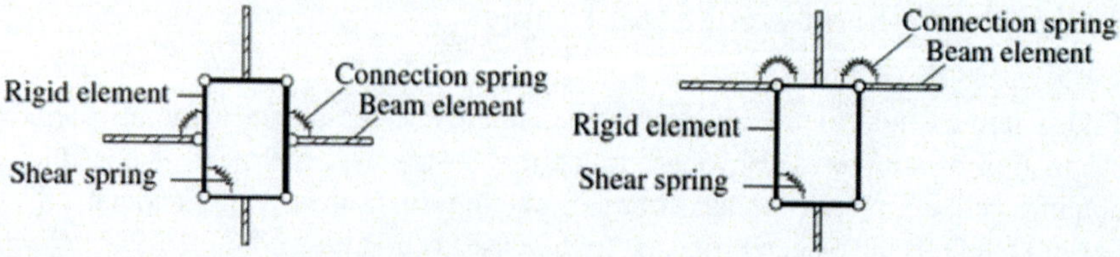

Fig. 8 FE beam-to-column joint models [18]

Table 2 Ways to apply semi-rigid joints in software packages

Software name	Linear spring FE	Nonlinear spring FE	Direct joint modeling	Easy-to-use for civil engineering
ANSYS	+	+	+	–
ABAQUS	+	+	+	–
SAP2000	+	–	+/–	+
Autodesk Robot	+	–	+/–	+
SCAD	+	–	+/–	+
Lira SAPR	+	–	+/–	+

4 Conclusion

The existing approaches to the study of the behavior of steel structure joints based on engineering and regulatory methods, experimental and numerical modeling are reviewed. Their advantages and disadvantages are described. The feature of the behavior of steel joints is highlighted, which is taking into account the flexibility, required for more accurate calculations of load-bearing systems.

When designing steel frames of multi-storey buildings, to evaluate the distribution of forces and movements it is important to determine the rigidity of the beam-to-column connections, which depends directly on the design solution of the joints. As research in this field shows, it is possible to reliably estimate stiffness using numerical modeling of joint structures. In this aspect, due to the large variety of design solutions for the connections of steel beams with columns, it seems reasonable to expand and use the data banks of unified standard joints with certain characteristics and classification by their stiffness. This will not only eliminate errors in the calculation of the frames, but will also reduce the complexity of the design process itself.

References

1. Chen WF, Kishi N, Komuto M (2011) Semi-rigid connections handbook. J. Ross Publishing, pp 365–378

2. Haapio J, Heinisuo M (2010) Minimum cost steel beam using semi-rigid joints. Rakenteiden Mekaniikka (J Struct Mech) 43(1):1–11
3. Farani de Souza LA, Vanalli L, Bueno de Luz A (2020) Numerical-computational model for nonlinear analysis of frames with semirigid connection. Math Probl Eng 11
4. Kartal ME, Basaga HB, Bayraktar A, Muvafık M (2010) Effects of semi-rigid connection on structural responses. Electron J Struct Eng 10:22–35
5. Xin-Wu W (2007) Experimental research and finite element analysis on behavior of steel frame with semi-rigid connections. In: 5th Wseas international conference on environment, ecosystems and development, pp 80–84
6. EN 1993-1-8 (2005) Eurocode 3: Design of steel structures
7. ANSI/AISC 360-16 (2016) Specification for Structural Steel Buildings
8. Goverdhan AV (1984) A collection of experimental moment–rotation curves and valuation of prediction equations for semi-rigid connections, Master thesis, Vanderbilt University, Nashville (TN)
9. Nethercot DA (1985) Steel beam-to-column connections: a review of test data and its applicability to the evaluation of joint behaviour in the performance of steel frames. CIRIA report
10. Kishi N, Chen WF (1986) Steel connection data bank program. Struct Eng 2
11. Cruz PJS, Simões da Silva LAP, Rodrigues DS, Simões RAD (1998) SERICON II — database for the semi-rigid behaviour of beam-to-column connections in seismic regions. J Constr Steel Res 46(1–3):233–234
12. Pucinotti R (2007) Prediction of cyclic moment-rotation behaviour for top and seat and web angle connections by mechanical model. Adv Steel Constr 3(2):530–552
13. Tusnina VM (2017) Semi-rigid steel beam-to-column connections. Mag Civ Eng 5:25–39
14. Kong Zh, Kim S-E (2017) Moment-rotation model of single-web angle connections. Int J Mech Sci 9:851 854
15. Bahaz A, Amara S, Jaspart JP, Demonceau JF (2018) Analysis of the behaviour of semi rigid steel end plate connections. In: MATEC Web of Conferences, vol 149
16. Jabłońska-Krysiewicz A, Gołowczyc J (2018) Numerical simulation of beam-to-column joints with flange cleats. In: MATEC Web of Conferences, vol 174
17. Nguyen P-C, Kim S-E (2013) Nonlinear elastic dynamic analysis of space steel frames with semi-rigid connections. J Constr Steel Res 84:72–81
18. Savadkoohi AT, Molinari M, Bursi OS, Friswell MI (2011) Finite element model updating of a semi-rigid moment resisting structure. Struct Control Health Monitoring 18:149–168

Finite Element Analysis of 3D Thin-Walled Beam with Restrained Torsion

Tesfaldet Gebre, Vera Galishnikova, Evgeny Lebed, Evgeniya Tupikova, and Zinah Awadh

Abstract This paper considers a modified formulation of finite element modeling for thin-walled sections with restrained torsion, including the deformation and angle of twisting caused by combined loading conditions. The unknown parameters of thin-walled cross-section are defined as a function of section properties that represent warping behavior along the beam axis. Thin-walled structures are widely used in various structural engineering applications due to their advantage of high bearing strength when compared to self-weight and when subjected to combined loadings. When a thin-walled section is subjected to a combined load with restrained torsion, they are ineffective at resisting, and it leads to reduce the capacity of the beam due to the torsion and additional warping stresses. A finite element calculation can be used to analyze a 3D bar of thin-walled structural sections. Different commercial software and studies commonly consider six degrees of freedom at each node of a member for a space frame without considering the effect of warping restraint at the member's ends. This paper presents a finite element calculation for thin-walled sections with restrained torsion using the 14 × 14 member stiffness matrix, which includes warping as an additional degree of freedom and is commonly used for open thin-walled sections. In this study, we considered two different methods to take into consideration the additional degree of freedom for the stiffness matrix, which are very close to each other for small values of characteristics number.

Keywords Thin-walled structures · Non-uniform warping · Open section · Stiffness matrix · Restrained torsion

---

T. Gebre (✉) · E. Tupikova · Z. Awadh
Peoples' Friendship University of Russia, Miklukho-Maklaya, 6, Moscow 117198, Russian Federation
e-mail: tesfaldethg@gmail.com

E. Tupikova
e-mail: tupikova-em@rudn.ru

V. Galishnikova · E. Lebed
Moscow State University of Civil Engineering, Moscow 129337, Russian Federation

P. Akimov et al. (eds.), *Proceedings of FORM 2022*, Lecture Notes in Civil Engineering 282, https://doi.org/10.1007/978-3-031-10853-2_34

1 Introduction

Nowadays, steel members are manufactured as thin-wall sections due to its high strength, highly flexible, ductility, fast construction, and effective space partition hence it is widely used in different engineering structures. Those mainly subjected to bending are called thin-walled beams. When a thin-walled section is subjected to a combined load, they are ineffective at resisting, and it leads to reduce the capacity of the beam. The behavior is poorly described by elementary formulations for which the mechanical components are reduced to stretching, bending and uniform torsion (i.e. the simplest case of a uniform distribution of the cross-sectional warping along the beam axis) [1]. Warping effects occur mainly at the points of action of the concentrated torsional moments (except for free beam ends) and at sections with free-warping restrictions and it included through an additional degree of freedom at each nodal point in the form of the first derivative of the angle of twist of the cross-section of the beam [2–4].

The analysis for extension, bending and flexure is rather straight- forward, but the analysis for the coupled deformations of torsion, warping and distortion poses a major challenge [5–8]. Currently, most design specifications do not provide clear guidance for combined bending and torsion design and the need exists for a simple design equation. The variation of the displacement over a section of a member is expressed with a common function for stretching, torsion and bending [9]. I-shaped steel beams are widely used as structural elements because of their flexural efficiency about the strong axis. It considers the cross section as completely rigid in its own plane, and the effect of shearing deformations is neglected [10, 11]. The solutions for thin-walled section with nonuniform torsion were developed as initial works and also there are studies considered to be as a design aids for simple cases [12, 13]. This is limited for a slender beam and the shear deformation in middle surface is negligible but for short-deep beam and closed thin-walled beams, the shear deformation should be considered [14]. However, in many applications beams are eccentrically loaded and as a result experience torsional loads in combination with bending. The importance of restrained torsion of thin-walled section has grown significantly as the deformations and stresses caused by torsion affects the behavior of the structures with open as well as closed section [15, 16]. Like all open sections, I-shaped steel beams are very inefficient at resisting torsion and the interaction effects due to torsion acting in combination with bending can significantly reduce the capacity of the beam.

Many design methods have been developed to deal with combined bending and torsion, but none have been universally adopted by design standards. In the past decades, many relevant research have been conducted and different commercial software commonly consider six degrees of freedom at each node of a member for a space frame without considering the effect of warping restraint at the ends of the member [9, 17, 18]. Modern software packages for structural analysis use finite element types which consider up to six degrees of freedom at the structural nodes, which corresponds to the linear and angular displacements in these nodes as for the rigid bodies [19]. The warping part of the first derivative of the twist angle has been considered as the additional degree of freedom in each node at the element ends which can be regarded as part of the twist angle curvature caused by the warping moment [17, 20, 21]. Numerous studies developed the 14 × 14 member stiffness matrix including warping as an additional degree of freedom and commonly with open thin-walled section [18, 19, 22, 23].

In this paper, a space frame stiffness matrix will be presented which is more convenient for advanced structural analysis of 3D beam structures. The structures are analyzed or designed by using only the effect of Saint Venant torsion resistance thus the analysis may ignore the torsion part in the members and the design may be underestimated. To overcome this inaccuracy, several researchers tried to develop stiffness matrix with seven degrees of freedom at each node of a member for a space frame. This additional stiffness matrix considers the warping degree of freedom at the ends of the member with thin-walled section. This study deals with the Space frame finite element method regarding the first order theory based on the assumption is that the resulting deformations are small, and that the equilibrium may be formulated for the undeformed structure as an approximation. This is done by considering beam element and equation which are necessary for the computing deformations will be derived thus to calculate the displacements and internal forces and moments for frame structures.

2 Methods

Considering the Governing Equation for non-uniform torsion, as the rate of change of the total twisting moment in a bar is in equilibrium with the applied load. The general solution for the homogeneous equation of the governing equation is satisfied by the following assumed twisting angle function $\beta_1(y_1)$ and it yields to the exact solutions of the angle of twists as shown in expression 1.

A cantilever beam is considered, loaded with bending moments, lateral forces from the lever, suspension and load and a section of 10Б is considered. Thus, the design scheme is adopted as follows: At the first step of loading (the weight of the beam itself is considered automatically, the force is equal to the sum of the weights of the lever and suspension, the moments from the weight of the lever and suspension (Fig. 1).

The governing equations for a member and frame are derived by applying the principle of virtual work to the frame. The differential governing equations for the generalized member displacements are satisfied for arbitrary virtual displacements and expressed as follows:

$$\begin{aligned} EAv_{1,1} + q_1 &= 0 \\ EJ_3v_{2,1111} - q_2 + m_{3,1} &= 0 \\ EJ_2v_{3,1111} - q_3 - m_{2,1} &= 0 \\ EJ_\omega\beta_{1,1111} - GJ_T\beta_{1,11} - m_1 - m_{\omega,1} &= 0 \end{aligned} \quad (1)$$

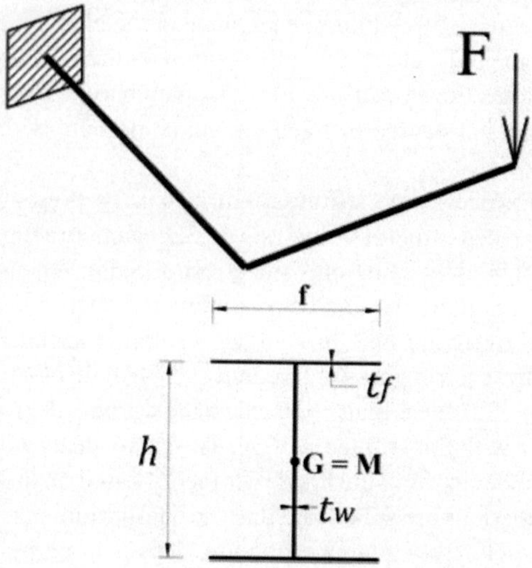

Fig. 1 Cantilever beam loaded with bending moments and lateral forces

The displacement variation over the length of a member is related to the nodal displacements by solving the differential equations the differential governing equations for the generalized member displacements such that the values of the displacement functions at the nodes equal the unknown nodal displacement values. For non-uniform torsion, a trigonometric interpolation of rotation β_1 is used as an initial parameter and finally compared with the approximation solution. To consider the warping of the restrained member, additional degrees of freedoms are introduced at the nodes and added to member displacement vector. An interpolation function containing hyperbolic functions of y_1, which satisfies the governing differential equation for torsion considered:

$$\begin{aligned} \beta_1(y_1) &= \mathbf{g}(y_1)^T \mathbf{b} \\ \mathbf{g}^T &= \begin{bmatrix} g_1(y_1) & g_2(y_1) & g_3(y_1) & g_4(y_1) \end{bmatrix} \\ \mathbf{b}^T &= \begin{bmatrix} \beta_{1A} & \beta_{1,1A} & \beta_{1B} & \beta_{1,1B} \end{bmatrix} \end{aligned} \tag{2}$$

$$\beta_1 = h_\omega^T C$$

$$h_\omega^T = \begin{bmatrix} \sinh \theta z \\ \cosh \theta z \\ z \\ 1 \end{bmatrix} \qquad C = \begin{bmatrix} C_1 & C_2 & C_3 & C_4 \end{bmatrix}$$

The interpolation functions are substituted into the left-hand side of expression (1) and the integration over the length of the member is performed for axial and bending loads but separately considered for torsion as it developed based on the two different methods.

$$\begin{aligned} E\,A \int_0^a \delta v_1\, v_{1,1}\, dy_1 &= \delta \mathbf{v}_1^T\, \mathbf{K}_1 \mathbf{v}_1 \\ E\,J_3 \int_0^a \delta v_2\, v_{2,11}\, dy_1 = \delta \mathbf{v}_2^T\, \mathbf{K}_2 \mathbf{v}_2\; E\,J_2 \int_0^a \delta v_3 v_{3,11}\, dy_1 &= \delta \mathbf{v}_3^T\, \mathbf{K}_3 \mathbf{v}_3 \end{aligned} \quad (3)$$

$$K_1 = \frac{EA}{a}\begin{bmatrix} 1 & -1 \\ -1 & 1 \end{bmatrix} = \begin{bmatrix} k_1 & k_2 \\ k_2 & k_1 \end{bmatrix}$$

$$K_2 = \frac{EJ_2}{a^3}\begin{bmatrix} 12 & 6a & -12 & 6a \\ 6a & 4a^2 & -6a & 2a^2 \\ -12 & -6a & 12 & -6a \\ 6a & 2a^2 & -6a & 4a^2 \end{bmatrix} = \begin{bmatrix} k_3 & k_4 & k_6 & k_4 \\ k_4 & k_5 & k_7 & k_8 \\ k_6 & k_7 & k_3 & k_7 \\ k_4 & k_8 & k_7 & k_5 \end{bmatrix}$$

$$K_3 = \frac{EJ_3}{a^3}\begin{bmatrix} 12 & -6a & -12 & -6a \\ -6a & 4a^2 & 6a & 2a^2 \\ -12 & 6a & 12 & 6a \\ -6a & 2a^2 & 6a & 4a^2 \end{bmatrix} = \begin{bmatrix} k_9 & k_{10} & k_{12} & k_{10} \\ k_{10} & k_{11} & k_{13} & k_{14} \\ k_{12} & k_{13} & k_9 & k_{13} \\ k_{10} & k_{14} & k_{13} & k_{11} \end{bmatrix}$$

The contribution of torsion to the internal virtual work of the governing differential Eq. (1) is given as the following expressions:

$$\int_0^a (E\,C_\omega \delta\beta_{1,11}\beta_{1,11} + G\,J_T\,\delta\beta_{1,1}\beta_{1,1})\, dA = \delta \mathbf{b}^T (\mathbf{K}_{\omega 1} + \mathbf{K}_{\omega 2})\, \mathbf{b} \quad (4)$$

$\mathbf{K}_{\omega 1}$ *warping* stiffness matrix
$\mathbf{K}_{\omega 2}$ stiffness matrix for torsion with out warping restraint

Stiffness matrices $\mathbf{K}_{\omega 1}$ *and* $\mathbf{K}_{\omega 2}$ are added to the member stiffness matrix $\mathbf{K}_m$ in the usual manner.

$$\mathbf{K}_T = \frac{EC_\omega}{a^3}\begin{bmatrix} k_{T1} & k_{T2} & k_{T3} & k_{T4} \\ k_{T2} & k_{T6} & k_{T7} & k_{T8} \\ k_{T3} & k_{T7} & k_{T11} & k_{T12} \\ k_{T4} & k_{T8} & k_{T12} & k_{T16} \end{bmatrix}$$

$$K_{T1} = K_{T11} = S * \theta \sinh\theta, \quad K_{T6} = K_{T16} = S * (\cosh\theta - \frac{\sinh\theta}{\theta}) * a^2$$

$$K_{T2} = K_{T4} = S * (\cosh\theta - 1) * a, \quad K_{T8} = S * \left(\frac{\sinh\theta}{\theta} - 1\right) * a^2$$

$$S = \left(\frac{\theta^2}{Q}\right), \quad Q = 2(1 - \cosh\theta) + \theta\sinh\theta, \qquad K_{T3} = -K_{T1}, \quad K_{T7} = K_{T12} = -K_{T2} \tag{5}$$

The above element stiffness matrix for torsion with restrain warping can be used by divided into two matrices. The parameters K_{T1}, K_{T2}, K_{T6} and K_{T8} can be replace by approximation as shown below:

$$\begin{aligned} K_{T1a} &= 12 + \tfrac{6}{5} * \theta^2 \; K_{T2a} = 6 + \tfrac{1}{10} * \theta^2 \\ K_{T6a} &= 4 + \tfrac{2}{15} * \theta^2 \; K_{T8a} = 2 - \tfrac{1}{30} * \theta^2 \end{aligned} \tag{6}$$

Considering the above series expressions, the alternative matrices can express as shown below:

$$\mathbf{K}_{Ta} = \frac{EC_\omega}{a^3}\begin{bmatrix} 12 & -6a & -12 & 6a \\ 6a & 4a^2 & 6a & 2a^2 \\ -12 & -6a & 12 & 6a \\ 6a & 2a^2 & 6a & 4a^2 \end{bmatrix} + \frac{GJ}{30a}\begin{bmatrix} 36 & -3a & -36 & -3a \\ -3a & 4a^2 & 3a & -a^2 \\ -3a & 3a & 36 & 3a \\ -3a & -a^2 & 3a & 4a^2 \end{bmatrix} \tag{7}$$

3 Results and Discussion

Trigonometric and approximation solutions of an interpolation function are considered to express the stiffness matrix for the non-uniform torsion by considering an additional degree of freedom at each node. Comparing both methods, we can conclude that both are similar for small value of θ and this is commonly considered for open thin-walled section as their value of θ is small. The % of error between the two methods of element stiffness matrix for torsion with restrain warping is given graphically as shown below. As the variation of the total torsional components depends on the value of θ and we can consider different section types. For $\theta = 1$

and 2 the errors range between 6.7 to 9.7% which is considered reasonable and both methods are acceptable for open thin-walled sections as sown in Fig. 2.

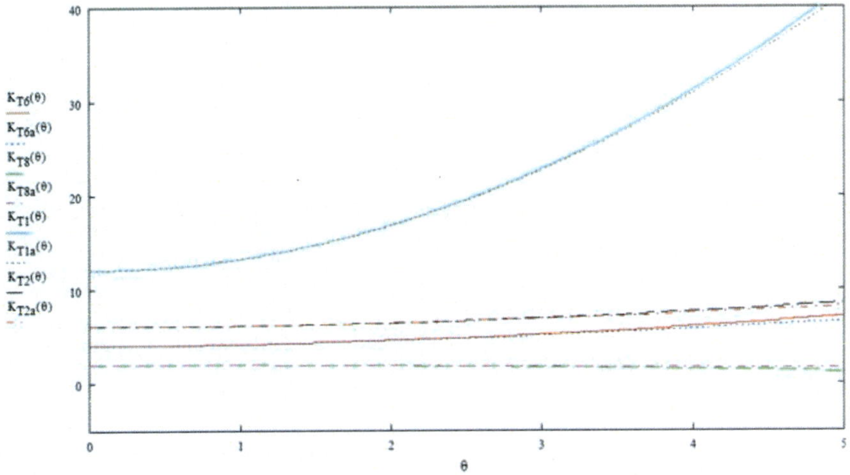

Fig. 2 Evaluation of exact and approximate methods for various values of θ

The stiffness matrix as it is known, the relationship between the generalized force vector q_m and the generalized displacement vector v_m is established by the stiffness matrix K_m of the element.

$$\mathbf{q}_m = \mathbf{K}_m \quad \mathbf{v}_m \tag{8}$$

In the local coordinate system, the nodal displacement and load vector are shown below:

$$\mathbf{v}_e = \begin{bmatrix} v_{1A} & v_{2A} & v_{3A} & \beta_{1A} & \beta_{2A} & \beta_{3A} & \beta_{1,1A} & v_{1B} & v_{2B} & v_{3B} & \beta_{1B} & \beta_{2B} & \beta_{3B} & \beta_{1,1A} \end{bmatrix}^T$$

$$\mathbf{q}_m = \begin{bmatrix} n_{1A} & n_{2A} & n_{3A} & m_{1A} & m_{2A} & m_{3A} & m_{\omega A} & n_{1B} & n_{2B} & n_{3B} & m_{1B} & m_{2B} & m_{3B} & m_{\omega B} \end{bmatrix}^T$$

The results are compared with different studies in both methods to include the aadditional degrees of freedom and are introduced at the nodes and added to member displacement vector [22, 23]. The member variables are collected in member displacement vector V_m and member load vector q_m and the matrices are arranged correspondingly in member stiffness matrix k_m.

$\mathbf{K}_m$ =

k_1							k_2						
	k_3				k_4			k_6				k_4	
		k_9		k_{10}					k_{12}		k_{10}		
			k_{T1}			k_{T2}				k_{T3}			k_{T4}
		k_{10}		k_{11}					k_{13}		k_{14}		
	k_4				k_5			k_7				k_8	
			k_{T5}			k_{T6}				k_{T7}			k_{T8}
k_2							k_1						
	k_6				k_7			k_3				k_7	
		k_{12}		k_{13}					k_9		k_{13}		
			k_{T9}			k_{T10}				k_{T11}			k_{T12}
		k_{10}		k_{14}					k_{13}		k_{11}		
	k_4				k_5		k_7					k_8	
			k_{T13}			k_{T14}				k_{T15}			k_{T16}

Based on the analytical method and the simplified approach presented in the preceding sections, an example of an open thin-walled beam section with external combined torque and point load applied on shear center is considered to demonstrate the method's applicability and fidelity. Furthermore, a commercial software, ANSYS, is used to provide finite solid element analysis and is graphically compared as shown in the figures below. Simulations have been carried out in the Ansys program with beam189 elements, considering an additional degree of freedom (Fig. 3).

Fig. 3 Finite element module of the thin-walled beam section

Table 1 shows the percentage difference in values of deflection and non-uniform angle of twist (restrained torsion) on the axis of an I-section of a bar using both methods.

Table 1 The percentage difference between the two methods

Deflection (m)			Angle of twisting (m)		
Ansys	Mathcad 14 × 14 DOF FEM	% of difference	Ansys	Mathcad 14 × 14 DOF FEM	% of difference
−7.25E–08	−8.48E–08	**−14.50**	−1.70E–05	−1.59E–05	**6.95**
−2.94E–06	−3.48E–06	**−15.50**	−2.49E–03	−2.23E–03	**11.73**
−8.57E–06	−9.96E–06	**−14.00**	−8.71E–03	−7.66E–03	**13.64**
−1.73E–05	−1.92E–05	**−10.00**	−1.81E–02	−1.59E–02	**13.64**
−2.81E–05	−3.09E–05	**−9.00**	−3.00E–02	−2.72E–02	**10.50**
−3.95E–05	−4.46E–05	**−11.50**	−4.41E–02	−4.01E–02	**9.89**
−5.22E–05	−6.00E–05	**−13.00**	−5.98E–02	−5.44E–02	**9.89**
−7.22E–05	−7.68E–05	**−6.00**	−7.68E–02	−7.18E–02	**6.95**
−8.32E–05	−9.46E–05	**−12.00**	−9.48E–02	−8.48E–02	**11.73**
−9.83E–05	−1.13E–04	**−13.00**	−1.13E–01	−1.03E-01	**10.50**
−1.17E–04	−1.30E–04	**−10.00**	−1.30E–01	−1.19E–01	**9.89**

The variation of the non-uniform angle of twist (restrained torsion) and deformation on the axis of an I-section of the bar are shown in Fig. 4. The proposed beam elements were created using the mathematical program Mathcad 14 × 14 node FEs and ANSYS commercial software. Figure 4 depicts the solution results for deflection and angle of twisting, with 14 × 14 DOFs, which are validated by a comparison with the corresponding values from Mathcad 14 × 14 DOFs as well as the results obtained from commercial software (Fig. 4).

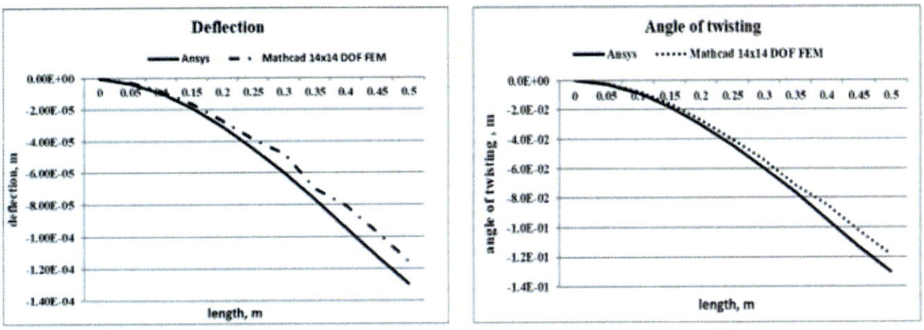

(a) (b)

Fig. 4 The deflection **a** and angle of twisting **b** distribution along the beam's longitudinal direction of thin-walled section

4 Conclusions

According to this study the following conclusions are drawn.

1. The stiffness matrix for 3D beam of thin-walled sections is presented which is more convenient for advanced structural analysis bar elements.
2. The stiffness matrix for thin-walled sections with restrained torsion using the 14×14 member stiffness matrix including warping as an additional degree of freedom and commonly with open thin-walled section is considered.
3. To include the additional degree of freedom for the stiffness matrix, in this study we considered two different methods and their results are very close to each other for small value of characteristics number.
4. This stiffness matrix is more applicable for open thin-walled sections because the value of characteristics number for open section is very small comparing to the closed thin-walled sections.
5. To include the additional degree of freedom both trigonometric and approximate methods are considered and for characteristics number $(\theta) = 1$ and 2 the errors range between 6.7 to -9.7% which is considered reasonable and both methods are acceptable for open thin-walled sections.
6. The length of the member is limited based on the section type and with maximum value of characteristics number (θ) less than 2.

References

1. Saadé K, Warzée G, Espion B (2006) Thin Walled Struct 44(7):808–821. https://doi.org/10.1016/j.tws.2006.04.003
2. Murín J et al (2014) Eng Struct 59:153–160. https://doi.org/10.1016/j.engstruct.2013.10.036
3. Saadé K, Espion B, Warzée G (2004) Thin Walled Struct 42(6):857–881. https://doi.org/10.1016/j.tws.2003.12.003
4. Sapountzakis EJ (2013) Bars under torsional loading: a generalized beam theory approach. In: ISRN Civil Engineering, pp 1–39. https://doi.org/10.1155/2013/916581
5. Kim JH, Kim YY (2000) Int J Numer Methods Eng 49(5):653–668. https://doi.org/10.1002/1097-0207(20001020)49:5<653::AID-NME973>3.0.CO;2-I
6. Jönsson J (1999) Thin Walled Struct 33(4):269–303. https://doi.org/10.1016/S0263-8231(98)00050-0
7. Put BM, Pi Y-L, Trahair NS (1999) J Struct Eng 125(5):9445. https://doi.org/10.1061/(ASCE)0733-9445(1999)125
8. Galishnikova VV (2020) A theory for space frames with warping restraint at nodes. In: 1st IAA/AAS SciTech forum on space flight mechanics and space structures and materials, pp 763–784
9. Vatin N, Havula J, Martikainen L, Sinelnikov A (2013) Thin-walled cross-sections and their joints : tests and FEM-modelling. In: Lellep J, Puman E (eds) Proceedings of the 2nd International conference, University of Tartu Press, Tartu. https://doi.org/10.4028/www.scientific.net/AMR.945-949.1211
10. Paul AS, Charles JC (2003) AISC Design Guide: Torsional Analysis of Structural Steel Members_Revision, Steel Design Guide Series, p 116

11. Vlasov VZ (1984) Thin-walled Elastic Beams, National Technical Information Service, Springfield, Virginia, p 493
12. Alwis WAM, Wang CM (1996) Eng Struct 18(2):125–132. https://doi.org/10.1016/0141-0296(95)00112-3
13. Gebre TH, Galishnikova VV (2020) J Phys Conf Ser 1687:012020. https://doi.org/10.1088/1742-6596/1687/1/012020
14. Banić D, Turkalj G, Brnić J (2016) Trans Famena 40(2):71–82. https://doi.org/10.21278/TOF.40206
15. Murín J, Kutiš V (2002) Comput Struct 80(3–4):329–338. https://doi.org/10.1016/S0045-7949(01)00173-0
16. Jönsson J (1998) Comput Struct 68(4):393–410. https://doi.org/10.1016/S0045-7949(98)00070-4
17. Perelmuter A, Yurchenko V (2014) Metal Constr 20:179–190
18. Wang ZQ, Zhao JC, Zhang DX, Gong JH (2012) Restrained torsion of open thin-walled beams including shear deformation effects. J Zhejiang Univ Sci A 13(4):260–273. https://doi.org/10.1631/jzus.A1100149
19. Wang Z-Q, Zhao J-C (2014) Warping behavior of open and closed thin-walled sections with restrained torsion. J Struct Eng 140(11):040140899. https://doi.org/10.1061/(asce)st.1943-541x.0001010
20. Tusnin A (2016) Finite element for calculation of structures made of thin-walled open profile rods. Procedia Eng 150:1673–1679. https://doi.org/10.1016/j.proeng.2016.07.149, http://dx.doi.org/10.1016/j.proeng.2016.07.149
21. Xiao-Feng W, Qi-Lin Z, Qing-Shan Y (2010) A new finite element of spatial thin-walled beams. Appl Math Mech 31(9):1141–1152. https://doi.org/10.1007/s10483-010-1349-7
22. Murín J, Kutiš V, Královič V, Sedlár T (2012) 3D Beam Finite element including nonuniform torsion. Procedia Eng 48:436–444. https://doi.org/10.1016/j.proeng.2012.09.537
23. Sapountzakis EJ, Mokos VG (2004) 3-D beam element of variable composite cross section including warping effect. Acta Mechanica 171(3–4):151–169. https://doi.org/10.1007/s00707-004-0148-4

Impact of Joint Quality on Stress–Strain State and Stability of Bureyskaya Concrete Dam

Viktor Tolstikov and Yara Waheeb Youssef

Abstract This research is aimed at investigating the impact of grouting quality of the inter–columnar joint as well as the impact of quality of horizontal construction joints on stress–strain state and anti–slide stability of Bureyskaya concrete gravity dam. The calculations are carried out using the software complex "CRACK" based on finite element method under design loads. The results of the numerical modelling calculations show that the grouting quality of the inter–columnar joint significantly affects the stress–strain state of Bureyskaya concrete gravity dam under design loads. However, the stress–strain state of Bureyskaya concrete gravity dam is insignificantly affected by the quality of horizontal construction joints under design loads. In addition, it is found that in comparison with the non–grouted inter–columnar joint, the grouting percentages of the inter–columnar joint 38%, 53 and 100% lead to increasing the dam's anti–slide stability coefficient value at the contact joint by 12.2%, 13.7% and 14.1%, respectively. Furthermore, the results show that in consideration of good–quality horizontal joints, the dam's anti–slide stability coefficient value at the first horizontal joint (at the level of 151 m) is greater by more than 2 times than the case of poor–quality horizontal joints, and greater by 24% than the case of average–quality horizontal joints.

Keywords Concrete gravity dam · Inter–columnar joint grouting · Horizontal construction joints · Horizontal joint quality · Stress–strain state · Anti-slide stability coefficient · Joint Shear characteristics · Angle of internal friction · Cohesion · Software complex CRACK

V. Tolstikov · Y. W. Youssef (✉)
Moscow State University of Civil Engineering, Moscow 129337, Russian Federation
e-mail: yarayoussef2811@hotmail.com

V. Tolstikov
e-mail: TolstikovVV@mgsu.ru

P. Akimov et al. (eds.), *Proceedings of FORM 2022*, Lecture Notes in Civil Engineering 282, https://doi.org/10.1007/978-3-031-10853-2_35

1 Introduction

During the construction and operation of massive concrete gravity dams, various violations of the continuity of their material occur: cracks form, construction joints open and close, etc. [1]. In severe weather conditions, majority of high concrete gravity dams, including Bureyskaya concrete gravity dam, are constructed by the method of cutting the structure into separately constructed columnar missives. Then the temporary inter–columnal temperature–shrinkage joints in the concrete gravity dam's body are grouted using a pipe system with outlets for cement mortar injection [2, 3].

Construction joints are defined as the spaces between adjacent blocks and lifts in conventional vibrated concrete (CVC) and roller compacted concrete (RCC) dams. These discontinuities into the concrete gravity dam body are the weakest zones, which formed because of the construction technology of the dam [4, 5]. Construction joints lead to a degradation of dam continuity and an increase in seepage between blocks and lifts, which leads to reducing the tensile–shear strength [6, 7]. Furthermore, the presence of construction joints inside the body of concrete gravity dam has a great impact on displacements and forces at the points of dam [1]. In addition, construction joints can create special conditions in stability of concrete gravity dams [8].

One of the main problems in concrete gravity dam design and safety monitoring is to investigate the impact of construction joints on structural behavior of dams [8]. The results of experimental studies have shown that increasing shear characteristics (angle of internal friction and cohesion) of construction joints leads to increasing the ultimate bearing capacity and improving the stability of the concrete gravity dam [9, 10]. Shear characteristics of construction joints are increased by carring out grouting woks, which provide the monolithic behavior of the concrete gravity dam and help to avoid dam cracking. Therefore, joint grouting plays an important role in improving dam stability [9, 11]. According to the calculation results from the action of the hydraulic load, it was concluded that account of the disclosure joints leads to a significant increase in the calculated stresses in the concrete gravity dam body [1].

At present, numerical methods are used in different fields of engineering and science because of their computational speed and high efficiency [8]. Actually, finite element method is considered to be an important method for modelling and investigating the structural behavior of concrete gravity dams [12].

Primary grouting works of inter-columnar joints of Bureyskaya concrete gravity dam began in 1995 in the station part of the dam. In 2008, they were fully completed in all parts of the dam. Studies have shown that carrying out grouting works of intercolumnar joints of Bureyskaya concrete gravity dam had a positive effect on the stress-strain state of the dam [13].

The purpose of this research is to investigate the impact of grouting quality of the inter–columnar vertical joint as well as the impact of quality of horizontal construction joints on stress–strain state and anti–slide stability of Bureyskaya concrete gravity dam. The research is based on finite element method using the software complex "CRACK". The calculations are carried out under static loading condition.

2 Materials and Methods

Finite Element Modelling of «Bureyskaya Concrete Gravity Dam-Rock Foundation» System

The studied dam in this research is Bureyskaya concrete gravity dam, which is located on Bureya river in Russian Federation, with a dam crest length of 789 m. Dimensions of the investigated right–bank non-overflow section №16 of the dam are as follows: dam height 122 m, dam crest width 14 m. The concrete gravity dam of Bureyskaya hydroelectric power plant (HPP) is consisted of a conventional vibrated concrete (CVC) upper column and a roller compacted concrete (RCC) lower wedge–shaped part. At positive temperatures in winter, firstly the upper column of Bureyskaya concrete gravity dam was constructed of conventional vibrated concrete (CVC-B15W8) under a tent. Then during the warm period, the internal zone of the dam of roller compacted concrete (RCC-B10) as well as the protection layer on the downstream slope of conventional vibrated concrete (CVC-B15F200) were constructed with a lag in height, in comparison with the height of the dam's upper column. Due to temperature deformations during shrinkage and cooling of concrete, an inter–columnar vertical joint formed between conventional vibrated concrete and roller compacted concrete. This inter–columnar joint requires grouting works [5, 13].

A two–dimensional (2D) finite element model of «Bureyskaya concrete gravity dam–rock foundation» system is constructed for the right–bank non-overflow section №16 of the dam, as shown in Fig. 1. The two–dimensional model mesh has got 2697 nodes and 2562 finite elements.

Table 1 shows physical and mechanical characteristics of the materials of «Bureyskaya concrete gravity dam–rock foundation» system. Shear characteristics

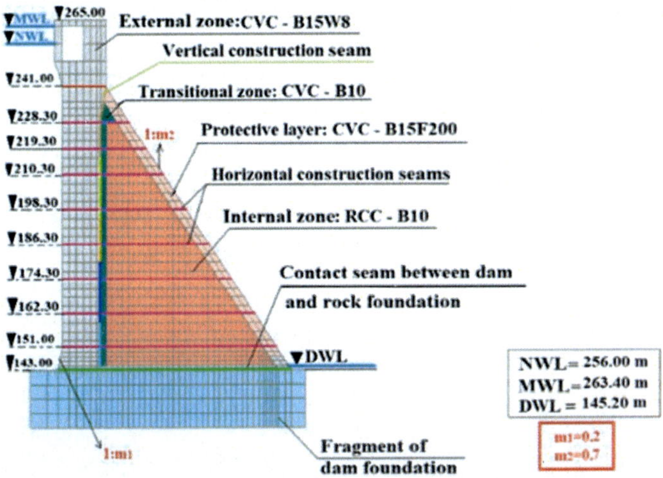

Fig. 1 Two–dimensional (2D) finite element model of «Bureyskaya concrete gravity dam–rock foundation» system for the right–bank non-overflow section №16 of the dam

Table 1 Physical and mechanical characteristics of the materials of «Bureyskaya concrete gravity dam–rock foundation» system [13]

Material	Volumetric Weight (KN/m^3)	Poisson ratio, ν	Elastic Modulus, E (MPa)	Compressive Strength, R_c (MPa)	Tensile strength, R_t (MPa)
CVC–B15W8	24.0	0.15	34,500	11.3	1.15
CVC–B15F200	24.5	0.15	34,500	11.3	1.15
CVC–B10	24.0	0.15	30,000	7.5	0.85
RCC–B10	23.5	0.15	30,500	7.5	0.78
Foundation rock	20.0	0.24	17,000	9.0	0.25

of the contact joint between the dam and the rock foundation are: angle of internal friction $\varphi = 45°$, cohesion C = 1 MPa [13].

The numerical modelling calculations are carried out using the software complex "CRACK" developed at the Department of Hydraulics and Hydrotechnical Engineering of Moscow State University of Civil Engineering, based on finite element method [14–16]. The calculations are carried out considering design loads on the dam: the self–weight of the concrete gravity dam, the hydrostatic pressure on the upstream dam face and the seepage uplift pressure on the base of the dam. The calculations are carried out with 70 iterations and at the upstream water level corresponding the normal water level (NWL = 256 m).

Scenarios of Grouting Quality of Bureyskaya Dam's Inter–Columnar Joint
In order to investigate the impact of grouting quality of the inter–columnar joint on stress–strain state and anti–slide stability of Bureyskaya concrete gravity dam, five different scenarios of inter–columnar joint grouting are studied. Scenarios №1 and №2 take into account a non–grouted inter–columnar joint. Scenarios №3 and №4 consider a partially grouted inter–columnar joint (grouting works in scenario №3 are carried out from the dam base to the level of 180.3 m, while in scenario №4 grouting works are carried out from the dam base to the level of 195.3 m). However, scenario №5 is characterized by a fully grouted inter–columnar joint. Characteristics of dam's inter–columnar joint according to each scenario are shown in Table 2 [13].

The five scenarios consider that all the horizontal construction joints of Bureyskaya concrete gravity dam are poor–quality (angle of internal friction $\varphi = 35°$, cohesion C = 0.25 MPa).

Scenarios of Quality of Bureyskaya Dam's Horizontal Construction Joints
The impact of horizontal construction joint quality on stress–strain state and anti–slide stability of Bureyskaya concrete gravity dam is studied as follows. Scenario №5, which is characterized by a fully grouted inter–columnar joint and poor–quality horizontal construction joints, is studied for two additional qualities of horizontal construction joint quality: average–quality and good–quality horizontal construction joints. Shear characteristics (angle of internal friction and cohesion) of the poor–quality horizontal construction joints, average–quality horizontal construction joints and good–quality horizontal construction joints are shown in Table 3 [13].

Table 2 Characteristics of the inter–columnar joint of Bureyskaya concrete gravity dam according to each scenario [13]

Inter–Columnar joint characteristics	Scenario №1	Scenario №2	Scenario №3	Scenario №4	Scenario №5
Grouting percentage (%)	0	0	38	53	100
Ultimate allowed closure (mm)	1	2	1	1	1
Shear stiffness (MPa/m)	0.1	0.1	100	100	100
Normal stiffness (MPa/m)	0.5	0.5	300	300	300
Cohesion, C (MPa)	0	0	0.3	0.3	0.3
Angle of internal friction, φ (°)	1	1	35	35	35
Tensile strength (MPa)	0.01	0.01	0.1	0.1	0.1

Table 3 Shear characteristics of the three qualities of horizontal construction joints of Bureyskaya concrete gravity dam [13]

Shear characteristics of horizontal construction joints	Poor–Quality horizontal joints	Average–Quality horizontal joints	Good–Quality horizontal joints
Angle of internal friction, φ (°)	35	45	50
Cohesion, C (MPa)	0.25	0.75	1

3 Results and Discussion

Impact of Grouting Quality of the Inter–columnar Joint on Stress–Strain State and Anti–Slide Stability of Bureyskaya Concrete Gravity Dam
Figure 2 shows the stress state (normal stresses σ_y) of «Bureyskaya concrete gravity dam–rock foundation» system in consideration of five different scenarios of inter–columnar joint grouting.

Figure 3 shows the element state of Bureyskaya concrete gravity dam–rock foundation system in consideration of five different scenarios of inter–columnar joint grouting.

Table 4 shows the characteristics of the stress–strain state and the anti–slide stability of «Bureyskaya concrete gravity dam–rock foundation» system in consideration of five different scenarios of inter–columnar joint grouting.

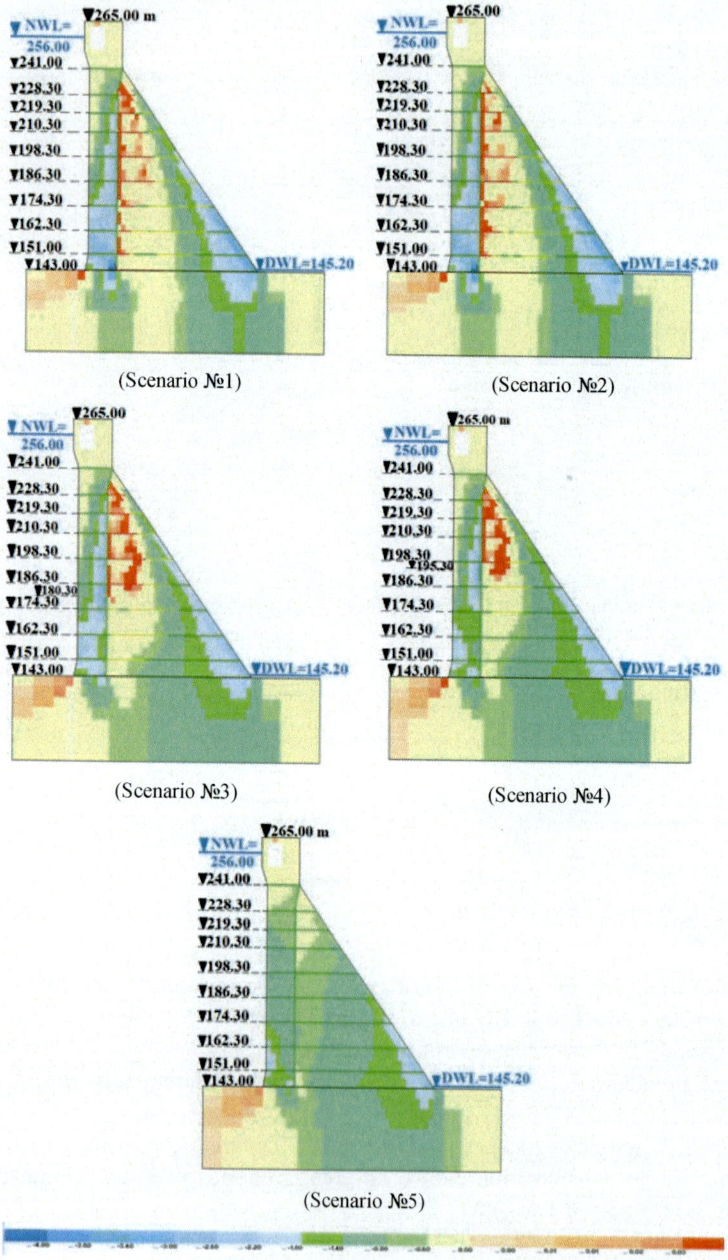

Fig. 2 Stress state (normal stresses σ_y) of «Bureyskaya concrete gravity dam–rock foundation» system in consideration of five different scenarios of inter–columnar joint grouting

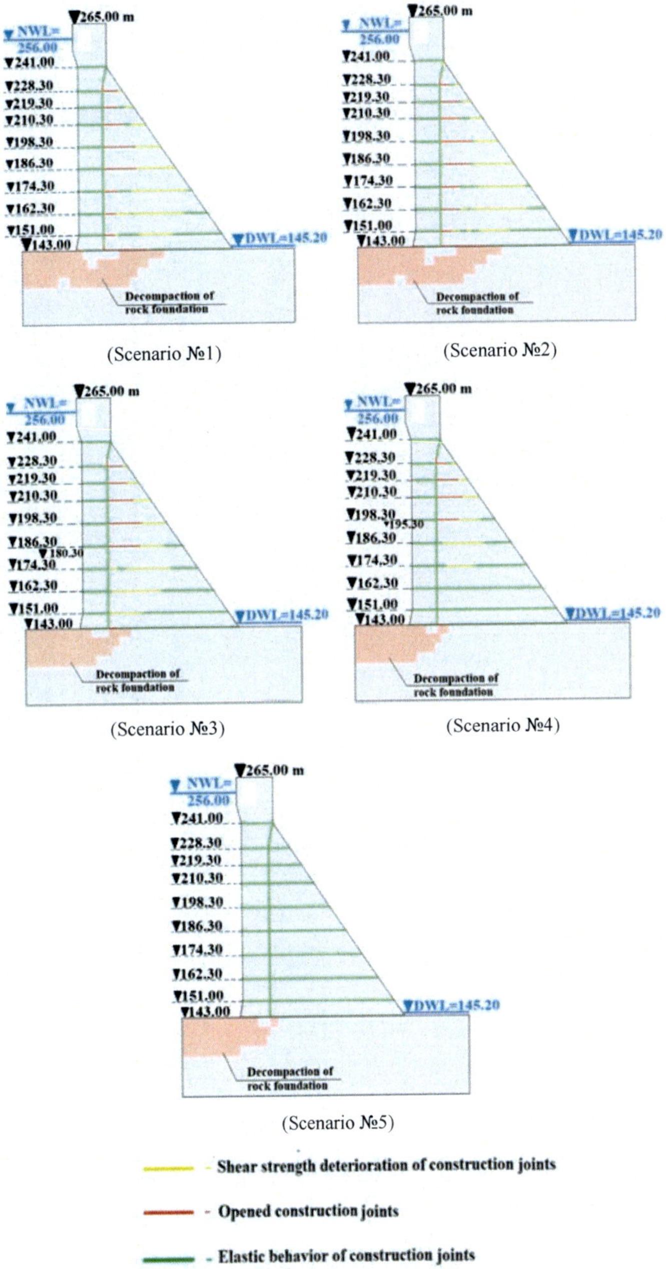

Fig. 3 Element state of «Bureyskaya concrete gravity dam–rock foundation» system in consideration of five different scenarios of inter–columnar joint grouting

Table 4 Characteristics of the stress–strain state and the anti–slide stability of «Bureyskaya concrete gravity dam–rock foundation» system in consideration of five different scenarios of inter–columnar joint grouting

Dam's characteristics	Scenario №1	Scenario №2	Scenario №3	Scenario №4	Scenario №5
Maximum Horizontal Displacement of the Dam Crest in X–Direction (cm):	6.5	6.6	5.1	4.6	2.7
Maximum Vertical Displacement of the Dam Crest in Y–Direction (cm):	1.7	1.3	1.5	1.4	1.1
Contact Joint Disclosure in the Upper Column of the dam (m):	2	2	0	0	2
Contact Joint Disclosure in the roller compacted concrete of the internal zone of the dam (m):	23.9	23.9	5.2	1.1	0
Length of Decompaction Zone of Rock Foundation Along the Dam Base (m):	47.4	48.8	27.9	21.7	19.7
Principal tensile stress, σ1 (MPa)	0.57	0.69	0.71	0.36	0.28
Principal tensile stress, σ2 (MPa)	− 5.34	− 5.4	− 4.48	− 4.35	− 4.28
Dam's Anti–Slide Stability Coefficient at the Contact Joint, K_S:	2.62 > 1.32	2.62 > 1.32	2.94 > 1.32	2.98 > 1.32	2.99 > 1.32

In consideration of the non–grouted inter–columnar joint of Bureyskaya concrete gravity dam (scenario №1 and scenario №2), it is observed that when the ultimate allowed closure of the inter–columnar joint is between [1, 2] mm, the stress–strain state of the dam and the state of the dam elements almost do not change.

In consideration of the partially grouted inter–columnar joint of Bureyskaya concrete gravity dam with a grouting percentage of 38% (scenario №3), it is observed that:

- The contact joint is closed in the upper column of the dam;

- The disclosure of the contact joint in the roller compacted concrete of the internal zone of the dam decreases in comparison with the two scenarios of the non–grouted inter–columnar joint;
- The distribution of decompaction zone of rock foundation along the dam base is less in comparison with the two scenarios of the non–grouted inter–columnar joint;
- The tensile stresses in the dam body are less in comparison with the two scenarios of the non–grouted inter–columnar joint;
- The coefficient of anti–slide stability along the contact joint increased by 12.2% (from 2.62 to 2.94) in comparison with the two scenarios of the non–grouted inter–columnar joint.

In consideration of the partially grouted inter–columnar joint of Bureyskaya concrete gravity dam with a grouting percentage of 53% (scenario №4), it is observed that:

- The disclosure of the contact joint in the roller compacted concrete of the internal zone of the dam decreases in comparison with the scenario of partially grouted inter–columnar joint with a grouting percentage of 38%;
- The tensile stresses in the dam body are less in comparison with the scenario of partially grouted inter–columnar joint with a grouting percentage of 38%;
- The coefficient of anti–slide stability at the contact joint increases by 13.7% (from 2.62 to 2.98) in comparison with the two scenarios of the non–grouted inter–columnar joint;
- The coefficient of anti–slide stability increases by 1.4% (from 2.94 to 2.98) in comparison with the scenario of partially grouted inter–columnar joint with a grouting percentage of 38%.

In consideration of the fully grouted inter–columnar joint of Bureyskaya concrete gravity dam (scenario №5), the stress–strain state of the dam improves as follows:

- The horizontal displacement of the dam crest in X–direction decreases;
- The contact joint in the roller compacted concrete of the internal zone of the dam is closed;
- The inter–columnar joint and the horizontal construction joints are completely closed;
- The length of decompaction zone of rock foundation along the dam base is less;
- The coefficient of anti–slide stability at the contact joint increases by 14.1% in comparison with the two scenarios of the non–grouted inter–columnar joint;
- The stress state of «Bureyskaya concrete gravity dam–rock foundation» system is characterized by the absence of tensile stresses as well as the absence of concentration of compressive stresses at the downstream toe of the dam.

It is observed that in all the five scenarios, the principal tensile stress (σ_1) values are less than the concrete tensile strength ($R_t = 0.78$ MPa). In addition, the principal compressive stress (σ_2) values are less than the concrete compressive strength ($R_C = 11.3$ MPa).

Table 5 Values of anti–slide stability coefficient of Bureyskaya concrete gravity dam at the first horizontal joint (at the level of 151 m) according to the quality of this horizontal joint

Quality of the dam's first horizontal joint (at the level of 151 m)	Dam's Anti–Slide stability coefficient value at the first horizontal joint (K_S)
Poor–Quality Joint	1.74 > 1.25
Average–Quality Joint	3.05 > 1.25
Good–Quality Joint	3.78 > 1.25

The normative documentation determines the minimum allowable value for the anti–slide stability coefficient at contact joint of the concrete dam as $K_{S,\ norm} = 1.32$ [13, 17]. It is observed that in all the five scenarios, the values of dam's anti–slide stability coefficient at the contact joint are greater the normative value ($K_{S,\ norm} = 1.32$).

Impact of Quality of Horizontal Construction Joints on Stress–Strain State and Anti–slide Stability of Bureyskaya Concrete Gravity Dam
It is observed that in consideration of scenario №5 with average–quality horizontal construction joints and good–quality horizontal construction joints under design loads acting on the dam, the stress state of Bureyskaya concrete gravity dam matches the stress state of the scenario №5 with poor–quality horizontal construction joints (Fig. 2). Furthermore, it is observed that in consideration of scenario №5 with average–quality horizontal construction joints and good–quality horizontal construction joints under design loads , the element state of Bureyskaya concrete gravity dam matches the element state of the scenario №5 with poor–quality construction joints (Fig. 3). Therefore, it is concluded that the quality of horizontal construction joints insignificantly affects the stress-state and the element state of the dam. It is worth noting that in another research, it was concluded that the quality of the horizontal construction joints significantly affects the stress–stress state and the bearing capacity of the concrete gravity dam under dam overloading [4, 5].

Table 5 shows the values of anti–slide stability coefficient of Bureyskaya concrete gravity dam at the first horizontal joint (at the level of 151 m) according to the quality of this horizontal joint.

The normative documentation determines the minimum allowable value for the anti–slide stability coefficient at the horizontal construction joints of the concrete dam as $K_{S,\ norm} = 1.25$ [13, 17]. It is observed that all the values of the anti–slide stability coefficient of Bureyskaya concrete gravity dam at the first horizontal joint are greater the normative value ($K_{S,\ norm} = 1.25$) regardless of the quality of this joint. Furthermore, it is noticed that the value of the dam's anti–slide stability coefficient increases with increasing the quality of horizontal construction joints as follows. In consideration of good–quality horizontal joints, the dam's anti–slide stability coefficient value at the first horizontal joint (at the level of 151 m) is greater by more than 2 times than the case of poor–quality horizontal joints, and greater by 24% than the case of average–quality horizontal joints.

Model Verification
Analysis of the field data of Bureyskaya concrete gravity dam shows that the dam's calculated stress–strain state values using the software complex "CRACK" in the different studied scenarios match well the measured values on the field using control and measuring instruments [13].

4 Conclusion

The main conclusions from this research are as follows:

1. The grouting quality of the inter–columnar joint significantly affects the stress–strain state of Bureyskaya concrete gravity dam under design loads.
2. The quality of horizontal construction joints insignificantly affects the stress–strain state of Bureyskaya concrete gravity dam under design loads.
3. In comparison with the non–grouted inter–columnar joint, grouting percentages of the inter–columnar vertical joint 38%, 53% and 100% lead to increasing Bureyskaya dam's anti–slide stability coefficient value at the contact joint by 12.2%, 13.7% and 14.1%, respectively.
4. In consideration of good–quality horizontal construction joints, Bureyskaya dam's anti–slide stability coefficient value at the first horizontal joint (at the level of 151 m) is greater by more than 2 times than the case of poor–quality horizontal joints, and greater by 24% than the case of average–quality horizontal joints.

References

1. Kolosova GS, Lalin VV, Kolosova AV (2013) Mag Civ Eng 5:76–85
2. Argal ES, Ryzhankova LN (2017) Gidrotekhnicheskoe Stroitel'stvo 9:40–45
3. Argal ES (1987) Monolithing of concrete dams through cementation of construction joints. Energoatomizdat, Moscow
4. Tolstikov VV Youssef Y (2021) Gidrotekhnicheskoe Stroitel'stvo 11:24–28
5. Tolstikov V, Youssef YW (2022) Impact of construction seams on the bearing capacity of a CVC-RCC combined dam. In: Akimov P, Vatin N (eds) Proceedings of FORM 2021. LNCE, vol 170, pp 371–382. Springer, Cham. https://doi.org/10.1007/978-3-030-79983-0_35
6. Zhang M et al (2020) Constr Build Mater 263:120248
7. Al Baghdady S, Khan L (2018) Designing roller compacted concrete (RCC) dams, MSc thesis, KTH School of ABE
8. Armaghani DJ, Bayat V, Koopialipoor M, Pham BT, Bull J (2021) Eng Geol Environ 80:55–70
9. Belchenko KP (1984) Impact of horizontal weak seams on the strength and stability of concrete gravity dams, Ph.D. thesis, Moscow Institute of Civil Engineering named after V.V. Kuibyshev (MISI)
10. Ibrahim BB (1991) Impact of columnar cutting on the static work of concrete gravity dams, Ph.D. thesis, Moscow Institute of Civil Engineering named after V.V. Kuibyshev (MISI)

11. Buffi G, Manciola P, Lorenzis LD, Gambi A (2018) Influence of construction joints in arch–gravity dam modelling: the case of Ridracoli. In: Proceedings of the 26th international congress on large dams, Vienna, Austria, pp 1047–1062
12. Ren QW, Jiang YZ (2011) Sci China Tech Sci 54:509–515
13. Declaration of hydraulic structure safety of OJSC RusHydro branch (2013) «Bureyskaya HPP», p 217, Talakan, Russia
14. Zertsalov MG, Ivanov VA Tolstikov VV (1988) Soil Mech Found Eng 2
15. Tolstikov VV (2006) Vestnik MGSU 2:123–132
16. Zertsalov MG Tolstikov VV (1988) Gidrotekhnicheskoe Stroitel'stvo 8
17. Yudelevich AM (2017) Assessment of reliability of concrete gravity dams on rock foundations at the stages of design, construction and operation, DSc thesis, JSC All–Russian Research Institute of Hydrotechnical Engineering named after B.E. Vedeneev

Influence of Polycarboxylate Superplasticizer and Silica Fume on the Properties of Self-compacting Concrete

Aleksandr Smirnov, Lev Dobshits, and Sergey Anisimov

Abstract Self-compacting concrete is a highly flowing concrete that compacts under self-weight without any vibration effort. To obtain self-compacting concretes with reduced cement consumption and high physical, mechanical and performance characteristics, the use of superplasticizers and mineral additives is required. This paper discusses the effect of a complex additive based on polycarboxylate superplasticizer and silica fume on the properties of self-compacting concretes. According to the study results, self-compacting concrete mixtures with high segregation resistance, high fluidity with a cone flow diameter of more than 700 mm and increased consistency retention of more than 3 h within the slump-flow class SF2 were obtained. The hardening kinetics of self-compacting concrete with a complex additive is characterized by high rates of strength development, even despite the long-term consistency retention of self-compacting concrete mixes. At 1 day, the strength of self-compacting concrete was 31.6 MPa, at 2 days—56.5 MPa, at 28 days—92.5 MPa with a cement consumption of 500 kg/m^3. According to the X-ray phase analysis results, it was found that the complex use of polycarboxylate superplasticizer and silica fume at 1 day leads to an acceleration of the hydration processes of alite by 13%, tricalcium aluminate—by 24% and an increase in the content of portlandite by 5%, ettringite—by 14%. At 28 days, the use of complex additive leads to the formation of a cement stone structure with a reduced content of portlandite by 40%, ettringite by 11% and increased content of amorphous low-basic hydrated calcium silicates by 18%, significantly compacting the structure of cement stone and improving its physical and mechanical characteristics.

Keywords Self-compacting concrete · Polycarboxylate superplasticizer · Silica fume · Sand-aggregate ratio · Slump-flow · Consistency retention · Compressive strength · Water absorption · Phase composition

A. Smirnov (✉) · S. Anisimov
Volga State University of Technology, 3 Lenin Square, Yoshkar-Ola 424000, Russian Federation
e-mail: smiralex93@gmail.com

L. Dobshits
Russian University of Transport (MIIT), 9 b 9 Obrazcova Street, Moscow 127994, Russian Federation

P. Akimov et al. (eds.), *Proceedings of FORM 2022*, Lecture Notes in Civil Engineering 282, https://doi.org/10.1007/978-3-031-10853-2_36

1 Introduction

Self-compacting concretes are increasingly used in constructing various objects in the world. Self-compacting concrete mixes are characterized by high workability, due to which they can spread under their own weight and fill heavily reinforced structures without compaction [1–4].

Obtaining a new type of concrete has become possible due to modern polycarboxylate superplasticizers. Unlike superplasticizers based on naphthalene and melamine sulfonates, polycarboxylate superplasticizers are characterized by a spatial structure of molecules with branched side chains, which contributes to more efficient dispersion of cement floccules due to the steric effect. These superplasticizers have a significant water-reducing and plasticizing ability and provide highly mobile and self-compacting concrete mixes with high retention of rheological properties [5–8].

However, the use of superplasticizers does not provide sufficient expansion of aggregate grains, in which the self-compacting concrete mixture is resistant to segregation. Reducing bleeding and segregation of the self-compacting concrete mix is achieved by increasing cement consumption and the use of mineral additives in the concrete mixture [1–4]. The use of mineral additives in the composition of self-compacting concrete mixtures is one of the priority areas for reducing cement consumption, improving their technological and performance characteristics. Industrial by-products such as fly ash, blast furnace slag, and silica fume are widely used as mineral additives [9–11].

In high-performance concretes, much attention is paid to the ferroalloy industry by-product—silica fume [12, 13]. Due to its large specific surface area and amorphous structure, silica fume has a high pozzolanic activity and is an effective micro filler. The cement stone structure formation in the presence of silica fume is based on the interaction of calcium hydroxide formed during the cement hydration with the active silicon dioxide. The formation of low-basic hydrated calcium silicates instead of portlandite crystals leads to the cement stone structure densification, increasing its strength and resistance to aggressive environments [14, 15].

Thus, the development and use of complex additives based on silica fume and polycarboxylate superplasticizer are one of the priority areas for modifying cement systems, which can significantly improve the technological, physical, mechanical and performance characteristics of self-compacting concrete.

The research goal was to study the effect of a complex additive based on a polycarboxylate superplasticizer and silica fume on the technological properties of self-compacting concrete mixes and the physical and mechanical characteristics of concrete.

2 Materials and Methods

For the preparation of self-compacting concrete mixes, Portland cement (PC) CEM I 52.5 N according to GOST 31,108-2020 produced by LLC «Akkermann Cement» (Novotroitsk) with a specific surface area of 400 m^2/kg and a normal consistency of 28.5% was used as a binder. The clinker of this cement had the following mineralogical composition: C_3S—62.1%; C_2S—15.8%; C_3A—5.0%; C_4AF—13.2%. Crushed stone (CS) with a fraction of 5–20 mm (5–10 mm—40% and 10–20 mm—60%) with a true density of 3.0 g/cm^3, a bulk density of 1.56 g/cm^3, and an intergranular porosity 48% was used as a coarse aggregate. Local fine-grained quartz sand (S) with a fineness modulus of 1.9, a true density of 2.65 g/cm^3, a bulk density of 1.51 g/cm^3, and an intergranular porosity of 43% was used as a fine aggregate. Polycarboxylate superplasticizer (PCE) Sika ViscoCrete 25 HE-C (an aqueous solution with a density of 1.07–1.09 g/cm^3) and condensed silica fume (SF) produced by PJSC «NLMK» with a bulk density of 155 kg/m^3 and mass content of silicon oxide SiO_2 of 92% was used as an additive. Dosages of superplasticizer and silica fume in the composition of self-compacting concrete mixtures were taken based on previous studies and amounted to 1 and 10% of the binder weight [16]. To compare the properties of self-compacting concretes with a complex additive, the organomineral modifier MB 10-01 (MB) GOST R 56,178-2014 was used. This modifier MB is a powdered product with a bulk density of 800 kg/m^3. The mineral part of the modifier (90%) is represented by silica fume, and the organic part (10%) contains a superplasticizer based on sulfonated naphthalene formaldehyde polycondensates (PNS) [17].

The water content (W) in the compositions was selected from the condition of obtaining self-compacting concrete mixtures with a slump-flow class SF2 according to GOST R 59,714-2021/EN 206:2013, which does not require vibration compaction of concrete mixtures. The consistency of concrete mixtures was determined by the slump-flow test in accordance with GOST R 58,002-2017/EN 12,350-8:2010. The viscosity of self-compacting concrete mixtures was determined by the flow time of concrete mixtures to a diameter of 500 mm (t_{500}) in accordance with GOST R 58,002-2017/EN 12,350-8:2010. Sample cubes 100 × 100 × 100 mm in size were made from concrete mixtures. At the age of 1, 2 and 28 days of hardening under normal conditions, the samples were subjected to mechanical tests. The strength of the samples was determined in accordance with GOST 10,180-2012, the density was determined in accordance with GOST 12,730.1-2020, and the water absorption was determined in accordance with GOST 12,730.3-2020.

3 Results and Discussion

One of the main tasks of optimizing concrete composition is determining the ratio of aggregates, ensuring a minimum cement consumption [18].

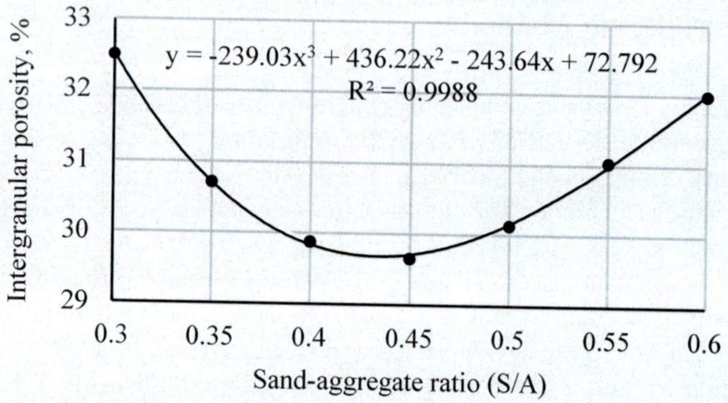

Fig. 1 Intergranular porosity of aggregates mixture with different sand-aggregate ratio

The consumption of coarse and fine aggregates in the composition of self-compacting concrete mixtures was selected from the condition of ensuring the minimum intergranular porosity of the aggregates and achieving the best concrete mixture workability with a fixed volume of cement paste [19].

Figure 1 shows the results of determining the intergranular porosity of aggregates mixture depending on the sand-aggregate ratio.

It has been established that for the aggregates used, the maximum packing density and minimum intergranular porosity of the aggregates mixture are achieved with the sand-aggregate weight ratio of 0.45 (S/A volume ratio 0.48). In this case, the minimum voids volume in the aggregates mixture is 297 l/m^3. With an increase in the S/A ratio to a value of 0.50, the voids volume increases by 1.6%. With a decrease in the S/A ratio to a value of 0.40, the intergranular porosity of the aggregate mixture increases by 0.7%. An increase in the intergranular porosity of aggregates will lead to the need to increase the mortar component, which will lead to overspending of the remaining components of the concrete mixture and an increase in its cost.

To clarify the content of coarse and fine aggregates, the properties of self-compacting concrete mixtures with a fixed volume of cement paste but a different aggregates ratio were studied. The sand-aggregate ratio by weight varied within 0.40; 0.45; 0.50 (by volume 0.43; 0.48; 0.53). Binder consumption in concrete mixtures was taken as 550 kg/m^3, water consumption—170 kg/m^3 (W/B = 0.31). At the same time, the volume of cement paste in the compositions was 350 l/m^3. This volume of cement paste is sufficient to fill the voids between aggregate grains and form the necessary layer of cement paste on their surface, which reduces friction between aggregate particles and increases the ability of the concrete mixture to self-compact. Reducing the volume of cement paste to less than 350 l/m^3 will reduce the expansion of aggregate grains and increase the friction between them. The workability of the system will be insufficient to obtain self-compacting concrete mixtures. As a consequence, increased superplasticizer consumption will be required to increase workability.

Table 1 Compositions and properties of concrete mixes and concretes with different sand-aggregate ratios

№	S/A	Concrete mix proportion, kg/m^3					W/B	Slump-flow, mm	Viscosity t_{500}, s
		PC	S	CS	W	PCE			
1	0.40	550	736	1104	170	5.5	0.31	725	7.9
2	0.45		828	1012				730	8.0
3	0.50		920	920				680	9.2

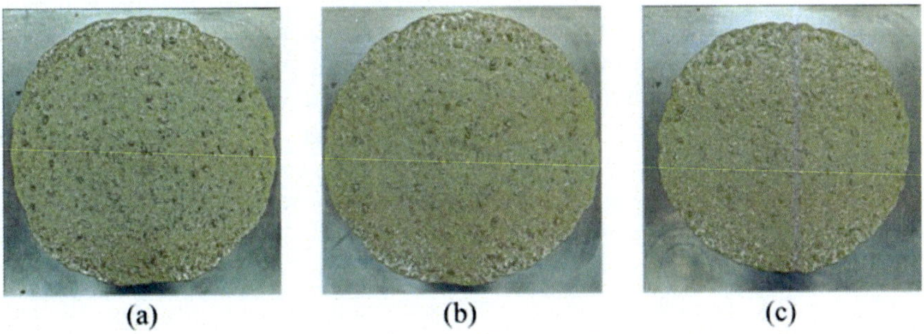

Fig. 2 Appearance of self-compacting concrete mixes with different sand-aggregate ratios: **a**—0.40; **b**—0.45; **c**—0.50

The studied compositions of self-compacting concrete mixtures and their properties are shown in Table 1.

Figure 2 shows the appearance of self-compacting concrete mixes with different sand-aggregate ratios.

According to the results of the workability determination, it was found that the resulting concrete mixtures have high fluidity with a cone flow diameter of 680–730 mm, which corresponds to the slump-flow class SF2, which does not require vibration compaction of concrete mixtures in accordance with GOST R 59,714-2021/EN 206:2013. It is noted that at the sand-aggregate ratio of 0.45, the best workability of the concrete mixture is observed, which is consistent with the results of determining the minimum intergranular porosity of the aggregate mixture. Reducing the amount of sand to a value of 0.40, despite a decrease in the specific surface area of aggregates and a decrease in their water demand, does not increase the workability of the concrete mixture. At the same time, a decrease in the amount of fine fraction leads to slight bleeding of the concrete mixture. With an increase in the sand-aggregate ratio to the value of 0.50, the thickness of the cement gel layer between the grains of aggregates decreases, which causes an increase in viscosity and a decrease in the workability of the concrete mixture.

Next, the influence of a complex additive based on a polycarboxylate superplasticizer and silica fume on the properties of self-compacting concrete mixes and concretes was studied. To compare the properties of self-compacting concretes with a complex additive based on a polycarboxylate superplasticizer and silica fume, a

Table 2 Compositions and properties of concrete mixes with modifying additives

№	Mix ID	Concrete mix proportion, kg/m^3							W/B	Slump, mm	Slump-flow, mm	Viscosity t_{500}, s
		PC	MB	PCE	SF	W	S	CS				
1	Control	550	–	–	–	248	621	1012	0.45	180	–	–
2	MB (PNS + SF)	500	111	–	–	220	725	886	0.36	–	710	5.8
3	PCE	550	–	5.5	–	170	828	1012	0.31	–	730	8.0
4	PCE + SF	500	–	5.5	50	170	819	1001	0.31	–	690	8.2

concrete mixture composition was prepared with an organomineral modifier MB, including silica fume (90%) and a superplasticizer based on sulfonated naphthalene-formaldehyde polycondensates (10%). Also, a control composition of the concrete mix without modifying additives with an S4 slump class was prepared.

The compositions and properties of concrete mixes are given in Table 2.

According to the results of the workability determination of concrete mixtures, it was found that obtaining self-compacting mixtures is possible only with the use of plasticizing additives. In the composition without additives, even with a high cement paste content of 425 l/m^3, the concrete mixture has a cone slump of only 180 mm (slump class S4). Using the organomineral modifier MB based on SF and PNS makes it possible to achieve self-compaction of the concrete mixture only at a very high dosage of the additive 22%, twice as much as when using SF and PCE. At the same time, using a PNS superplasticizer does not lead to a significant decrease in the water content of the concrete mix. The use of PCE superplasticizer makes it possible to obtain self-compacting concrete mixtures even at a low W/B = 0.31 due to the admixture's strong plasticizing and water-reducing effect. The use of highly dispersed silica fume together with polycarboxylate superplasticizer, due to its high water-retaining capacity, allows increasing the homogeneity and segregation resistance of the self-compacting concrete mixture without significantly reducing its workability.

Figure 3 shows the results of determining the consistency retention of the obtained self-compacting concrete mixes with modifying additives. Figures 4, 5 and 6 show the appearance of self-compacting concrete mixes with modifiers over time.

It has been established that the use of an organomineral modifier MB based on SF and PNS leads to a rapid loss of workability of concrete mixtures even at a high dosage of the additive. The consistency retention of self-compacting concrete mix with MB is less than 30 min within the slump-flow class SF2, which does not allow the use of this concrete mix during long-term transportation and the construction of monolithic structures without compaction.

Using a polycarboxylate superplasticizer ensures long-term consistency retention of concrete mixes even at low W/B values due to the strong steric effect of the admixture. At the same time, after repeated mixing of the self-compacting concrete mix with PCE, a slight increase in its flowability was observed after 60 min. This may be due to the peculiarities of the molecular structure of the used polycarboxylate superplasticizer. Researchers note that PCEs with a high grafting degree (low amount of carboxylic groups on the main polymer chain) are slowly adsorbed on cement

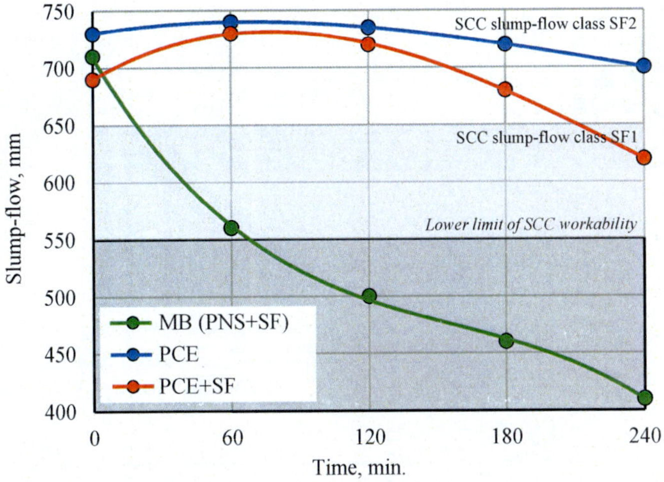

Fig. 3 Consistence retention of self-compacting concrete mixes with modifying additives

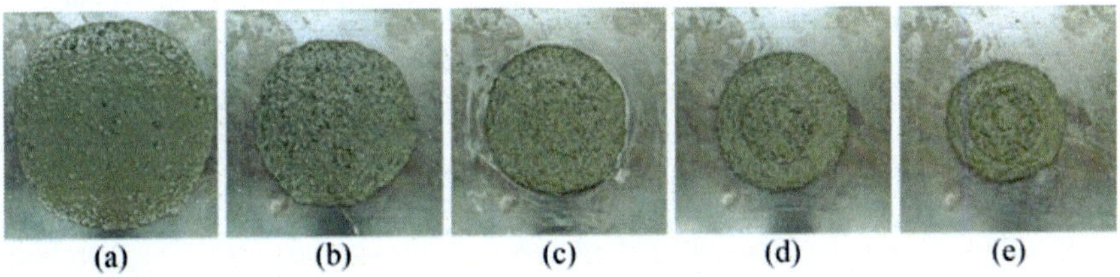

Fig. 4 Appearance of self-compacting concrete mixes with MB modifier over time: **a**—5 min; **b**—60 min; **c**—120 min; **d**—180 min; **e**—240 min

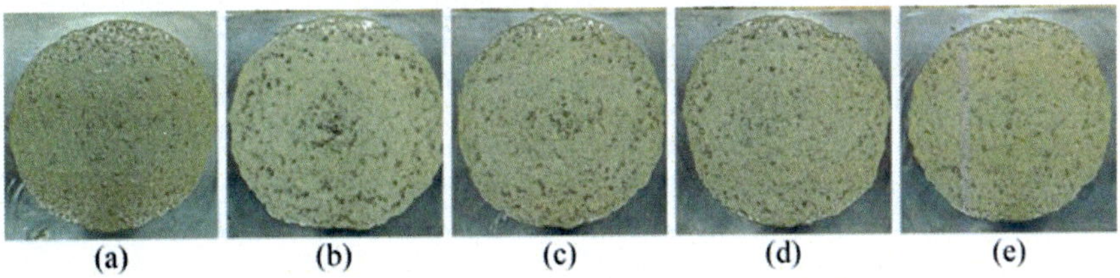

Fig. 5 Appearance of self-compacting concrete mixes with PCE superplasticizer over time: **a**—5 min; **b**—60 min; **c**—120 min; **d**—180 min; **e**—240 min

particles and have the delayed plasticization effect [8]. It is also noted that self-compacting concrete mix with PCE without stabilizing and water-retaining additives is prone to bleeding and segregation.

The use of highly dispersed silica fume due to its high water-retaining capacity increases the homogeneity and segregation resistance of the self-compacting concrete

Fig. 6 Appearance of self-compacting concrete mixes with PCE + SF additive over time: **a**—5 min; **b**—60 min; **c**—120 min; **d**—180 min; **e**—240 min

Table 3 Concretes properties with modifying additives

№	Mix ID	Compressive strength, MPa			Strength class	Density, kg/m^3	Water absorption, %	
		1 day	2 days	28 days			W_M	W_O
1	Control	12.0	25.1	48.0	B35	2353	6.6	15.5
3	MB (PNS + SF)	15.4	36.5	66.8	B50	2379	3.7	8.8
4	PCE	26.5	52.5	74.9	B55	2498	3.6	9.0
5	PCE + SF	31.6	56.5	92.5	B70	2530	1.9	4.8

mixture. It was also found that replacing cement with silica fume increases the viscosity and reduces the consistency retention of the concrete mixture. Despite this, the resulting self-compacting concrete mixture with the complex use of PCE and SF has high fluidity with a cone flow diameter of more than 700 mm and is characterized by increased consistence retention of more than 3 h within the slump-flow class SF2, which does not require vibration compaction of concrete mixtures.

The effect of modifying additives on the physical and mechanical properties of self-compacting concretes has been studied. The results of the study are presented in Table 3. Figure 7 shows the hardening kinetics of self-compacting concretes with modifying additives. Figure 8 shows the results of determining the open capillary porosity of self-compacting concretes.

It has been established that the strength of the control sample without additives at 28 days is 48.0 MPa, which corresponds to strength class B35. The use of the organomineral modifier MB makes it possible to increase the strength of concrete at 28 days by 39% compared to the control sample without additives and obtain self-compacting concrete of strength class B50. At the same time, the open capillary porosity of concrete decreases by 43%.

The use of polycarboxylate superplasticizer, due to a significant reduction in the water demand of the concrete mixture, allows to increase the strength of self-compacting concrete at 1 day by 72%, at 2 days—by 44%, at 28 days—by 12% compared with the MB modifier. At the same time, the volume of open capillary pores of concrete remains at the same level.

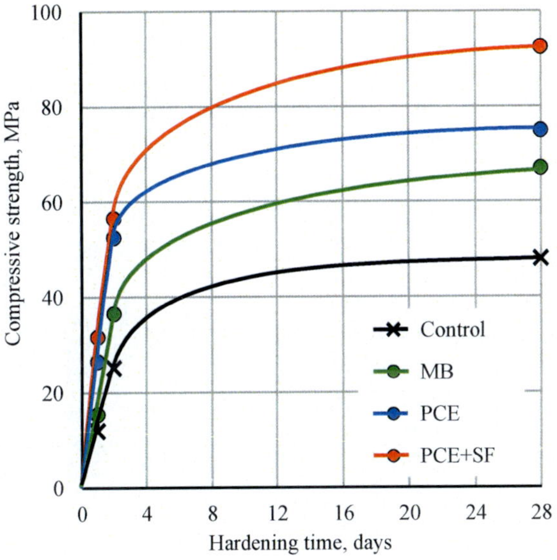

Fig. 7 Hardening kinetics of self-compacting concretes with modifying additives

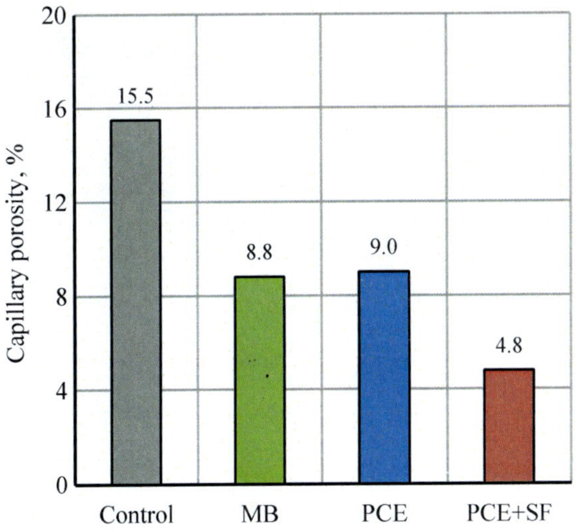

Fig. 8 Capillary porosity of self-compacting concretes with modifying additives

The use of a complex additive based on a polycarboxylate superplasticizer and silica fume, due to its microfilling and pozzolanic action, leads to an increase in the strength of self-compacting concrete at 1 day by 19%, at 2 days—by 8%, at 28 days—by 23% compared with the composition using PCE. At the same time, the cement stone structure is compacted, and the capillary porosity of concrete is reduced by

47%. The hardening kinetics of self-compacting concrete with a complex additive is characterized by high rates of strength development, even despite the long-term consistency retention of self-compacting concrete mixes. At 1 day, the complex use of additives makes it possible to obtain self-compacting concrete with a strength class of B20, at 2 days—B40, at 28 days—B70 at a cement consumption of 500 kg/m^3.

To identify the pattern of changes in the phase composition during hydration and structure formation of cement systems with polycarboxylate superplasticizer and silica fume, an X-ray phase analysis of cement stone samples with modifying additives was performed [20]. Figures 9 and 10 show X-ray diffraction patterns of cement stone with PCE (a) and with PCE + SF (b) at 1 and 28 days of hardening.

It has been established that the complex use of polycarboxylate superplasticizer and silica fume accelerates the hydration processes of cement clinker minerals at 1 day (Fig. 9). In the composition with silica fume, there is a decrease in the content of unreacted minerals of alite d = [3.03; 2.97; 2.77; 2.74; 2.61; 2.18 Å]—by 13%, tricalcium aluminate d = [2.70; 1.91 Å]—by 24%. At the same time, the acceleration of cement hydration leads to an increase in the content of hydrated phases in the

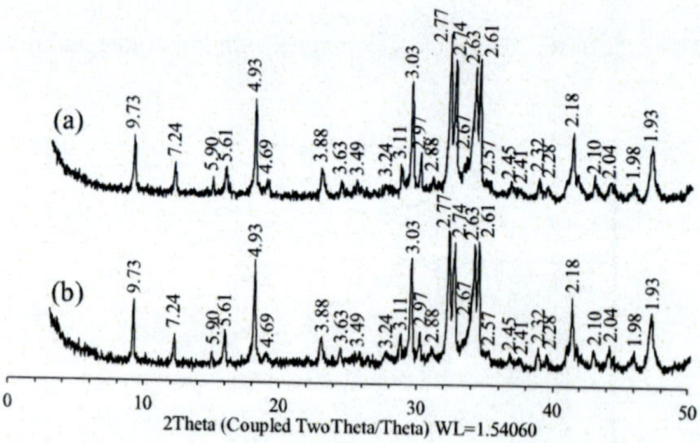

Fig. 9 X-ray diffraction patterns of cement stone at 1 day: **a**—PCE; **b**—PCE + SF

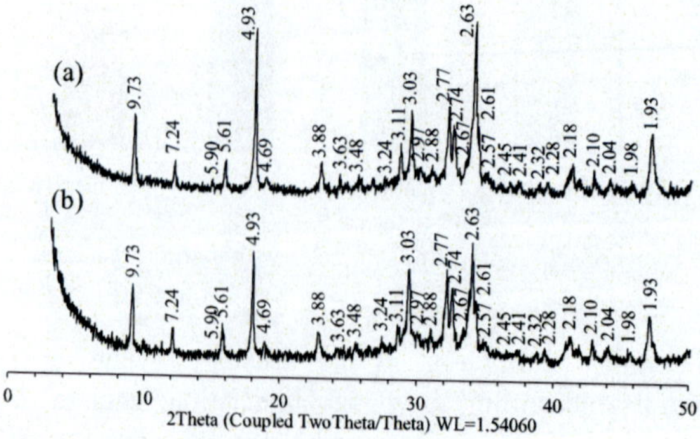

Fig. 10 X-ray diffraction patterns of cement stone at 28 days: **a**—PCE; **b**—PCE + SF

composition of cement stone at 1 day: portlandite d = [4.93; 3.11; 2.63; 1.93 Å]—by 5%, ettringite d = [9.73; 5.61; 3.88 Å]—by 14% compared with a cement stone sample using only superplasticizer.

At 28 days (Fig. 10), the complex use of polycarboxylate superplasticizer and silica fume leads to the formation of a cement stone structure with a reduced content of portlandite d = [4.93; 3.11; 2.63; 1.93 Å] by 40% and ettringite d = [9.73; 5.61; 3.88 Å] by 11%. At the same time, there is an increase in the content of the amorphous phase by 18%, which indicates the formation of a structure with increased content of low-basic hydrated calcium silicates C-S-H, significantly compacting the structure of cement stone and improving its physical and mechanical characteristics.

4 Conclusion

1. The consumption of coarse and fine aggregates in the composition of self-compacting concrete mixtures was selected using crushed stone with a fraction of 5–20 mm and quartz sand with a fineness modulus of 1.9. It has been established that the most effective ratio of these aggregates in the composition of self-compacting concrete mixtures, at which the minimum intergranular porosity of the aggregates is observed and the best concrete mixture workability is achieved, is the sand-aggregate ratio of 0.45.
2. The effect of a complex additive based on a polycarboxylate superplasticizer and silica fume on the properties of self-compacting concrete mixtures has been studied. Using a polycarboxylate superplasticizer makes it possible to obtain self-compacting concrete mixtures even at a low W/B = 0.31, and also ensures long-term consistency retention of concrete mixtures due to the strong steric effect of the admixture. However, self-compacting concrete mixtures with PCE without stabilizing and water-retaining additives are prone to bleeding and segregation. The use of highly dispersed silica fume, due to its high water-retaining capacity, allows increasing the homogeneity and segregation resistance of self-compacting concrete mixtures without significantly reducing their workability. With the complex use of PCE and SF additives, stable self-compacting concrete mixtures with high segregation resistance, high fluidity with a cone flow diameter of more than 700 mm and increased consistency retention of more than 3 h within the slump-flow class SF2 were obtained.
3. The effect of a complex additive based on a polycarboxylate superplasticizer and silica fume on the physical and mechanical properties of concrete has been studied. It has been established that the use of silica fume in self-compacting concretes with polycarboxylate superplasticizer, due to its microfilling and pozzolanic action, leads to an increase in the strength of concrete at 1 day by 19%, at 2 days—by 8%, at 28 days—by 23% and a decrease in their capillary porosity by 47%. The hardening kinetics of self-compacting concrete with a complex additive is characterized by high rates of strength development, even despite the long-term consistency retention of self-compacting concrete

mixes. At 1 day, the complex use of additives makes it possible to obtain self-compacting concrete with a strength class of B20, at 2 days—B40, at 28 days—B70 at a cement consumption of 500 kg/m^3.

4. According to the X-ray phase analysis results, it was found that the complex use of polycarboxylate superplasticizer and silica fume at 1 day leads to an acceleration of the hydration processes of alite by 13%, tricalcium aluminate—by 24% and an increase in the content of hydrated phases in the composition of cement stone: portlandite—by 5%, ettringite—by 14%. At 28 days, the complex use of polycarboxylate superplasticizer and silica fume leads to the formation of a cement stone structure with a reduced content of portlandite by 40%, ettringite by 11% and increased content of amorphous low-basic hydrated calcium silicates by 18%, significantly compacting the structure of cement stone and improving its physical and mechanical characteristics.

Acknowledgements The research was supported by the Ministry of Science and Higher Education of the Russian Federation (Grant № 075-15-2021-674) and Core Facility Centre «Ecology, biotechnologies and processes for obtaining environmentally friendly energy carriers» of Volga State University of Technology, Yoshkar-Ola.

References

1. Okamura H, Ouchi M (2003) J Adv Concr Technol 1:5–15
2. Brouwers HJH, Radix HJ (2005) Cem Concr Res 35:2116–2136
3. Shi C, Wu Z, Lv K, Wu L (2015) Constr Build Mater 84:387–398
4. Kaprielov S, Sheynfeld A, Arzumanov I, Chilin I (2021) Bull Sci Res Cent Constr 3(30):30–40. (in Russian)
5. Yamada K, Takahashi T, Hanehara S, Matsuhisa M (2000) Cem Concr Res 30:197–207
6. Plank J, Sakai E, Miao CW, Yu C, Hong JX (2015) Cem Concr Res 78:81–99
7. Felekoğlu B, Sarikahya H (2008) Constr Build Mater 22:1972–1980
8. Flatt R, Schober I (2012) Underst Rheol Concr 144–208
9. Gesoğlu M, Güneyisi E, Özbay E (2009) Constr Build Mater 23:1847–1854
10. Poon CS, Kou SC, Lam L (2006) Constr Build Mater 20:858–865
11. Juenger MCG, Siddique R (2015) Cem Concr Res 78:71–80
12. Shi C, Wu Z, Xiao J, Wang D, Huang Z, Fang Z (2015) Constr Build Mater 101:741–751
13. Zhou M, Wu Z, Ouyang X, Hu X, Shi C (2021) Cem Concr Compos 124:104242
14. Muller ACA, Scrivener KL, Skibsted J, Gajewicz AM, McDonald PJ (2015) Cem Concr Res 74:116–125
15. Rossen JE, Lothenbach B, Scrivener KL (2015) Cem Concr Res 75:14–22
16. Smirnov AO, Dobshits LM, Anisimov SN (2020) IOP Conf Ser Mater Sci Eng 896:012095
17. Kaprielov SS, Sheinfeld AV, Kardumyan GS (2017) Stroitel'nye Materialy [Constr Mater] 12:58–63. (in Russian)
18. Lin WT (2020) Constr Build Mater 242:118046
19. Smirnov A, Dobshits L, Anisimov S (2020) IOP Conf Ser Mater Sci Eng 869:032039
20. Smirnov A, Dobshits L, Anisimov S (2022) Akimov P, Vatin N (eds) Proceedings FORM 2021. LNCE, pp 111–121. Springer, Cham. https://doi.org/10.1007/978-3-030-79983-0_11

Granular Foam-Glass-Ceramic Thermal Insulation Based on Natural Quartz Sand

Ivan Vedyakov, Vladimir Vaskalov, Nikolai Maliavski, Andrey Nezhikov, and Mikhail Vedyakov

Abstract The article describes the modification of the production technology of granular alkali-silicate thermal insulation materials. The modification lies in the fact that natural quartz sand is used as a silica precursor, instead of amorphous silica, due to its greater availability. A mixture of glassy sodium silicate with soda and/or sodium hydroxide additives acts as an alkaline precursor. The intermediate material (pregranulate) is obtained by mixing the sand with a binder solution simultaneously with granulation followed by drying at 200 °C. Another feature of the technology is the high temperature of the pregranulate firing (850–930 °C). The closed nature of porosity, along with a high silicate modulus, significantly increases the water resistance of the foam silicate and also significantly reduces its thermal conductivity and water absorption. As a result, a granular foam material with a bulk density of 170–440 kg/m^3, compression strength in a cylinder of 0.5–6.3 MPa, thermal conductivity of 0.046–0.084 W/(m·K), volume water absorption of 7.8–13.5%, weight loss when boiling in water 0.12–0.33%. According to its consumer properties, the foam material is not inferior to that obtained with amorphous silica, but differs in cheaper logistics.

Keywords Alkali-silicate thermal insulation · Thermal foaming · Quartz sand · Sodium silicate · Water resistance

1 Introduction

Alkali-silicate thermal insulation materials, or foam silicates, are highly porous materials of a glassy or glass-crystalline structure. The basis of their composition are silicates of alkali metals, primarily sodium or potassium. They may be prepared by cold or thermal foaming of alkali silicates in the form of aqueous solutions (water glasses) or hydrogels.

In recent decades, the demand for such materials has been continuously growing, which contributes to the development of their production technologies. The reason for

I. Vedyakov · V. Vaskalov · N. Maliavski (✉) · A. Nezhikov · M. Vedyakov
Central Research Institute of Building Structures (TSNIISK), 2nd Institutskaya Street, 6, 109428 Moscow, Russia
e-mail: nikmal08@yandex.ru

P. Akimov et al. (eds.), *Proceedings of FORM 2022*, Lecture Notes in Civil Engineering 282, https://doi.org/10.1007/978-3-031-10853-2_37

this is a unique combination of consumer properties, both common to mineral thermal insulation (incombustibility, heat resistance, biostability) and individual, inherent in foam silicates (ease of regulation of foaming processes, the possibility of achieving very low values of density and thermal conductivity). The main disadvantage of foam silicates, insufficient water resistance, is successfully overcome by the correction of the composition and the introduction of water-strengthening additives, which makes it possible to obtain higher water resistance than that of Portland cement-based foams with an equal density.

Depending on the conditions of the foaming process, various substances can act as a foaming agent:

Cold foaming:

- mechanical foaming (mechanical stirring energy is a foamer, air is a bubble filler);
- chemical foaming (active metals or hydrogen peroxide are foamers, hydrogen or oxygen are foaming agents).

Medium temperature (150–550 °C) foaming (chemically bound water is a foamer [1], water vapor is a bubble filler).

High temperature (>550 °C) foaming (foaming agents—carbon-containing compounds are formers, carbon dioxide is a bubble filler).

At the same time, in the case of medium-temperature foaming, mainly open porosity is formed, and in other cases closed one [2].

In 2017, a patent was published for the manufacture of granular foam silicate from a raw material mixture including anhydrous sodium silicate and carbonate, amorphous silica, water and a high-temperature gas-forming agent [3]. The proposed technology was based on the rejection of the use of ready-made water glass and the synthesis of a binder solution at the preparatory stage of the process, during the grinding of a part of the silica component together with alkaline components. At the same time, in comparison with traditional foam silicates of medium-temperature foaming, the foaming temperature of raw granules was increased to 700–900 °C which made it possible to obtain granules mainly with closed porosity. As a result, a significant increase in the mechanical and chemical strength, as well as the water resistance of the foam material, was achieved. In a subsequent publication [4], it was also described the production of thermal insulation granules based on waste from the processing of apatite-nepheline ores as a silica component using the same technology. The results of determining the thermal conductivity and water resistance of granulates of various compositions obtained using this technology, were also presented.

Despite the quite satisfactory quality of the granules obtained by the technology described above, the latter can be used in the production of building materials, mainly in areas close to amorphous silica deposits. Away from these deposits, such production will be unprofitable due to the high cost of transporting raw materials to the place of production. This factor has a particularly strong effect on the logistics of amorphous silica which makes up 75–85% of the finished product mass. Unfortunately, its deposits are located on the territory of Russia extremely unevenly and in the northern regions they are practically absent at all. Thus, among the 22 Russian regions whose territory is located in the arctic or subarctic zones, only four have developed deposits of amorphous silica or nepheline-containing raw materials [5].

The replacement of amorphous silica with crystalline silica could reduce the cost of production logistics. First of all, this refers to natural quartz sand, industrial deposits of which are widespread everywhere. In particular, out of 22 Russian arctic and subarctic regions, quartz sand deposits are being developed in 21 ones [5, 6]. In addition, many rocks containing easily separable quartz, such as sandstone or marshalite, can be used as silica components.

A significant disadvantage of quartz as an object of alkaline dissection, is its low chemical activity. Therefore, for the complete silicatization of quartz, for example in the processes of obtaining water glass, temperatures of the order of 1300 °C and above, are required [7]. However, if there is no need to obtain an absolutely transparent and colorless glassy material as a result of synthesis, as is the case in the production of foam glass ceramics, the requirements of phase purity and the absence of inhomogeneities are not so important. Therefore we can significantly reduce the maximum temperature of heat treatment without the risk of serious deterioration in the quality of the product.

Examples of similar technical solutions proposed over the past 20–25 years can be considered:

- direct synthesis of water glass by hydrothermal reaction of finely ground quartz sand with an aqueous solution of NaOH at a temperature of 170–190 °C and a pressure of 0.8–1.2 MPa [8];
- direct synthesis of anhydrous sodium polysilicate by heating an alloy of quartz sand with NaOH at a temperature above 550 °C at atmospheric pressure [9];
- preparation of binding compositions by reaction of finely ground quartz sand with NaOH solution at a temperature of 85–90 °C in a steaming chamber at atmospheric pressure [10];
- production of granular foam silicate by foaming at 720–900 °C pregranules prepared on the basis of water glass and quartz sand with additives [11];
- production of granular foam silicate by foaming at 880–1050 °C pregranules prepared from a mixture of finely ground marshalite with a mixture of Na_2CO_3 and CaO [12].

In most of the above-mentioned technologies, high-energy grinding (wet or dry) to a particle size of about 0.1 mm or lower, is used as an additional factor activating the silica component. But the resulting degree of increase in the chemical activity of quartz cannot be explained only by an increase in its total specific surface area. Direct physico-chemical studies in recent years, aimed at finding methods to protect concrete from alkaline corrosion, have shown that prolonged high-energy mechanical effects on quartz grains lead to severe deformation and partial amorphization of their surface layers.

In particular, the paper [13] presents the results of a study of the effect of deformation of quartz crystals in aggregates (quartzite, mylonite and granite) on the development of alkaline silica reaction (LCR) in concrete by transmission electron microscopy, visible-field spectrophotometry and tests on mortar rods. The authors concluded that the intensity of the LCR is directly dependent on the degree of deformation (stress) of the quartz crystal lattice. This conclusion was later confirmed by the authors of [14] when studying the Portland cement-granite system using petrographic analysis. Finally, in the same year, the results of a study of the quartz sand—Portland cement system using a whole range of materials science and physico-chemical methods, were presented in [15]. According to the authors, the defect of the quartz crystal lattice, called "Dauphine twinning", has the greatest influence on the rate of LCR and consequently on the growth of the chemical activity of quartz in an alkaline medium.

The main purpose of this work, is to develop the chemical foundations of the technology for producing granular foam glass ceramics from raw mixtures based on sodium silicate, carbonate and hydroxide, on the one hand, and natural quartz sand, on the other.

2 Experimental

Samples of granular foam-glass–ceramic thermal insulation were made according to the method described in [4], with the replacement of amorphous silica with crystalline quartz in the composition of natural quartz sand. The maximum firing temperature of raw granules ranged from 850 to 930 °C. The bulk density of the finished granulate was determined according to GOST 9758-2012, its compressive strength when squeezed in a cylinder—according to GOST 32,495-2013. Thermal conductivity of granulates in backfill—according to GOST 32,497-2013, water absorption and frost resistance—according to GOST 9758-2012. The water resistance of foam materials was estimated by the percentage of mass loss by granules of a fraction of 54–10 mm as a result of boiling in water according to GOST 32,496-2013. The content of volatile components (mainly H_2O and CO_2) was determined by thermogravimetry using a thermal analyzer SDT Q600 (TA Instruments). Using the same device, the content of crystalline quartz in the samples was determined by the area of the endothermic effect on the DSC curve at 573 °S associated with the phase transition α-$SiO_2 \leftrightarrow \beta$-$SiO_2$.

Chemical analysis of the components of the raw materials and the resulting thermal insulation was carried out by atomic emission spectrometry using an induction-coupled plasma spectrometer of the iCAP-6200 Duo (Thermo Scientific) type, according to GOST R 57165-2016. The transfer of solid samples into solution was carried out according to the method described in the monograph [16] by 24-h dissolution of the ground sample in 45% HF followed by rapid 50-fold dilution with distilled water.

3 Results

The samples for the study were obtained at an experimental plant for the production of light concrete aggregates operating in the TSNIISK (Fig. 1). Three types of quarry sand, different in origin, were used as a silica component for the preparation of granulates: sand-1 (RF, Chechen Republic, Shelkovskaya district), sand-2 (RF, Leningrad region, Vsevolzhsky district), and sand-3 (RF, Tula region, Zaoksky district).

The chemical compositions of the silica components of the raw materials, are presented in Table 1. As for the alkaline component, in all cases it consisted of a mixture of sodium carbonate, glassy sodium silicate with a molar modulus of 2.5–3.0 and water in a mass ratio of 10:2:15. The external high-temperature foamer was finely ground silicon carbide SiC in an amount of 0.5% by weight of the final product.

The technology of production of granular foam glass ceramics based on quartz sand includes the following essential stages:

- fine grinding and sieving of the silica component;
- preparation of the binder solution by wet grinding at 100–110 °C of soda and soluble glass together with a part of quartz sand;
- mixing of metered amounts of binder solution, quartz sand and high-temperature gas-forming agent with simultaneous granulation;
- drying of the prepared granules at 200 °C to obtain a pregranulate capable of long-term storage;
- foaming of pregranules in a rotary kiln at 850–930 °C with subsequent cooling and sieving.

The most important features of the used method for obtaining granular foam silicate from natural quartz sand are:

Fig. 1 General view of the experimental plant for the production of foam granulates

Table 1 Chemical composition of used silica components of raw mixtures

Component	Content in mass.%		
	Sand-1	Sand-2	Sand-3
SiO_2	67.10	82.41	78.21
Al_2O_3	10.29	8.62	3.51
Fe_2O_3	2.77	0.77	2,02
MgO	0.99	0.74	1.74
CaO	5,69	1,96	6,45
Na_2O	2,78	2,43	2,20
K_2O	2,18	0,84	1,12
TiO_2	0,36	0,21	0,07
P_2O_5	0,15	0,05	0,11
LOI	7,39	1,97	3,76

- preliminary thorough grinding of quartz sand to a specific surface of 3500–4000 cm^2/g (for example, wet grinding on a ball mill or dry grinding on a disintegrator);
- high temperature of final foaming (open pores formed at medium temperarures, by the time of reaching 750 °C and above are closed, and new pores are already initially formed closed).

Natural sands containing silica mainly in the form of finely dispersed crystalline α-quartz, are used as silica components of the raw material mixtures. Sands of any type can be used, both according to mineralogical (alluvial, deluvial, marine, Aeolian, etc.) and commercial (river, quarry, alluvial, ground sandstone, ground marshalite, etc.) classifications. For use in this technology as silica components, sands containing at least 60 mass % SiO_2, no more than 12 mass % Al_2O_3, no more than 5 mass % Fe_2O_3 and no more than 15 mass % (CaO + MgO), are suitable.

In contrast to foam granulates based on amorphous silica, in the case of the foam materials studied here, there is a possibility of experimental monitoring of the kinetics of the silicatization process of quartz introduced as part of the raw material mixture, that is, the process of foam silicate synthesis. For this, as already noted above, one of the standard methods of thermal analysis can be applied, namely, differential scanning calorimetry [17]. The measured object of the study was the endothermic effect at 573 °C. Knowing the quartz content in the raw mixture and determining its residual content in the finished granulate by the area of the DSC peak, it is easy to determine the proportion of quartz that has entered into the silicatization reaction during the synthesis of the foam granulate.

Figure 2 shows the course of the TGA and DSC curves of the raw mixture with sand-1 as a silica component, as well as the DSC curves for fired foams prepared from this mixture at three firing temperatures. It is easy to notice that the peak area of the DSC at 573 °C is inversely dependent on the firing temperature (maximum in the case of the raw mixture after drying and practically zero for the granulate fired at 930 °C).

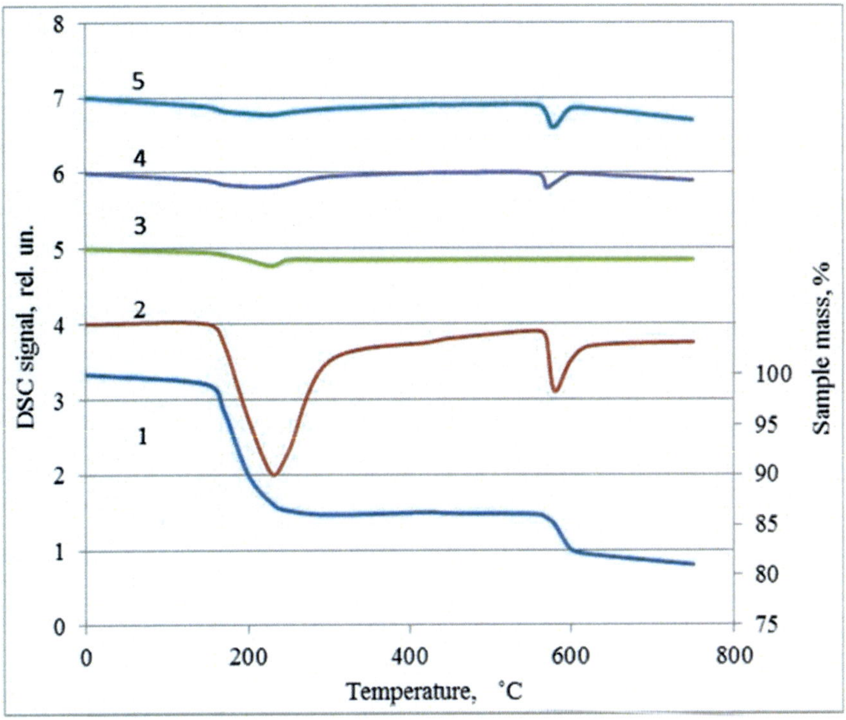

Fig. 2 Experimental curves of TGA and DSC for the pregranulate, containing sand-1, and its firing products (1 and 2—raw mixture, respectively, TGA and DSC, 3, 4, 5—DSC of final foam obtained at, 930, 900 and 850 °C respectively)

In the Table 2 shows the most significant properties of the obtained samples of foamed granulates, including the degree of quartz silicatization. The test results of samples of foamed granulates obtained from sand-1 indicate that with an increase in the foaming temperature, the average values of bulk density, thermal conductivity and water absorption decrease, and the average indicators of water resistance and the degree of silicatization of natural quartz increase.

As for the water resistance of foamed granules, it is determined primarily by the foaming temperature (samples 1-850—1-930) and to a much lesser extent by the composition of quartz sand (samples 1-900—3-900). In the latter case, among the samples with a foaming temperature of 900 °C, sample 2-900 showed the best water resistance, obviously due to the maximum content of ($MgO + CaO$).

Table 2 The most important operational properties of the obtained granules (sample indication: the sand number—maximum foaming temperature)

Item No	Property	Samples of foam granulate				
		1-930	1-900	1-850	2-900	3-900
1	Bulk density, kg/m^3	170–230	190–240	230–290	160–200	170–210
2	Compressive strength, MPa	1,1–1,6	1,1–1,6	1,1–1,7	1,1–1,5	1,6–2,1
3	Thermal conductivity, W/(m·K)	0,047–0,057	0,052–0,056	0,057–0,065	0,046–0,052	0,047–0,053
4	Mass loss during boiling, %	0,12–0,17	0,20–0,31	1,2–1,9	0,13–0,19	0,16–0,30
5	Volumetric water absorption, %	0,8–1,2	1,1–1,9	1,8–3,5	1,0–1,7	0,8–1,4
6	Frost resistance, brand	F25	F25	F25	F25	F25
7	Quartz silicatization extent, %	100	87	65	65	68

4 Discussions

A comparison of the water resistance of foam silicates made using amorphous and crystalline silica showed that the differences in properties between the samples of these two groups do not exceed the values of the dispersion within each of the groups. As an example, let's illustrate this situation with the example of water resistance of samples. Table 3 shows the experimentally found values of mass loss during boiling for five granulate samples made using two amorphous and three crystalline varieties of silica as silica components. All samples were prepared using the same technology

Table 3 Found parameters of water resistance of samples of foam silicates

Item No	Silica component	w, %	ρ, kg/m^3	h, %	H, mg/cm^3
1	Diatomite (Ulyanovsk region) [4]	0.37	483	0.33	1.59
2	Tripoli (Oryol region) [4]	0,83	294	0.22	0.64
3	Sand-1	0.93	425	0.26	1.10
4	Sand-2	0.69	374	0.17	0.64
5	Sand-3	0.85	380	0.25	0.95

and contained the same concentrations of SiO_2 and Na_2O. In addition to the values of mass loss during boiling (h), the values of loss during calcination (w) and the average density of granules (ρ) were also determined. The results are shown in Table 3.

The last column of the table shows the calculated values of effective water resistance H, which allows you to compare the water resistance of foams with different values of w and ρ, which is necessary for an objective assessment of the water-strengthening effect of additives. The value of H can be calculated by the formula [2]:

$$H = 0.01\rho h\left(1 - \frac{w}{100}\right) \qquad (1)$$

At $H \leq 3$, the foam silicate is considered highly water-resistant, at $3 < H \leq 10$—acceptable for non-extreme conditions.

All the values of H given in Table 3 fluctuate with small deviations within the range of 0.5–1.6, which indicates that the corresponding foam silicates belong to alkaline silicate insulation with the highest water resistance, such as (in parentheses—the values of H): Foam glass and Bisipor (1–2), Aerosteklo (0.6–4), Steklopor (1.5–2.5).

The use of quartz sand with an increased (up to 10–12%) content of Al_2O_3 significantly increases the water resistance of the foam material in an alkaline environment and makes it possible to obtain high-water-resistant granules at a reduced foaming temperature (see Table 2, Model 1-850). The same effect can be achieved by introducing additives of other glass-forming oxides, for example, in B_2O_3 or ZnO. An alternative option for obtaining granulates with increased water resistance, but without the flux effect, is the use of sands with a significant content of carbonate rocks, that is, a moderately high concentration of (CaO + MgO) in the glass phase (sample 3-900 in Table 2).

Experimental batches of granular thermal insulation of the proposed composition, obtained in 2020–2021 at TSNIISK, showed that in terms of its physico-mechanical and physico-chemical characteristics, the material is at the level of the best world analogues and will undoubtedly find wide application in many areas of construction production, in particular:

- as aggregates for ultralight concretes with a density of 500–600 kg/m^3, compressive strength of 3.5–5.0 MPa and thermal conductivity of 0.12–0.14 W/(m * K) for wall and other enclosing structures (monolithic and prefabricated panels, complete factory-ready house kits);
- as aggregates in structural concrete with a density up to 1500 kg/m^3 and compressive strength up to 35 MPa for load-bearing reinforced concrete structures;
- thermal insulation backfills in foundations, floors, attics, etc. in individual housing construction;
- as well as in some areas of non-construction production, for example, in the oil and gas industry for the production of grouting solutions with low cement consumption, thermal insulation mixtures and flame retardant paints.

The advantages of the proposed method of obtaining thermal insulation are:

- significant reduction in the cost of production logistics;
- the possibility of increasing the degree of foaming and water resistance, reducing the density and volumetric water absorption of foam;
- partial solution of the problem of disposal of silica-containing waste from various industries (foundry, chemical, as well as construction and production of building materials).

5 Conclusions

Alkali-silicate thermal insulation materials have recently become widespread in industrial and civil construction. They have such advantages over other types of mineral thermal insulation as the availability of raw materials, low cost, but most importantly—the ease of regulating the parameters of the production process and the possibility of obtaining a very low density and thermal conductivity. Recently, there has been a dynamic development of technologies for the production of granular and block alkali-silicate foams of high-temperature foaming, which are characterized by closed porosity, increased strength and reduced water absorption. As a silica precursor, natural or technogenic amorphous silica is usually used, which in many regions of Russia and other countries is characterized by limited availability. Therefore, the paper proposes its replacement with natural quartz sand, which has universal availability. The technological scheme of its production remains basically the same as when obtaining foam silicates based on amorphous silica, with the exception of the foaming temperature increased by 30–100 °C. The resulting foam silicate is characterized by increased strength, water resistance and reduced water absorption.

References

1. Korneev VI, Danilov VV (1996) Soluble and water glass. Stroyizdat St. Petersburg, p 216
2. Malyavsky N Alkaline-silicate heaters. Properties and chemical bases of production. Russ Chem J 47(4):39–45
3. Vaskalov V, Vedyakov I, Nezhikov A et al (2017) Patent RU 2605982 C2
4. Vedyakov I, Vaskalov V, Maliavski N, Vedyakov M (2012) E3S Web of Conferences 263:01017
5. Russian Federal Geological Fund. Objects of registration of the State cadastre of deposits and manifestations of minerals. https://www.rfgf.ru/gkm/
6. Biryulev GN (2015) Georesources 4:21–24
7. Iler RK Silica chemistry. Part 1. Mir, Moscow, p 416
8. Lotov VA, Lotov VA, Vereshchagin VI, Kosintsev VI, Pasechnikov YuV (1999) Patent RU 2132817
9. Lazaro AL, Rodriguez-Valadez FJ, Machorro Lopez JJ, Espejel-Ayala F (2020) Mater Res Express 7(4):045503
10. Mitina NA, Vereshchagin VI (2009) Bull Tomsk Polytechnic Univ 314(3):11–14
11. Ketov AA, Puzanov IS, Puzanov SI et al (2007) Patent RU 2291126

12. Orlov GA (2019) Patent RU 2701838
13. Tiecher F, Gomes MEB, Dal Molin DCC, Hasparyk NP, Monteiro PJM (2017) Materials 10:1022
14. Antolik A, Jozwiak-Niedzwiedzka D (2021) Constr Build Mater 295:123690
15. Franke Portella K, Evangelista Lagoeiro L, Luis Bronholo J (2021) IBRACON Struct Mater J 14(3):e 14308
16. Bock R (1984) Decomposition methods in analytical chemistry. Chemistry, p 432
17. Labus M (2017) J Thermal Anal Calorimetry 129:965–973

Resistance of Vertical Joints During Torsion of Multistorey Buildings

Valery Lyublinskiy and Vladislav Struchkov

Abstract Floor slabs, vertical and horizontal joints determine the spatial interaction of the entire load-bearing system of panel, frame-panel buildings. Under the action of wind loads and earthquake loads, torsion of the bearing system may occur for a number of reasons. In the asymmetry of the load-bearing system in the plan building rotates around the center of stiffness. The assessment of torsion conditions in building codes based on the parameters of elastic systems with an asymmetric plan has been the subject of numerous studies in the past. Vertical butt joints of panels, referred to as shear connections, in this case are affected in two planes. In this work, a numerical study of the welded joint on vertical shear force and horizontal torsional action is carried out. The object of the study was a fragment of a vertical wall of large-panel residential buildings, the thickness of which was 180 mm, and the total height was 700 mm. The stress–strain state of the vertical butt joint is determined, data on the asymmetric deformation of the considered welded joint are obtained. The test results can be taken into account when designing load-bearing systems of large-panel buildings, taking into account the proposed algorithm for nonlinear stiffness of welded joints.

Keywords Joints · Torsion · Panel buildings · Shear

1 Introduction

The main forces that arise in the vertical joints of large-panel buildings and frame-panel buildings are bending, stretching, compression and shear forces [1]. Mathematical models of load-bearing systems of multi-story buildings, which are the basis of software systems, focus on elastic deformation of butt joints [2, 3, 21]. Multi-story buildings are affected by horizontal loads, such as wind and earthquakes. Damage during earthquakes has shown that one of the causes of damage to load-bearing reinforced concrete structures is the reaction of the building to torsion [4–8].

V. Lyublinskiy (✉) · V. Struchkov
Moscow State University of Civil Engineering, Moscow 129337, Russian Federation
e-mail: lva_55@mail.ru

P. Akimov et al. (eds.), *Proceedings of FORM 2022*, Lecture Notes in Civil Engineering 282, https://doi.org/10.1007/978-3-031-10853-2_38

Torsion of bearing systems of buildings can occur for various reasons. The main reasons are the asymmetry of the load-bearing system in the plan, the asymmetry of the load-bearing system in the height of the building, the uneven distribution of stiffness elements. Vertical and horizontal joints are one of the stressed elements of large-panel buildings. Vertical joints perceive shear, stretching, compression and bending forces. There are good reasons to believe that the work of these joints is far from linear, which has been the subject of numerous studies on linear and nonlinear deformation of various joints [2, 9–12, 21]. Diagrams of joint deformation depending on shear forces are obtained. At the same time, the work of the joint in the plane of the panel walls that it connects is considered.

When the center of mass (CM) and the center of stiffness (CS) do not coincide under the action of horizontal load, torsion occurs in multi-story buildings. In asymmetric load-bearing systems, in buildings with eccentricities between the centers of mass and stiffness, under the action of wind and seismic influences, significant damage may occur in structures that are most remote from the center of mass [13–17]. The appearance of torsional influences is also possible with progressive destruction as a result of a local change in the bearing system of the building. In building codes of many countries contain recommendations for the design of buildings with uniform distributions of stiffness along the axes of symmetry of the building, there are recommendations for limiting the ratio (ξ) of the maximum horizontal displacement of the corner point of the floor to the average horizontal displacement. When ξ is more than 1.8, deformations in structures increase sharply under torsional action. If the difference in eccentricity between the CS and the CM, according to our data, is more than one meter, a building rotation occurs, leading to significant additional movement and effort in structures removed from the center of the building.

The resistance of reinforced concrete structures under the action of torsion and bending is a difficult task. Questions arise when calculating reinforced concrete elements under the influence of longitudinal forces and torques. In a number of works proposed a computational model of the complex resistance of reinforced concrete structures in buildings and structures during torsion with bending [18–20]. These proposals are very interesting, but have not been brought for inclusion in software packages.

Previous studies show that the stress-strain state of reinforced concrete structures and the operational qualities of frame-panel and panel buildings largely depend on the rigidity of the butt joints of load-bearing walls used [1–3]. Flat wall panels work in their own plane. In the joints connecting adjacent panels or panels of perpendicular directions and removed from the center of mass, forces in the horizontal plane may arise from the torsional load. An algorithm was proposed in [2] to include experimental deformation curves of tight shear bonds when assessing the stress-strain state of load-bearing reinforced concrete structures.

The building's resistance to external load is described by the system of differential equations [2]:

$$\mathrm{N}'' - R\,\mathrm{N} = \mathrm{F} \tag{1}$$

where N—matrix of unknowns, size $(m + 1) \times n$, m—number of vertical load-bearing elements; n—the number of design sections for the height of the supporting system; R—square matrix of stiffness coefficients, size $(m + 1) \times (m + 1)$; F—load matrix, size $(m + 1) \times n$.

The last differential equation of system 2 represents the bimoment T [1] as a function of torsion in the height of the building. Bimoment T combines bimoments of all groups of vertical load-bearing elements forming open and closed contours.

The algorithm for calculating the spatial load-bearing system of multi-story buildings in a nonlinear formulation was taken as a basis [2]. The nonlinear operation of vertical load-bearing elements in the algorithm was taken into account using shear deformation diagrams. In this case, an iterative process is performed, during which the main system of the discrete-continuum model is solved together with systems of equations that describe the nonlinear character of column deformation and shear relations:

$$\begin{cases} \mathrm{N}'' - R\,\mathrm{N} = F \\ \mathrm{S} = \mathrm{F(N)} \end{cases} \quad (2)$$

The nonlinear system that describes the change in the compliance of shear bonds:

$$\mathrm{S} = \mathrm{F(N)} \quad (3)$$

where S—shear bond deformability matrix, size $\mathrm{l} \times \mathrm{k}$;

l—number of vertical shear bonds;

k—the number of intervals along the height of the building, within which deformability is constant.

And although large panels work in their own plane, vertical connections between them can experience a complex stress-strain state. Horizontal forces may occur at the joints connecting the panels of perpendicular directions. Experimental data on the deformation of dense joints exposed to shear and torsional action are extremely necessary to include these diagrams in (3).

In order to prepare experimental tests in this paper, numerical modeling was used to consider the problem of estimating the stress-strain state of the vertical joint of a panel building that resists shear and rotation.

2 Methods

It was supposed to research a flat butt joint [2, 21] on the influence of vertical shear forces and horizontal torsion to obtain the rigidity of the shear bonds of the system (3). The results of such a shear test are shown in Fig. 1. However, the issues of planned experimental tests, the stability of the test sample and the possibility of applying loads in two planes determined the spatial variant of the interface of panel walls.

Fig. 1 Destruction of the connecting plate during shear

A fragment of a large-panel building consisting of the intersection of walls connected by welded seams is selected. Instead of a flat connecting plate, the corners welded to the embedded parts of the walls was used. In the proposed work, vertical joints and part of panel buildings were modeled by the finite element method.

The total height of the sample is 70 cm. The sample was loaded with its own weight, a planned vertical concentrated shear load of 33 ts and two oppositely directed horizontal forces of 5 ts along the X axis, simulating the torsional effect from a horizontal load on the building. The size of the finite element is assumed to be 2 cm. The finite element type is taken as a flat shell.

Figures 2 and 3 show a fragment of wall panels connected by a welded joint. Reinforced concrete panels are made of 180 mm thick heavy concrete of class B20.

The nonlinear processor of the Lira 10.12 software package was used to implement the step-by-step method of load application. For the numerical experiment, 10 steps of load application were provided. For concrete, a two-line deformation diagram was set. According to [22], when taking into account the nonlinear resistance of steel structures, the calculated deformation diagram of steel is adopted with generalized parameters σ_1 and ε_1. In accordance with the accepted class of steel structures was set diagram of nonlinear metal work (Fig. 4).

The embedded parts are made of C255 steel. Non-linear material was also set for them [22]. For steel C255 with a rolled thickness of 4 to 10 mm, the value of the ultimate stress Ryn = 245 MPa is accepted.

The loading process was carried out in proportion to the expected breaking load.

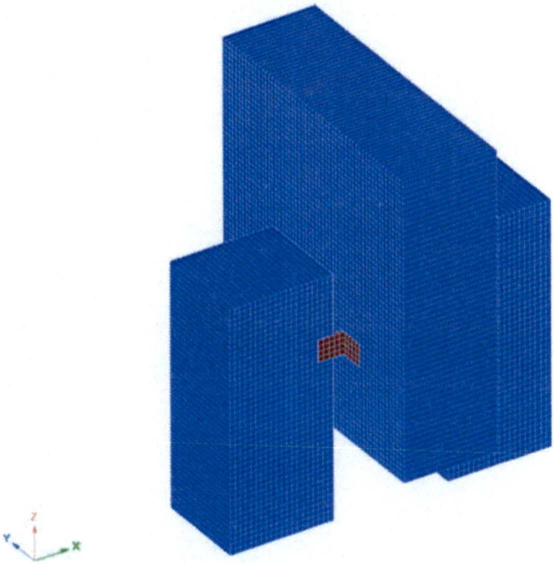

Fig. 2 General view of the calculation scheme

3 Results and Discussion

As a result of sequential loading, the destruction of the test sample numerically occurred (Figs. 5 and 6). The maximum vertical load on the sample was 231 kN, and the maximum horizontal loads creating torsion were 35 kN. The joint collapsed on the metal joint. The maximum compressive stresses in the metal linings reached the values of σ, which were set in the schedule of the lining material (Fig. 4). From horizontal loads that create torsion of the panel and the inserts, maximum compressive stresses with a minus sign arise in two metal linings (near left and far right linings), and in the other two tensile stresses (Fig. 5). The destruction designation in Lira 10.12 is graphically displayed on the plates in circles.

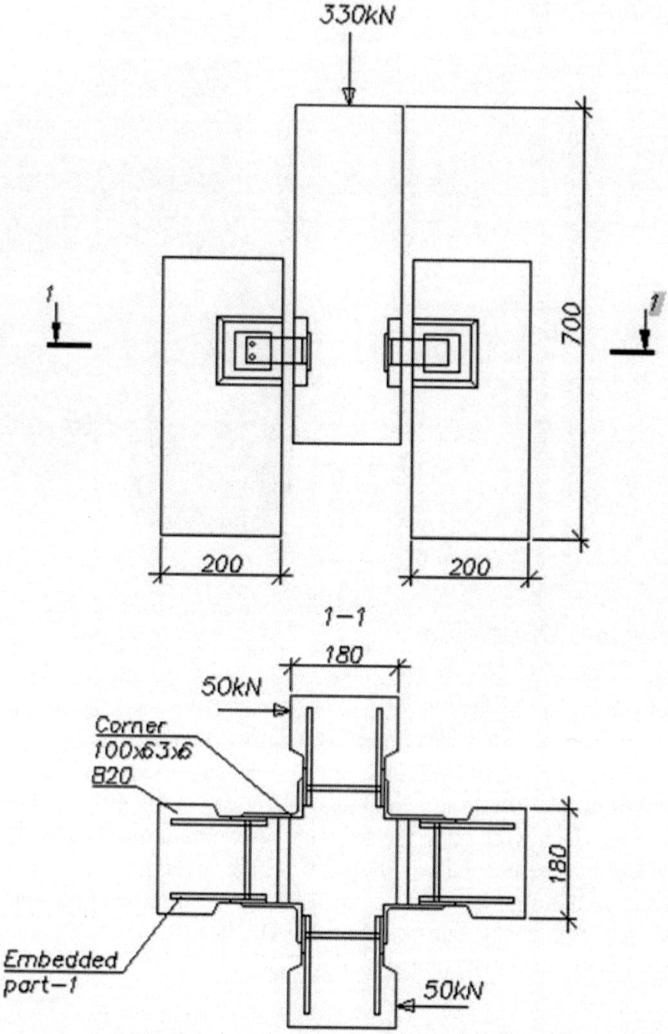

Fig. 3 Sample loading scheme

The stiffness of the shear joints throughout the loading was assumed to be linearly elastic. The joint metal was deformed non-linearly and reached extreme stresses. It was assumed that the test sample would withstand an external impact of 330 kN. With a sequential load, the limiting stresses in the steel shear connecting elements reached 231 kN at a load. This needs to be discussed. It is subject to verification of the correctness of the accepted finite element, which should provide for the possibility of resistance in two planes. At the same time, it is necessary to take into account the complex stress state of these elements.

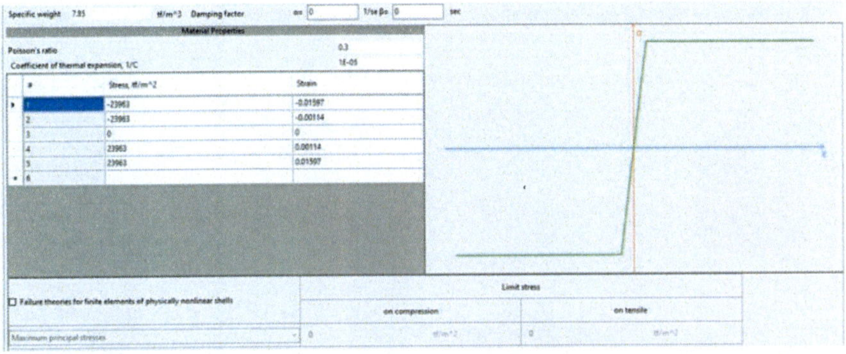

Fig. 4 Graph of non-linear operation of the metal lining material

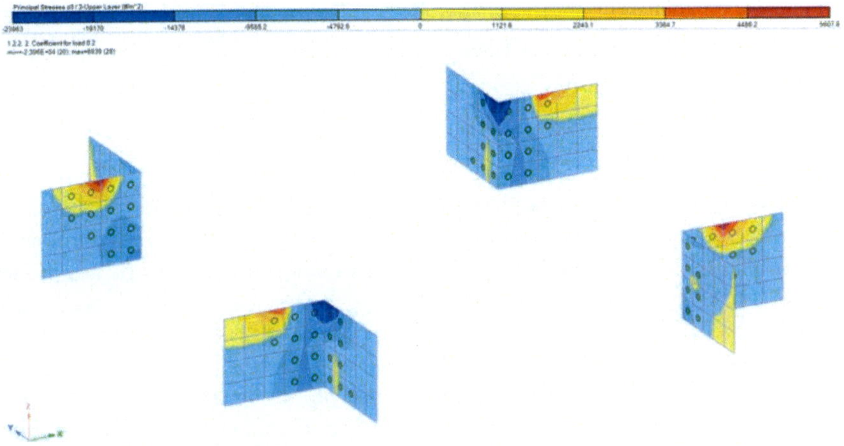

Fig. 5 Stresses **σ** in vertical joints

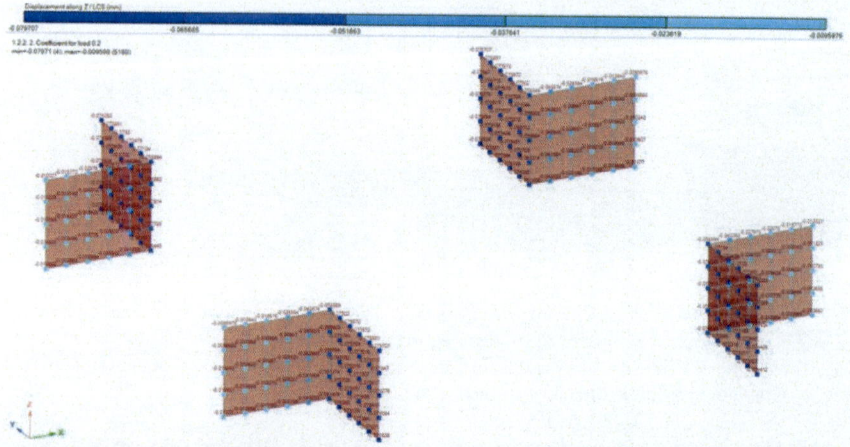

Fig. 6 Movement of nodes in vertical joints

4 Conclusion

According to the numerical model, the destructive forces of a dense joint exposed to shear and torsional load are obtained. A difference was found in the stress-strain state of vertical joints under the action of torques of different signs. The destruction of the welded joints begins in the corner zone. Cracking in the concrete of wall panels has not been recorded.

The final element modeling the stress state of the joint metal, the method of constructing a finite element solid-state model, the boundary conditions corresponding to the loading of the fragment and the coupling of concrete panels need to be clarified with such a complex loading. Obtaining the planned experimental data on the resistance of a dense joint to shear and torsional horizontal action will allow creating a more accurate numerical model of the operation of the structure under study and verifying it.

Acknowledgements This work was financially supported by the Ministry of Science and Higher Education of the Russian Federation (grant # 075-15-2021-686). All tests were carried out using research equipment of The Head Regional Shared Research Facilities of the National Research Moscow State University of Civil Engineering.

References

1. Drozdov P (1977) Design and Calculation of Load-bearing Systems of Multistorey Buildings and their Elements. Stroyizdat, Moscow, p 223
2. Lyublinskiy V, Ubysz A (2021) E3S Web of Conferences, vol 263, p 02015. https://doi.org/10.1088/1757-899X/869/5/052046

3. Lyublinskiy V (2019) J Phys Conf Ser 1425:012159. https://doi.org/10.1088/1742-6596/1425/1/012159
4. Rossley N, Aziz F, Chew HC, Farzadnia N (2014) Aust J Basic Appl Sci 8(1):370–380
5. Sharma J, Singh A, Sehgal R (2015) Int J Eng Res Technol 3(10):1–4
6. Kewalramani M, Syed Z (2017) Eurasian J Anal Chem 13(3):289–292
7. Naresh Kumar BG, Punith N, Bhyrav D, Arpitha T (2017) Buildings. Int J Eng Res Technol 6(5):801–804
8. Lim H, Kang JW, Pak H, Chi H, Lee Y, Kim J (2018) Appl Sci 8:479. https://doi.org/10.3390/app8040479
9. Shuvalov A, Gorbunov I, Kovalev M, Faizova A (2018) MATEC Web of Conferences, vol 196, p 02049. https://doi.org/10.1051/matecconf/201819602049
10. Blazhko V (2017) Hous Constr 3:17–21
11. Tamrazyan A, Avetisyan L (2016) MATEC Web of Conferences, vol 86, p 01029. https://doi.org/10.1051/matecconf/20168601029
12. Tamrazyan A (2014) Appl Mech Mater 475–476:1563–1566. https://doi.org/10.4028/www.scientific.net/AMM.475-476.1563
13. Ajay T, Parthasarathi N, Prakash M (2021) Mater Today Proc 40:1. https://doi.org/10.1016/j.matpr.2020.03.499
14. Akis T, Tokdemir T, Yilmaz C (2009) J Asian Archit Build Eng 8(2):531–538. https://doi.org/10.3130/jaabe.8.531
15. Khatiwada P, Lumantarna E (2021) Civ Eng 2:290–308. https://doi.org/10.3390/civileng2020016
16. Hussein G, Eid N, Khaled H (2019) IOSR-JMCE 16(5):40–55. https://doi.org/10.9790/1684.1605044055
17. Botis M, Cerbu C (2020) Appl Sci 10:5555. https://doi.org/10.3390/app10165555
18. Demyanov AI, Kolchunov VII, Pokusaev AA (2017) Struct Mech Eng Constr Build 6:37–44. https://doi.org/10.22363/1815-5235-2017-6-37-44
19. Salnikov A, Kolchunov Vl, Yakovenko I (2015) Appl Mech Mater 725–726:784–789. https://doi.org/10.4028/www.scientific.net/AMM.725-726.784
20. Kolchunov VII, Demyanov AI, Naumov NV (2019) Int J Comput Civ Struct Eng 15(4):66–82. https://doi.org/10.22337/2587-9618
21. Lyublinskiy V (2019) Build Reconst 5:17–22. https://doi.org/10.33979/2073-7416-2019-85-5-17-22
22. Russian Building Code SP 16.13330.2017 Steel structures. Design rules

Effective Polymer-Modified Bitumen Based on PET Waste

Dmitry Vorobyev, Yuri Borisenko, Dmitry Shvachev, and Stepan Rudak

Abstract The research results of polymer-modified bitumen (PMB) based on recycled polyethylene terephthalate for road asphalt concrete are presented. A more efficient and economical technology for the thermochemical destruction of recycled polyethylene terephthalate (PET) is proposed. Efficient compositions of Polymer-Modified Bitumen with different content of recycled PET with an increased range of plasticity have been developed. Regularities for the influence of the recycled PET concentration in the modifier and in PMB on the plasticity range of the binder have been established. The optimal (from a technical and environmental point of view) content of a modifying additive based on recycled PET at various ratios of polyethylene terephthalate and glycerin in the modifier has been established. Studies of the aging processes' kinetics of the proposed polymer-bitumen compositions based on recycled PET showed a high structural stability of PMB over time.

Keywords Bitumen · Polymer · Waste · Polyethylene terephthalate · Modification

1 Introduction

The condition of asphalt concrete road surfaces in many regions of the Russian Federation is assessed as critical or close to it. Often the reason is the low quality of the asphalt materials. Used road construction materials need modernization and new technological solutions to ensure durability, shear resistance, heat and crack resistance of road surfaces. The determining factors influencing the physical, mechanical and operational properties of asphalt concrete are the content and quality of the bituminous binder. An effective way to solve the problem of improving the quality of applied road bitumen is the use of bituminous binders modified with polymers. Currently, for the road bitumen modification, the following are successfully used: thermosetting polymers (duroplastics); thermoplastic polymers (thermoplastics or plastomers); thermoplastic elastomers (thermoplastic rubbers); rubber-like polymers

D. Vorobyev (✉) · Y. Borisenko · D. Shvachev · S. Rudak
Federal State Autonomous Educational Institution for Higher Education "North-Caucasus Federal University", 1, Pushkin Street, Stavropol 355017, Russian Federation
e-mail: davorobev@ncfu.ru

P. Akimov et al. (eds.), *Proceedings of FORM 2022*, Lecture Notes in Civil Engineering 282, https://doi.org/10.1007/978-3-031-10853-2_39

(elastomers). However, the use of modifying additives based on the above polymers leads to a significant increase in the cost of the resulting polymer-modified bitumen (PMB), complicates the technology of its preparation and the production of asphalt concrete itself.

One of the promising areas for reducing the cost and improving the production technology of PMB is the use of household plastic waste as a bitumen modifier, in particular, waste based on recycled polyethylene terephthalate (PET) (used containers for the food industry; plastic bottles, canisters, containers and other types of packaging). Polyethylene terephthalate has become widespread in the past few decades as a safe, strong, durable material. But its reuse in the food industry is prohibited. As a result, this valuable material accumulates in dump sites and landfills. It accounts for about 11% of all plastic waste [1].

In recent years, a number of research, devoted to the possibility of effective use of PET as a modifying polymer additive for road bitumen and bitumen-mineral compositions, have appeared [2–6]. These studies show that the modifier based on PET plastic improves the physical–mechanical and operational characteristics of asphalt concrete and opens the way for the recycling of plastics.

Thus, according to the results of studies [7–11], the use of plastic waste as a modifying additive makes it possible to obtain polymer-modified bitumen for asphalt concrete mixtures with a wider range of plasticity (the range of plasticity is calculated as the difference between the softening temperature Ts and the brittle temperature Tb of the binder). The higher this indicator, the better the bitumen resists thermal and mechanical influences. The use of asphalt concrete mixtures containing PMB with an increased plasticity range will significantly increase the operating temperatures and the main operational properties of road surfaces (durability, thermal stability, heat and crack resistance, shear resistance, etc.) in a wide range.

However, the known methods and technologies for the use of polymer waste for bitumen modification have a number of significant drawbacks: high cost and energy consumption in the production of PMB, the complexity of the technology and the instability of the resulting PMB; use of high-speed mixers; high mixing time when preparing compositions, etc.

Reducing the cost, energy consumption, simplification and optimization of the technology for obtaining effective PMBs based on household PET waste is an urgent task, the solution of which will create the preconditions for its wider industrial introduction.

2 Materials and Methodology

Cleaned crushed PET bottles (taken from household waste) with a particle size of 2–10 mm were accepted for research.

Preliminarily, before introducing into bitumen, thermochemical destruction of PET was carried out by the method of alcoholysis [12]. The method of thermochemical destruction proposed by A.M. Syroezhko, V.V. Vasiliev and others [13] was used.

For the processing of polyethylene terephthalate, glycerin $C_3H_8O_3$ was chosen as one of the most common representatives of polyhydric alcohols.

Thermochemical destruction of PET was carried out in a reactor with access to atmospheric air. Crushed clean plastic with glycerin was mixed in a reactor and heated to 220–240 °C, this temperature regime was kept for 1–1.5 h. A sample weighing 0.5 kg was taken from the resulting modifier into a porcelain glass. The sample was mixed for 5 min, after that the solution was poured through a bitumen sieve and uniformity was visually checked (no particles of undissolved PET plastic). After that, the modifier was poured into a container for storage and left for further research.

The modifier was made in 4 different ratios of PET:glycerin (% wt.), respectively: 50:50 (composition 1); 55:45 (composition 2); 60:40 (composition 3) and 65:35 (composition 4). It has been established that when the content of glycerin in the mixture is 40% wt. and less the resulting mixture is not homogeneous. In further studies, formulations 1–3 were used.

The study of the influence of modifiers obtained on the basis of PET was carried out on bitumen BND 60/90. Bitumen properties are presented in Table 1.

In order to obtain the optimal composition of the polymer-bitumen binder, the influence of the bitumen modifier content on the properties of the obtained PMB was studied. To achieve this the obtained PET modifiers (compositions 1–3) were introduced into bitumen in an amount of 0.5–10% wt. with an interval of 0.5% wt.

Bitumen and PET modifiers were heated to temperatures of 130–150 °C, then dosed. The dosed bitumen was loaded into a laboratory mixer, then the dosed modifier was added to the bitumen and mixed at a speed of 60–100 rpm for 1 min.

Table 1 Properties of bitumen BND 60/90 (PJSC "Saratovskiy NPZ")

Name of indicator	Test method	Actual value	Norm according to Russian State Standart (GOST) 22,245-90 for BND 60/90
Needle penetration depth, 0.1 mm at 25 °C; at 0 °C	GOST 11501	87 35	61–90 At least 20
Ring-and-ball softening temperature, °C	GOST 11506	48	Not lower than 47
Extensibility, cm at 25 °C; at 0 °C	GOST 11505	71 5.1	At least 55 At least 3.5
Fraas Brittle temperature, °C	GOST 11507	Minus 19	Not higher than minus 15
Measurement of softening temperature after warming up, °C	GOST 18180	4	No more than 5
Penetration index	GOST 22245	– 0.3	From –1.0 to +1.0

The prepared PMBs were cooled to room temperature. After a day, PMB was tested – penetration (P), ring-and-ball softening temperature (Ts), ductility (extensibility) (D), Fraas brittleness temperature (Tb), softening temperature after heating (Tr) were measured, penetration index (PI) was calculated.

The penetration index of the studied PMBs was determined by GOST 22245-90 according to the formula:

$$PI = \frac{30}{1 + 50A} - 10, \tag{1}$$

where A is calculated by the formula:

$$A = \frac{2,9031 - \log P}{T_s - 25}, \tag{2}$$

where P is the penetration depth of the needle at 25 °C, 0.1 m;

Ts is the softening temperature, °C.

3 Results

The results of studies of the main properties of polymer-modified bitumen with modifiers based on PET (PMB-1 with modifier composition 1; PMBV-2 with modifier composition 2; PMB-3 with modifier composition 3) are respectively presented in Tables 2, 3 and 4.

Tests of PMB-1 with a modifier based on PET showed that an increase in the content of the modifier leads to an increase in the viscosity of PMB-1 (Table 2), penetration and ductility indicators decrease, and softening temperature increases. With an increase in the content of the modifier, the brittleness temperature steadily decreases. The minimum value of Tb–24 °C was recorded at the modifier content in the composition of 7.5–10% wt. In the same range of modifier content, a decrease (by 1 °C) in the softening temperature of PMB after heating was recorded.

Analyzing the PMB-2 test data (Table 3), the following should be noted: with an increase in the content of the modifier in PMB (as well as with the use of the modifier composition 1), the penetration and ductility indicators decrease, and the ring-and-ball softening temperature increases. However, an increase in the content of PET in the modifier (composition 2) by 5% wt. contributes to a more intensive increase in the viscosity of PMB, as evidenced by a more intensive decrease in penetration and ductility and an increase in Ts. The introduction of the modifier also makes it possible to reduce the brittleness temperature to –26 °C (with an additive content of 8.5–10 wt. % bitumen). A decrease in the softening point of PMB after heating was recorded at a modifier content in the polymer-bitumen binder of 3% wt. However, with an increase in the content of the modifier more than 8.5% wt. PMB inhomogeneity is observed.

Table 2 Properties of PMB-1 with a modifier based on PET (composition 1, PET:glycerin ratio (% wt.)—50:50)

Indicator name	Test method	Value of the indicator when adding a modifier based on PET (composition 1), % of the mass of bitumen																			
		0,5	1,0	1,5	2,0	2,5	3,0	3,5	4,0	4,5	5,0	5,5	6,0	6,5	7,0	7,5	8,0	8,5	9,0	9,5	10
Needle penetration depth, P, 0.1 mm at 25 °C	GOST 11501	87	87	87	87	87	87	87	87	87	86	86	85	85	84	83	83	82	82	81	81
0 °C		35	35	35	35	35	35	35	35	35	35	34	34	34	33	32	32	31	31	30	30
Ring-and-ball softening temperature, Ts, °C	GOST 11506	48	48	48	48	48	48	48	48	48	48	49	49	49	49	49	50	50	50	50	50
Extensibility, D, cm at a temperature of 25 °C	GOST 11505	71	71	71	71	71	71	70	70	70	69	69	69	69	68	68	67	67	66	66	66
0 °C		5,1	5,1	5,1	5,0	5,0	4,9	4,9	4,9	4,8	4,8	4,8	4,7	4,7	4,6	4,5	4,5	4,4	4,4	4,4	4,3
Fraas brittle temperature, Tb, °C	GOST 11506	–19	–19	–19	–19	–19	–20	–20	–20	–21	–21	–22	–22	–22	–23	–24	–24	–24	–24	–24	–24
Measurement of the softening temperature after warming up, Tr, °C	GOST 18180, GOST 11506	4	4	4	4	4	4	4	4	4	4	4	4	4	4	3	3	3	3	3	3
Homogeneity	GOST 52056	Homogeneous																			

Table 3 Properties of PMB-2 with a PET-based modifier (composition 2, PET:glycerin ratio (% wt.)—55:45)

Indicator name	Test method	Value of the indicator when adding a modifier based on PET (composition 1), % of the mass of bitumen																			
		0,5	1,0	1,5	2,0	2,5	3,0	3,5	4,0	4,5	5,0	5,5	6,0	6,5	7,0	7,5	8,0	8,5	9,0	9,5	10
Needle penetration depth, P, 0.1 mm at 25 °C	GOST 11501	87	86	86	85	85	84	83	83	82	82	81	80	79	78	78	77	75	74	74	73
0 °C		35	35	35	34	34	33	32	32	31	30	30	29	28	28	27	27	26	25	25	24
Ring-and-ball softening temperature, Ts, °C	GOST 11506	48	48	48	48	48	48	49	49	49	49	50	50	50	50	51	51	51	52	52	52
Extensibility, D, cm at a temperature of 25 °C	GOST 11505	71	70	70	70	68	68	67	67	67	65	64	64	63	63	62	61	61	60	60	59
0 °C		5,0	5,0	4,9	4,9	4,8	4,8	4,7	4,7	4,6	4,6	4,6	4,5	4,5	4,4	4,3	4,3	4,2	4,2	4,0	3,9
Fraas brittle temperature, Tb, °C	GOST 11506	– 19	– 19	– 20	– 20	– 21	– 21	– 22	– 22	– 22	– 23	– 24	– 24	– 24	– 24	– 25	– 25	– 26	– 26	– 26	– 26
Measurement of the softening temperature after warming up, Tr, °C	GOST 18180, GOST 11506	4	4	4	4	4	3	3	3	3	3	3	3	3	3	3	3	3	3	3	3
Homogeneity	GOST 52056	Homogeneous																Non-homogeneous			

Table 4 Properties of PMB-3 with PET-based modifier (composition 3, PET:glycerin ratio (% wt.)—60:40)

Indicator name	Test method	Value of the indicator when adding a modifier based on PET (composition 1), % of the mass of bitumen																			
		0,5	1,0	1,5	2,0	2,5	3,0	3,5	4,0	4,5	5,0	5,5	6,0	6,5	7,0	7,5	8,0	8,5	9,0	9,5	10
Needle penetration depth, P, 0.1 mm at 25 °C	GOST 11501	85	84	82	80	79	77	77	76	75	73	73	71	70	69	67	66	65	64	64	63
0 °C		33	32	30	29	27	27	26	26	26	25	24	24	23	23	22	22	18	18	16	14
Ring-and-ball softening temperature, Ts, °C	GOST 11506	48	49	50	50	51	52	52	52	53	54	54	54	54	54	55	55	55	55	55	55
Extensibility, D, cm at a temperature of 25 °C	GOST 11505	71	70	69	69	68	67	66	65	65	64	64	63	62	61	61	60	59	58	58	57
0 °C		5.0	4,9	4,9	4,8	4,7	4,6	4,6	4,5	4,4	4,4	4,3	4,2	4,1	4,0	4,0	3,9	3,8	3,6	3,6	3,5
Fraas brittle temperature, Tb, °C	GOST 11506	–19	–19	–20	–21	–22	–23	–24	–25	–26	–28	–28	–27	–27	–26	–25	–24	–24	–23	–23	–22
Measurement of the softening temperature after warming up, ITr, °C	GOST 18180, GOST 11506	4	4	3	3	3	3	3	2	2	2	2	2	2	2	2	2	2	2	2	2
Homogeneity	GOST 52056	Homogeneous														Non-homogeneous					

Tests of PMB-3 (Table 4) revealed the following: the nature of the change in the viscosity of the binder PMB-3 is similar to that in the tests of PMB-1 and PMB-2. The viscosity of PMB-3 increases with increasing content of the modifier, and the indicators characterizing it (penetration, ductility, ring-and-ball softening point) have the highest values compared to the corresponding indicators of PMB-1 and PMB-2. This indicates that the viscosity of PMB depends on the content of polyethylene terephthalate in the binder modifier. The change in the brittleness temperature with an increase in the modifier content in the binder is extreme and is minimal (–28 °C) at a modifier content of 5–5.5% wt. This corresponds to the maximum recorded plasticity interval of the binder, corresponding to 82 °C, which is 21% higher than the plasticity interval of the original BND 60/90 bitumen. With an increase in the content of the modifier more than 7.5% wt. PMB inhomogeneity is observed.

It should be noted that when using the proposed modifiers based on PET (compositions 1–3), with an increase in its content in PMB, the softening point change index after heating decreases by 1–2 °C (Tables 2, 3 and 4). This fact may indirectly indicate an increase in the durability of the proposed PMBs.

The thermal sensitivity of bitumen components characterizes the penetration index (PI) more differentiated, depending on bitumen penetration at 25 °C and bitumen softening temperature Ts. Between the value of PI and the properties of bitumen, there is the following relationship – the higher the penetration index, the higher the heat resistance of the binder. In order to avoid obtaining bitumen with low ductility, the upper value of PI is limited. Specific limits for changing the PI are set depending on the conditions for the use of bitumen. So for BND 60/90 bitumen, the PI values range from −1.0 to +1.0. The introduction of the proposed modifiers into the composition of bitumen affects the increase in PI values.

The higher the PI, the lower the temperature sensitivity of the bituminous binder and the lower the tendency to embrittlement.

The influence of the content of modifiers in PMB on the change in the value of the penetration index (PI) was studied (Table 5).

Table 5 Influence of the content of modifiers based on PET in PMB on the change in the penetration index (PI)

Index	The content of the modifier based on PET (composition 1) in PMB-1, wt %.																			
	0,5	1,0	1,5	2,0	2,5	3,0	3,5	4,0	4,5	5,0	5,5	6,0	6,5	7,0	7,5	8,0	8,5	9,0	9,5	10,0
PI	-0,31	-0,31	-0,31	-0,31	-0,31	-0,31	-0,31	-0,31	-0,31	-0,34	-0,06	-0,09	-0,09	-0,13	-0,16	0,11	0,07	0,07	0,04	0,04
Index	The content of modifier based on PET (composition 2) in PMB-2, % wt.																			
	0,5	1,0	1,5	2,0	2,5	3,0	3,5	4,0	4,5	5,0	5,5	6,0	6,5	7,0	7,5	8,0	8,5	-	-	-
PI	-0,31	-0,34	-0,34	-0,37	-0,37	-0,41	-0,16	-0,16	-0,20	-0,20	0,04	0,00	-0,04	-0,07	0,19	0,15	0,08	-	-	--
Index	The content of modifier based on PET (composition 3) in PMB-3, % wt.																			
	0,5	1,0	1,5	2,0	2,5	3,0	3,5	4,0	4,5	5,0	5,5	6,0	6,5	7,0	7,5	-	-	-	-	-
PI	-0,37	-0,13	0,07	0,00	0,23	0,41	0,41	0,37	0,58	0,74	0,74	0,66	0,62	0,58	0,53	-	-	-	-	

As a result of the tests of the proposed modifiers, it was found that the highest PI values were obtained: for PMB-1 with a modifier content of 8% wt. (PI = + 0.11); for PMB-2 with a modifier content of 7.5% wt. (PI = +0.19); for PMB-3 with a modifier content of 5–5.5% wt. (PI = +0.74).

Taken together, the results of studying the properties of polymer-modified bitumen with PET-based modifiers show that the proposed technology for the preparation of PMB improves almost the entire range of technical properties of bitumen, and the following compositions of PMB with a modifier based on PET: PMB-1 with modifier content (composition 1) 8 wt %; PMB-2 with modifier content (composition 2) 7.5% wt. and PMB-3 with modifier content (composition 3) 5–5.5% wt.

To assess the kinetics of PMB aging processes, studies of the influence of the time factor on the change in the properties of the proposed PMB were carried out. The selected PMB samples of optimal compositions were stored under normal temperature and humidity conditions and tested at intervals of 1 month for six months. Penetration (P), softening temperature (Ts), ductility (D), Fraas brittleness temperature (Tb), change in softening temperature after heating (Tr) of PMB from the manufacture moment were determined. The test results are presented in Table 6. According to the data obtained, the development of aging processes in the studied compositions of PMB is practically absent within half a year.

Thus, the proposed compositions of polymer-modified bitumen can be recommended for use during the road construction season, which, as a rule, does not exceed the studied time interval.

Table 6 Change in the physical and mechanical properties of PMB over time, depending on the optimal amount of the modifier

Name	Time, month	P, 0.1 mm at 25°C/0°C	Ts, °C	D, cm at 25°C/0°C	Tb, °C	Tr after warming up, °C
PMB-1 (content of PET modifier 8% wt.)	1	81/30	50	66/4,3	-24	3
	2	80/30	50	66/4,3	-24	3
	3	80/30	50	66/4,2	-24	3
	4	81/30	50	65/4,2	-24	3
	5	81/30	50	66/4,3	-24	3
	6	81/30	50	66/4,3	-24	3
PMB-2 (content of PET modifier 7.5% wt.)	1	77/27	51	61/4,3	-25	3
	2	76/28	51	61/4,3	-25	3
	3	77/27	51	61/4,3	-25	3
	4	77/27	51	61/4,3	-25	3
	5	77/27	51	61/4,3	-25	3
	6	77/27	51	61/4,3	-25	3
PMB-3 (content of PET modifier 5.5% wt.)	1	73/24	54	64/4,3	-28	2
	2	72/26	54	64/4,3	-28	2
	3	74/24	54	66/4,2	-28	2
	4	73/24	54	65/4,2	-28	2
	5	73/24	54	64/4,3	-28	2
	6	73/24	54	64/4,3	-28	2

Analysis of the results of the studies of polymer-modified bitumen based on recycled polyethylene terephthalate showed the following:

4 Conclusion

1. A more efficient and economical technology for the thermochemical destruction of secondary polyethylene terephthalate with glycerin in a reactor by PET alcoholysis with access to atmospheric air has been proposed.
2. Compositions of polymer-modified bitumen with different content of recycled PET have been developed: PMB-1 (BND 60/90 bitumen with PET modifier composition 1—PET:glycerin ratio—50:50); PMB-2 (bitumen BND 60/90 with PET modifier composition 2—PET:glycerin ratio—55:45); PMB-3 (bitumen BND 60/90 with PET modifier composition 3—PET:glycerin ratio—60:40);
3. The use of the proposed modifiers based on recycled PET significantly expands the plasticity range of the bituminous binder. It has been established that with an increase in the content of destructured PET in a polymer-bitumen binder, its range of plasticity steadily increases. The interval of plasticity of PMB increases the more intensively, the higher the ratio of the polyethylene terephthalate concentration in the modifier.
4. As a result of the analysis of the data obtained from the studies carried out, the optimal (from a technical and environmental point of view) content of the modifying additive based on PET was established at various ratios of PET:glycerin in the modifier, which amounted to: for PMB-1—8 wt %; for PMB-2—7.5% wt.; for PMB-3—5–5.5% wt.
5. The study of the aging processes' kinetics of optimal polymer-bitumen compositions based on recycled PET showed a high structural stability of the proposed PMBs over time.

Acknowledgements The work was performed using the Center for Collective Use of the North Caucasus Federal University with the financial support of the Ministry of Education and Science of Russia, unique project identifier RF—2296.61321X0029 (agreement № 075-15-2021-687).

References

1. Materials from the portal of the Federal Service for Supervision of Natural Resources. Analytical note on the management of solid household and industrial waste and on the implementation of pilot projects for waste processing in the constituent entities of the Russian Federation, p 14. [Electronic resource]. rpn.gov.ru/sites/all/files/documents/doklady/analiticheskaya_zapiska2.doc/
2. Lapian FEP, Ramli MI, Pasra M, Arsyad A (2021) The performance modeling of modified asbuton and polyethylene terephthalate (PET) mixture using response surface methodology (RSM). Appl sci 11:6144. https://doi.org/10.3390/app11136144

3. Sojobi AO, Nwobodo SE, Aladegboye OJ, Pratico FG (2016) Recycling of polypthylene terepthalate (PET) plastic bottle wastes in bituminous asphaltic concrete. Cogent Eng 3(1):1133480
4. Köfteci S, Ahmedzade P, Gunay T Evaluation of bitumen physical properties modified with waste plastic pipes. Can J Civ Eng 45(6):469–477. https://doi.org/10.1139/cjce-2017-0672
5. Mishra B, Gupta MK (2021) Performance study on bituminous concrete mixes using varying content and sizes of shredded polyethylene terephthalate. J Mater Civ Eng 33(1):1–13. https://doi.org/10.1061/(ASCE)MT.1943-5533.0003460
6. Xia T, Zhang A, Xu J et al (2021) Rheological behavior of bitumen modified by reclaimed polyethylene and polypropylene from different recycling sources. J Appl Polym Sci 138:e50435. https://doi.org/10.1002/app.50435
7. Singh B et al (2013) Polymer-modified bitumen of recycled LDPE and maleated bitumen. J Appl Polym Sci 127(1):67–78. https://doi.org/10.1002/app.36810, https://search.ebscohost.com/login.aspx?direct=true&db=aci&AN=82300128&site=ehost-live
8. Levchenko SI, Pen VR, Kolyada KA (2019) The use of modified bitumens to improve the quality of asphalt concrete coatings, Actual problems of aviation and cosmonautics: Collection of materials of the V International scientific and practical conference dedicated to the Day of Cosmonautics. In: Loginova YuYu (ed) 3 volumes, Krasnoyarsk, Krasnoyarsk: Federal State Budgetary Educational Institution of Higher Education Siberian State University of Science and Technology named after Academician M.F. Reshetnev, vol 3, pp 497–499, 08–12 April 2019
9. Yagafarova GG, Latypov VM, Moskovets AV et al (2012) New road mixes based on large-tonnage wastes of the oil and gas industry. In: Proceedings of the Samara Scientific Center of the Russian Academy of Sciences, vol 14, no 5–3, pp 871–873
10. Moldakhmetova AN (2021) The use of municipal solid waste of polyethylene terephthalate in road construction/A. N. Moldakhmetova, M. A. Elubay, Science and technology of Kazakhstan 1:49–54. https://doi.org/10.48081/LJWJ8316
11. Zhen L, Anand S, Rabindra KP, Zhifei T (2018) Value-added application of waste PET based additives in bituminous mixtures containing high percentage of reclaimed asphalt pavement (RAP). J Clean Prod 196:615–625. https://doi.org/10.1016/j.jclepro.2018.06.119. ISSN 0959-6526
12. Balakin VM, Krasilnikova MA, Starodubtsev AV (2013) Chemical methods of utilization of polyethylene terephthalate. Plastic Masses 1:57–64
13. Syroezhko AM, Vasiliev VV et al (2015) Patent 2573012 Rus. Federation: IPC51 C08L 95/100; C08J 11/24. Polymer-bitumen binder [Text], Applicant and patent holder Federal State Budgetary Educational Institution of Higher Professional Education St. 19/01/2015, publ. 20/01/2016, Bull. No. 2

Properties of 1915T Aluminum Alloy at Low Temperatures

Oleg Kornev, Aleksandr Shuvalov, and Eugenia Sokolova

Abstract Many characteristics of some alloys based on aluminum are not studied profoundly enough or information about them is lacking at all, especially about 1915T which is considered in this study. Aluminum alloy 1915T is suggested to be structural material for vertical cylindrical tanks exploited in conditions of temperatures below 0 °C. To confirm that considered material corresponds structure operation requirements tests with static and low-cycle loading were carried out. As a result of tensile tests mechanical properties such as ultimate strength, limit of proportionality, offset yield strength and elastic modulus were obtained for temperatures +20, –40, –70 and –100 °C. It was noticed that values of mechanical properties increased at low temperatures. Alloy deformability grows insignificantly within temperature decrease which means that toughness of the material almost does not change. Low-cycle loading was performed with stress ratio of r = 0,02 to determine low-cycle fatigue strength of aluminum alloy 1915T for the same temperatures. During static tests change of elastic modulus was noticed after the load was removed in a specimen which deformation reached plastic stage. Strain amplitude dependence on number of cycles was observed to fit an approximation curve for every testing temperature. Strength characteristics and deformability in conditions of low temperatures showed that 1915T not only strengthens but also still goes through plastic stage of deformation.

Keywords Aluminum alloy · 1915T · EN AW-7005 · Low temperature · Mechanical properties · Fatigue strength · Vertical cylindrical tank

O. Kornev · A. Shuvalov · E. Sokolova (✉)
Moscow State University of Civil Engineering, Moscow 129337, Russian Federation
e-mail: jane10.08.98@yandex.ru

O. Kornev
e-mail: Kornev-O@live.ru

A. Shuvalov
e-mail: edic@mgsu.ru

P. Akimov et al. (eds.), *Proceedings of FORM 2022*, Lecture Notes in Civil Engineering 282, https://doi.org/10.1007/978-3-031-10853-2_40

1 Introduction

Due to low unit weight and high strength aluminum alloys have found specific application in aircraft and automotive industry, shipbuilding, and rocket technology [1, 2]. Also, aluminum alloys have advantages of ease of processing, low maintenance, corrosion resistance and low-temperature resistance [3–6]. Owing to some characteristics such as low elastic modulus and higher cost in comparison with steel alloys based on aluminum do not meet many requirements of construction industry as good as steel does. Even though practice in construction established standard base of applied materials for specific structures, it is worth paying attention to a combination of some aluminum alloys characteristics which gives an opportunity of brand-new field of aluminum alloy application.

Taking into account that elastic modulus of aluminum alloys is 3 times lower than the one of the steel, aluminum alloys can absorb 3 times higher elastic energy. For structures subjected to significant stresses and containing big amount of elastic energy, both of which may cause sudden failure, it is important for structural material to have optimal combination of high strength and low elastic modulus. Vertical cylindrical tanks are considered such structures. Being main structural material in aircraft and space industry aluminum alloys can withstand loads in conditions of significantly low temperatures. Such behavior of the material is caused by toughness being slightly changed within range of decreasing temperature [7], which is essential feature for structures subjected to cyclic loading. Previously mentioned type of loading is typical for tanks exploited in product storage mode of more than 100 turns per year and is especially worth attention because this type of loading causes brittle failure at low temperatures [8–14].

Unfortunately, many characteristics of some alloys based on aluminum are not studied profoundly enough or information about them is lacking at all, especially about aluminum alloy considered in this study [15–17]. Mostly it concerns low temperature influence on strength, elastic modulus, and endurance limit of aluminum alloy. In this article test results of mechanical properties and low-cycle fatigue strength at low temperatures for aluminum alloy 1915T are offered to be considered. According to international grades system for wrought aluminum alloys 1915T is one of the 7xxx series, representatives of which are the strongest among other alloys based on aluminum. A full analogue of 1915T is EN AW-7005 [18]. According to [7] EN AW-7005 has the most beneficial combination of toughness and strength for low temperature application, that is why Russian analogue was taken as a test object.

2 Methods and Materials

To determine mechanical characteristics of 1915T such as ultimate strength, offset yield strength, limit of proportionality, and elastic modulus tensile tests were carried out according to GOST 1497-84 "Metals. Methods of tension test" [19]. Low cycle fatigue testing was held according to GOST 25.502-79 "Strength analysis and testing in machine building. Methods of metals mechanical testing. Methods of fatigue testing" [20]. Number of cycles did not exceed $5 \cdot 10^4$. Type of cyclic loading was tension-compression, stress-controlled. In this research Instron 8802 and LabTest 6.500H were used as testing machines. Shape and dimensions of specimens are shown in Fig. 1.

The content of each series of specimens and their loading type are shown in Table 1.

Specimens for static and cyclic loading are shown in testing machine clamps in Fig. 2.

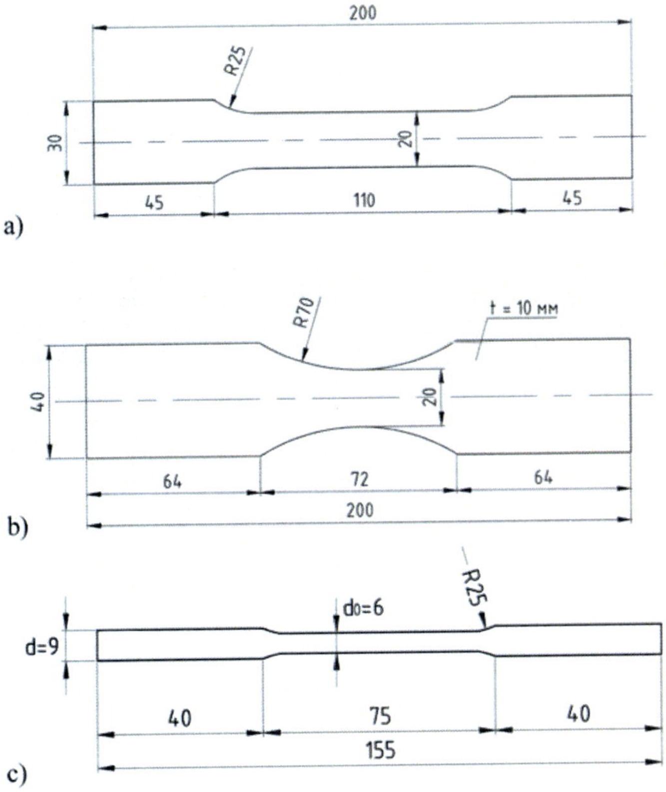

Fig. 1 Specimen dimensions for **a** static testing, **b** cyclic testing and **c** testing to determine elastic modulus change, mm

Table 1 Characteristics of specimen series

Series mark	Number of specimens	Test temperature, °C	Stress ratio, r
Static loading (tension)			
1s	5	+20	–
2s	5	–40	–
3s	5	–70	–
4s	5	–100	–
Low-cycle loading			
1c	10	+20	0,02
2c	10	–40	0,02
3c	10	–70	0,02
4c	10	–100	0,02

(a)

(b)

Fig. 2 Specimen in testing machine clamps for **a** static and **b** cyclic loading

3 Results and Discussion

According to results obtained from static testing "mechanical properties—temperature" diagrams for 1915T alloy were gained for such characteristics as limit of proportionality, offset yield strength, ultimate strength (Fig. 3), elastic modulus (Fig. 4) and both area reduction and elongation as well (Fig. 5).

According to results given in Figs. 3, 4 and 5 it is seen that with temperature decrease from +20 °C down to –100 °C strength parameters increase by 6% for limit of proportionality, by 8,9% for offset yield strength and by 8,8% for ultimate strength. Elastic modulus grows by 10%. For metals with face-centered cubic crystal structure, including aluminum alloys, it is a typical feature to have deformability unchanged or slightly growing with temperature decrease [21] which is seen in Fig. 5.

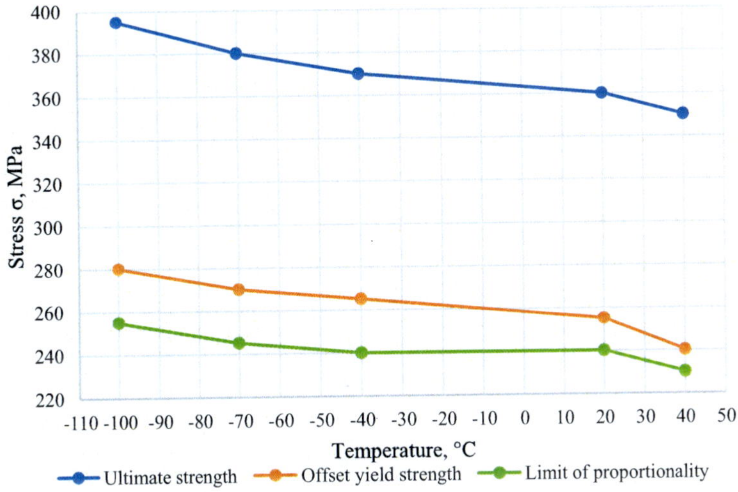

Fig. 3 "Offset yield strength/ultimate strength/limit of proportionality–temperature" diagram for 1915T alloy

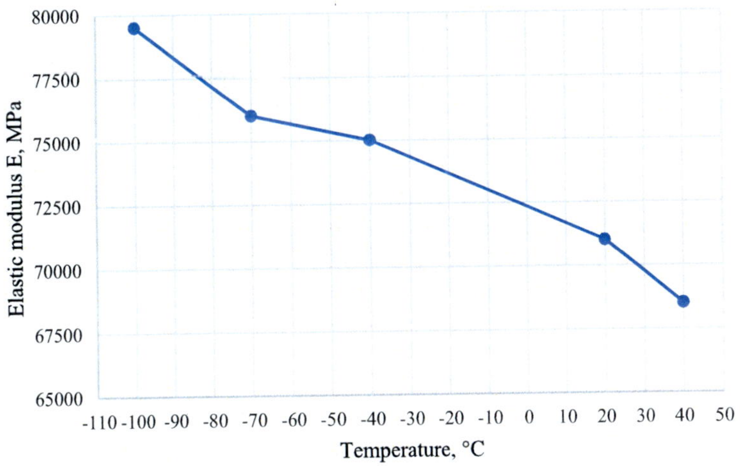

Fig. 4 "Elastic modulus-temperature" diagram for 1915T alloy

During static tests at +20 and –70 °C some of the specimens being in plastic region already were subjected to several cycles of loading-unloading in order to determine elastic modulus change (Fig. 6).

Considering obtained results (Fig. 6) elastic modulus of 1915T alloy reaching some point in plastic deformation decreases after unloading at both positive and negative temperatures. Stabilization of elastic modulus was observed at 2% of initial plastic deformation at +20 °C and at 3% at –70 °C. Despite elastic modulus growth

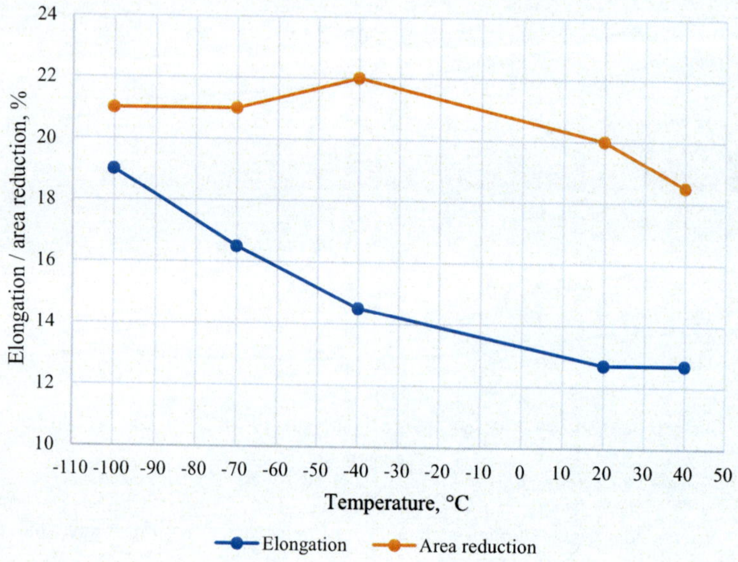

Fig. 5 "Elongation/area reduction-temperature" diagram for 1915T alloy

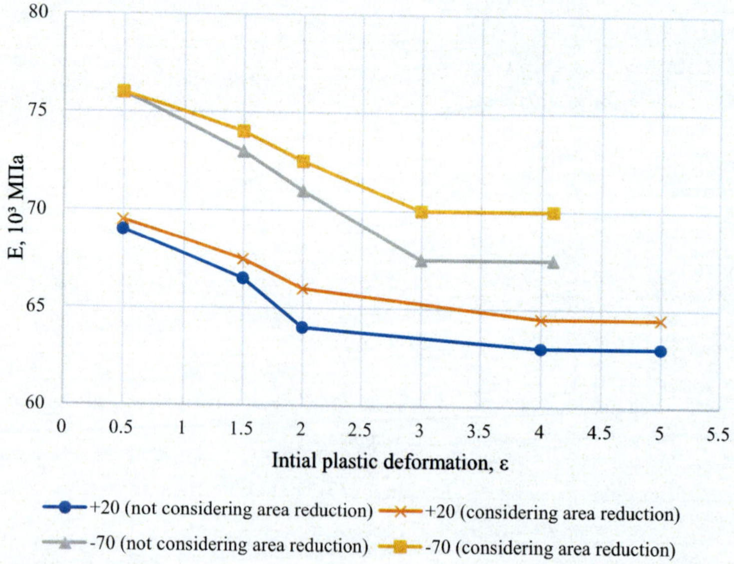

Fig. 6 "Elastic modulus-initial plastic deformation" diagram for 1915T alloy at +20 и –70 °C temperatures

at temperatures below 0 °C multiple loading and unloading cycles lead to mentioned property decrease which means that embrittlement effect does not affect material much in conditions of considered loading as it does in a static one.

Low-cycle fatigue test results are shown in Figs. 7 and 8.

In Fig. 7 it is seen that with temperature decrease low-cycle fatigue strength of 1915T increases. Depending on temperature the material at maximum stress of 350–380 MPa can withstand 10^4 numbers of cycles which corresponds to minimum operational cyclic loading of a tank with service life of 100 years.

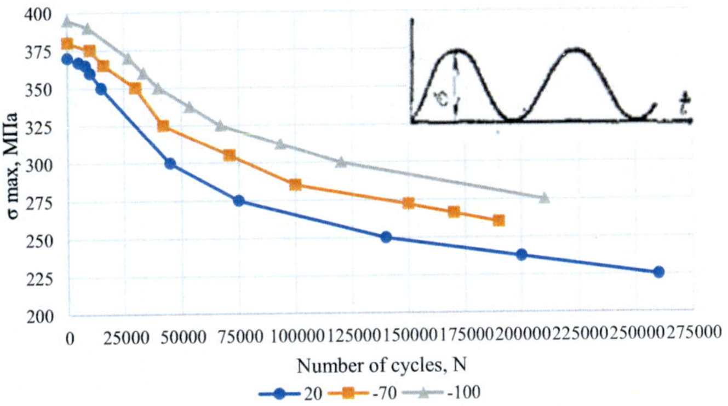

Fig. 7 "Maximum stress—number of cycles" diagram for 1915T at +20, –70 and –100 °C

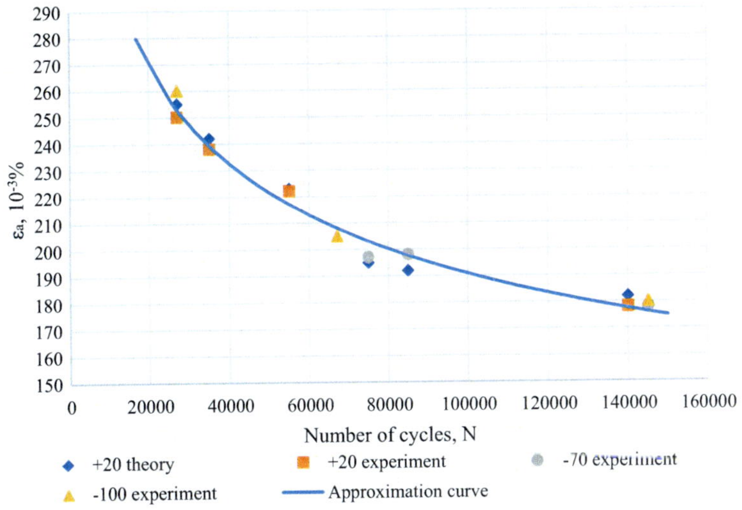

Fig. 8 "Strain amplitude—number of cycles" diagram for 1915T at +20, –70 and –100 °C

"Strain amplitude—number of cycles" diagram is shown in Fig. 8. Values obtained during experiments at different temperatures lie along the approximation curve with coefficient of determination $R^2 = 0{,}98$.

4 Conclusions

Aluminum alloy 1915T testing with static and cyclic loading at room and low temperatures, viz. +20, –40, –70 and –100°C, showed that values of strength parameters such as offset yield strength, limit of proportionality, ultimate strength and elastic modulus increase in the material. Alloy deformability grows insignificantly within temperature decrease which means that toughness of the material almost does not change. Fatigue strength increases and change of elastic modulus after unloading a specimen which deformation reached plastic region indicates that the material can absorb bigger amount of elastic energy with every "loading-unloading" cycle and as a result that early prediction of material failure may become more possible. At 10^4 numbers of cycles 1915T alloy can withstand significant amplitude of stress change with stress ratio $r = 0{,}02$.

By and large strength characteristics and deformability in conditions of low temperatures showed that 1915T not only strengthens but also still goes through plastic stage of deformation. Therefore, considered alloy can be potentially used as structural material for vertical cylindrical tanks subjected to low-cycle loading at low temperatures. In addition to that it was proven in empirical way that there is a correlation between strain amplitude and number of cycles despite of temperature conditions and this correlation can be described with some dependence corresponding to real values distribution with high accuracy.

Acknowledgements This work was financially supported by the Ministry of Science and Higher Education of the Russian Federation (grant # 075-15-2021-686). All tests were carried out using research equipment of The Head Regional Shared Research Facilities of the National Research Moscow State University of Civil Engineering.

References

1. Luts AR, Suslina AA (2018) Aluminum and its alloys, Textbook, Samara, p 81
2. Davis JR (2001) Aluminum and aluminum alloys. In: Alloying: Understanding the Basics, ASM International, Ohio, pp 351–416
3. Zhao X, Li H, Chen T, Cao B, Li X (2019) Mechanical properties of aluminum alloys under low-cycle fatigue loading. Materials 12. https://doi.org/10.3390/ma12132064
4. Shuvalov A, Katanina A, Kornev O, Kovalev (2019) Working capacity of aluminum alloys in structure elements. In: E3S Web of Conferences 97, FORM-2019 Proceeding
5. Davis JR (1999) Corrosion of Aluminum and Aluminum Alloys, ASM International, p 313
6. Korgin AV, Odesskiy PD, Ermakov VA, Seid Kilani LZ, Romanez VA, Korolyova EA (2019) Aluminum alloys strength for bridge construction. Deformation Fract Mater 8:10–19

7. Kaufman JG, Wanderer ET (1983) Mechanical properties of structural materials at low temperatures. In: Fridlyander JN (ed) Metallurgiya, Moscow, pp 163–175
8. Zaharova MI (2015) Analysis and estimation of risks of tank and gas holder failures at low temperatures, Ph.D. thesis, V.P. Larionov Institute of the Physical-Technical Problems of the North of the Siberian Branch of the RAS
9. Kondrashova OG, Nasarova MN (2004) Cause-and-effect analysis of vertical steel tanks failures. Oil Gas Bus 2
10. Volkov VN, Popova NV, Burmistrova ON (2014) Working capacity assessment of oil tanks in environmental conditions of the Komi Republic. Modern Probl Sci Educ 4. https://science-education.ru/ru/article/view?id=13855
11. Bolshakov AM, Andreev YM (2015) Failure analysis of metal structures operating in north environmental conditions. Aviat Mater Technol S1:27–31
12. Bolshakov AM, Tatarinov LN (2009) Reliability of main gasline after 30 years of operation in north environmental conditions. Gas Ind 2:28–31
13. Bolshakov AM, Andreev YM (2012) Defects nature and failure types of tanks operating in north environmental conditions. Gas Ind 3:90–92
14. Konovalov NN (2006) Regulation of defects and reliability of non-destructive methods of weld joint control, STC Industrial Safety, Moscow, p 111
15. Ma J, Wang P, Fang H (2021) Fatigue life of 7005 aluminum alloy cruciform joint considering welding residual stress. Materials 14. https://doi.org/10.3390/ma14051253
16. Kaufman JG (2008) Properties of aluminum alloys: Fatigue data and the effects of temperature, product form, and processing, ASM International, Ohio, p 557
17. Osintsev OYe, Konkevich VYu (2010) High-strength rapidly solidified Al-Zn-Mg and Al-Zn-Mg-Cu aluminum alloys. Technol Light Alloys 1:157–163
18. GOST 4784-2019 Aluminium and wrought aluminium alloys. Grades
19. GOST 1497-84 Metals. Methods of tension test
20. GOST 25.502-79 Strength analysis and testing in machine building. Methods of metals mechanical testing. Methods of fatigue testing
21. Park D-H, Choi S-W, Kim J-H, Lee J-M (2015) Cryogenic mechanical behavior of 5000- and 6000-series aluminum alloys: Issues on application to offshore plants. Cryogenics 68:44–58

Bearing Capacity of Anchors at Multi-cyclic Dynamic Loads

Oleg Kabantsev and Mikhail Kovalev

Abstract Most frequently used types of anchors were subjected to experimental studies to obtain valid data on post-installed anchors, such as mechanical anchors (wedge expansion anchors, undercut anchors, screw anchors, push-in anchors, etc.); bonded anchors (epoxy resin, polyurethane, polyester anchors with quartz sand, cement mixtures added, etc.), and plastic anchors (plastic dowels, plate anchors) and cast-in-place anchors (bent anchors, anchors with an anchor plate, with a conical end, with a bar hook; compound anchors, etc.). The authors studied the effect of an earthquake-induced damage (plastic deformation) of a concrete base and multi-cyclic dynamic loads, similar to seismic ones, on the bearing capacity of anchors and scatter of acquired results. Plastic deformations of the concrete base are assumed to be in the form of cracks located directly at the point of anchoring. The results of the research show that decrease in bearing capacity in cracked concrete under multi-cyclic dynamic loads should be taken into account during design of anchorage under seismic loads depending on failure mechanisms, e.g. for steel failure reduction of bearing capacity does not exceed 6% whereas for failure at the interface between the adhesive composition and the base it can reach up to 50%. Values of variation coefficient depend on type of failure mechanisms, e.g. for steel failure its value does not exceed 5.2%, whereas for interface failures it can reach up to 17.4%. Acquired values of variation coefficients should be taken into account for calculation of safety factors for different types of anchors under seismic loads action.

Keywords Anchor · Dynamic loading · Seismic · Failure mechanism · Pullout test

1 Introduction

Nowadays anchors are in common use as structural elements made for fastening other structures and machinery to the concrete base. Anchors are used to fasten a broad array of structures and elements from individual structural members, that

O. Kabantsev · M. Kovalev (✉)
Moscow State University of Civil Engineering, Moscow 129337, Russian Federation
e-mail: kovalyov.mike@gmail.com

P. Akimov et al. (eds.), *Proceedings of FORM 2022*, Lecture Notes in Civil Engineering 282, https://doi.org/10.1007/978-3-031-10853-2_41

are parts of a building bearing system, to various types of engineering machinery, such as pipelines and control panels, and non-bearing elements of buildings, such as suspended facades, room partitions, and translucent structures.

Anchors are divided into two groups by the method of their embedment into structures:

- post-installed anchors, such as mechanical anchors (wedge expansion anchors, undercut anchors, screw anchors, push-in anchors, etc.); bonded anchors (epoxy resin, polyurethane, polyester anchors with quartz sand, cement mixtures added) and plastic anchors (plastic dowels, plate anchors);
- cast-in-place anchors, installed in the base during its casting (bent anchors, anchors with an anchor plate, anchors with a conical end, and anchors with a bar hook, compound anchors, etc.).

Mechanical characteristics of mechanical expansion wedge anchors [1], undercut anchors [2] and bonded anchors [3] under static loads have been fairly studied. The studies of anchors, exposed to dynamic loads and cyclic crack opening conditions, are presented in [4, 5].

However, analysis of these studies results shows that reduction of bearing capacity under the seismic loading or the loading simulating it (e.g. multi-cyclic dynamic loadings) and scatter of ultimate pullout loads, needed to assess the anchor behavior under seismic loading, have not been presented.

A traditional approach to design of buildings under the action of earthquake is based on the theory of development of the plastic deformation phase, which in the case of concrete base can be described as a system of cracks with different width. The most conservative case is crack occurrence at the point of anchor embedment. The probability of cracks passing through the point of anchor embedment is confirmed in a number of studies [6–10]. According to [6], the width of residual crack openings after earthquake in columns and beams of different cross-section, alternated from 0.26 to 1.04 mm. Other studies [11] show that national regulatory documents (e.g. [12]) give requirements for the maximum acceptable crack width in concrete after the action of seismic loads up to 1.5 and 3 mm.

It is evident that anchors, installed in concrete with cracks simulating earthquake damage, may have bearing capacity that severely differs from behavior under static loads in uncracked concrete. However, currently there is almost no data on the anchor behavior in cracked concrete under dynamic loads.

The research findings can be used to determine bearing capacity of anchors under seismic load.

2 Methods and Materials

Research was carried out for anchors, installed in uncracked concrete and concrete with a crack at the anchor embedment point. The crack opening width was equal to 0.8 mm which corresponds to the research results [6].

Fig. 1 Control of crack width

Cracks were made by driving steel wedge-shaped elements in preformed holes in concrete base. Crack opening width was controlled using dial gauge displacement indicators (Fig. 1).

Table 1 shows types of tested anchors.

Concrete blocks with compressive strength classes C25/30 and C30/37 were used for the embedment of anchors. Dimensions of concrete blocks were determined depending on the embedment depth and the anchor diameter to prevent any obstacles to the formation of concrete cone during loading. Cast-in-place anchors (types 1–3) were installed in the large holes, that had been made during the concrete casting. After the embedment of anchors, the holes were filled with expanding concrete with compressive strength class of C40/50. The installation of post-installed anchors was performed into pre-drilled holes. The diameter and depth of a hole were defined according to manufacturer's recommendations. The anchor nut was tightened with the tightening torque up to the value specified by the manufacturer.

The experimental study of anchors consisted of static and multi-cyclic dynamic pullout tests. Multi-cyclic dynamic tests simulated seismic loads action. For each load pattern series of specimens were tested. The series consisted of 5 specimens, exposed to static loading, and 5 specimens, exposed to multi-cyclic dynamic loading. Figure 3 shows multi-cyclic loading pattern used in tests according to European regulatory documents [15].

Figure 4a shows scheme of experimental setup used in tests. Figure 4b shows anchor specimen during tests. Test load was transferred from the loading frame to the anchor through a steel plate connected to the hydraulic cylinder by steel threaded studs. The specimens were loaded by the reconfigurable loading frame equipped with hydraulic cylinders. Two clamps and studs were used to fasten the concrete block to the reinforced floor.

Table 1 Anchor types

№	Anchor type	Anchor	Anchor diameter d, mm	Hole diameter, mm	Embedment depth h_{ef}, mm	Crack opening width, mm
1	Cast-in-place anchors	Bent anchor bolts, made according to [13] (Fig. 2a)	M16	82	200–300	0; 0.4; 0.8
			M24	122	250–450	
			M36	152	410–580	
2	Cast-in-place anchors	Bolts with anchor plate, made according to [13] (Fig. 2b)	M20	82	120–220	0; 0.4; 0.8
			M30	122	210–270	
			M48	152	340–430	
3	Cast-in-place anchors	Anchors with bar hook, made according to [14] (Fig. 2c)	M10	82	90	0; 0.4; 0.8
			M16	122	140–200	
			M24	152	220–290	
			M36	152	320–480	
4	Post-installed anchors	Mechanical undercut anchors Hilti HDA-T, made according to the manufacturer's documentation (Fig. 2d)	M12	22.3	125	0; 0.4; 0.8
			M16	30.3	190	
			M20	37.4	250	
5	Post-installed anchors	Mechanical expansion wedge anchor Hilti HST3, made according to the manufacturer's documentation (Fig. 2e)	M10	10.3	60	0; 0.4; 0.8
			M16	16.3	85	
6	Post-installed anchors	Bonded cementitious adhesive anchor Hilti HIT-HY200 with threaded studs, made according to the manufacturer's documentation (Fig. 2f)	M16	18.3	60	0; 0.4; 0.8
7	Post-installed anchors	Epoxy resin bonded anchor Hilti HIT-RE500V3 with threaded studs, made according to the manufacturer's documentation (Fig. 2f)	M12	14.3	70	0; 0.4; 0.8
			M20	22.4	100	
			M24	28.4	120	

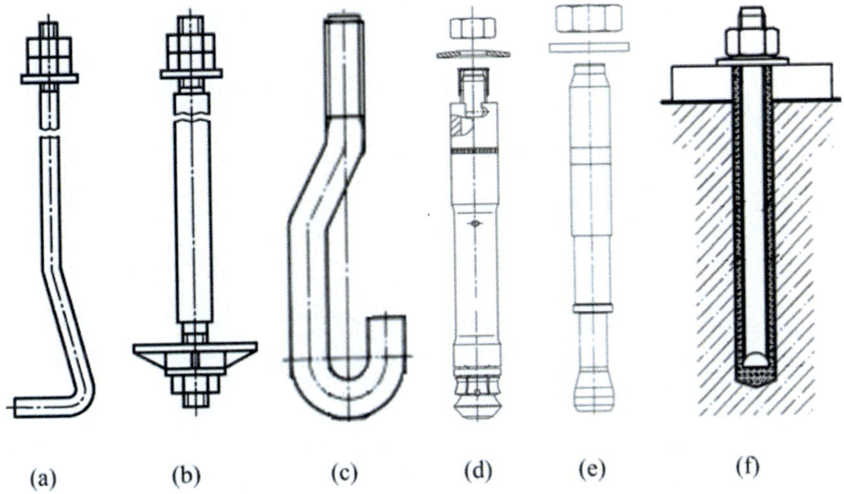

Fig. 2 Types of tested anchors: **a** bent anchors; **b** anchors with an anchor plate; **c** anchors with a bar hook; **d** undercut anchors; **e** wedge expansion anchors; **f** bonded anchors

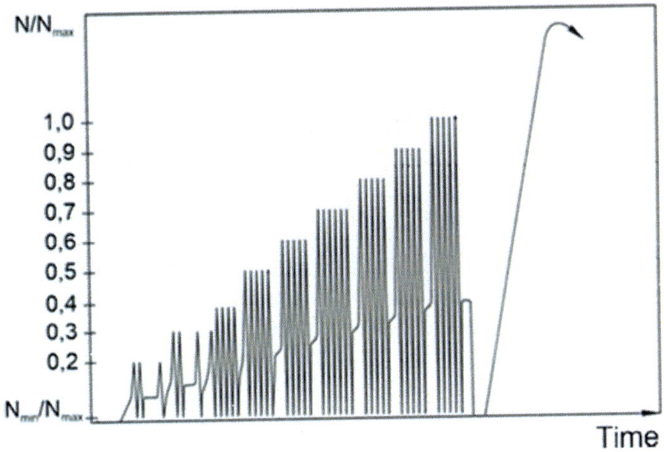

Fig. 3 Cyclic loading pattern

3 Results and Discussion

Table 2 shows test results for different types of tested anchors ($N_{u,st}$)—mean value of ultimate pullout load acquired in static tests with crack width $\Delta w = 0.8$ mm; v ($N_{u,st}$)—variation coefficient for $N_{u,st}$; $N_{u,seism}$—mean value of ultimate pullout load acquired in dynamic multi-cyclic tests with crack width $\Delta w = 0.8$ mm; v ($N_{u,seism}$)—variation coefficient for ($N_{u,seism}$). Figure 5 shows failure mechanisms for different types of anchors.

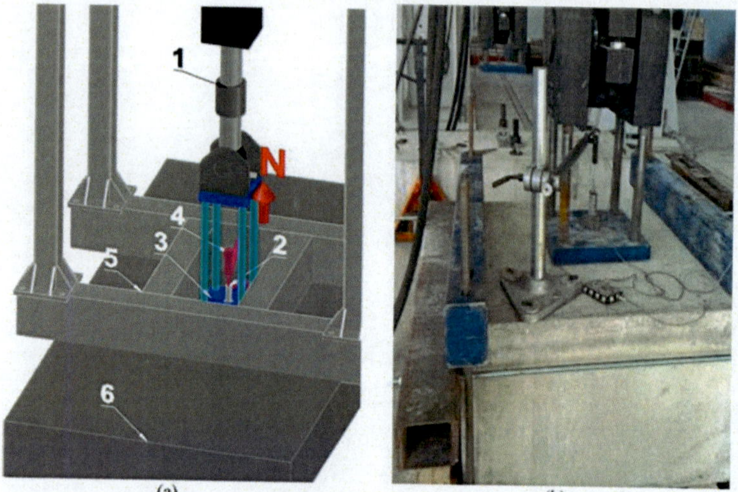

(a) (b)

Fig. 4 Test setup: **a** 1—hydraulic cylinder, 2—anchor, 3—fastened element, 4—displacement gauge, 5—loading frame, 6—concrete base; **b** anchor specimen during tests

Figure 6 shows the comparison of the ultimate pullout load featuring different values of the base crack opening width (0, 0.4, 0.8 mm) and loading patterns (static and dynamic multi-cyclic) depending on the anchor type (the load is presented as the percentage relative to the ultimate load level, applied to the anchor installed in uncracked concrete). The values of the ultimate load for anchors in case of both uncracked concrete and cracked concrete with crack width of 0.4 mm, are taken from article [16].

In case of steel failure cracks in the concrete and dynamic loads do not have significant effect on anchor bearing capacity. Its reduction does not exceed 6% as seen in Fig. 6. In case of concrete cone and slippage failure cracks in concrete and dynamic loads have a major effect on anchor bearing capacity. Its reduction was up to 26%. In case of failure at the interface between the adhesive composition and the base bearing capacity reduction depends on type of adhesive compound—for cementitious adhesive anchors decrease in bearing capacity reaches 50%, while for epoxy resin decrease is only up to 34%. This corresponds with different load-transfer mechanisms from concrete base to anchor for different types of adhesives. For cementitious adhesive anchors main load-transfer mechanism is micro-keying between adhesive compound and concrete base, whereas for epoxy-resin main load-transfer mechanism is adhesion of epoxy resin to concrete base.

Variation coefficient for ultimate pullout load in dynamic tests is on average 1.6 higher than for static tests in the same width of crack as Table 2 shows. Variation coefficient corresponds with failure mechanism and reaches values of 5.2% in case of steel failure, 13.1% in case of concrete cone failure, 11.3% in case of slippage failure and 17.4% in case of failure at the interface between the adhesive composition and the base. High variation coefficient for concrete cone failure can refer to higher scatter of

Table 2 Test results

Anchor type	d, mm	h_{ef}, mm	Steel strength, MPa	Concrete class	Failure mechanism	$N_{u,st}$, kN	$v(N_{u,st})$, %	$N_{u,seism}$, kN	$v(N_{u,seism})$, %
1	16	150	450	C25/30	Concrete cone failure	59.4	5.2	53.1	9.7
	16	200	450	C25/30	Steel failure	78.7	2.4	75.8	3.8
	24	250	450	C25/30	Concrete cone failure	151.4	6.1	139.3	10.4
	24	300	450	C25/30	Steel failure	160.7	1.9	156.1	2.9
	36	370	450	C25/30	Concrete cone failure	361.3	7.2	303.5	11.5
	36	440	450	C25/30	Steel failure	373.4	2.3	360.2	3.0
2	20	120	450	C25/30	Concrete cone failure	99.5	5.7	87.5	8.9
	20	180	450	C25/30	Steel failure	111.5	1.4	108.2	2.3
	30	210	450	C25/30	Concrete cone failure	206.3	6.3	196.0	10.1
	30	270	450	C25/30	Steel failure	266.3	1.8	256.3	2.7
	48	340	450	C25/30	Concrete cone failure	636.0	6.8	551.4	12.4
	48	430	450	C25/30	Steel failure	662.9	2.1	652.5	3.4
3	10	70	450	C25/30	Concrete cone failure	26.4	5.7	23.8	13.1
	10	90	450	C25/30	Steel failure	28.5	3.4	27.1	4.2
	16	140	450	C25/30	Concrete cone failure	69.8	5.6	62.9	12.4
	16	170	450	C25/30	Steel failure	78.6	3.7	75.3	4.7
	24	220	450	C25/30	Concrete cone failure	166.5	6.1	147.2	11.8
	24	270	450	C25/30	Steel failure	176.8	3.5	170.2	4.9
	36	320	450	C25/30	Concrete cone failure	305.0	7.2	274.8	10.3
	36	410	450	C25/30	Steel failure	410.2	4.1	399.4	5.2
4	12	125	Bolt 8.8	C25/30	Steel failure	74.3	2.6	72.7	4.8

(continued)

Table 2 (continued)

Anchor type	d, mm	h_{ef}, mm	Steel strength, MPa	Concrete class	Failure mechanism	$N_{u,st}$, kN	$v(N_{u,st})$, %	$N_{u,seism}$, kN	$v(N_{u,seism})$, %
	16	190	Bolt 8.8	C25/30	Steel failure	135.7	3.3	130.9	5.1
	20	250	Bolt 8.8	C25/30	Concrete cone failure	209.7	5.2	189.7	7.1
	20	250	Bolt 8.8	C30/37	Steel failure	219.7	1.2	218.0	3.1
5	10	60	Bolt 8.8	C25/30	Slippage + concrete cone failure	16.1	8.7	13.8	10.4
	10	60	Bolt 8.8	C30/37	Slippage	21.4	7.2	18.5	9.7
	16	85	Bolt 8.8	C25/30	Slippage + concrete cone failure	54.2	9.1	46.9	11.3
	16	85	Bolt 8.8	C30/37	Slippage	63.6	7.6	55.7	10.1
6	16	60	Bolt 8.8	C30/37	Failure at the interface between the adhesive composition and the base	45.2	11.4	35.9	15.3
7	12	70	Bolt 8.8	C30/37	Failure at the interface between the adhesive composition and the base	49.7	11.5	46.1	15.5
	20	100				102.4	10.8	93.0	16.0
	24	120				152.9	9.6	137.3	17.4

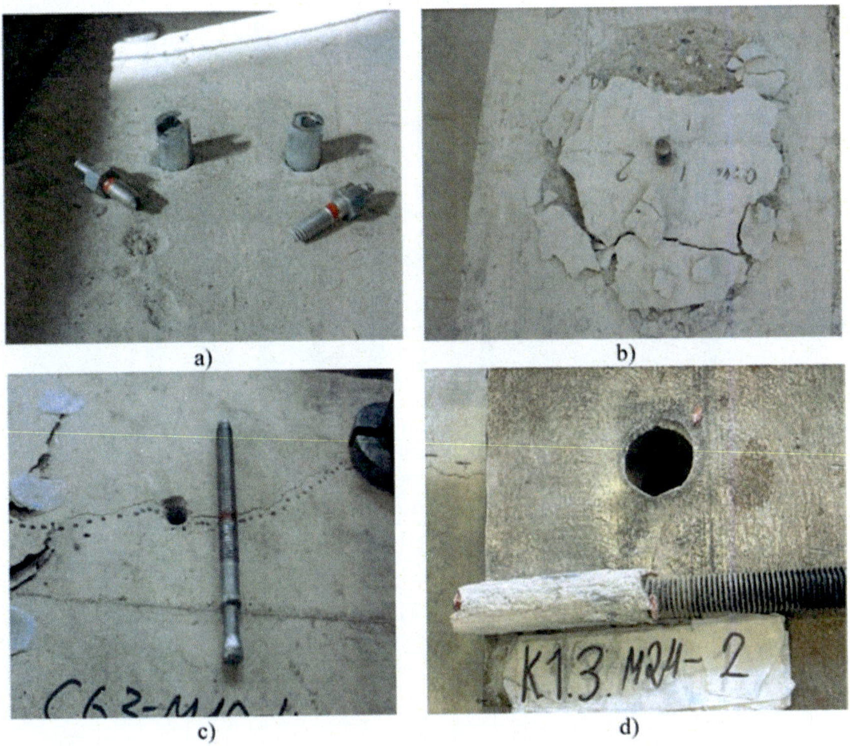

Fig. 5 Anchor failure mechanisms: **a** steel failure; **b** concrete cone failure; **c** slippage failure; **d** failure at the interface between the adhesive composition and the base

concrete ultimate tension loads. High variation coefficient for interface failures can refer to higher number of factors that influence bearing capacity of anchors compared to other failure cases such as crack route in concrete base and quality of hole cleaning before anchor embedment. Acquired values of variation coefficients should be taken into account for calculation of safety factors for different types of anchors under seismic loads action.

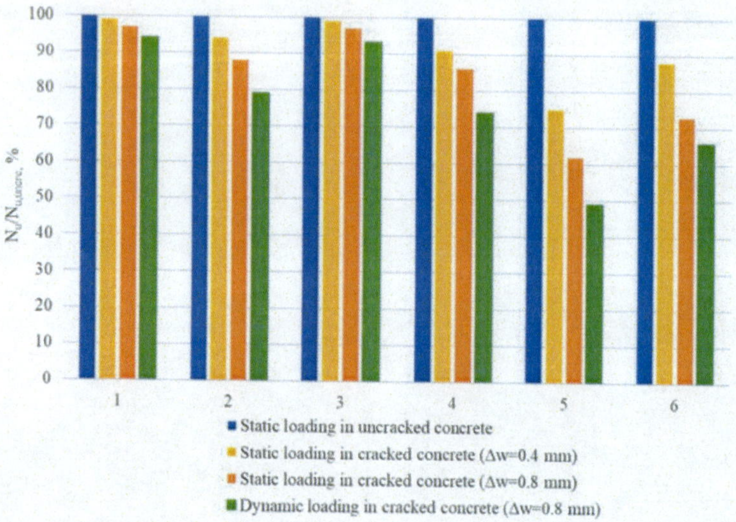

Fig. 6 Reduction of ultimate pullout load for anchors depending on the crack width and loading pattern: 1—cast-in-place anchors in case of steel failure; 2—cast-in-place anchors in case of concrete failure; 3—mechanical undercut anchors; 4—mechanical expansion wedge anchors; 5—bonded cementitious adhesive anchors; 6—epoxy resin bonded anchors

4 Conclusions

The results of the research show that decrease in bearing capacity in cracked concrete under multi-cyclic dynamic loads should be taken into account during design of anchorage under seismic loads depending on failure mechanisms, e.g. for steel failure reduction of bearing capacity does not exceed 6% whereas for failure at the interface between the adhesive composition and the base it can reach up to 50%. Values of variation coefficient depend on type of failure mechanisms, e.g. for steel failure its value does not exceed 5.2%, whereas for interface failures it can reach up to 17.4%. Acquired values of variation coefficients should be taken into account for calculation of safety factors for different types of anchors under seismic loads action.

Acknowledgements This work was financially supported by the Ministry of Science and Higher Education of the Russian Federation (grant # 075-15-2021-686). All tests were carried out using research equipment of The Head Regional Shared Research Facilities of the National Research Moscow State University of Civil Engineering.

References

1. Chen Z (2021) A review of current research progress on tensile behavior of expansion anchors in concrete. Structures 34:2276–2287. https://doi.org/10.1016/j.istruc.2021.08.128

2. Eligehausen R, Mallée R, Silva J (2012) Anchorage in concrete construction. Ernst und Sohn. ISBN 9783433011430
3. Karmokar T, Mohyeddin A, Lee J, Paraskeva T (2021) Concrete cone failure of single cast-in anchors under tensile loading – a literature review. Eng Struct 243. https://doi.org/10.1016/j.engstruct.2021.112615
4. Shafei E, Tariverdilo S (2021) Seismic pullout behavior of cast-in-place anchor bolts embedded in plain concrete: damage plasticity based analysis. Structures 34:479–486. https://doi.org/10.1016/j.istruc.2021.07.085
5. Mahrenholtz P, Eligehausen R (2016) Anchor displacement behavior during simultaneous load and crack cycling. ACI Mater J 113:645–652. https://doi.org/10.14359/51689109
6. Hoehler MS (2006) Behavior and testing of fastening to concrete for use in seismic applications. Ph.D. thesis, IWB University
7. Mahrenholtz P (2012) Experimental performance and recommendations for qualification of post-installed anchors for seismic applications. Ph.D. thesis, IWB University
8. Hoehler MS, Mahrenholtz P, Eligehausen R (2011) Behavior of anchors in concrete at seismic-relevant loading rates. ACI Struct J 108:238–247
9. Mahrenholtz P, Eligehausen R (2015) Post-installed concrete anchors in nuclear power plants: performance and qualification. Nucl Eng Des 287:48–56. https://doi.org/10.1016/j.nucengdes.2015.03.004
10. Rodriguez M, Lotze D, Gross JH, Zhang Y-G, Klingner RE, Graves HL (2001) Dynamic behavior of tensile anchors to concrete. ACI Struct J 98:511–524
11. Perelmuter A, Kabantsev O (2020) On conceptual provisions of design standards for earthquake-resistant construction. Vestnik MGSU 15:1684. https://doi.org/10.22227/1997-0935.2020.12.1673-1684
12. Federal Emergency Management Agency FEMA 273. NEHRP guidelines for the seismic rehabilitation of buildings
13. Standardinform Russian State Standard GOST 24379.1-2012 Foundation Bolts. Structure and dimensions
14. ANSI DIN 529:2010 Masonry and Foundation Bolts
15. European organisation for technical approvals guideline for European technical approval of metal anchors for use in concrete ETAG 001 annex E
16. Shuvalov A, Gorbunov I, Kovalev M (2020) Anchorage strength and ductility at various loading conditions. Proc IOP Conf Ser Mater Sci Eng 869

Modified Clay Gypsum Materials for Facade Systems

Andrey Ushakov, Anton Pilipenko, Maria Kaddo, and Marat Asamatdinov

Abstract Existing trends in the development of technology for gypsum-containing materials include expansion of their application possibilities in systems of façade coatings, the usage of by-products of other industries and local gypsum-containing raw materials. Considering the low durability of gypsum materials in wet operating conditions, system solutions are being developed based on the use of various methods of gypsum modification, including the usage of polymers. The purpose of the research described in the article was to develop a method for selecting the composition of modified clay-gypsum binder, optimizing its compositions and assessing the properties of the materials obtained under conditions of high humidity. The experiment to assess the influence of the composition of the modified clay-gypsum binder on its properties was carried out according to the general methodology of planning the experiment and analytical optimization of its results. The research results showed that the compressive strength of the clay gypsum samples with 15% modified melamine–formaldehyde resin in-creases by 15% after 80 days of storage in air. Products made of modified clay-gypsum have a fairly high frost resistance of 75 cycles and a softening coefficient of 0.95. The vapor permeability of the material is 0.088 mg/(m·h·Pa), which determines the favourable moisture regime of the load-bearing walls with lining with this material.

Keywords Gypsum binder · Clay gypsum · Analytical optimization · Moisture resistance · Softening coefficient

A. Ushakov · A. Pilipenko (✉) · M. Kaddo
Moscow State University of Civil Engineering, Yaroslavskoye shosse 26, Moscow 129337, Russian Federation
e-mail: pilipenko.ans@gmail.com; pilipenkoas@mgsu.ru

M. Asamatdinov
Karakalpak State University, Ch. Abdirov Street 1, Nukus, Republic of Karakalpakstan 230112, Uzbekistan

P. Akimov et al. (eds.), *Proceedings of FORM 2022*, Lecture Notes in Civil Engineering 282, https://doi.org/10.1007/978-3-031-10853-2_42

1 Introduction

Gypsum materials and products are mainly used for interior decoration. For this purpose, gypsum plasterboard or gypsum fiber sheets, stucco decoration, self-leveling floors, various dry plaster, putty and adhesive gypsum mixtures, etc., are used. Existing trends in the development of the technology of gypsum-containing materials include the expansion of their application possibilities, including in the systems of facade coatings, the use of by-products of other industries, as well as local gypsum-containing raw clay-based materials. Taking into account the low durability of gypsum materials in wet operating conditions, system solutions are being developed based on the use of various methods of gypsum modification [1–3]. Methods for increasing the water resistance of gypsum are based on the introduction of various mineral additives, lime, and finely ground slag. Gypsum-cement-pozzolanic and gypsum-slag-cement-pozzolanic binders have been created and comprehensively studied. New technologies have been developed for the production of gypsum binders, composite gypsum binders and water-resistant gypsum binders of low water demand, which can be effective in monolithic construction and allow work in the winter without heating [4–6]. Noteworthy are studies aimed at the use of secondary products from other industries, and, in particular, modified phosphogypsum in gypsum-containing materials technologies [7–9]. Studies of the possibility of using clay gypsum as the main raw material for the manufacture of plaster mixes and products have shown a positive result.

The studies carried out by the authors suggest that the clay-gypsum binder may well be considered as an independent type of air binder, and under certain modification conditions and hydraulic hardening. Clay gypsum belongs to natural composites containing sufficient mass fractions of calcium sulfates and aluminosilicates. The properties of such composite systems, including those based on gypsum, are determined by the type of structuring additives and the peculiarities of their technological processing [10, 11].

Clay gypsum binder is a product obtained by moderate firing of natural clay gypsum to convert the dihydrate gypsum contained in it into semi-aqueous form. The next components of clay gypsum in terms of quantitative content are silica oxides (SiO_2) and sesquioxides ($Al_2O_3 + Fe_2O_3$). The content of $CaSO_4{\cdot}2H_2O$ is the most important component of the gypsum, and the higher the percentage of gypsum content, the higher the binding properties of the gypsum binder after firing. The content of gypsum in the modified building gypsum to obtain a binder must be at least 40% in terms of $CaSO_4$. With further modification to obtain plaster solutions with the addition of Portland cement in an amount of 10%, a content of at least 30% for $CaSO_4$ is allowed. The optimum heat treatment interval is 180–230 °C.

An effective method for modifying the properties of mineral binders is the introduction of polymers, synthetic resins and their derivatives [12, 13]. Polymers not only make it possible to change the mobility and workability of solutions, but also to regulate the operational characteristics of materials, in particular, to improve the water resistance of gypsum binder and its derivatives. In building materials science, composite building materials are considered as systems that allow, based on the integration of the properties of individual components, to obtain systems with the highest quality indicators. Taking into account the existing experience in the research, a method was adopted for modifying the clay-gypsum binder with various polymers in the form of aqueous solutions or emulsions. The influence of the type of polymer and its consumption on the properties of the composite was studied, as well as the influence of temperature and humidity conditions on these properties.

2 Methods and Materials

To reach the goal of research, it was necessary to study the effect of various polymers in the form of aqueous solutions or emulsions on the properties of the clay-gypsum-based composite. For this purpose, preliminary tests were made using different poly-condensation resins of various origins in forms of aqueous and organic solvent-based solutions. Tests shown that using polycondensation resins, the curing of which results in the release of low molecular weight products, water in particular, could be preferable to obtain dense structure of clay-gypsum matrix. For the chemical binding of the released water, a structuring additive based on polyisocyanates was introduced into the composition. An experiment to assess the effect of the composition of the modified plaster of clay-gypsum binder on its properties was carried out according to the general methodology for planning the experiment and analytical optimization of its results [14–16] based on the matrix of a full quadratic three-factor experiment. The resulting regression equations were tested against all statistical hypotheses. On the fact of com-parison with confidence intervals (Δ_{bj}), only significant factors were left, and as a result of checking by Fisher's criterion, a conclusion was made about the adequacy (or inadequacy) of the obtained models.

In the experiment, the following factors were taken as variable factors: the strength of the hardened clay gypsum binder (X_1), the consumption of melamine–formaldehyde resin (X_2) and the consumption of the structuring additive (X_3). The strength of the gypsum polymer after 7 days of specimen hardening (Y_1) was taken as a response function, and the softening coefficient of the gypsum polymer samples was taken as the optimization parameter according to the results of climatic tests (Y_2). The experimental conditions are given in Table 1.

Table 1 Intervals of variation of factors

Factor's name	Coding symbol, Xi	Average factor value, $\overline{X}_i$	Variation interval, ΔX_i	Factor values at variation levels	
				−1	+1
Hardened clay gypsum strength, MPa	X_1	0.5	0.2	0.3	0.7
Consumption of thermosetting resin, %	X_2	13	3	10	16
Consumption of structuring additive, %	X_3	0.6	0.2	0.4	0.8

3 Results and Discussion

Mathematical processing of the experimental results made it possible to obtain the regression equations for the compressive strength of the modified clay gypsum specimens (Y_1) and its softening coefficient (Y_2). The significance of the coefficient was checked by confidence intervals, respectively, the confidence interval for strength was $\Delta b_1 = 0.6$ MPa, and the softening coefficient was $\Delta b_2 = 0.02$. The following mathematical models (algebraic polynomials) are obtained:

- for compressive strength:

$$Y_1 = 32 + 7X_1 + 3X_2 + 3X_3 + 2X_1X_2 + X_1X_3$$

- for the softening coefficient:

$$Y_2 = 0.74 + 0.04X_1 + 0.10X_2 + 0.05X_3 - 0.03X_1X_2 - 0.03X_3^2$$

The resulting models were tested for adequacy according to Fisher's criterion. The calculated values of the Fisher's criterion are equal for the compressive strength model $F_1 = 16.2$ and for the model for the softening factor $F_2 = 15.9$. The tabular values of the criteria, respectively, are 19.2 and 19.3. The calculated values of the F-criterion do not exceed the tabular one, and with the corresponding confidence level (98%) the model can be considered adequate. This fact will be taken into account in the analytical optimization of mathematical models.

The analysis of the coefficients of the equation $Y_1 = f_1(X_1, X_2, X_3)$ shows that the strength of the gypsum polymer increases with an increase in the values of all factors (in the regions of their definition). The strength of gypsum has the greatest effect on strength (coefficient at X_1 is equal to 15). The influence of the consumption of resin and structuring is also significant (the coefficients at X_2 and X_3 are equal to 7 and 6, respectively). At the same time, there are significant paired influences, for X_1X_2 and X_1X_3. This behavior of the factors makes it possible to assume the possibility of synergistic effects of the combined influence of the strength of gypsum and the consumption of resin and structuring additive, which suggests the development of further research in this direction.

Analysis of the coefficients of the equation $Y_2 = f_2(X_1, X_2, X_3)$ shows that the greatest influence on the increase in the softening coefficient is exerted by the consumption of thermosetting resin (coefficient at X_2 equal to 0.14). The strength of gypsum affects the optimization parameter significantly, but not significantly (coefficient at X_1, equal to 0.06). With an increase in the consumption of the structuring additive, first there is an increase in the softening coefficient, and then, at high flow rates, a decrease is observed (coefficients at X_3 and X_3^2, equal to 0.09 and –0.05, respectively). This allows us to assume that the function $Y_1 = f_1(X_1, X_2, X_3)$ has a local extremum along X_3,. and analytical optimization can be applied. The "–" sign in the case of a pair interaction X_1X_2 suggests the possibility of an antagonistic effect based on the insignificant influence of the gypsum strength on the softening coefficient.

The research methodology is based on the cybernetic system, called the "black box". In this regard, the evaluation of the results is limited only to external interactions and obtaining dependencies connecting the result and the variable factors; the internal processes occurring in such systems are not considered. Analysis of a polynomial describing the relationship between compressive strength. The softening coefficient and variable factors show that this function (which is essentially a function of several variables) for one of these variables, namely, the polymer consumption (X_3), has a local extremum. Therefore, we can use the mathematical apparatus of analytical local optimization.

Analytical optimization is based on the fact that the functions for strength and density $Y_1 = f_1(X_1, X_2, X_3)$ and $Y_2 = f_2(X_1, X_2, X_3)$ are mathematical and methods of mathematical analysis can be applied to them, provided that the condition of adequacy is not violated. In the case under consideration, the following scheme is adopted:

- the equation $Y_2 = f_2(X_1, X_2, X_3)$ is differentiated by X_3 and equated to zero, determining the extremum of the function Y_2 by X_3;
- solve the functions $Y_1 = f_1(X_1, X_2, X_3)$ and $Y_2 = f_2(X_1, X_2, X_3)$ with $X_3 = \text{opt}$ and carry out local optimization.

Analytical optimization includes the following sequence of actions:

1) Determine the value of the local extremum of the function $Y_2 = f_1(X_1, X_2, X_3)$ by X_3:

$$\frac{\partial Y_2}{\partial X_3} = 0.09 - 0.10X_3 = 0 \rightarrow X_3 \frac{0.09}{0.1} = 0.9$$

2) Calculation of the natural value of the plasticizer consumption (corresponding to the possible obtaining of the maximum compressive strength of polymer concrete), using the decoding factor formula and the experimental conditions (Table 1): Consumption of structuring additive $= 0.6 + 0.2 \times 0.9 = 0.78\%$

3) Calculation of mathematical models (polynomials) for the optimized value of the factor X3 = 0.9:

- for compressive strength:

$$Y_1 = 35 + 8X_1 + 3X_2 + 2X_1X_2$$

- for the softening coefficient:

$$Y_2 = 0.77 + 0.04X_1 + 0.10X_2 - 0.03X_1X_2$$

The graphical interpretation of the obtained dependencies consists in the representation of the obtained dependences $Y_2 = f_3(X_1, X_2)$ with an optimized value of X_3 in the Cartesian coordinate system (X_1, X_2), which makes it possible to evaluate the influence of factors on the properties (softening coefficient) of the modified gypsum plaster and develop a nomogram for assessing material properties and determining the consumption of polycondensable thermosetting resin.

Graphical interpretation of the dependence $Y_2 = f_3(X_1, X_2)$ with an optimized value of X_3 equal to 0.78% with softening coefficient values: 1—0.6; 2—0.7; 3—0.8; 4—0.9 (see Fig. 1).

The method for modifying the composition of the clay-gypsum mixture with water-soluble polymers has a number of advantages. The introduction of organic additives into the clay-gypsum mixture leads to creation of a framework of crystalline intergrowths of dihydrate, upon hydration of gypsum with aluminum–silicate phase in composition. In the same time, curing resin forms a continuous polymer protective surface around gypsum crystalline framework and both components form matrix of the material. The pores in the gypsum body are filled with vitreous-like

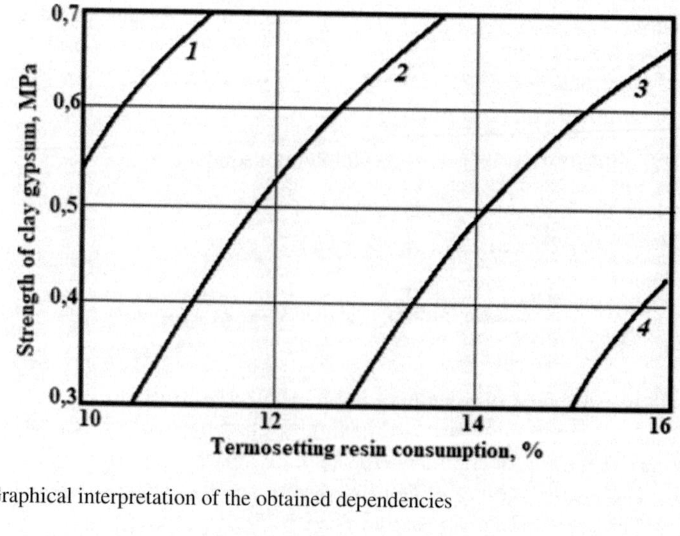

Fig. 1 Graphical interpretation of the obtained dependencies

substance. Formed protective screen of a polymer film around the gypsum crystals prevents water from reaching the highly soluble calcium sulfate, water and vapor permeability of the gypsum-based material is significantly reduced. It was proven that the compressive and bending strength of samples with addition of 15% modified melamine–formaldehyde resin (with 80 days of storage in air) increases by ~15 and ~10%, respectively. Clay-gypsum-based samples with addition of 20% melamine–formaldehyde resin could withstand ~75 cycles of freezing and thawing. The vapor permeability of clay-gypsum-polymer is equal to 0.088 mg/(m·h·Pa), which determines a favorable moisture regime for load-bearing walls made of brickwork with lining from this material.

4 Conclusions

Conducted research shows that developed decorative materials based on modified clay-gypsum binder could become effective materials not only for interior design, but could also be used on facade of buildings under construction and renovation.

To obtain an effective clay-gypsum binder, the content of gypsum in the modified clay gypsum must be at least 40% (calculated on $CaSO_4$). With further modification to obtain plaster solutions with the addition of Portland cement in an amount of 10%, a content of at least 30% calculated on $CaSO_4$ is allowed. The optimal heat treatment interval is 180–230 °C. Plasters based on modified clay-gypsum binder have a sufficiently long setting interval and can be applied both manually and by machine.

It was found that the compressive and bending strength of samples with addition of 15% modified melamine–formaldehyde resin (with 80 days of storage in air) increases by ~15 and ~10%, respectively. Clay-gypsum-based samples with addition of 20% melamine–formaldehyde resin could withstand ~75 cycles of freezing and thawing. The vapor permeability of clay-gypsum-polymer is equal to 0.088 mg/(m·h·Pa), which determines a favorable moisture regime for load-bearing walls made of brickwork with lining from this material. The effective field of application of developed clay-gypsum-based materials could be expanded to interior materials in Southern regions of Russian Federation or Karakalpak area of Uzbekistan Republic, where sufficient amounts of raw material are located.

Acknowledgements This work was financially supported by the Ministry of Science and Higher Education of the Russian Federation (grant # 075-15-2021-686). All tests were carried out using research equipment of The Head Regional Shared Research Facilities of the National Research Moscow State University of Civil Engineering.

References

1. Buryanov AF (2005) Build Mat 9:46–48
2. Ferronskaya AV, Korovyakov VF, Melnichenko SV, Chumakov LD (1992) Build Mat 5:34–35
3. Panchenko AI, Buryanov AF, Kozlov NV, Soloviev VG, Pashkevich SA (2014) Build Mat 12:72–74
4. Korolev EV, Danilov AM, Garkina IA (2006) Build Mat 7:55–57
5. Pustovgar AP, Buryanov AF, Vasilik PG (2010) Build Mat 12:61–64
6. Sapelin AN, Zhukov AD, Bessonov IV, Naumova NV, Chkunin AS (2014) Italian Sci Rev 11:155–157
7. Bozhenov PI, Meshheryakov YuG (1976) Einflub der Beimengungen and die technischen Eigenschaften son Gipsbinderu. Int Baustoff Sieikattagung Weimar
8. Meshheryakov YuG, Fedorov SV (2008) Industrielle verarbeitung von Phosphogips and Phospho Energiesparende technologien. Int Kongress Fachmesse Euro ECO. Hannover
9. Yakovlev G, Polyanskikh I, Fedorova G, Gordina A, Buryanov A (2015) Proc Eng 108:13–21
10. Kodzoev M-BKh, Isachenko SL, Kosarev SA, Basova AV, Skvortzov AV, Asamatdinov MO, Zhukov AD (2018) MATEC Web Conf 170:03022
11. Asamatdinov MO, Medvedev AA, Zhukov AD, Zarmanyan EV, Poserenin AI (2018) MATEC Web Conf 193:03045
12. Pyataev ER, Pilipenko ES, Burtseva MA, Mednikova EA, Zhukov AD (2018) IOP Conf Ser Mater Sci Eng 365:032041
13. Pyataev E, Zhukov A, Vako K, Burtseva M, Mednikova E, Prusakova M, Izumova E (2019) E3S Web Conf 97:02032
14. Zhukov A, Shokodko E, Bobrova E, Bessonov I, Dosanova G, Ushakov N (2019) Adv Intel Syst Comput 983
15. Zhukov A, Shokodko E (2020) Adv Intel Syst Comput 1116
16. Bessonov I, Zhukov A, Shokodko E, Chernov A (2020) E3S Web Conf 164:14016

Relationship Between Strength and Deformation Characteristics of High-Strength Self-compacting Concrete

Igor Bezgodov, Simon Kaprielov, and Andrey Sheynfeld

Abstract The paper provides data on the strength and deformation characteristics of heavy self-compacting concrete of classes B30–B100 with a cubic compressive strength of 36.5–114.8 MPa. It has been established that the values of the concrete prism compressive strength (36.2–104.2 MPa) are 42–64% higher than the normalized values given in the building code of the Russian Federation SP 63.13330.2018. The values of the static modulus of elasticity for high-strength concretes of classes B80–B100 are 44.1–48.1 GPa and exceed by 5–12% the values given in SP 63.13330.2018. The ultimate compressive strains of concrete of classes B30–B100 are in the range from 261×10^{-5} to 326×10^{-5} and exceed the value of 200×10^{-5} given in SP 63.13330.2018.

Complete deformation diagrams of self-compacting concretes of classes B30–B100 have been constructed. The nonlinearity of these ones decreases with increasing concrete strength. The descending branch of the σ-ε diagram is observed only for concrete of a class below B55 with a total relative compressive strain of 403.3×10^{-5} under a loading level of $0.85R_b$. Concrete of classes B55-B100 has no descending branch. Previously established dependencies are refined for the analytical description of strains and stresses at any stages of loading structures.

Keywords High-strength self-compacting concrete · Complete strain diagrame · Deformation characteristics · Poisson's ratio · Ultimate compressive strains · Compressive strength · Static modulus of elasticity

I. Bezgodov (✉)
Moscow State University of Civil Engineering, Moscow 129337, Russian Federation
e-mail: ibezgodov52@yandex.ru

S. Kaprielov · A. Sheynfeld
«Research Institute for Concrete and Reinforced Concrete» named after A.A. Gvozdev, JSC Research Center of Construction, Moscow, Russian Federation
e-mail: kaprielov@masterbeton-mb.ru

A. Sheynfeld
e-mail: sheynfeld@masterbeton-mb.ru

P. Akimov et al. (eds.), *Proceedings of FORM 2022*, Lecture Notes in Civil Engineering 282, https://doi.org/10.1007/978-3-031-10853-2_43

1 Introduction

All regulatory documents for the calculation of reinforced concrete structures both in Russia (set of rules SP 63.13330.2018) and abroad (international standard Model Code MC 2010, European standard EN 1992-1-1: 2004 Eurocode 2) are based on the relationship between the strength and deformation characteristics of concrete.

In Russia, the generalized strength parameter is the compressive strength class of concrete determined by the values of cubic (R) and prism (R_b) compressive strength. Among several normalized deformation characteristics of concrete, the most significant is the static modulus of elasticity (E_b), which according to SP 63.13330.2018 "Concrete and reinforced concrete structures. General Provisions", depends mainly on the classes and types of concrete. The remaining deformation characteristics are either depends on the elastic modulus (shear modulus ($G = 0{,}4 \cdot E_b$), or are taken constant (transverse strain coefficient—Poisson's ratio $\nu_b = 0.2$; ultimate strain for axial compression under short-term load $\varepsilon_{b0} = 0.002$).

At the same time, foreign standards give strength and deformation characteristics (except for Poisson's ratio) for each class of concrete. .

An analysis of the existing concrete strain diagrams under compression in Russian and foreign regulatory documents has shown that the ultimate compressive strain of concrete in many countries is taken as a variable. And non-linear dependences of strain diagrams are proposed, considering the ascending and descending branches [1]. In addition, the values on the ascending branch of the corresponding cylindrical strength and the total strain, taking into account the descending branch, are determined for each class.

However, the application of the dependencies given in foreign standards and literature is difficult in Russia, since the methods for determining the main characteristics of concrete have significant differences regarding the shape and size of the samples, the conditions for their maintenance, the speed and discreteness of loading, etc.

A number of studies [2–9] showed that the experimentally obtained dependences between the strength and deformation characteristics of ordinary concrete with a compressive strength not exceeding 40 MPa cannot always be used for high-strength concrete with a strength of 60–120 MPa.

Therefore, the study of the relationship between the strength and the complex of deformation characteristics of both ordinary and high-strength concretes with the construction of complete strain diagrams (including the descending branch) and the refinement of the previously obtained dependencies is an urgent task. This is of particular interest for concretes made of self-compacting mixtures, the volume of which is growing.

The purpose of the research was to determine the relationship between the strength and deformation characteristics of concrete classes from B30 to B100, prepared from self-compacting mixtures, with the refinement of previously obtained dependencies.

To achieve the purpose, the following tasks were solved:

– determination of strength (cubic and prismatic compressive strength) and deformation (static modulus of elasticity, Poisson's ratio, ultimate compressive strains)

characteristics of five compositions of self-compacting concrete of classes B30–B100;
- construction of complete concrete deformation diagrams, including the descending branch;
- refining of the previously established in [8, 9] dependences of the deformation characteristics of concretes on their prism compression strength;
- comparing of the obtained results with the standard values given in the building code of the Russian Federation SP 63.13330.2018.

2 Materials and Test Methods

Applied Materials

The materials (cement, modifier, microfiller, sand and crushed stone), which satisfy the standards of the Russian Federation and applied in the production of self-compacting concrete mixes for construction projects at the Moscow-City MIBC, were used for the preparation of concrete in laboratory conditions.

The characteristics of the materials were as follows:

- Portland cement CEM I 52.5 N, corresponding to GOST 31,108;
- organomineral concrete modifier MB10-50C A-II-2, including microsilica (45%), fly ash (45%) and superplasticizer (10%) [10], corresponding to GOST R 56,178;
- micro-filler—non-activated mineral powder MP-1 (ground limestone) with a particle size of less than 1.25 mm, corresponding to GOST R 52,129 and GOST R 56,592;
- superplasticizer SikaPlast E4 based on a mixture of modified lignosulfonates and polycarboxylate esters, corresponding to GOST 24,211;
- class I quartz sand with fineness modulus $M_{cr} = 2.55$, corresponding to GOST 8736;
- crushed granite fraction 5–10 mm, corresponding to GOST 8267;
- water for mixing concrete mixtures, corresponding to GOST 23,732.

Compositions and Properties of Concrete Mixes

In laboratory conditions, 5 concrete compositions were prepared using self-compacting mixtures with a cement consumption of 287 to 482 kg/m^3 with the addition of MB modifier and micro-filler at a water-binding ratio W/(C + MB) from 0.25 to 0.69.

The compositions of self-compacting concrete mixtures are presented in Table 1.

Table 1 Composition of self-compacting concrete mixtures

No composition	Composition of self-compacting concrete mixes, kg/m^3						
	C	MB	MP-1	FA	CA	SP	W
1	287	–	148	822	871	3,53	198
2	305	29	167	836	836	–	177
3	349	65	150	818	838	–	161
4	423	70	101	826	846	–	161
5	482	131	50	733	904	–	151

Note C—Portland cement; MB—modifier; MP-1—microfiller; FA—sand; CA—crushed stone; SP—superplasticizer; W—water

Concrete mixtures were prepared in a 25-L forced-action mixer with mixing of each batch for 5 min. The results of tests of concrete mixtures showed that their mobility, determined by the spread of a normal cone [11], is in the range from 55 to 65 cm.

Object of Research and Test Methods

Samples were formed from the prepared concrete mixtures: 3 cubes with a size of 100 × 100 × 100 mm to determine the cubic compressive strength of concrete (R) according to Russian standards GOST 10,180 and GOST 31,914; 6 prisms with a size of 100 × 100 × 400 mm to determine the prism compressive strength of concrete, the static modulus of elasticity, Poisson's ratio according to the standards of the Russian Federation GOST 24,452 and GOST 31,914; 3 prisms with size of 70.7 × 70.7 × 280 mm to establish complete strain diagrams of concrete according to the method [13, 14].

Control specimens were stored for 28 days under normal conditions (temperature plus 20 ± 2 °C, humidity 95 ± 5%) before testing. The loading of the prisms was carried out in steps equal to 0, $1R_b$ with holding at each step for 5 min until the destruction of the specimens. The static modulus of elasticity and Poisson's ratio were determined at a loading level of 30–40% of the prism strength.

3 Test Results and Discussion

Table 2 presents concrete test results in terms of: concrete class (B), cubic (R) and prism (R_b) compressive strength, static modulus of elasticity (E_b), Poisson's ratio (ν_b), ultimate compressive strain (ε_{b0}). And Fig. 1 shows complete strain diagrams of concrete.

Compressive Strength

The cubic compressive strength (R) for all concretes at the age of 28 days is in the range from 36.5 to 114.8 MPa and corresponds to concrete compressive strength

Table 2 Strength and short-term deformation characteristics of concrete

No	Strength and deformation characteristics of concrete						
	B, MPa	R, MPa	R_b, MPa	E_b, MPa	ν_b	$\varepsilon_{b0} \times 10^5$	$\varepsilon_{b0}{}^{-0,85} \times 10^5$
1	B30	36.5	36.2	32.5	0.206	261	403
2	B55	61.8	56.0	39.2	0.214	276	–
3	B70	80.2	74.5	40.5	0.198	299	–
4	B80	92.5	85.3	44.1	0.205	294	–
5	B100	114.8	104.2	48.1	0.230	326	–

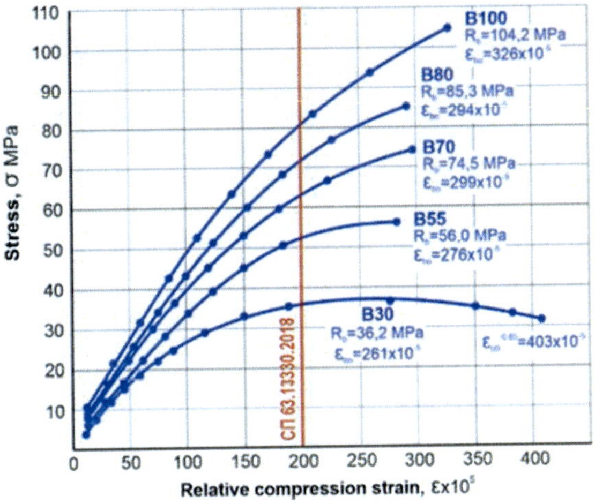

Fig. 1 Complete concrete compressive strain diagrams σ–ε

classes from B30 to B100. The prismatic compressive strength of concrete (R_b) at the age of 28 days is in the range from 36.2 to 104.2 MPa. Evaluation of the above results according to the criterion of the prism strength coefficient, determined by the ratio of the prismatic compressive strength of concrete to cubic strength ($K_{pp} = R_b/R$) shows that its actual values are in the range from 0.91 to 0.99 and significantly exceed the values of this coefficient calculated according to the parameters given in the building code of Russian Federation SP 63.13330.2018 (from 0.71 to 0.73).

Static Modulus of Elasticity

The static modulus of elasticity of concretes of classes B30–B100 with prismatic strength from 36.2 to 104.2 MPa is in the range from 32.5 to 48.1 GPa (see Table 2). Figure 2 shows the dependence between the static modulus of elasticity and the prismatic compressive strength of concrete.

The obtained results (see Fig. 2) show that the static modulus of elasticity of concrete is proportional to the prismatic strength and can be determined by the corrected formula [8] in the form:

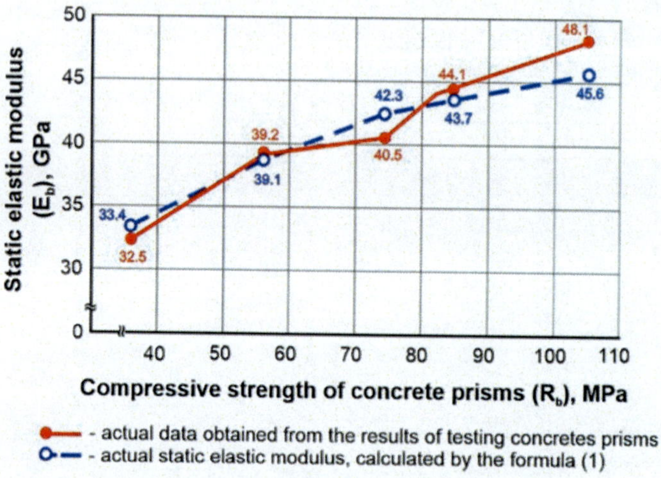

Fig. 2 Static elastic modules vs compressive strength of concretes prisms

$$E_b = \frac{52000 \cdot R_b}{23 + 0.92 R_b} \tag{1}$$

Comparison of the calculated results and those obtained experimentally shows (see Fig. 2) that this formula can be used to determine the static modulus of elasticity of concrete, since the calculation error does not exceed 5%.

Comparison of the experimentally obtained values of E_b with normalized values showed that the static modulus of elasticity of high-strength concretes of classes B80-B100 exceeds by 5 … 14% the values given in SP 63.13330.2018, corresponds to Model Code MC 2010 and is consistent with previously obtained results [4–6, 15, 16].

Poisson's Ratio

Poisson's ratio of heavy concretes of classes B30–B100 is in the range from 0.198 to 0.230 (see Table 2) and, in general, corresponds to the normalized value of the coefficient of transverse deformations $\nu_{b,P} = 0.2$ according to the building code of the Russian Federation SP 63.13330.2018.

Ultimate Compressive Strain

Ultimate compressive strains of concrete of classes B30–B100 with prismatic strength from 36.2 to 104.2 MPa are in the range from 261×10^{-5} to 326×10^{-5} (see Table 2) and exceed the value of 200×10^{-5} given in SP 63.13330.2018. The complete strain diagrams of concrete under compressive show (see Fig. 1) that the descending branch is observed only in class B30 concrete with a prism strength of 36.2 MPa, where the total relative compressive strain was 403.3×10^{-5} at the level of $0.85R_b$ of the descending branch. For concretes with higher strength, there is no descending branch, which corresponds to the results obtained in [9].

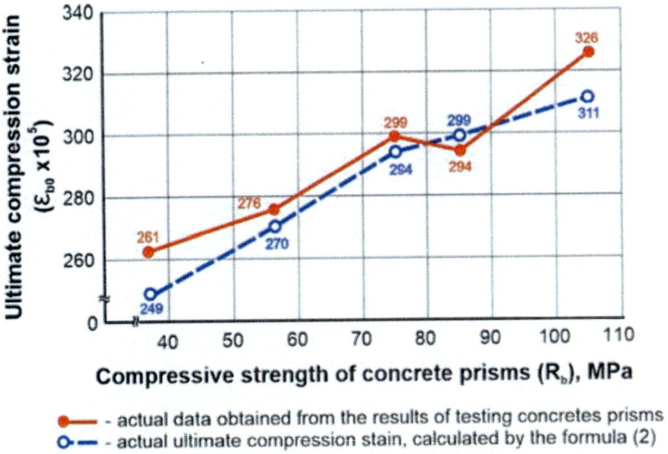

Fig. 3 Ultimate compressive strain vs compressive strength of concretes prisms

Figure 3 shows the dependence between the ultimate strain and the prismatic compression strength of concrete.

The obtained results (see Fig. 3) show that the ultimate strain of concrete increase with an increase in prismatic strength and can be determined by the corrected formula [8, 9] in the form:

$$\varepsilon_{b0} = 0.024 \sqrt[3]{\frac{R_b}{E_b}} \tag{2}$$

Comparison of the calculated results and those obtained experimentally shows (see Fig. 3) that this formula can be used to determine the limiting relative deformations of concrete, since the calculation error does not exceed 5%.

Relationship Between Strains and Stresses

It is necessary to determine the values of strains or stresses at any stage of loading or deformation of structures when calculating or testing.

For an analytical description of relative strains and stresses at any stages of loading, it is proposed to use the refined equations [8, 9] in the form:

$$\varepsilon_\eta = \varepsilon_{b0} \cdot \left(1 \pm \sqrt[n]{1 - \frac{\sigma}{R_b}}\right) \tag{3}$$

$$\sigma_\eta = R_b - \left|1 - \frac{\varepsilon_\eta}{\varepsilon_{b0}}\right|^n \cdot R_b \tag{4}$$

where:

ε_n is strain at a given stress level;

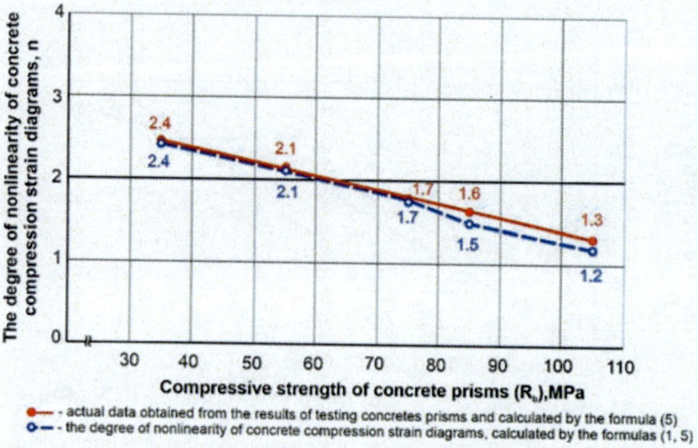

Fig. 4 The degree of nonlinearity of concrete compressive strain diagrams vs compressive strength of concretes prisms

σ_η is the stress at given strain, MPa;
η is the loading level equal to the ratio of stresses to prismatic strength (σ/R_b);
ε_{b0} is ultimate compressive strain;
R_b is the prismatic compressive strength of concrete, MPa;
n is the degree of non-linearity of the concrete strain diagram, which depends on its strength and can be determined by the formula:

$$n = 3.5 - \frac{R_b \cdot 10^3}{E_b} \quad (5)$$

The dependence between the degree of nonlinearity of deformation diagrams and the prism compressive strength of concrete is shown in Fig. 4, from which it can be seen that with increasing strength, the degree of nonlinearity decreases and tends to unity. This confirms the results of previous studies [8, 17].

It should be noted that when determining the strain in the descending branch of the diagram σ–ε, the sign in front of the root of the second polynomial in formula (3) should be replaced from minus to plus.

Comparison of the strainf of self-compacting concretes of classes B30–B100, obtained experimentally, with the results calculated from the analytical dependence (3) showed satisfactory convergence, which makes it possible to use Eqs. (3) and (4) to assess strains and stresses at all stages of loading structures.

4 Conclusions

1. Strength (cubic and prismatic compressive strength) and deformation (static modulus of elasticity, Poisson's ratio, ultimate compressive strains) characteristics of five compositions of self-compacting concretes of classes B30–B100 were determined using standard and special methods.
2. The obtained results show that the values of the strength and deformation characteristics of high-strength self-compacting concrete of classes B80–B100 exceed the standard values given in the building code of the Russian Federation SP 63.13330.2018.
3. Complete strain diagrams of self-compacting concretes of classes B30–B100 are constructed. The nonlinearity of these diagrams decreases as the strength of concrete increases. The descending branch of the $\sigma - \varepsilon$ diagram is observed only for concrete of classes below B55 with a prism strength of 36.2 MPa, while it is absent for concrete of classes B55–B100 with a prism strength of 56.0–104.2 MPa.
4. Previously established dependences for determining the static modulus of elasticity, ultimate compressive strains and analytical expressions for strains and stresses at any stages of loading the structures made of self-compacting concrete of classes B30–B100 have been refined.

References

1. Panfilov DA, Pischulev AA, Dimadetdinov KI (2013) Review of existing diagrams of deformation of concrete under compression in domestic and foreign regulatory documents. Ind Civil Constr (3):34–36
2. Sviridov NV, Kovalenko MG, Chesnokov VM (1991) Mechanical properties of especially strong cement stone. Concr Reinforced Concr (2):7–9
3. Beshr H, Almusallam AA, Maslehuddin M (2003) Effect of coarse aggregate quality on the mechanical properties of high strength concrete. Constr Build Mater (97–103):97–103
4. Kaprielov S, Karpenko N, Sheinfeld N, Kouznetsov E (2003) Influence of multicompo-nent modifier containing silica fume, fly ash, superplasticizer and air-entraining agent on structure and deformability of high-strength concrete. In: Seventh CANMET/ACI International Conference on Superplasticizers and other chemical admixtures in concrete, Berlin, Germany, pp 99–107
5. Kaprielov S, Karpenko N, Sheinfeld A (2004) On controlling modulus of elasticity and creep in high-strength concrete with multicomponent modifier. In: Eighth CANMET/ACI international conference on fly ash, silica fume, slag and natural pozzolans in concrete, Las Vegas, 23–29 May 2004, Supplementary Papers, pp 405–421
6. Karpenko NI, Kaprielov SS, Kuznetsov EN, Sheinfeld AV, Bezgodov IM (2004) Creep measures for high-strength concretes based on MB. RAASN Bull Dept Build Sci (8):203–214
7. Lee BJ, Kee S-H, Oh T, Kim Y-Y (2015) Effect of cylinder size on the modulus of elasticity and compression strength of concrete from static and dynamic tests. Adv Mater Sci Eng 12 p. No. ID 580638
8. Bezgodov IM (2015) On the issue of evaluation, the ultimate compression strain for various classes of concrete. Concr Reinforced Concr (15):9–11

9. Bezgodov IM, Dmitrenko EN (2019) Improvement of curvilinear diagrams of concrete deformation. Ind Civil Constr (8):99–104
10. Kaprielov S, Sheinfeld A (2000) Influence of silica fume/fly ash/superplasticizer combinations in powder-like complex modifiers on cement paste porosity and concrete properties. In: Sixth CANMET/ACI International conference on superplasticizers and other chemical admixtures in concrete, Nice, France, October 2000. Proceedings, pp 383–400
11. EN 12350-8: 2010 Testing fresh concrete - Part 8: Self compacting concrete - Slump-flow test
12. EN 206: 2013 Concrete - Specification, performance, production and conformity
13. Bezgodov IM, Levchenko PYu (2013) On the issue of the method of obtaining complete strain concrete diagrams. Tekhnologii betonov (10):34–36
14. Bezgodov IM (2020) Methodological features of the study of complete strain diagrams and stress relaxation in concrete. Tekhnologii betonov (11–12):39–44
15. Sviridov NV (1991) Extra strong cement concrete. Energy Constr (8):21–29
16. Sviridov NV, Kovalenko MG (1990) Concrete with a strength of 150 MPa on ordinary Portland cements. Concr Reinforced Concr (2):21–22
17. Mkrtchan AM, Aksenov VM (2013) Analytical description of the diagram of deformation of high-strength concretes. Inzhenerny Bull Don (3). http://www.ivdon.ru/magazin/archive/n3y2013/1818.

Optimization of Mixture Compound for Additive Technologies

A. O. Korneeva, B. A. Bondarev, A. A. Kosta, A. A. Meshcheryakov, and A. B. Bondarev

Abstract The planned properties of the material are regulated by changing the formulation and technological factors, the level of knowledge of concrete mixtures and methods of their production, as well as the equipment of the enterprise, is of great importance. Usually, methods of mathematical planning of the experiment are used, thanks to which regression equations are derived for the determined indicators. With this approach, applied tasks are solved and it becomes possible to identify patterns of formation of new materials suitable for use in 3-D printing.

The purpose of the research is to study concrete mixtures of various compositions and their rheological properties, including conditional viscosity, workability, consistency of concrete dough, as well as the selection of the optimal composition of the mixture suitable for additive technologies.

This article presents the results of optimizing the composition of building mixes used for 3-D printing for the Lipetsk region, since it includes blast furnace slag, which is a by-product of Novolipetsk Metallurgical Plant. The resulting mixture, selected with the help of mathematical planning of the experiment, has the necessary rheological properties and optimal setting times and may well be used for the construction of buildings and structures using additive technologies.

In the course of these studies, the optimal composition of the concrete mixture was found, which allows reducing cement consumption to 150 kg/m^3 of the mixture. At the same time, the compressive strength is 51 MPa and allows the use of this mixture for the construction of buildings and structures using additive technologies.

Keywords 3-D printing · Concrete mix · Setting time · Rheological properties · Flowability of concrete · 3-D printer · Additive technologies · Microsilica · Blast furnace slag

---

A. O. Korneeva (✉) · B. A. Bondarev · A. A. Meshcheryakov · A. B. Bondarev
Federal State Budgetary Educational Institution of Higher Education «Lipetsk State Technical University», Moskovskaya Street, 30, Lipetsk 398060, Russian Federation
e-mail: 2010anasta@mail.ru

A. A. Kosta
Moscow State University of Civil Engineering, Moscow 129337, Russian Federation
e-mail: KostaAA@mgsu.ru

P. Akimov et al. (eds.), *Proceedings of FORM 2022*, Lecture Notes in Civil Engineering 282, https://doi.org/10.1007/978-3-031-10853-2_44

1 Introduction

Equipment for additive technologies has made a breakthrough both in prototyping and in mass production. To date, three-dimensional printing is used in the manufacture of household appliances, the space industry, medicine, mechanical engineering, aircraft construction, and the construction industry is no exception [1, 2].

Currently, the construction industry needs to reduce material and technical, energy, labor, and as a result, financial costs, reduce the risk of injury in the production process and improve the quality of finished products through mechanization and automation, increases interest in the possibilities of additive technologies in the construction of structures of buildings and structures. Every year, the volume of work on the construction of facilities using 3-D technologies is growing quite intensively and it is becoming more and more relevant [3–5].

For construction 3D printers, concrete mortars are used as "ink", the deformative properties of which should provide the possibility of layer-by-layer growth of building structures. Such properties are characterized by rheological indicators [6], as which in the construction industry, instead of real rheological characteristics (viscosity, ultimate shear stress, etc.), some generalized indicators are used: conditional viscosity, consistency of the binder dough, workability of mortar or concrete mixture, etc. Simultaneously with the rheological characteristics, the setting time of concrete mortars is monitored.

For the use of a concrete mixture in 3-D printing in accordance with experimental studies based on literature data [7], a condition for obtaining concrete with the required characteristics is necessary. I.e., the concrete composition must have important properties corresponding to additive technology. At the stage of molding (extrusion) and hardening of the mixture in the field, this is:

- dispersion;
- rheological properties (formability, possibility of transportation through pipes, plastic strength);
- high adhesion between the layers of the mixture, tight fit of the layers;
- no fractures of the mixture;
- no cracking, low shrinkage;
- uniformity of hardening (setting);
- high setting speed after extrusion;

For the finished product:

- required strength
- high uniformity and stability of properties;
- low density and thermal conductivity;
- high adhesion strength;
- frost resistance.

The required characteristics are formed already at the stage of substantiating the composition of concrete mixtures and developing their compositions [8]. To ensure the formability of the mixture, it is rational to use fine-grained or powdered concrete mixtures based on Portland cement and its varieties with different types of fillers, fillers and modifying additives [9].

2 Methods

In this work, studies were carried out to determine the rheological properties of concrete mixtures that allow them to be used for additive technologies. One of the most significant characteristics is the density of the cement paste. But since it is important to study not only the rheology, but also the rate of strength gain, the setting time of the concrete mixture was also investigated.

The study was carried out on a ready-made concrete mixture and on samples-cubes with dimensions of 100 × 100 × 100 mm and beams with dimensions of 40 × 40 × 160 mm. Taking into account the presented data, an experimental program was developed and implemented. At the first stage of the research, concrete samples were made (Fig. 1), at the second stage, physical and mechanical properties were determined.

The tests were carried out on 10 compositions of a concrete mixture consisting of the following components in various proportions:

1) Cement: portland cement CEM I 42.5H;
2) Blast furnace slag;
3) Sand: fine-grained, 0.315 mm fraction + 0.16 mm sieve residue; Sentsovsky deposit;
4) Fly ash: coal industry waste;

Fig. 1 Components for the production of concrete mix

5) Microsilicon;
6) Metacaolin (VMK): highly active (VMK-45), white;
7) Slaked lime;
8) Polypropylene fiber: fiber length 10–12 mm, thickness 12 microns;
9) Superplasticizer Cemmix Cem Plast.

The setting time was determined using the Vic device with a needle according to Russian National Standard 310.3-76* «Cements. Methods for determining the normal density, setting time and uniformity of volume change». The flowability of the concrete mixture is characterized by the depth of loading of the reference cone into it measured in centimeters (Russian National Standard 5802-86). The compressive and bending strength was also determined.

The analysis of the data obtained made it possible to determine the components of the concrete mixture suitable for use in construction 3-D printing. This composition showed a high strength index in the 28-day hardening period, optimal mobility and setting time.

Further, by the method of mathematical planning of the experiment by changing the ratio of components to different properties, in order to identify the optimal composition of the concrete mixture was studied for additive technologies. Thus, the algebraic dependence of the characteristics of the resulting mixture on the consumption of components, their properties and the influencing variables was established.

An experiment planning matrix was compiled, the number of experiments was determined based on the number of factors, as well as in accordance with the conditions of the problem being solved. The results of the experiments were processed using methods of mathematical statistics. As a result, regression equations reflecting the relationship between the studied properties and the influencing factors were obtained. Based on these algebraic equations, graphs were constructed, which made it possible to quickly determine the value of the output parameter when each factor changes.

For concrete mixtures, it is advisable to optimize the formulation and preparation technology according to strength. The remaining quality indicators are not optimized, because it is assumed that their values should vary in a given range.

Results

For each of the compositions, the following were determined:

- the setting time according to the Vika device with a needle (Russian National Standard 30,744-2001);
- flowability by immersion of the cone (Russian National Standard 5802-86);
- compressive strength of samples at the age of 28 days (Russian National Standard 30,744-2001);
- bending strength of samples at the age of 28 days (Russian National Standard 30,744-2001) (Table 1).

Table 1 Test results

№ composition	Density, kg/m^3	Setting time		Lowability by immersion of the cone, mm	Strength of 28 days, MPa	
		Beginning, h	End, h		Compressive	Bending
1	2000	4 h	6 h	7,95	41,2	5,0
2	2150	4 h 30 min	6 h	3,00	48,9	7,9
3	1950	4 h	6 h 30 min	3,30	42,0	5,8
4	1980	3 h	4 h	3,10	35,1	6,3
5	2050	4 h	5 h в	6,50	53,4	8,0
6	1970	3 h	3 h 40 min	5,40	28,7	4,8
7	1965	2 h 30 min	3 h 50 min	3,90	38,5	4,7
8	1850,6	2 h 20 min	3 h 20 min	3,05	28,7	4,5
9	2005,8	2 h 50 min	3 h 40 min	6,07	51,7	7,5
10	1940,3	3 h 20 min	4 h	3,40	30,4	4,7

The results of the experiment planning tests are shown in Fig. 2. Based on the data obtained, a regression equation is derived that characterizes the dependence of the properties of the concrete mixture on the level of variable factors, which made it possible to construct the "compressive strength" response surface (Fig. 3).

After excluding insignificant coefficients, the regression equation has the form:

$$R(x_1, x_2, x_3) = 51,85 + 0,47x_1 + 0,29x_2 - 0,22x_3 - 1,22x_1^2 - 0,61x_2^2 - 0,65x_3^2 - 0,15x_2x_3$$

Microsoft Excel software was used to calculate the coefficients of regression equations and verify the adequacy of the obtained model.

3 Results

During the transition from coded factors to natural ones, the consumption of materials per 1 m^3 of concrete mixture suitable for additive technologies, kg was found: cement—385; water—520; blast furnace slag—70; plasticizer—13; sand—1170; microsilica—150; metacaolin (WMC)—80; slaked lime—195; polypropylene fiber—1.2.

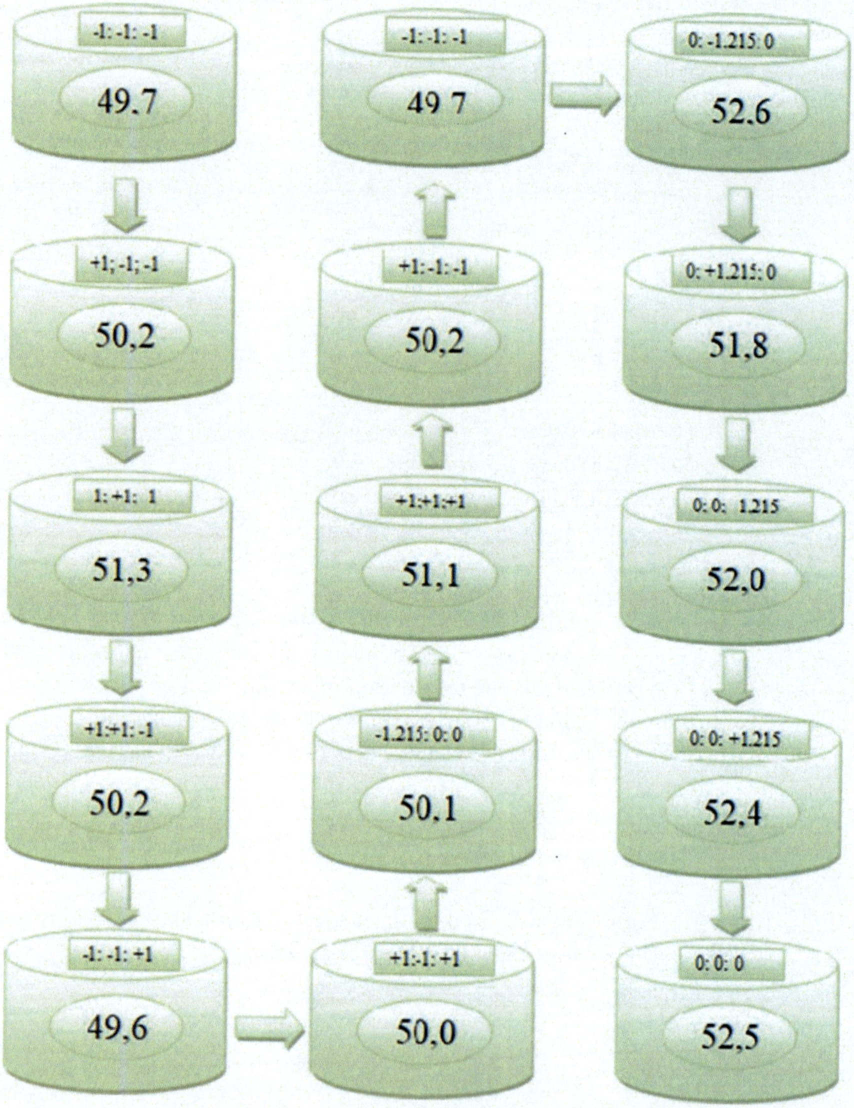

Fig. 2 Components for the production of concrete mix

In addition, the problem of waste disposal of the metallurgical industry is being solved, which will have a positive impact on the environmental situation of the region. This experimental composition of the concrete mixture is suitable for use in construction 3-D printing.

Fig. 3 Response surface «compressive strength»

4 Conclusion

The optimal composition of the concrete mixture, which was found in the course of these studies, allows achieving a sufficiently high compressive strength (51 MPa). At the same time, the introduction of blast furnace slag into the composition of the concrete mixture used for additive technologies allows reducing cement consumption up to 150 kg per 1 m^3 of the mixture.

Acknowledgements This work was financially supported by the Ministry of Science and Higher Education of the Russian Federation (grant # 075-15-2021-686). All tests were carried out using research equipment of The Head Regional Shared Research Facilities of the National Research Moscow State University of Civil Engineering.

References

1. Pahomova EG, Monastyrev PV, Mishchenko ES, Ivanov IA, Balthazar AD, Yezerskiy VA (2019) House-building analysis when using additive technologies: classification, advantages and disadvantages. J Appl Eng Sci 17(4):449–456
2. Glagolev ES, Lesovik VS, Zagorodnyuk LH, Podgornyi DS (2021) Composite binders and dry building mixes for 3d additive technologies. In: Lecture notes in civil engineering, vol 147, pp 229–235
3. Osmanov SG, Manoilenko AYu, Litovka VV (2019) The choice of options for the mechanization of concrete work in monolithic-frame construction. Eng Bull Don (1). ivdon.ru/ru/magazine/archive/n1y2019/5507/
4. Rael R, Fratello VS (2017) Printing architecture: innovative recipes for 3D printing. Princeton Architectural Press, New York, 176 p

5. Kravchenko GM, Trufanova EV, Danileiko IYu (2019) Investigation of the principles of shaping of objects of parametric architecture. Eng Bull Don (1). ivdon.ru/ru/magazine/archive/n1y2019/5513/
6. 3D printing in construction, Vatin NI, Chumadova LI, Goncharov IS et al (2017) Construction of unique buildings and structures 1(52):27–46
7. Canessa E (2013) Accessible 3D printing for science, education and sustainable development. In: Fonda K, Zennaro G (eds) International center for theoretical physics Abdus Salam, 192p
8. Bazhenov YuM (2002) Concrete technology. Izd-vo ASV, 500 p
9. Kopanitsa NO, Sorokina EA (2016) Features of the formation of requirements for the construction and technical characteristics of concrete mixes for 3d printing. In: Proceedings of the III international scientific conference of students and young scientists: youth, science, technology: new ideas and prospects (MNT 2016), pp 407–410
10. Naboni R, Paoletti I (2015) Advanced customization in architectural design and construction. Springer, Berlin, p 170

Numerical Simulation of the Temperature Field of an Office Space with Three Types of Heating Systems

Robert Akhverdashvili, Aleksandr Gulkanov, Sergey Saiyan, and Konstantin Modestov

Abstract The modern development of engineering systems design and production of climate equipment strives to increase the level of comfort of human stay and ensure favourable microclimate parameters of the serviced area of the room. Existing approaches to the use of various fundamental solutions for certain space-planning conditions often allow us to provide normalized microclimate parameters, but cannot guarantee them with a certain accuracy. In this regard, numerical modelling of heat and mass transfer processes for non-standard design conditions makes it possible to identify zones whose parameters do not correspond to the normalized ones. Choosing one of the types of heating systems that provide a microclimate of the room, it is impossible to check all areas of the working area for compliance with the requirements of regulatory documents on air temperature. The tasks of checking the thermal regime of the room were considered in many studies and scientific and technical works, however, heat exchange by convection was not taken into account in these works, and the air temperature was assumed constant throughout the working area, which is not a physical solution. In this paper, computational studies were expanded by taking into account the temperature fields of the airspace for three types of heating systems using numerical modelling methods implemented in the ANSYS Fluent software package. The changes in air temperature along the height of the working area were revealed, and the "unfavourable" areas of the working area were checked for compliance with regulatory requirements. The results obtained will allow design organizations to more appropriately approach the choice of a building heating system. Visual identification of problem areas and phenomena will eliminate further errors and misconceptions in the design.

Keywords Temperature field · Heating systems · Computational fluid dynamics

R. Akhverdashvili · A. Gulkanov · S. Saiyan (✉) · K. Modestov
Moscow State University of Civil Engineering, Moscow 129337, Russian Federation
e-mail: berformert@gmail.com

K. Modestov
e-mail: ModestovKA@mgsu.ru

A. Gulkanov
Politecnico di Milano, Piazza Leonardo da Vinci, 32, 20133 Milan, Italy

P. Akimov et al. (eds.), *Proceedings of FORM 2022*, Lecture Notes in Civil Engineering 282, https://doi.org/10.1007/978-3-031-10853-2_45

1 Introduction

The development of engineering systems design methods and the production of climate equipment are aimed at increasing the comfort level of human stay and ensuring favourable microclimate parameters of the serviced area of the room [1, 2]. The requirements for the rational use of energy resources are increasing, especially in the design and operation of engineering systems that ensure the microclimate of rooms [1]. The existing approach to the use of various principal solutions for certain space-planning conditions often allows you to provide normalized microclimate parameters, but cannot guarantee them with a certain accuracy. In this regard, numerical modelling of heat and mass transfer processes for non-standard design conditions makes it possible to identify zones whose parameters do not correspond to the normalized ones.

Choosing one of the types of heating systems, it is impossible to check all zones of the working area for compliance with the requirements of regulatory documents on air temperature. Problems of this kind have been considered in studies [3] and in papers [4–6]. In the calculations, only radiation temperatures were determined at various points of space during radiator heating, which took into account only heat exchange using thermal radiation. Convection was not considered in these studies. The air temperature in the volume of the working area was assumed to be constant, which is not a physical solution. The action of the convective jet from the heater and the intensification of the movement of air masses leads to a stratification of the air temperature along the height and space of the room volume.

Based on this, in this paper, computational studies were expanded by taking into account the temperature fields of the airspace for three types of heating systems using numerical modelling methods implemented in the ANSYS Fluent software package. Changes in air temperature along the height of the working area were revealed, and the "unfavourable" areas of the working area were checked for compliance with regulatory requirements [7]. The aim is to study the asymmetry of the distribution of temperature and velocity fields affecting the comfort of people staying indoors.

The results obtained will allow design organizations to more appropriately approach the choice of a building heating system. Visual identification of the problematic areas will eliminate further errors and misconceptions in the design. The solution to this problem will ensure in the future the creation of a comfortable thermal environment for a person to stay and will exclude inappropriate waste of thermal energy.

2 Materials and Methods

Characteristics and Design Parameters of Heating Systems

An office space for 35 people of a real object located in the Volgograd city is accepted as the investigated room. The estimated winter design temperature –22 °C [8].

The heat transfer resistance of the outer wall is 2.33 (m^2·°C)/W, and windows 0.59 (m^2·°C)/W [9]. The heat transfer resistance of the internal partition between adjacent rooms is 0.63 (m^2·°C)/W, the overlap resistance between the underlying and the overlying rooms is 0.73 (m^2·°C)/W. The emissivity of the surfaces are equal to: window 0.92; wall, floor and ceiling surfaces 0.9; the white paint coating of the heating device 0.8. The estimated temperature of the internal air is assumed to be 18 °C [7]. The dimensions of the room are given in Fig. 1.

Characteristics and design parameters of heating systems:

1. Water with panel radiators [10]. Dimensions of the heater: length 1600 mm, height 600 mm, depth 50 mm. The distance from the floor to the bottom surface is 100 mm. The surface temperature is 59 °C. The scheme with boundary conditions is shown in Fig. 2.
2. Air heating combined with a ventilation system [11]. Air distributors—ceiling diffusers rectangular 600 × 600 mm. The balance between supply and exhaust air is zero. The mass air consumption per diffuser is 0.265 kg/s. The supply air temperature is 21 °C. The scheme with boundary conditions is shown in Fig. 3.
3. The "Underfloor heating" system. The area of the heat-emitting surface of the floor is 62.32 m^2, the surface temperature is 23 °C. The surface temperature is accepted according to the requirements [2]. The scheme with boundary conditions is shown in Fig. 4.

Description of the Technique of Numerical Simulation of Temperature-Convective Fields of Air Space

To write down mathematically a complete system of equations, it is necessary to consider three main mechanisms of heat transfer in a room: thermal conductivity, convection and radiation [12, 13].

Convection can be described as follows:

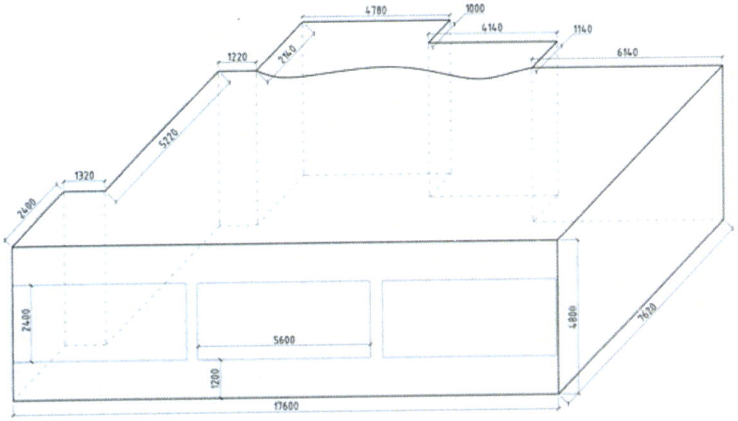

Fig. 1 Scheme with the dimensions of the room under study

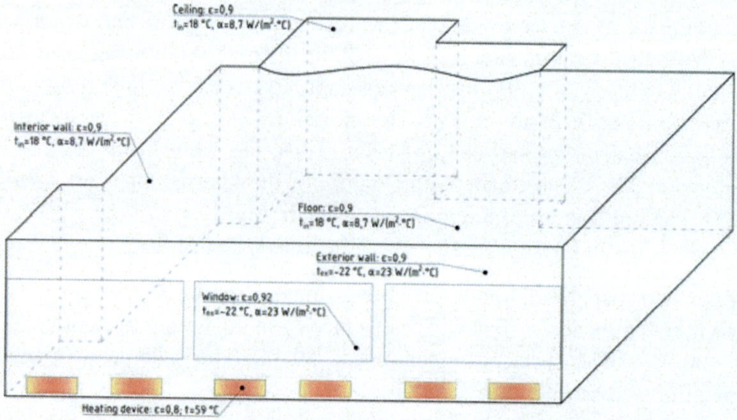

Fig. 2 Scheme of boundary conditions of a heating system with panel radiators

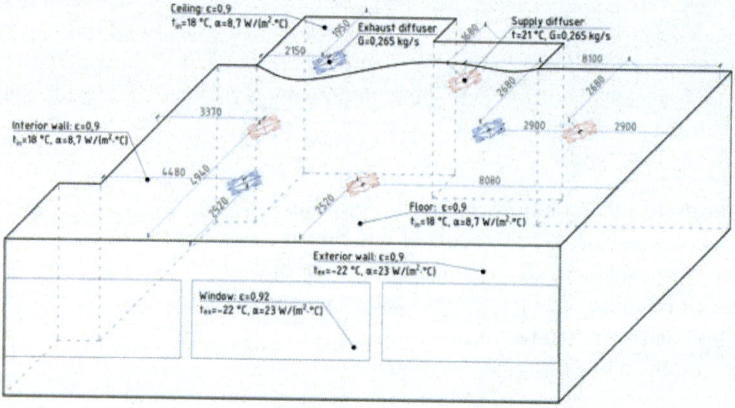

Fig. 3 Scheme of the boundary conditions of the air heating system

1. Let's write down the differential equation of convective heat transfer taking into account thermal conductivity:

$$\rho \cdot c\left(\frac{\partial T}{\partial t} + \left(\vec{v} \cdot \vec{\nabla}\right)T\right) = div[\lambda \cdot grad(T)] + q_v + \mu \cdot \Phi - p \cdot div(\vec{v}), \tag{1}$$

where: ρ—function expressing density, $\left[kg/m^3\right]$; c—specific heat function c = const, $\left[J/(kg \cdot K)\right]$; T—temperature function, $[K]$; t—time function, $[s]$; $\vec{v}$—air velocity vector, $[m/s]$; λ—coefficient of thermal conductivity of air λ = const, $\left[W/(m \cdot K)\right]$; q_v—the source term, which expresses the

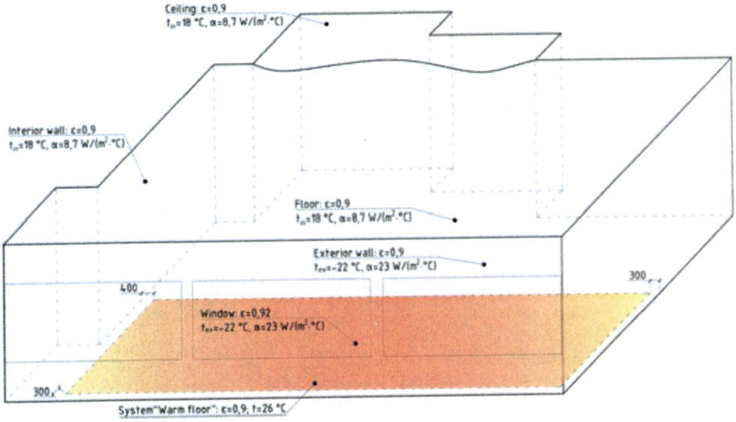

Fig. 4 Scheme of boundary conditions of the heating system "Underfloor heating"

inflow/outflow of energy under the action of internal heat sources; μ—dynamic viscosity coefficient; Φ—dissipative function, $[W]$;

$\rho \cdot c \cdot \frac{\partial T}{\partial t}$—an unsteady term that expresses the unsteadiness of the heat exchange process, in our case $\rho \cdot c \cdot \frac{\partial T}{\partial t} = 0$ because the temperature field is constant;

$\rho \cdot c \cdot \left(\vec{v} \cdot \vec{\nabla}\right) T$—convective term, which expresses the transfer of heat during the movement of the medium;

$div[\lambda \cdot grad(T)]$—a conductive term that expresses the transfer of heat by thermal conductivity;

$\mu \cdot \Phi$—dissipative term, which expresses the heating of the medium during the dissipation of kinetic energy during motion;

$-p \cdot div(\vec{v})$—the term of thermal compression/expansion, which expresses the change in the energy of the air during its compression/expansion.

Thus, the equation in our case will take the form:

$$\rho \cdot c\left(\vec{v} \cdot \vec{\nabla}\right) T = \lambda \Delta T + q_v + \mu \cdot \Phi - p \cdot div(\vec{v}). \tag{2}$$

2. The Navier–Stokes equation.

   Since air is a compressible medium, therefore it is necessary to write down the Navier–Stokes equation for a compressible fluid. We will write it down under the condition that the η—coefficient of dynamic viscosity and the ζ—"second" viscosity are constant in vector form:

$$\rho\left(\frac{\partial \vec{v}}{\partial t} + \left(\vec{v} \cdot \vec{\nabla}\right)\vec{v}\right) = -\vec{\nabla} p + \eta \Delta \vec{v} + \left(\zeta + \frac{\eta}{3}\right)\vec{\nabla} div(\vec{v}). \tag{3}$$

3. The law of conservation of mass and the equation of state of an ideal gas.

For the system of equations to be complete, it is necessary to introduce two more equations.

The law of conservation of mass:

$$\frac{\partial \rho}{\partial t} + div(\rho \vec{v}) = 0. \quad (4)$$

The equation of state of an ideal gas:

$$pV = NRT, \quad (5)$$

where: N—number of moles; $R = 8.31\ J/(mol \cdot K)$—universal gas constant.

We have chosen the ideal gas model, because it is known that the state of the air in the room space is well described by the ideal gas model.

The heat flux of radiation can be described according to the Stefan-Boltzmann law, then the balance relative to the j surface will look like this:

$$P_j = \sigma \varepsilon_j \sum\nolimits_{i=1}^{n} \varepsilon_i T_i^4 A_i F_{i \to j} - \sigma \varepsilon_j T_j^4 A_j, \quad (6)$$

where: $\sigma = 5.67 \cdot 10^{-8}\ W/(m^2 \cdot K^4)$—Stefan-Boltzmann constant; ε_i—emissivity of radiating surface; ε_j—emissivity of surface affected by radiation; A—radiation area, $[m^2]$; T_i—radiating surface temperature, $[K]$; T_j—surface temperature affected by radiation, $[K]$.

An analytical solution to the problem is generally impossible, due to the instability and complexity of the solution of Navier–Stokes equation, therefore, to simplify the problem and obtain a solution, we will use methods based on the use of numerical methods for solving systems of partial differential equations. The finite volume method was considered as the numerical method used.

Since direct numerical simulation (DNS) requires an extremely small mesh size (on the order of the Kolmogorov scale), and computational costs are proportional to the number of nodes in each direction and the number of time steps, we will use methods based on the use of turbulence models to simplify the numerical solution. To date, there is no universal turbulence model, and choosing a specific model from the many existing ones is a difficult task [14–17]. A two-parameter k-epsilon turbulence model was used as the computational model of this work. The selected turbulence model refers to the RANS (Reynolds Averaged Navier–Stokes) models [13]. RANS models are quite well studied, their applications are described, and they can be used to model both two and three-dimensional flows, which is suitable for the internal aerodynamics of a room.

The radiation model is adopted by DO (Discrete Ordinates). It allows the calculation model to take into account the radiation from the surface of the grey body, which are all surfaces inside the room. Also, this model is the most suitable for taking into account local heat sources.

The implementation of mathematical models, numerical scheme and simulation parameters for solving the problem of temperature fields of the airspace are embodied in the ANSYS Fluent software package. The boundary conditions, as well as the simulation parameters are taken in accordance with the initial data in the subsection "Characteristics and design parameters of heating systems".

Finite Volume Models of Heating Systems

In the ANSYS Meshing module, finite volume models of the room under study were developed for various heating options of the room. The geometric model was divided into several components related to heating devices, necessary for local refinement of the finite volume mesh in places with high gradients of change in the solution fields.

The characteristic dimensions of the elements inside the volumes of the models were: 0.9 m (for the model with underfloor heating), 0.2 m (for the model with panel radiators) and 0.08 m (for the model with air heating). Such a large variation in the characteristic sizes of elements inside the airspace is associated with a different degree of gradient of temperature fields for three different heating systems obtained from the results of preliminary numerical modelling. For a model with an air heating system, it turned out to be an order of magnitude higher than for the other two systems, therefore, in order to resolve such a large gradient of change in the solution fields, the size of the final volumes inside the airspace was significantly reduced.

The characteristic dimensions of the elements on the surfaces of heating devices were: 0.08 and 0.04 m (for the model with underfloor heating), 0.05 and 0.1 m (for the model with panel radiators), 0.01, 0.03 and 0.05 m (for the model with air heating). Hexagonal shapes of finite volume elements were used in the airspace (type 1), in the areas of rapid changes in air parameters (temperature and velocity), the shape of the elements was used in the form of tetrahedra with a higher density (type 2). Such areas were adopted: the airspace near the heating devices in the water heating system with panel radiators and the space near the supply openings in the system with air heating. The transition regions connecting the two types of mesh were split using pyramidal finite volumes (type 3). Trigonal prismatic finite volumes (type 4) were used to create a boundary layer in the zones of heating devices.

A boundary layer with the following parameters was created on all internal surfaces: 5 elements in thickness, the total thickness of the boundary layer is 0.04 and 0.02 m, the layer growth coefficient is 1.05. The need for a boundary layer for internal aerodynamics problems is associated with providing a correct numerical solution in the wall layer to obtain more physical simulation results. Finite volume model for the variant with air heating is shown in Fig. 5.

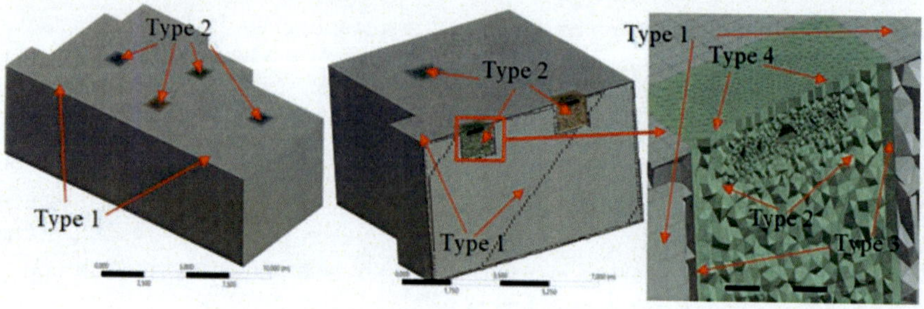

Fig. 5 Finite volume model for an version with air heating

3 Results and Discussion

Water System with Panel Radiators

For a water system with panel radiators, the most intense temperature change in height occurs. The maximum temperature difference between the planes is 1.76 °C. There is a clear stratification of the air space by temperature. The increased temperature in the upper area of the room causes additional heat losses, while it does not bring thermal efficiency to the working area, limited by a height of two meters from the floor surface. The convective jet from the heater, moving along the surface of the window and the outer wall, also leads to additional inappropriate heat losses. At the border of the working area of the frontal outer wall and the window at a height of 0.5 m from the floor surface, the air temperature is: opposite the heating device 17.9 ± 0.2 °C; opposite the section of the wall without the device 18.3 ± 0.2 °C. The temperature field in sections is shown in Fig. 6.

Air Heating

The intensity of air temperature changes in height is less, compared to panel radiators. The maximum temperature difference between the planes is 1.08 °C. Similarly, stratification is observed, while the airspace is explicitly divided into two areas. The upper area, similar to the system with radiators, has a higher temperature, due to the floating of part of the supply jet to the ceiling of the room, which leads to insufficient heating of the working area and additional thermal costs. Proceeding from this,

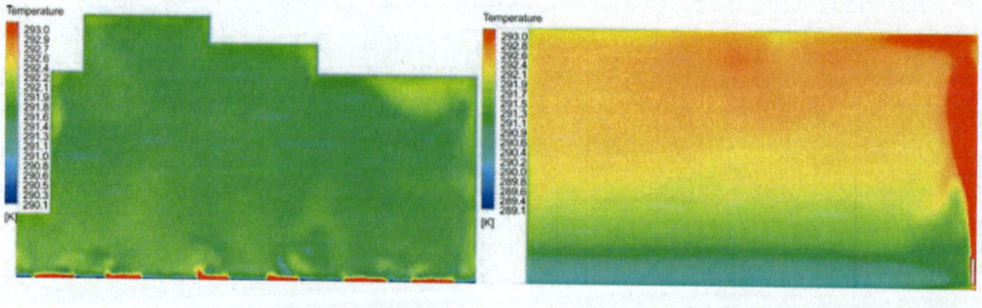

Fig. 6 The temperature field in the plan at a height of 1.1 m from the floor surface and in a cross section, perpendicular to the outer wall

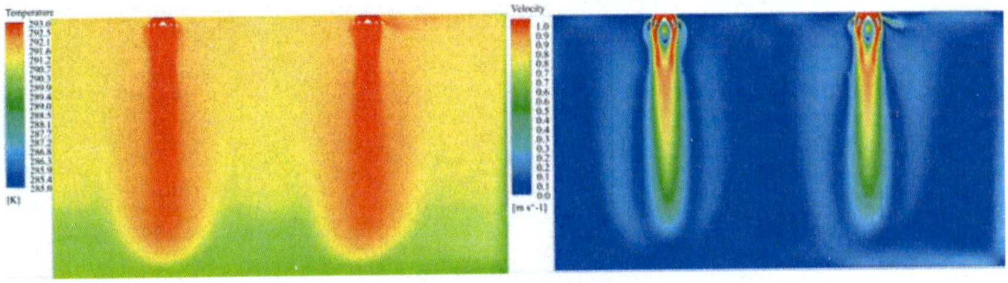

Fig. 7 The temperature field and the velocity field in the cross section, perpendicular to the outer wall

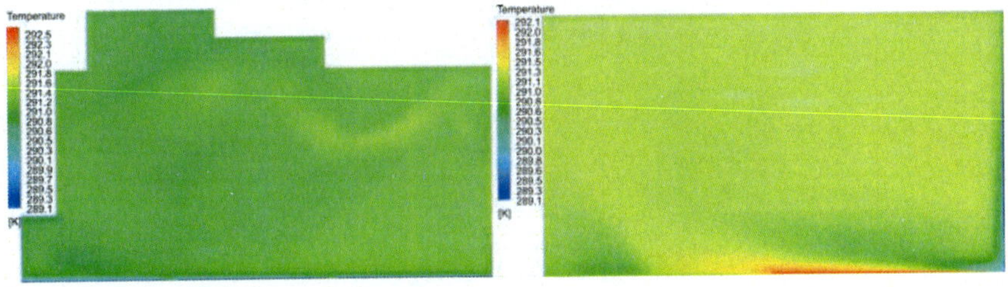

Fig. 8 The temperature field in the plan at a height of 1.1 m from the floor surface and in a cross section, perpendicular to the outer wall

regarding the operation of the ventilation system, compensation of heat losses with the help of supply air, respectively, having a temperature above the working area, leads to a decrease in the efficiency of air exchange. At the border of the working area of the frontal outer wall and window at a height of 0.5 m from the floor surface, the air temperature is 17.5 ± 0.2 °C. The temperature field and velocity field are shown in Figs. 7 and 8.

Underfloor Heating

The intensity of the change in air temperature along the height has the smallest value. The maximum temperature difference between the planes is 0.18 °C. The temperature field throughout the working area has a uniform distribution. There is no overheating of the air in the upper area of the room. At the border of the working area, the front of the outer wall and the window at a height of 0.5 from the floor surface, the air temperature is 18.1 ± 0.1 °C. At a height of 0.1 m from the floor, the action of a cold convective jet from the window is interrupted by a convective flow from the heat-emitting surface of the floor. The temperature has a value of 17.5 ± 0.2 °C. The temperature field is shown in Fig. 9.

Table 1 Average air temperatures in the plane at a height Z_i from the floor surface

The position of the plane parallel to the floor at a height of Z_i, m	The average air temperature in the Z_i plane for the following types of heating systems:		
	Panel radiators	Air heating	Underfloor heating
0.1	17.45	17.04	18,02
0.2	17.61	17.10	17,99
0.3	17.64	17.17	18,00
0.4	18.03	17.23	18,02
0.5	18.05	17.28	18,03
0.6	18.40	17.33	18,05
0.7	18.33	17.37	18,08
0.8	18.54	17.42	18,11
0.9	18.56	17.42	18,14
1.0	18.73	17.48	18,15
1.1	18.74	17.55	18,17
1.2	18.87	17.64	18,17
1.3	18.88	17.75	18,18
1.4	18.98	17.85	18,18
1.5	18.99	17.93	18,19
1.6	19.07	17.99	18,19
1.7	19.07	18.04	18,19
1.8	19.15	18.09	18,20
1.9	19.15	18.09	18,20
2.0	19.21	18.12	18,20
Standard deviation	0.56	0.36	0,08

The obtained results of air temperature changes in the height of the working area for three variants of the heating systems are shown in Table 1.

The smallest standard deviation of the air temperature in the height of the working area is observed in the "Underfloor heating" system and is 0.08. The average mass air temperature by room volume in the calculation model for these three heating systems is 18.15 ± 0.05 °C. Consequently, maintaining the same air temperature throughout the room, we get a different temperature distribution, which occurs due to the convective movement of air masses.

4 Conclusion

Numerical modelling is an effective method for studying temperature and velocity fields, allowing designing systems for a comfortable indoor microclimate. Combining accurate mathematical models and effective numerical methods, it is possible to take into account effects previously unavailable according to existing engineering and regulatory methods.

Using numerical modelling, a different distribution of air temperature over the height of the working area was revealed at the same average room temperature. The air temperature at the border of the working area, at the points indicated above, during the operation of all three heating systems meets the requirements (the local deviation of the air temperature from the calculated value in the space of the working area of the room should not exceed: for optimal parameters—2 °C, for permissible—3 °C) [7].

The smallest standard deviation of the air temperature along the height of the working area is observed in the "Underfloor heating" system and is 0.08, which indicates the most uniform distribution of air temperature along the height of the room. The average air temperature by room volume in the calculated model for these three heating systems is 18.15 ± 0.05 °C. Consequently, maintaining the same air temperature throughout the room, we get a different temperature distribution, which occurs due to the convective movement of air masses.

Based on the results of numerical simulation of convective-radiant heat exchange, on the example of an office space for 35 people, it can be concluded that the most effective heating system is the "Underfloor heating" system. This system has the smallest asymmetry in the distribution of temperature and velocity fields, which favourably affects the comfort of people staying indoors.

In the process of designing radiator and air heating systems, it is necessary to take into account the occurrence of air overheating zones in the upper area of the room. It is impractical to use these types of systems for rooms with a high ceiling height.

Acknowledgements This work was financially supported by the Ministry of Science and Higher Education of the Russian Federation (grant # 075-15-2021-686). All tests were carried out using research equipment of The Head Regional Shared Research Facilities of the National Research Moscow State University of Civil Engineering.

References

1. Federal Law № 261 (2021) On energy saving and energy efficiency improvement and on amendments to certain legislative acts of the Russian Federation (as amended as of June 11, 2021), Russian newspaper, №226, 11/27/2009. Parliamentary newspaper, №63, 11/27/2009, Collection of Legislation of the Russian Federation, №48, 11/30/2009, art. 5711
2. Code of Practice 60.13330.2020 (2021) Heating, ventilation and air conditioning sanitary code 41-01-2003 (as amended). Standartinform
3. Bogoslovsky VN (1982) Building thermophysics (thermophysical fundamentals of heating, ventilation and air conditioning) Textbook for universities, 2nd edn, revised. And additional – (Higher. School), p 415
4. Malyavina EG, Landyrev SS (2022) Verification of compliance with the requirements of GOST 30494-2011 to the parameters of the internal environment at the border of the serviced area. AVOK 2:14–16

5. Malyavina EG, Frolova AA, Landyrev SS (2021) Distribution of local asymmetry of the resulting temperature over the room. S.O.K. Plumb Heat Air Cond 10:36–39
6. Malyavina EG, Samarin OD (2018) Building thermophysics and microclimate of buildings. Ministry of Education and Science Ros. Federation, National research. Moscow state builds. un-t. Publishing house MISI – MGSU, Moscow, p 288
7. GOST 30494-2011 (2019) Buildings residential and public. Indoor Climate Parameters (Revised Edition). Standartinform, Moscow
8. Code of Practice 131.13330.2020 (2021) Building climatology Building Norms and Rules 23-01-99*. Standartinform, Moscow
9. Code of Rules 50.13330.2012 (2012) Thermal protection of buildings. Updated version of the Sanitary Norms and Rules 23-02-2003 (with Amendment №1). Ministry of Regional Development of Russia, Moscow
10. Makhov LM (2015) Heating, Proc. for universities. 2nd edn, corrected and enlarged. ASV Publishing House, Moscow, p 400
11. Kamenev PN, Tertichnik EI (2008) Ventilation. Tutorial. Publishing House of ASV, Moscow, 624 p
12. Landau LD, Lifshits EM (2015) Hydrodynamics, 6th edn. Theoretical Physics, vol 6. Fizmatlit, Moscow, p 728
13. Ansys Fluent, Release 2021R2, help system, Theory Guide, ANSYS, Inc.
14. Dubinsky SI (2010) Numerical modelling of wind effects on high-rise buildings. Diss. for the degree of Ph.D. MGSU, Moscow, p 199
15. Belostotsky AM, Akimov PA, Afanas'eva IN (2017) Computational aerodynamics in construction problems. DIA Publishing House, Moscow, p 720
16. Methodological manual "Mathematical (numerical) modeling of wind loads and impacts" (to SP 20.13330.2016), M., FAU FTS of the Ministry of Construction of Russia, 2020 (developed by CJSC NRC StaDiO)
17. Roach P (1980) Computational fluid dynamics. Mir, Moscow, p 618

Integrating BIM in Operation and Maintenance Stage

Nikolai Bolshakov, Alberto Celani, and Liliya Azhimova

Abstract This study examines the integration of building information modelling (BIM) technologies in operation & maintenance stage in the system of managing real estate that helps to reduce transaction costs. The approach and method are based on Digital Twin technology and Model Based System Engineering (MBSE) approach. The results of the development of a service for digital facility management and digital expertise are presented. The connection between physical and digital objects is conceptualized.

Keywords BIM · Operation and maintenance · Facility management · Digital twin · MBSE · Asset management

1 Introduction

Information technologies (IT) inevitably penetrate construction market digitizing not only design stage of investment construction projects but also construction and operation & maintenance stages [1]. According to recent studies [2] building information modelling technologies have a great potential in applying to operation & maintenance stage. The object of this study is existing capital construction facility with all the systems that are included in it.

Subjects of study are properties of the existing construction facility that determine the operation process and are presented in digital form. Therefore we are going to analyze maintenance activities that are going to be executed in facility management.

N. Bolshakov (✉)
Peter the Great Saint Petersburg Polytechnic University, Saint Petersburg, Russia
e-mail: nikolaybolshakov7@gmail.com

A. Celani
ABC Department, Politecnico di Milano, Milan, Italy
e-mail: alberto.celani@polimi.it

L. Azhimova
Kazan State University of Architecture and Engineering (KazanSUAE), Kazan, Russia

P. Akimov et al. (eds.), *Proceedings of FORM 2022*, Lecture Notes in Civil Engineering 282, https://doi.org/10.1007/978-3-031-10853-2_46

Building and structure data in actual construction often differs from the designed one [3], therefore it is necessary to obtain actual as-is BIM data which can be possible with laser scanning technology [4].

The introduction of information technologies into the operation phase is an issue at the forefront of the development of the construction industry [5]. Participants understand that this is inevitable [6], but they are generally afraid to be the first to implement BIM in operation & maintenance, which is also supported by the fact that in government standards (GOST) and codes of practice (CP) the operational information model is only advisory. The introduction of IT is a big risk and many rightly consider it a wrong move in the short term [7]. This is due to the lack of open information and qualified personnel [8].

However, the same thing has already happened with the introduction of BIM into design stage [9]. 5–10 years ago, companies were skeptical about new technology, preferring proven CAD [10]. BIM in design started to work when a huge amount of information appeared in the public domain.

As the practice of BIM in design shows, now is the right time to use it in operation, since, later, only those companies that have managed to fully implement information technologies will remain afloat [11].

One of the barriers on the way of implementing BIM in operation & maintenance is the fact that the relationship of the properties of the elements of existing buildings is unknown [12]. These relationships as well as the actual properties of the building and structural elements need to be digitized.

The purpose of the study is development of a method for the effective operation of existing capital construction projects through the formation and maintenance of an information model of the building. Relationship between systems and subsystems is the key factor of this effective operation [13].

From economic point of view the overall goal is reducing transaction costs. Digitization turns out to be the way to reduce these transaction costs [14].

Identification of connections is also necessary during reconstruction and repair in order to perform effective management of real estate objects [15]. In the absence of connections, it is necessary to design buildings and structures again.

2 Method

Research method is focused on creating a digital passport [16], the data from which is used by various predictive analytics models. The result of applying the method is the creation of a digital asset [17] with real value for the owner.

Figure 1 represents the Model-Based System Engineering [18] approach which lies in the basement of developed methodology. Starting with conceptual design on customer level the designed buildings and structures need to be decomposed as systems which consist of subsystems. At the same time subsystems need to be decomposed in elements which have their own properties and connections between each other. Elements of systems include not only construction data but also physical

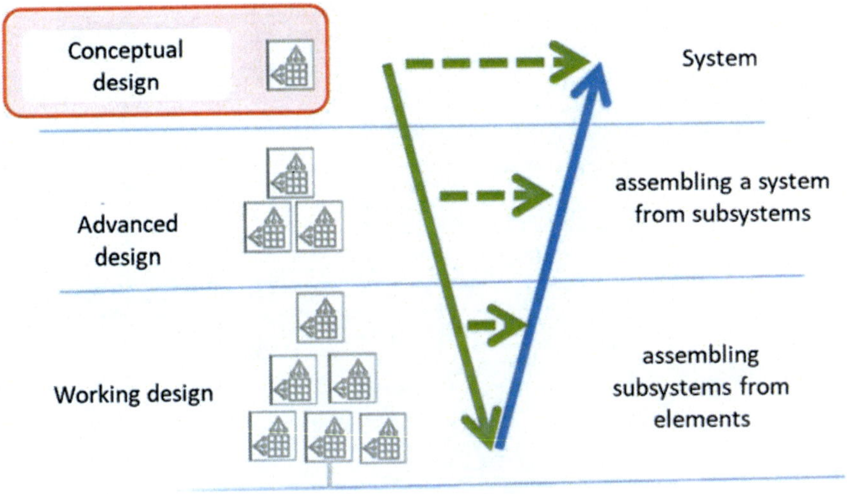

Fig. 1 Model-based system engineering approach

elements of production with their digital representation—digital twins. Therefore on this level integration of building information modelling and digital twin occurs.

It is necessary to develop adequate models that determine the rational operation of the capital construction facility and the transformation of its space-planning solutions. In order to clearer understand the maintenance processes, we divide the object into physical and digital. Load bearing capacity of physical object is maintained according to calculation models of digital object, energy efficiency of physical object is obtained through energy modelling of digital object and etc. Digital and physical subsystems are divided into digital and physical elements respectively.

Data and information from calculation models, energy modelling, sensors and Internet of Things (IoT), economic models, space management, digital management become an input for system information modelling (SIM) [19] and customer relationship management (CRM) [20]. SIM is the process of modelling complex connected systems. System information models are digital representations of connected systems, such as electrical instrumentation and control, power, and communication systems. CRM is application software for organizations designed to automate strategies for interacting with customers (clients), in particular to increase sales, optimize marketing and improve customer service by storing information about customers and the history of relationships with them, establishing and improving business processes and subsequent analysis results. Digital part of the construction facility should not be independent, but should be connected with management systems (Fig. 2).

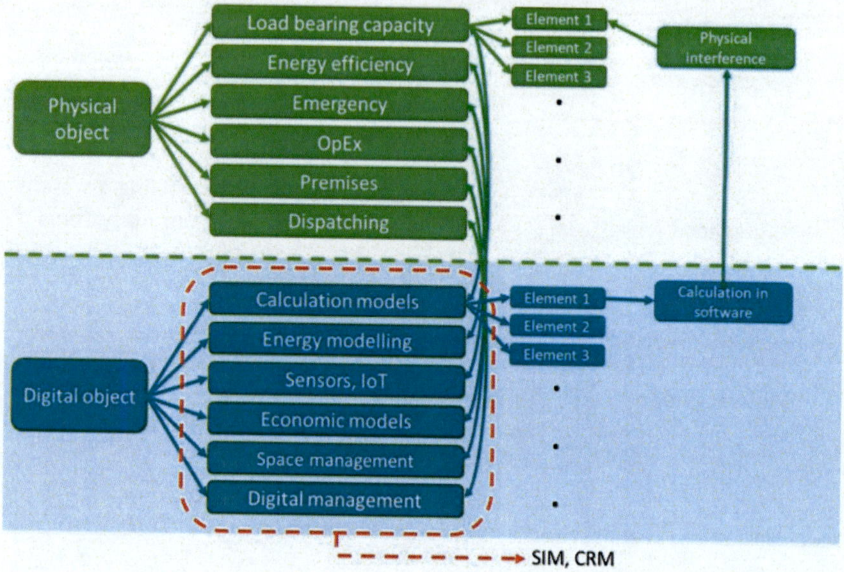

Fig. 2 Method of the research. Connection between physical and digital object

3 Results

Formation of a Digital Asset

The set of digital technologies and their bidirectional connections with the physical world is considered in this article as a digital asset. Digital asses is formed not only on the operation & maintenance stage but throughout the whole lifecycle of physical asset which includes also design, construction, commissioning and demolition.

Each stage includes several stakeholders. Interaction between these stakeholders can be digitized, therefore we identify 9 processes which can be areas of developed web-service application. Currently two of them are under development: interaction between design contractor and expertise and facility management.

These interactions include:

1. Customer—design contractor. Digitization of customer requirements regarding cost and time of design, properties of designed object, etc.
2. Design contractor—expertise. Digitization of application to expertise, check of BIM-model and documentation, submission of plans, sections, etc., formation of expertise notes. This interaction may also include the customer depending on the contract between customer and design contractor.
3. Customer—construction contractor. Digitization of customer requirements regarding cost and time of construction, subcontractors. This stage may implement modern approaches such as Advanced Work Packages (AWP).

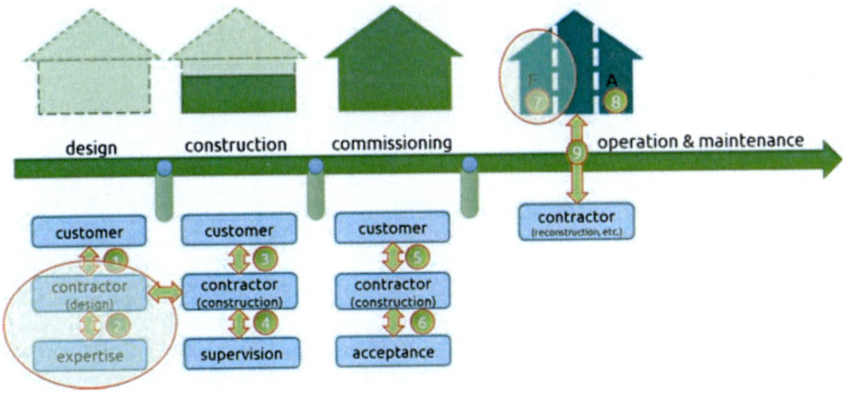

Fig. 3 Areas of application

4. Construction contractor—supervision. Digitization of interaction between supervision authorities and construction contractors and subcontractors, submission of work packages, etc.
5. Customer—construction contractor on commissioning stage. Digitization of commissioning stage.
6. Construction contractor—acceptance authorities. Digitization of application to acceptance authorities during commissioning stage.
7. Facility management
8. Asset management
9. Owner—reconstruction contractor. Digitization of reconstruction and minor construction activities during operation & maintenance stage (Fig. 3).

All the described interactions can be digitized and optimized using BIM technologies. In the approbation section below we discuss results of digitizing two of them: design contractor—expertise and facility management.

Approbation

The developed methodology has been approbated in a form of web-service. Like any web application, the developed service consists of a client and a server part. Among the advantages of developing the service in web we can mention:

- A browser is enough to use
- Many basic functions are available for free, more complex operations are paid for in proportion to use
- Simple, visual interface, understandable to allied professionals

Figure 4 represents the results of digitizing the interaction between expertise authorities and design contractor. Each of the stakeholders have their own personal login and password. After logging in design contractor is able to upload BIM model

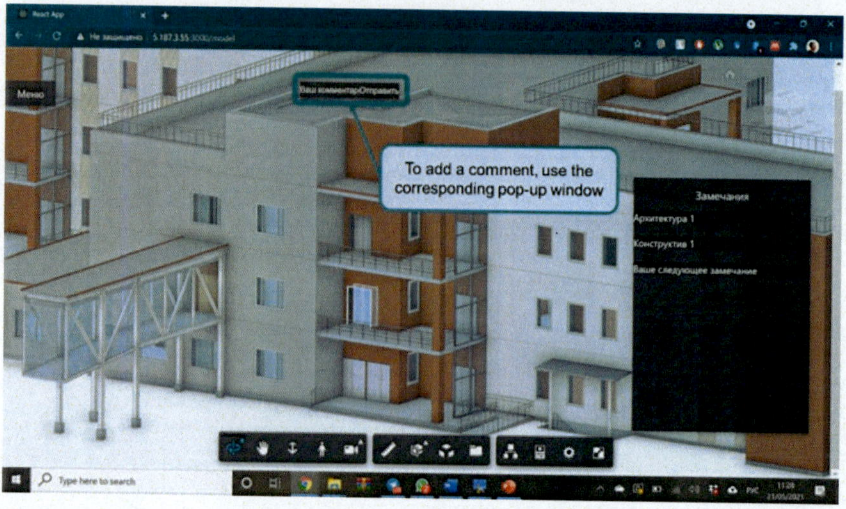

Fig. 4 Web-service interface

to the web service. BIM model contains not only geometrical data but also qualitative properties of systems, subsystems and elements such as material, chemical or physical properties, maintenance plans, drawings, etc.

Expertise authority after logging in through browser is able to check the model and it's properties. The main feature is that expertise authority is able to attach comments to the elements, writing remarks regarding design decisions. All the restored remarked are stored and become available for design contractor.

Another approbations that is currently under development is digitization of facility management. It includes:

- Navigation.
- Location of building components.
- Visualization and marketing.
- Maintainability check.
- Creation and updating of digital assets.
- Space management.
- Planning and feasibility studies for non-capital construction.
- Management in emergency situations.
- Energy control and monitoring.
- Staff training and development.
- Augmented reality.

4 Conclusion

Building information modelling can be implemented not only in design and construction stage of construction investment project but also in operation & maintenance. Applying BIM in operation & maintenance reduces operational expenditures. Interaction between different stakeholders such as customer, construction contractor, design contractor, expertise authorities, commissioning authorities, facility managers and asset managers can be digitized and therefore optimized. It is planned to integrate existing solutions into the developed web service. Intermediate results of web-service development show how the expertise and facility management procedures can be simplified. Future development of the project is going to be focused on digitizing asset management, construction stage and improvement of digital facility management solutions.

Acknowledgements The reported study was funded by RFBR according to the research project № 20-38-90055

References

1. Pishdad-Bozorgi P, Gao X, Eastman C, Self AP (2018) Planning and developing facility management-enabled building information model (FM-enabled BIM). Autom Constr 87:22–38
2. Pärn EA, Edwards DJ, Sing MC (2017) The building information modelling trajectory in facilities management: a review. Autom Constr 75:45–55
3. Jung J, Hong S, Jeong S, Kim S, Cho H, Hong S, Heo J (2014) Productive modeling for development of as-built BIM of existing indoor structures. Autom Constr 42:68–77
4. Badenko V, Fedotov A, Zotov D, Lytkin S, Volgin D, Garg RD, Liu M (2019) Scan-to-BIM methodology adapted for different application. Int Arch Photogram Rem Sens Spat Inf Sci 42:1–7
5. Gao X, Pishdad-Bozorgi P (2019) BIM-enabled facilities operation and maintenance: a review. Adv Eng Inform 39:227–247
6. McArthur JJ (2015) A building information management (BIM) framework and supporting case study for existing building operations, maintenance and sustainability. Proc Eng 118:1104–1111
7. Chen W, Chen K, Cheng JC, Wang Q, Gan VJ (2018) BIM-based framework for automatic scheduling of facility maintenance work orders. Autom Constr 91:15–30
8. Hu ZZ, Tian PL, Li SW, Zhang JP (2018) BIM-based integrated delivery technologies for intelligent MEP management in the operation and maintenance phase. Adv Eng Softw 115:1–16
9. Azhar S (2011) Building information modeling (BIM): trends, benefits, risks, and challenges for the AEC industry. Leadersh Manag Eng 11(3):241–252
10. Denzer AS, Hedges KE (2008) From CAD to BIM: educational strategies for the coming paradigm shift. In: AEI 2008: building integration solutions, pp 1–11
11. Liu R, Issa RR (2016) Survey: common knowledge in BIM for facility maintenance. J Perform Constr Facil 30(3):04015033
12. Yin X, Liu H, Chen Y, Wang Y, Al-Hussein M (2020) A BIM-based framework for operation and maintenance of utility tunnels. Tunn Undergr Space Technol 97:103252
13. Russell M (2012) Using MBSE to enhance system design decision making. Proc Comput Sci 8:188–193

14. Chen C, Tang L (2019) BIM-based integrated management workflow design for schedule and cost planning of building fabric maintenance. Autom Constr 107:102944
15. Becker S, Peter M, Fritsch D (2015) Grammar-supported 3d indoor reconstruction from point clouds for "as-built" BIM. ISPRS Ann Photogram Rem Sens Spat Inf Sci 2(3):17
16. Honic M, Kovacic I, Sibenik G, Rechberger H (2019) Data-and stakeholder management framework for the implementation of BIM-based Material Passports. J Build Eng 23:341–350
17. Bolshakov N, Plyako A, Celani A, Azhimova L, Akimov L (2021) Digital asset in the system of real estate management. In: E3S web of conferences, vol 263. EDP Sciences, p 04039
18. Madni AM, Madni CC, Lucero SD (2019) Leveraging digital twin technology in model-based systems engineering. Systems 7(1):7
19. Love PE, Zhou J, Matthews J, Luo H (2016) Systems information modelling: enabling digital asset management. Adv Eng Softw 102:155–165
20. Chen IJ, Popovich K (2003) Understanding customer relationship management (CRM): people, process and technology. Bus Process Manage J

Studying the Structure Formation of Cement Stone in the Presence of Fine-Disperse Additives

V. D. Cherkasov, V. V. Avdonin, D. N. Pronin, D. A. Varankina, and R. S. Rogozhkin

Abstract The study and creation of highly effective chemical additives and their complexes for the modification of cement systems is an urgent scientific and applied task of building materials science. The article discusses the role of the used finely dispersed additives in the process of structure formation of cement stone, provides data on the effect of these additives on the strength of the MZB. The possibility of creating an effective additive for hardening cement composites on average by 30%, with a dosage of not more than 1.5% by weight of cement based on fine carbon, is shown. This approach is distinguished by environmental safety, technological simplicity and economic efficiency.

Keywords Cement · Finely dispersed additives · Strength · X-ray diffractometry · Structure please · Carbon · Calcium titanate

1 Introduction

Obtaining concretes based on Portland cement binder with high performance properties and low cost is an important task of building materials science [1–4]. One of the ways to solve this problem is the addition of solid-phase finely dispersed additives (SDA) of a certain nature into the initial cement blend [5–15]. These are the TDD of particle surfaces, which, in our opinion, can interact with calcium ions in a solution of blended cement, either by an ion-exchange mechanism or by the principle of formation of colloidal micelles, depending on the nature of the additive.

V. D. Cherkasov · V. V. Avdonin (✉) · D. N. Pronin
N. P. Ogarev's Mordovian State University, Bolshevistskaya Street 68, Saransk, Russia
e-mail: rm2012@inbox.ru

D. A. Varankina · R. S. Rogozhkin
Vyatka State University, Moskovskaya Street 36, Kirov, Russia

P. Akimov et al. (eds.), *Proceedings of FORM 2022*, Lecture Notes in Civil Engineering 282, https://doi.org/10.1007/978-3-031-10853-2_47

2 Methodologies

According to the first mechanism, the adsorption of calcium ions can occur on the surface of such adsorbents as coals of various types, according to the following scheme [16]:

$$\text{coal}^{2-}2\text{H}^{+} + \text{Ca}^{2+} + 2\text{OH}^{-} \Leftrightarrow \text{coal}^{2-}\text{Ca}^{2+} + 2\text{H}^{+} + 2\text{OH}^{-}. \quad (1)$$

At the same time, calcium ions quite strongly bind to the surface of the coal; as a result, the formed charged particles can later play the role of additional crystallization centers for the products of primary hydration of cement (and this is mainly calcium hydroxide) from a supersaturated cement slurry. The uniform distribution of TDD particles in the original cement-containing system will ensure a more uniform flow of structure formation processes, and ultimately, an increase in the strength of the cement composite.

According to the second mechanism, the adsorption of calcium ions from an aqueous solution of lime can occur on the surface of hardly soluble solids, in the crystal structure of which there are calcium cations, in accordance with the Fajans-Panet rule. Such substances include, for example, natural mineral perovskite; calcium titanate. In this case the adsorption process will proceed according to the following scheme:

$$\begin{aligned} & m\text{CaTiO}_3(\text{perovskite macroparticle surface}) + n\text{Ca}^{2+} + 2n\text{OH}^{-} \\ & \rightarrow \left[(m\text{SaTiO}_3)n\text{Ca}^{2+}2(n - x)\,\text{OH}^{-}\right]2x\text{OH}^{-}. \end{aligned} \quad (2)$$

3 Results and Discussion

To confirm the above assumption the experiments were carried out to study the effect of fine particles of coals of various types and calcium titanate on the strength characteristics of fine-grained concrete, which are shown in Fig. 1.

From the presented data it follows that the additives used in relatively small amounts relative to the mass of the binder significantly increase the strength of cement concrete. So the maximum increase in strength is achieved with an additive amount of about 1% by weight of cement, and is: 34%—for carbon MUI 99; 28%—BAU-A coal; 23%—for calcium titanate. The most effective TDD tested was carbon MUI 99.

To establish the possibility of interaction of calcium ions with the surface of the particles of the additives used in the cement mortar, X-ray diffraction research [17] of cement composites modified with carbon, coal, and calcium titanate was carried out on an Empyrean Panalytical X-ray diffractometer in the filtered radiation of a copper anode with a Pixcel 3D semiconductor detector. Portland cement M500

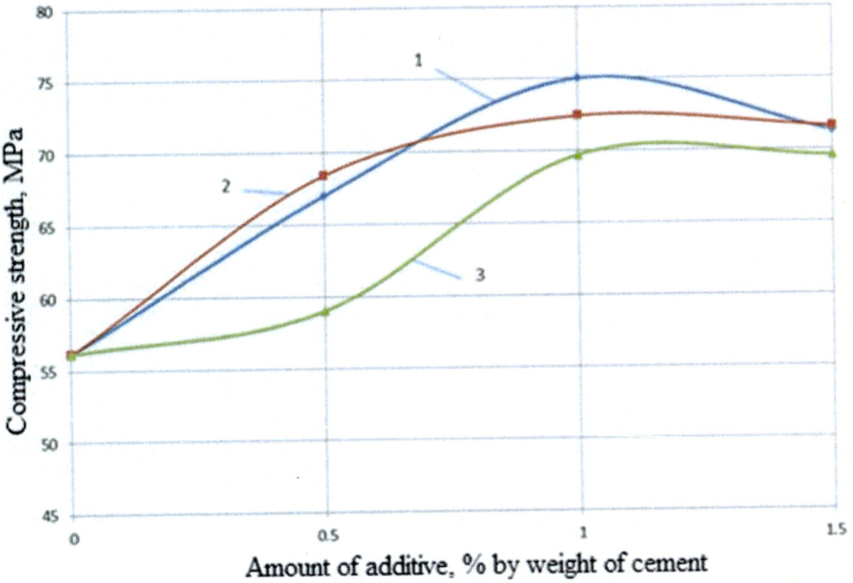

Fig. 1 The strength of the MZB depending on the amount of additives from the mass of cement, where: 1—carbon MUI 99; 2—BAU-A coal; 3—calcium titanate. Materials: cement—750 kg/m^3; sand (0.8–2 mm)—1000 kg/m^3; sand (0.16–0.63 mm)—500 kg/m^3; W/C = 0.3; GP (MasterGleniun) = 1%

was used as the initial binder. X-rays were taken after 28 days of hardening. The analysis of the obtained results showed that for the sample without additive (C-1) (Fig. 2) the intensity of $Ca(OH)_2$ reflections (interplanar distances d = 4.93; 3.11; 2.63; 1.79; 1,69 Å and double Bragg angles 2θ = 17.98; 28.60; 34.04; 50.75; 54.28°, respectively) are significantly larger than for samples with additives. The lowest values of the intensity of reflections of lime are noted for the sample with the addition of carbon (G-1). In samples with additions of charcoal and $CaTiO_3$ (Y-1 and K-1, respectively), these values are approximately equal, but higher than in the sample with carbon addition. Possibly, the lower reflection intensities of lime in the presence of carbon-containing additives compared to the control sample are associated with the adsorption of $Ca(OH)_2$ on carbon particles.

At the same time, the intensity of the diffraction lines of calcite (Fig. 3), in particular at d = 3.029 Å (2θ = 29.47°) is much greater for the sample with the addition of carbon, which may indicate the activation of the carbonization process in the presence of a large amount of lime F.

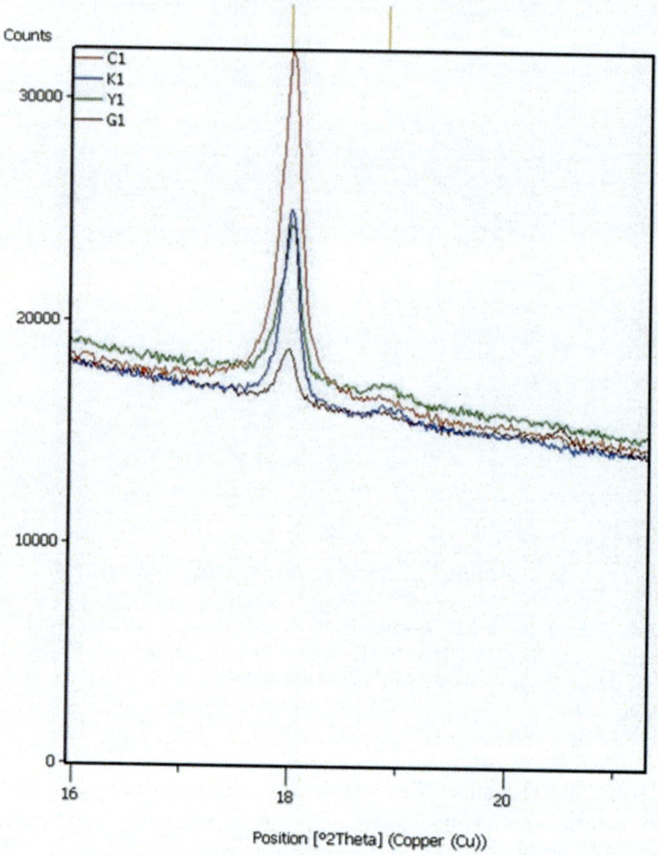

Fig. 2 Intensity of reflections of lime, where: C 1—cement without additives; K 1—calcium titanate; Y 1—BAU-A coal; G 1—MUI 99 carbon

For the sample without additive, the lowest value of the intensity of calcite reflections was noted. In samples with the addition of charcoal and $CaTiO_3$, the value of $CaCO_3$ reflections is higher than in the control sample, but significantly lower than in the sample with the addition of carbon. This allows us to assume that in samples with additives (and especially in the sample with the addition of carbon), the increase in strength can partially occur through the carbonate mechanism.

The value of the reflection intensities of anhydrous calcium silicates (Fig. 4) (d $= 2.706–2.607$ Å; $2\theta = 32.22–32.61°$) for all samples is approximately equal, but for samples with carbon and $CaTiO_3$ additions, it is slightly less than for the control composition and composition with the addition of charcoal. This may indicate that the hydration process in the samples with carbon additions proceeds in a more active form. The difference in the degree of conversion for the samples with the addition of carbon black and charcoal may be due to the difference in particle sizes and their specific surface area.

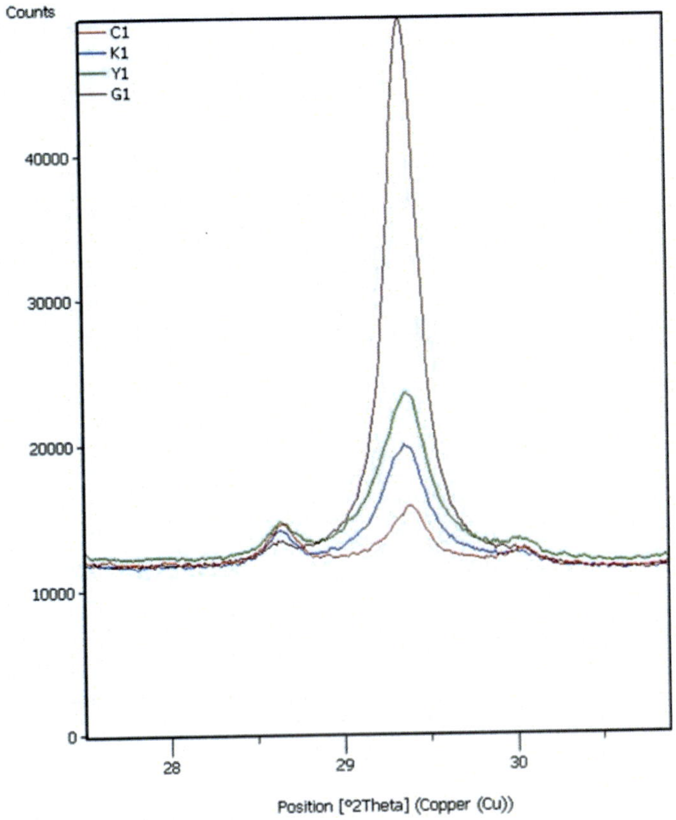

Fig. 3 Intensity of calcite reflections, where: C 1—cement without additives; K 1—calcium titanate; Y 1—BAU-A coal; G 1—MUI 99 carbon

At the same time a corresponding change in the intensity of lime reflections is clearly not observed. It can be assumed that part of $Ca(OH)_2$ binds to other hydrated phases (part of Ca^{2+} ions may be in the structure of ettringite formed at the early stages of hydration) or is adsorbed in crystalline form on particles of charcoal or carbon.

The intensity of ettringite reflections (Fig. 5), (d = 3.88; 2.29; 2.23 Å; 2θ = 22.9; 39.31; 40.42°) is somewhat higher in samples with the addition of carbon and charcoal, which probably indicates the activation of the formation of this phase.

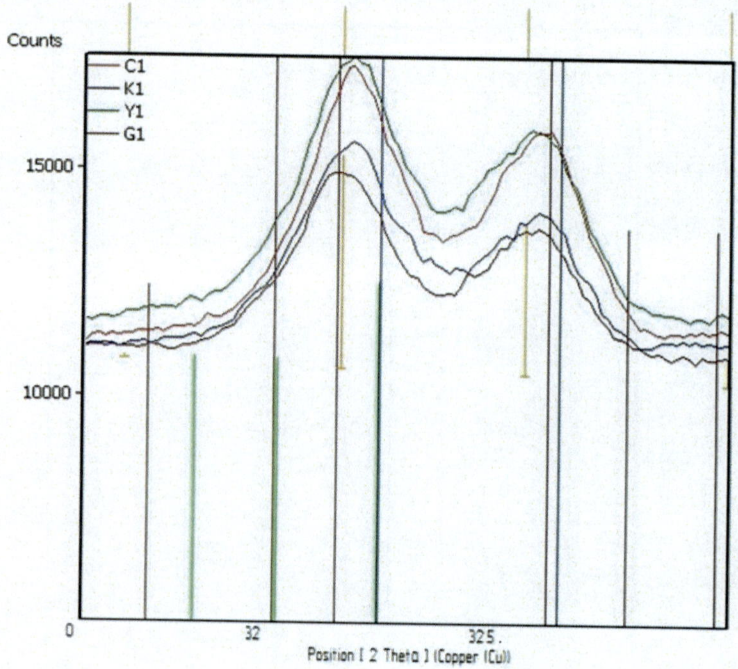

Fig. 4 Intensity of reflections of anhydrous calcium silicates, where: C 1—cement without additives; K1—calcium titanate; Y1—BAU-A coal; G1—carbon MUI 99

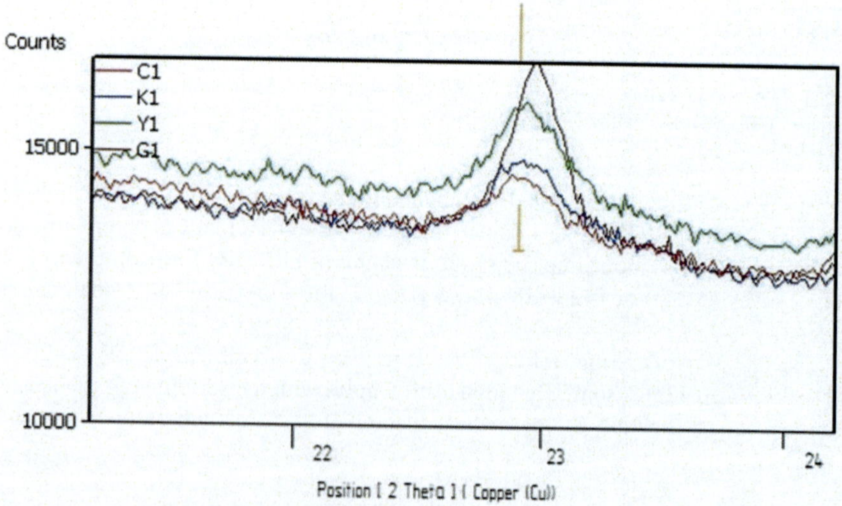

Fig. 5 The intensity of reflections of ettringite, where: C 1—cement without additives; K 1—calcium titanate; Y 1—BAU-A coal; G 1—MUI 99 carbon

4 Conclusions

The obtained results of X-ray phase analysis confirm, to a certain extent, the theoretical assumptions about the active participation of the used additives in interaction with calcium ions in the blended cement slurry in the primary processes.

References

1. Bazhenov YuM, Demyanova VS, Kalashnikov VI (2006) Modifitsirovannye vysokokachestvennye betony [Modified high-quality concretes]. Publishing House DIA, Moscow, 368 p
2. Kaprielov SS, Batrakov VG, Sheinfeld AV (1999) Modified concretes of a new generation: reality and prospects. Concr Reinforced Concr № 6:610
3. Batrakov VG (1998) Modified concretes. Stroyizdat, Moscow, 768s
4. Makridin NI, Korolev EV, Maksimova IN (2013) Structural formation and structural strength of cement composites. MGSU, Moscow, p 152
5. Cherkasov VD, Buzulukov VI, Pronin DN (2016) Modification of cement stone with a complex additive based on super-, hyperplasticizer and carbon powder. Reg Archit Constr (4):40–49
6. Cherkasov VD, Buzulukov VI, Pronin DN (2018) Study of the influence of various fine-disperse additives on physical and mechanical properties of cement composites. Reg Archit Constr (3):62–70
7. Cherkasov VD, Buzulukov VI, Emelyanov AI, Pronin DN (2018) Development of fine hardening additives for cement systems. Izvestia vuzov. Building 3:109–116
8. Scrivener KL, Nonat A (2011) Hydration of cementitious materials - present and future. Cem Concr Res 41(7):641–650
9. Ramachandran AR, Arutzeck MW (1986) Hydration of trical silicate at fixed pH. In: 8th international congress on the chemistry of cement, Rio de Janeiro, vol 3, pp 225–230
10. European Nanotechnology Geteway (2006) Nanotechnology and construction. NanoforumReport, pp 8–12, November 2006
11. Gaitero JJ, Campillo I, Guerrero A (2008) Reduction of the calcium paste by addition of silica nanoparticles. Cem Concr Res 38:1112–1118
12. Khasanov B, Irmuhamedova L, Firlina G, Mirzaev T (2020) Theoretical foundations of the structure formation of cement stone and concrete. 2020 IOP Conf Ser Mater Sci Eng 869:032032
13. Gospodinov P, Kazandjiev R, Mironovab M (1996) The effect of sulfate ion diffusion on the structure of cement stone. Cem Concr Comp 18(6):401–407
14. Viktor S, Lidiya P, Larisa B, Ekaterina L, Olga M (2017) The role of chemical admixures in the formation of the structure of cement stone. MATEC Web Conf 116:01018
15. Morozova M, Akulova M, Ayzenshtadt A, Frolova M (2022) Specifics of structure formation of cement stone with silica raw material - a substrate of photocatalytic composite material. Digit Technol Constr Eng 173:183
16. Kiselev VYa, Komarov VM (2005) Adsorption at the "solid - solution" interface. Tutorial. MITHT, Moscow, 81 p
17. Gorelik SS, Skokov YuA, Rastorguev LN (1994) Radiographic and electron-optical analysis. MISiS, Moscow, 328 p

Methodology of Information Modeling at the Design Stage of Railway Infrastructure Construction Facilities

Maksim Zheleznov, Liubov Adamtsevich, and Angelina Rybakova

Abstract The expansion of the transport system in Russia is one of the urgent tasks of economic development. The high level of transport infrastructure ensures the development of many business lines, simplifies the procedure of cargo transportation, and also creates comfortable conditions for the movement of passengers. Currently, a new stage of development has begun for railway transport, associated not only with its demand, but also with the introduction of new technologies and working methods. One of the most promising areas of work today is information modeling technologies, which include a large number of different functions and implementation possibilities. Therefore, the purpose of this work is to analyze the methodology of information modeling at the design stage of railway infrastructure construction facilities, and the task is to form an algorithm for designing and interacting with an information model. The article describes the main features of transport infrastructure facilities, the specifics of work based on information modeling throughout the life cycle, and also analyzes the functionality of information modeling at the design stage. As a result, an analysis of the effectiveness of using information modeling technology tools for the main tasks of designing railway infrastructure facilities is presented, and an algorithm for developing an information model at the design stage is formed.

Keywords Building information modeling · Complete modular units · Prefabrication · Design automation · Construction management · Modular structure · Modular construction

M. Zheleznov · L. Adamtsevich · A. Rybakova (✉)
Moscow State University of Civil Engineering, Moscow 129337, Russian Federation
e-mail: AngelinaRibakova@yandex.ru

M. Zheleznov
e-mail: ZheleznovMM@mgsu.ru

L. Adamtsevich
e-mail: AdamtsevichLA@mgsu.ru

Sirius University of Science and Technology, 1 Olympic Ave, Sochi 354340, Russian Federation

P. Akimov et al. (eds.), *Proceedings of FORM 2022*, Lecture Notes in Civil Engineering 282, https://doi.org/10.1007/978-3-031-10853-2_48

1 Introduction

Information modeling technologies (BIM) are a set of interrelated processes based on the use of intelligent 3D models that include the maximum amount of information about the future object. Information modeling covers all stages of the life cycle, from concept to operation and, increasingly, even to dismantling. The purpose of information modeling technology is to impoverish all the processes of urban planning. BIM technologies are one of the most promising areas not only in the construction industry, but in related engineering sciences [1].

Even though computing and graphics software has evolved over several years, the adoption of Information Modeling (BIM) in the construction industry is quite slow compared to other industries. Today, however, Building Information Modeling (BIM) covers the entire life cycle of a building. In addition to the 3D representation of the building, the model also includes other characteristics of the elements. A complex 5D model consists of time (4D) and cost (5D) parameters. The 6th dimension is currently being propagated: the report during the life cycle. However, full implementation is not yet possible due to the complexity of implementation and the low level of software functions in this matter [2, 3].

The development of each subsequent parameter (4-5-6D) is determined not only by the development of new tools and software capabilities, but also by a more complete and high-quality building information model (3D). There are a large number of approaches, methods and standards for developing the most detailed model of buildings or structures, however, the existing theoretical base is insufficient for developing a railway infrastructure model [4].

Unlike capital construction projects, the railway infrastructure is characterized by the following complexities of information modeling:

- lack of a special software package for railway facilities;
- minimum number of railway infrastructure families developed;
- low level of standardization and lack of a regulatory framework;
- shortage of specialists in the field of information modeling of the railway infrastructure;
- lack of education and training systems for specialists.

Despite the rapid growth and spread of information modeling technologies in transport infrastructure, the above obstacles are of great importance for slowing down the further development of both railway infrastructure and BIM technologies in this industry.

To overcome all the difficulties, it is advisable to consider foreign experience in the implementation of transport facilities based on information modeling technologies. [5] The success is based on a complex of factors: an effective design methodology, the use of experience in capital construction, close interaction with government agencies, focus on the next stages of the life cycle, a high level of collaboration (Fig. 1).

On the territory of Russia today there are no implemented objects from this list, which gives a lot of space for activity, development of own projects and new scientific

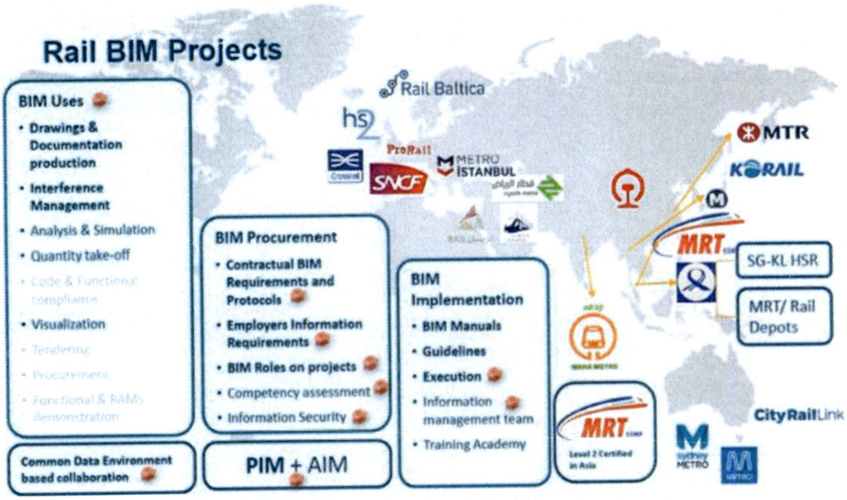

Fig. 1 Map of the largest BIM-projects of the railway infrastructure [5]

research. Based on the analysis of foreign experience, it is possible to identify the key factors for successful implementation, as well as the features and indicators of objects. Thus, it is advisable to form a methodology for the effective development of information models of railway infrastructure facilities [6, 7].

2 Methods

2.1 *Features of Railway Infrastructure Facilities*

The functional purpose of railway transport facilities is the transportation of passengers and goods. Railway transport infrastructure is a complex of interconnected structures that provide the entire transportation process. From an economic point of view, railway infrastructure is a separate industry that includes a number of facilities, processes, and specialists [8].

Thus, the railway infrastructure is a complex that includes linear railway tracks, wagons, stations, traffic control and regulation systems, information complexes, power supply devices, means of loading and unloading operations, as well as complexes for their technical operation. Consequently, railway infrastructure facilities are divided into two conditional groups: railway tracks and related structures.

Information modeling of objects of each group has significant differences. Features of objects form the features of their development and modeling. The formation of models of railway tracks includes the following features:

- availability of a model unit of the track sector in accordance with standards and norms;
- possibility of operational one-dimensional scaling;
- availability of splitters and connecting elements;
- the possibility of correcting the model taking into account changes in the input data;
- availability of additional accompanying elements;
- the possibility of automated construction of two-dimensional drawings.

Modeling of another group of railway structures is most comparable with capital construction projects [9, 10]. However, taking into account the peculiarities of the infrastructure, these objects have specific features:

- relatively small area and number of storeys;
- prefabricated;
- interaction with railway tracks;
- availability of standard projects and their systematization;
- high level of technological requirements.

When designing railway infrastructure facilities, it is necessary to take into account not only the above specifics, but also the peculiarity of the infrastructure as a whole: a large territorial extent, climatic differences, a high percentage of location outside the city [11, 12]. Therefore, unlike capital construction projects, a number of conditions are formed that complicate the design, even despite the small indicators of technical and economic indicators.

2.2 *Information Modeling Toolkit for Railway Infrastructure Objects*

In the process of developing information models of railway facilities, the same tools are used as for capital construction projects [13]. However, taking into account the peculiarities of the railway, some of the tools are used in a non-standard way or are developed additionally. The functionality of popular information modeling software systems in most cases is quite identical, which makes it possible to determine compliance with the elements performed and design tasks (Table 1).

Based on the results of the analysis of the comparison of the main tasks and the functionality of information modeling for railway infrastructure facilities, it is possible to determine the correspondence and the maximum possibility of performing any tasks. Based on the functionality of popular software systems, it is possible to fully develop an information model of absolutely any railway facility [14].

However, when considering the model in terms of time duration, it is advisable to consider not only the functionality, but also its effectiveness within the entire life cycle of the object. Some information modeling tools perform different functions

Table 1 Comparison of information modeling functionality for railway infrastructure facilities

№	Task group information modeling tool	Task group information modeling tool
1	Architecture	Architectural tools Concept modeling tools Material settings Importing objects
2	Constructive	Constructive tools Architectural tools Importing objects Family or feature tools
3	Engineering support	Engineering systems tools Add-ons and base from the manufacturer Family or feature tools with added types
4	Development of documentation and drawings	Documentation tools Visibility control tools
5	Development of a scalable sector model unit	Model groups Family tools Family element geometric modeling tools Editing tools
6	Development of splitters, connecting and additional elements	Family tools Family element geometric modeling tools Editing tools
7	Formation of a library of typical objects for reuse	Model groups Family tools Import and export tools
8	Fastening and docking of elements	Constructive tools Architectural tools Importing objects Family or feature tools
9	Grouping and joining	Model groups Family tools

at different stages of the life cycle. Thereby increasing the overall efficiency of information modeling.

2.3 *Application of Information Modeling Tools at Various Stages of the Life Cycle*

The design, construction and operation of railways is a complex task for all participants in the life cycle of an object. Projects for the construction of railway infrastructure facilities include a large number of issues related to both the existing infrastructure and the solution of local technical problems [15].

Information modeling technologies make it possible to have information about an object at all stages of its life cycle - from the development of technical specifications to its decommissioning [16]. In addition to concretizing the indicators of the object and its cost, simplifying communication and interconnection between specialists and tasks at different stages of the life cycle, new prospects for organizing and ensuring the safety of work appear, especially in projects for the reconstruction of existing railway infrastructure facilities [17] (Table 2).

With this method of dividing the life cycle into stages, one can see not only the functionality of solving certain tasks, but also the positive effect of information modeling [18]. Simultaneously with the simplification of the work of a particular interested specialist, the exchange of data, their updating, the change of employees, the preparation of reporting documents and visualization are accelerated. Considering that most new railway projects are closely interconnected with existing infrastructure, the presence of a digital model greatly simplifies the design work, minimizes errors and speeds up the entire construction process.

Table 2 Information modeling functions at life cycle stages

№	Stage	Information modeling function
1	Concept	- Quickly create multiple project options - Quick creation of a construction site using open data sources and geodetic surveys
2	Engineering survey	- Using the point cloud obtained during laser scanning as the initial data for building a model - Processing and integration into the BIM process of engineering surveys
3	Design	- Creation of information models of railways. Obtaining project documentation - Design of culverts and other artificial structures - Checking visibility distance
4	Working documentation	- Creation of working documentation - Obtaining the necessary statements and reports
5	Construction	- Creation of accurate estimates and bills of quantities for the object - Using up-to-date information from the model at the construction site
6	Coordination and interaction	- Construction simulation - Coordination of adopted design decisions and maintenance of a shared data environment - Distribution of roles in the project with appropriate levels of access to project data
7	Exploitation	- Maintaining an operational information model of culverts
8	Dismantling	- Integration of data from all stages of the life cycle; - Schedule control; - Simplification of the control procedure;

3 Results

As a result of the analysis of information modeling at the design stage of railway infrastructure facilities, several fundamental features are formed. Regardless of the purpose, each object interacts in one way or another with the existing railway infrastructure, which complicates the implementation of some tasks [19, 20]. However, the spread of information modeling tools for various purposes simplifies the work.

At the design stage, seven main tasks can be identified (paragraphs 3–4 of Table 2):

1. Import (use) data of the pre-project stage.
2. Development of an information model of the railway.
3. Engineering communications design.
4. Model verification.
5. Formation of statements and specifications.
6. Formation of documentation
7. Export the model for the next stages of the life cycle.

To solve each problem, previously developed information model data is used in one form or another. The efficiency of the design stage depends on the quality and usefulness of previous developments. Thus, the algorithm for information modeling of railway facilities at the design stage is a complex process interconnected on the model data (Fig. 2):

Interaction with the information model, including its adjustment, is performed at all stages of the life cycle [21]. Due to this, not only the development and improvement of the model takes place, but also the interaction and coordination between different specialists at different stages. However, the implementation of this technique is possible with the appropriate software (Autodesk Civil 3D, Autodesk InfraWorks Autodesk Recap Pro, Autodesk Revit, Autodesk Navisworks, etc.) and functionality (Table 1). As a result, along with the project documentation, an information model of

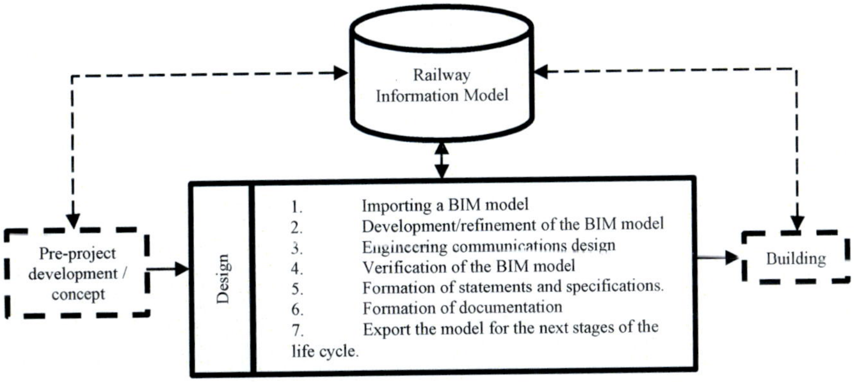

Fig. 2 Algorithm of information modeling of railway objects

Fig. 3 Railway information model

the railway is formed, which in the future is used not only as an object of operation, but also as initial data for analyzing and simplifying the development of new railway infrastructure facilities (Fig. 3) [22].

4 Conclusion

The results of this study are theoretically and practically significant for activities in the field of railway infrastructure construction and information modeling. In the first case, the main elements and tasks are identified, which can be significantly rationalized both for the future functioning of the railway, and to simplify the local work of designing certain objects. For information modeling technologies, a list of the most versatile functions and tools is being formed that allow solving non-standard problems in the field of transport construction, and at the same time form promising areas for developing tools in the field of all types of transport.

A promising direction based on this study is the analysis of the integration of the formed tasks and functions, their interaction and effectiveness within the framework of the railway infrastructure. That is, the next step is to analyze the methodology of information modeling at the stage of construction and operation of railway infrastructure facilities.

Acknowledgements The reported study was funded by RFBR, project number 20-38-51013.

References

1. Vignali V, Acerra EM, Lantieri C, Vincenzo FD, Piacentini G, Pancaldi S (2021) Building information Modelling (BIM) application for an existing road infrastructure. Autom Constr 128:103752

2. Abbondati F, Biancardo SA, Palazzo S, Capaldo FS, Viscione N (2020) I-BIM for existing airport infrastructures. Transp Res Proc 45:596–603
3. Chong HY, Lopez R, Wang J, Wang X, Zhao Z (2016) Comparative analysis on the adoption and use of BIM in road infrastructure projects. J Manag Eng 32(6):05016021
4. Ehrbar H (2016) Building information modelling–a new tool for the successful implementation of major projects of German railways. Geomech Tunnelbau 9(6):659–673
5. Building Information Modelling (BIM) in Railways for Design, Construction, Operation and Asset Management, 21 May 2019. https://events.development.asia/materials/20190521/buildinginformation-modelling-bim-railway-design-construction-operationand
6. Kuprijanovskij VP, Sinjagov SA, Dobrynin AP (2016) BIM Digital Economy. How to achieve success? A practical approach to the theoretical concept. Part 1. Approaches and main advantages of BIM. Int J Open Inf Technol 4(3):1–8
7. Aziz Z, Riaz Z, Arslan M (2017) Leveraging BIM and big data to deliver well maintained highways. Facilities 35(13):818–832
8. Fanning B, Clevenger CM, Ozbek ME, Mahmoud H (2015) Implementing BIM on infrastructure: comparison of two bridge construction projects. Pract Period Struct Des Constr 20(4):04014044
9. Latiffi AA, Mohd S, Kasim N, Fathi MS (2013) Building information modeling (BIM) application in Malaysian construction industry. Int J Construct Eng Manag 2:1–6
10. Leone M, D'Andrea A, Loprencipe G, Malavasi G, Bernardini L (2017) Building information modeling (BIM): prospects for the development of railway infrastructure industry. In: Proceedings of AIIT International Congress TIS Rome, 10–12 April 2017, Rome
11. Castañeda K, Sánchez O, Herrera RF, Pellicer E, Porras H (2021) BIM-based traffic analysis and simulation at road intersection design. Autom Constr 131:103911
12. Pasetto M, Giordano A, Borin P, Giacomello G (2020) Integrated railway design using infrastructure-building information modeling. The case study of the port of Venice. Transp Res Procedia 45:850–857
13. Zhao J, Knoop VL, Wang M (2020) Two-dimensional vehicular movement modelling at intersections based on optimal control. Transp Res B Methodol 138:1–22
14. Yu B, Zhang M, Wang Z, Bian D (2014) Dynamic translation for virtual machine based traffic simulation. Simul Model Pract Theory 47:248–258
15. Chao Q et al (2020) A survey on visual traffic simulation: models, evaluations, and applications in autonomous driving. Comput Graph Forum 39(1):287–308
16. Biancardo SA, Capano A, De Oliveira SG, Tibaut A (2020) Integration of BIM and procedural modeling tools for road design. Infrastructures 5(4):1–14
17. Zheleznov M, Adamtsevich L, Vorobiev P, Rybakova A (2021) Analysis of building information modeling technologies for transport infrastructure objects at the stages of the life cycle. E3S Web Conf 263(6):05030
18. Bongiorno N, Bosurgi G, Carbone F, Pellegrino O, Sollazzo G (2019) Potentialities of a highway alignment optimization method in an I-BIM environment. Periodica Polytech Civ Eng 63(2):352–361
19. Adibfar A, Costin A (2019) Next generation of transportation infrastructure management: fusion of intelligent transportation systems (ITS) and bridge information modeling (BrIM). In: Mutis I, Hartmann T (eds.) Advances in Informatics and Computing in Civil and Construction Engineering. Springer, Cham. https://doi.org/10.1007/978-3-030-00220-6_6
20. Costin A, Adibfar A, Hu H, Chen SS (2018) Building information modeling (BIM) for transportation infrastructure–literature review, applications, challenges, and recommendations. Autom Constr 94(7):257–281
21. Häußler M, Borrmann A (2020) Model-based quality assurance in railway infrastructure planning. Autom Constr 109:1–15
22. Autodesk: Programs for design. Architecture, design and construction (2022). https://autodesk.ru/

Comparative Analysis of Domestic and Foreign Methods for Processing Tests for Torsional Shear Device

A. Z. Ter-Martirosyan, V. V. Sidorov, E. S. Sobolev, and A. S. Almakaeva

Abstract There is discusses in the article the features of processing test reports on a ring shear device and studies the history of the emergence of the formulas used to determine shear stresses in the shear plane according to Russian and foreign standards. The purpose of the article is to determine the most correct method for processing test results. 9 tests were carried out on a ring shear device at three normal stresses: 100, 200 and 400 kPa. The tests were carried out at a constant speed of 0.1138°/min and a constant vertical pressure. Solid loams were used for testing. Soil samples were cut from monolithic cores. The soils were taken from the construction site of the city of Moscow. Based on the results of laboratory experiments, graphs of the dependence of shear stresses on linear shear strains $\tau = f(d)$ were plotted. Shear stresses were determined according to the formula regulated by GOST R 59,937-2021, however, it was found that when using the formula presented in ASTM D 6467 - 99 the shear stresses will be 2 times lower. There is in the article provides a derivation of the formula for determining shear stresses during torsion and describes an explanation that determines the reason for the differences in the formulas.

Keywords Torsional shear device · Comparative analysis · Processing tests

1 Introduction

In experimental studies of soils the correct laboratory testing and its processing are of great importance in determining their parameters and components of the stress–strain state. In this article in order to determine the effective shear stresses in a ring shear device laboratory tests were carried out in accordance with the current regulatory documents. However during the study of domestic and foreign regulatory literature, differences were revealed in the formulas for determining the shear stress specified in GOST R 59,937-2021 Soils. Methods for laboratory determination of strength characteristics with a ring shear device and in ASTM D 6467 - 99 Standard Test

A. Z. Ter-Martirosyan · V. V. Sidorov · E. S. Sobolev · A. S. Almakaeva (✉)
Moscow State University of Civil Engineering, Moscow 129337, Russian Federation
e-mail: totilas96@mail.ru

P. Akimov et al. (eds.), *Proceedings of FORM 2022*, Lecture Notes in Civil Engineering 282, https://doi.org/10.1007/978-3-031-10853-2_49

Method for Torsional Ring Shear Test to Determine Drained Residual Shear Strength of Cohesive Soils. At the same time, the value of the calculated shear stress calculated according to GOST R 59,937-2021 is 2 times higher than according to ASTM D 6467 - 99, which is a significant difference and requires detailed consideration.

2 Literature References

In world practice, there are several devices that allow determining the shear stresses and strength characteristics of soils: a triaxial compression device, a direct shear device, a ring shear device and a simple shear device. In Russia devices of direct shear and triaxial compression have received the greatest popularity and distribution, as well as the regulatory and technical base. They have their set of advantages and disadvantages. A ring shear and a simple shear devices are widely known and used abroad. However, with the release of the Russian standard GOST R 59,937-2021, which establishes requirements for laboratory methods for determining the resistance to torsional shear of dispersed soils, it became possible to conduct experiments on a ring shear device in domestic practice. This device has application features: firstly, the shear plane is constant and, unlike the direct shear device, the ring shear device allows to get more reliable results and secondly, this allows to provide large shear deformations including narrower formed shear plane. As a result, peak and residual characteristics of soil strength can be determined. The latter are of great importance for problems associated with the loss of stability of slopes because for this type of problem it is recommended to use residual strength characteristics. Nevertheless, the ring shear device is almost the first known design of the device developed in Russia to determine the strength characteristics of soils. For the first time, this device was mentioned in the work of the Russian scientist Nemilov N. in 1913 [1]. In addition, several Russian scientists were engaged in its modification and improvement: Ter-Stepanyan G.I. (1955), Meschyan S.R. (1962), Vyalov S.S. (1962), Lomize G.M. (1970), Goldstein M.N. (1979), as well as many foreign scientists. However for a long time in Russia, the ring shear device was not certified for laboratory testing and did not have a regulatory and technical base. This was probably due to the complexity of preparing the sample for testing and more complex processing of the results than in the same direct shear device, for example. In world practice a significant contribution to the development and adaptation of the device for use in scientific and construction purposes was made by Bishop A.W. in 1971 [2]. The design that he proposed is the most popular all over the world at the present time because Bishop A.W. eliminated the shortcomings of the previous designs of the ring shear device: provided a small gap between the rings, eliminated lateral friction of the soil sample against the rings, made it possible to test at low stress levels, etc. A few years later, in 1979, Bromhead [3] proposed a different design for the shear torsion tool, which served as the basis for the creation of the English standard for laboratory testing.

In addition, attempts have been made to study the applicability of these devices to different types of problems in order to select a device that allows to determine the most reliable results of the obtained strength parameters. Moreover, these studies had to explain the reasons why there are differences in the results. Similar studies were carried out by Hanzava et al. [4] who showed that a direct shear device for clay soils gives higher values of strength and stiffness while in a simple shear device, on the contrary, they are lower. The authors believe that this is due to the different mechanism of destruction of the soil sample in these devices. The same conclusions were also made by the authors Tsubakihara Y. and Kishida H. (1993) [5] during tests of clay soils. Comparative tests of sandstone in direct shear and ring shear devices showed that there is no significant difference in the values of the peak shear stress but the further behavior of the soil after the peak will differ depending on the test conditions [6]. The conducted studies indicate that the direct shear device gives unreliable results because during testing the shear plane changes and the distribution of shear stresses is uneven [7–9]. The error of values is 20% per 1 cm which is unacceptable in the design and constraction [10]. Previously, the authors [11] carried out comparative tests of clay soils in a direct shear device and ring shear device which showed that in the first case, the strength of soils is higher during the entire test. Hammoud F., Boumekik A. and Lemos L.J.L., Vaughan P.R. have been studying the nature of soil destruction at contact with solid materials on a ring shear device and determining the thickness of the shear zone involved in the work [12, 13]. They determined that the nature of soil failure and the thickness of the shear zone would depend on the roughness of the contact and the type of soil.

The residual strength characteristics are significantly affected by normal stress [2, 3]. The higher the normal stress, the lower the residual strength. The experiments that were carried out by the authors [14–16] showed that the residual strength is also affected by humidity and the size of the fractions of clay particles. Based on the results obtained on a ring shear device, the authors developed a correlation dependence that allows estimating the residual strength of soils depending on its moisture content [17].

3 Methods

The ring shear device allows to determine the parameters of the residual strength of soils which are important in solving problems related to the stability of slopes and retaining walls. Due to the continuous rotation of one part of the ring relative to the other it is possible to provide large shear deformations and the possibility of re-shearing along an already formed plane. Similarly, in nature there is a loss of stability of the slope.

In this work a ring shear device was used to test soils. This is combined with certified and calibrated control and measuring equipment manufactured by NPP Geotek LLC which included a set of GT 2.0.7 10 kN pneumatic loader and control units. The general view of the device is shown in Fig. 1.

This device allows to carry out kinematic torsional shearing by turning the lower part of the ring at a known speed. The shear resistance is measured with a horizontal

Fig. 1 General view of the ring shear device

force transducer connected by a lever to the upper part of the ring and preventing it from turning together with the lower part. The engagement of the sample, which excludes its rotation inside the ring during torsional shear, is provided with the help of blades located in the side parts of the rings as well as in its upper and lower parts. The view of the rings with blades is shown in Fig. 2.

The tests were carried out according to the consolidated-drained scheme in accordance with GOST R 59,937-2021. Solid loams were taken as soil samples, which were taken from the construction site of the Moscow Metro in the Fili-Davydkovo area from a depth of 34–36 m. They were cut from monolithic cores. The sample diameter and height are 71.4 and 35 mm, respectively, which correspond to the

Fig. 2 General view of the device rings

Table 1 Physical properties of soil

Type of soil	Humidity of the lower plastic limit W_P, %	Humidity of liquid limit W_L, %	Plasticity number I_P
Loam	28	44	0.16

dimensions of the inside of the device's rings. Before the beginning of each experiment the soil samples were kept under a vertical load corresponding to the normal pressure during the shear test: 100, 200 and 400 kPa, for at least 8 h. The shear was carried out at a constant speed of 0.1138°/min and a constant vertical load until a rotation angle of 17° was reached. The main physical properties of the soil are shown in Table 1.

4 Results

Based on the results of the experiments test reports were obtained containing an array of data on the measured stresses and strains. However the shear stress is not constant in the shear plane but it changes according to a linear law: in the center of the section, it is equal to zero and at the points of the outer contour it reaches maximum values. Therefore, it is necessary to determine the average shear stress in accordance with clause 9.1 of GOST R 59,937-2021 using the formula:

$$\tau = 10\frac{3M_t}{2\pi(\mathrm{R}_a^3 - \mathrm{R}_i^3)} \tag{1}$$

where M_t – the moment of the rotational force applied to the sample, kN cm;

R_a – the outer radius of the annular sample, cm;

R_i – the inner radius of the annular sample, cm.

Due to the fact that the tested soil samples had the shape of a solid cylinder, formula (1) took the following form:

$$\tau = 10\frac{3M_t}{2\pi\,\mathrm{R}_a^3} \tag{2}$$

A similar formula was described by Boldyrev [18]. For plots of the dependence of shear stresses on linear shear strains $\tau = f(d)$, the formulas presented in [11] were used to convert angular strains in degrees to equivalent linear strains in millimeters. The resulting graphs are shown in Fig. 3.

As shown in Fig. 4, increasing of vertical load is the reason of increasing of shear resistance: at a vertical pressure of 100 kPa, peak shear stresses vary from 90 to 120 kPa, residual - from 70 to 90 kPa, at a vertical pressure of 200 kPa, peak values of shear stresses vary in the range from 170 to 200 kPa, residual - from 140 to 160 kPa, and at a vertical pressure of 400 kPa peak stresses ranged from 240

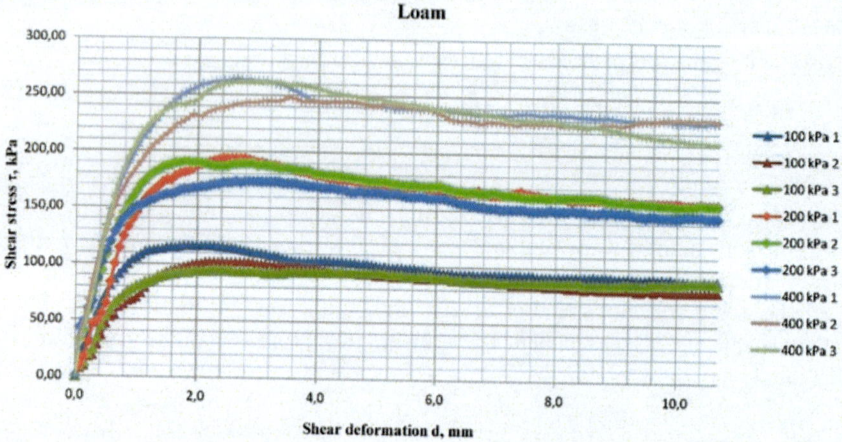

Fig. 3 Graphs of dependence of shear stresses on shear strains $\tau = f(d)$

Fig. 4 Tested soil sample (hard loam)

to 260 kPa, residual vary in the range from 210 to 230 kPa. On average, there is a decrease in strength by 25–30% after the shear stresses reach peak values while most of the plotted graphs have pronounced peaks. When shear strains reach values approximately equal to 6–7 mm shear stresses reach constant values, i.e. it is residual strength.

Figure 4 shows a tested sample of clay soil. In the shear plane it is possible to see furrows indicating the nature of the shear - rotation. In addition, the shear plane is not smooth because there are depressions and elevations which can be one of the reasons for the appearance of pronounced peaks in the graphs. A more pronounced effect was observed in clays when the presence of irregularities had a direct relationship with the appearance of one or even two pronounced peaks in the graphs [19].

During the study of the foreign standard ASTM D 6467 - 99 it was revealed that the formula for determining the average value of shear stress differs from that presented in GOST R 59,937-2021. According to a foreign source the difference lies in the numerical constant: in GOST R 59,937-2021 it is 3/2 and in ASTM D 6467 - 99 - 3/4:

$$\tau = \frac{3(F_1 + F_2)L}{4\pi(R_2^3 - R_1^3)} \tag{3}$$

where F_1 and F_2 – measured loads applied to the sample, kN;

L – leg length, cm;

R_2 – the outer radius of the annular sample, m;

R_1 – the inner radius of the annular sample, m.

Thus, there was a need for a detailed study of where the formula for determining the shear stress came from. According to the discipline of resistance of materials [20] it is known that the formula for determining shear stress is as follows:

$$\tau = \frac{M}{W_p} \tag{4}$$

where M – torsion moment, kN;

W_p – the polar moment of cross section.

In order to calculate the polar moment of a cross section of a circular or annular shape consider an elementary area of thickness dp located at a distance p from the center of rotation. If consider the entire section of the circle then the elementary areas form a ring with a thickness of dp (Fig. 5).

The polar moment of resistance of an elementary ring is defined as the sum of multiplying the areas of dp and their distances to the center of rotation taken over the entire area:

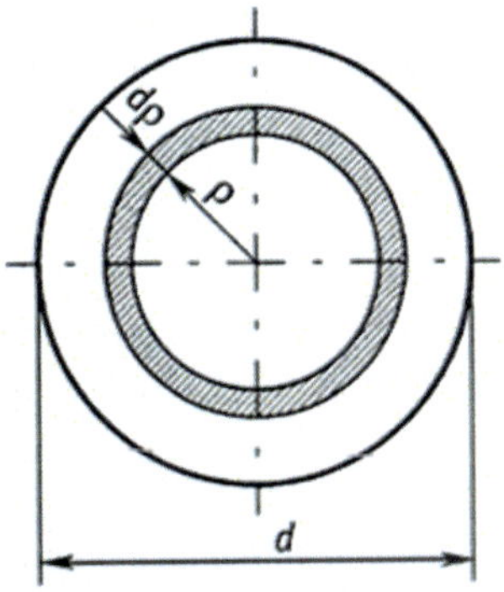

Fig. 5 To determine the polar moment of resistance of an elementary ring with a thickness *dp*

$$dW_{pk} = \int_S p \cdot dS = p \cdot S = p \cdot 2\pi pdp \tag{5}$$

where S – area of the elementary ring.

The polar moment of the entire circular section is determined by differentiating expression (5) with respect to the area of the circle:

$$dW_p = \int_0^R p \cdot 2\pi pdp = \int_0^R 2\pi p^2 dp = 2\pi \frac{R^3}{3} \tag{6}$$

where R – the radius of the circle.

It should be noted that these formulas are valid for a solid circular section. The polar moment of resistance of the annular section will be determined as the difference between the polar moments of resistance of the large and small circles.

Thus, if it substitutes the result obtained from expression (6) into expression (4), then the formula for determining the shear stress will take the following form:

$$\tau = \frac{3M}{2\pi R^3} \tag{7}$$

That is, the derived formula for determining the shear stress during torsion corresponds to the formula presented in GOST R 59,937-2021.

However, according to the description of the torsional shear device presented in the foreign standard ASTM D 6467 - 99, the measurement of the torque M_t is carried out using a torsion beam and two strain gauges. In this case, GOST R 59,937-2021 provides a formula for determining the torque:

$$M_t = (F_1 + F_2)L/2 \tag{8}$$

where F_1 и F_2 – the measured loads applied to the torsion beam, kN;

L – the distance between the points of application of forces F_1 and F_2, m.

If expression (8) for torque is substituted into expression (7), then the formula for determining shear stress will be obtained, presented in the foreign standard ASTM D 6467 - 99.

Thus, in the Russian regulatory document GOST R 59,937-2021 a formula is presented for determining the shear stress in ring shear in a general form and in the foreign standard ASTM D 6467 - 99 - in a particular form for a specific device design described in the same document. Therefore, a difference appeared in the formula for shear stresses. Due to the fact that there is a wide variety of designs of the ring shear device it is recommended that the torque M_t be determined each time individually as the sum of the moments of the shear forces applied to the soil sample relative to the center of rotation, i.e. the center of the cross section of the soil sample.

5 Conclusions

1. Laboratory tests were performed on a кштпl shear device for hard loam at three normal stresses: 100, 200 and 400 kPa. The torsional shear was carried out at a constant speed of 0.1138°/min.
2. Based on the test results plots of shear stresses versus shear strains $\tau = f(d)$ were constructed which showed that with an increase of normal load shear resistance increases and that after reaching the peak value shear stresses decrease by 25–30% when shear strains reach 6–7 mm.
3. According to the presence of notches and hills in the shear plane it can be assumed that they caused the appearance of pronounced peaks in the plots of shear stress versus shear strain.
4. During the processing of test reports and plotting a difference was revealed in the formulas for shear stresses presented in Russian GOST R 59,937-2021 and foreign ASTM D 6467 - 99. As a result of studying the origin of the formulas it was found that the difference arose due to design features of a ring shear device certified for testing. In GOST R 59,937-2021, the formula for determining shear stresses is given in a general form, and in ASTM D 6467 - 99 - for the particular case when the torque M_t is measured using a torsion beam and two strain gauges.
5. In general it is recommended that the torque be determined each time when a test is performed with using a new construction of device. It is according with the large number of existing designs of the ring shear device in the world.

References

1. Nemilov N (1913) On the theory of loose bodies. J Minist Commun Russia. Book 9
2. Bishop AW et al (1971) A new ring shear apparatus and its application to the measurement of residual strength. Geotechnique 21:273–328
3. Bromhead EN (1979) Mikasa's a simple ring shear apparatus. Ground Eng 15:40–44
4. Hanzawa H, Nutt N, Lunne T, Tang YX, Long M (2007) A comparative study between the NGI direct simple shear apparatus and the Mikasa direct shear apparatus. Soils Found 47:47–58
5. Tsubakihara Y, Kishida H (1993) Frictional behaviour between normally consolidated clay and steel by two direct shear type apparatuses. Soils Found 33:1–13
6. Liao CJ, Lin HM, Lee DH, Chen PY, Liao JJ (2007) Study on shear behavior of interbedded sandstone and mudstone slope using ring shear test. In: ISOPE Proceedings of the Seventeenth International Ofshore and Polar Engineering Conference, pp 1601–1606
7. Goldstein MN (1979) Mechanical Properties of Soils. Stress-Deformation and Strength Characteristics, p 269. Stroyizdat, Moscow
8. Maslov NN (1935) On the significance of some factors during shear experiments. In: Sat Svirstroy. L. No. 1U
9. Meschyan SR (1974) Mechanical Properties of Soils and Laboratory methods for their Determination, p 192. Nedra, Moscow
10. Bardet JP (1997) Experimental Soil Mechanics, p 584. Prentice Hall, News Jersey
11. Ter-Martirosyan AZ, Sidorov VV, Almakaeva AS (2019) Determination of the interface parameters on the contact of concrete and soil by different methods. IOP Conf Ser Mater Sci Eng 698:022071

12. Hammouda F, Boumekik A (2006) Experimental study of the behaviour of interfacial shearing between cohesive soils and solid materials at large displacement. Asian J Civ Eng 7:63–80
13. Lemos LJL, Vaughan PR (2000) Clay–interface shear resistance. Géotechnique 50:55–64
14. Skempton AW (1964) Long-term stability of clay slopes. Fourth Rankine lecture. Geotechnique 14:77–101
15. Tika TE, Hutchinson JN (1999) Ring shear tests on soil from the Vaiont landslide slip surface. Géotechnique 49:59–74
16. Yoshimine M, Kuwano R, Kuwano J, Ishihara K (1999) Dynamic properties of fine-grained soils in pre-sheared sliding surfaces. Slope Stab Eng 1:595–600
17. Eid HT, Amarasinghe RS, Rabie KH, Wijewickreme D (2015) Residual shear strength of fine-grained soils and soil-solid interfaces at low effective normal stresses. Can Geotech J 52:198–210
18. Boldyrev GG (2014) Methods for Determining the Mechanical Properties of Soils with Comments on GOST 12248-2010, p 812, 2nd ed., add. and correct. Prondo LLC, Moscow
19. Ter-Martirosyan AZ, Sidorov VV, Almakaeva AS (2019) Peculiarities and difficulties in determining the contact strength of soil and structural materials. J Geotech 11:30–41
20. Darkov AV, Shpiro GS (1975) Strength of Materials, p 654, Ed. 4th. Textbook for High Schools, Higher School, Moscow

Some Aspects of the Formation of the Cellular Structure of Slag-Silicate Porous Wood Chip Concrete

Ruslan Dvornikov and Svetlana Samchenko

Abstract The paper considers some aspects of the formation of the porous structure of slag-silicate porous wood chip concrete. To improve the quality of technical foam different additives stabilizers were used: bentonite clay, liquid glass and sodium humate. As a result of experiments, it was found that the properties of the resulting foam are affected to a greater extent by the addition of bentonite clay. When adding clay in an amount equal to 4% of the mass of the foaming solution, water separation decreases to 28.2% without changing the multiplicity relative to the control sample. At the second stage of the work were carried out studies of the kinetics of hydration of the binder by isothermal calorimetry. The results showed an insignificant slowing of the hydration processes and reduction of heat release for the sample with surface active substance in the sample. This fact may be related to the presence of surfactants, which due to their high adsorption capacity of the molecules are able to slow down the processes of structure formation, as well as can affect the morphology of the newly formed composite.

Keywords Lightweight concrete · Slag-alkali binders · Improved cellular structure · Porous wood chip concrete · Hydration kinetics

1 Introduction

Since the beginning of the twenty-first century, the growing volume of construction of residential and public buildings, as well as increasing demands for thermal insulation and economic efficiency of building structures gave impetus to research in the field of creating environmentally friendly, perfect building materials with improved physical, mechanical and construction and technical properties.

With the development of scientific and technological progress there are more and more new technologies and methods of production of building materials. In particular, technologies of production of highly efficient lightweight concretes including those

R. Dvornikov · S. Samchenko (✉)
Moscow State University of Civil Engineering, Moscow 129337, Russian Federation
e-mail: samchenko@list.ru

P. Akimov et al. (eds.), *Proceedings of FORM 2022*, Lecture Notes in Civil Engineering 282, https://doi.org/10.1007/978-3-031-10853-2_50

based on cementless binders such as, for example, slag-silicate binders are being developed.

The group of lightweight concretes includes, among others, autoclaved and non-autoclaved cellular concretes. The most promising and most frequently used wall materials in the construction of residential and public buildings are non-autoclave foam and gas concrete of D600-D700 average density grades. The advantage in the construction is more often given to the non-autoclave foam concrete because of the simpler technology of preparation, which eliminates the operation of removing the hump, as well as a more stable cellular structure obtained by the introduction of technical foam [1, 2].

Other promising material from the class of lightweight concrete known today is wood chip concrete. This material was invented in Holland in the 1930s and was called "Durisol". The resulting composite has proven itself as a reliable and durable material in Europe, Canada and the USA. In the 1960s, the Soviet Union produced an analogue of a foreign material called wood chip concrete. A variety of wood-cement composite is porous wood chip concrete. This material is obtained by mixing cement binder, organic aggregate, technical foam and chemical additives. The advantages of this material is a universal technology that allows to obtain this material on the usual molding equipment of existing enterprises for the production of precast concrete products without disturbing the basic technological process. In addition, the porosity of wood chip concrete mixture leads to a significant improvement of its physical and mechanical properties, as well as reducing the thermal conductivity [3–5].

However, despite the obvious advantages, cellular concrete based on technical foam has disadvantages. The main difficulty in the production of non-autoclaved concretes with porous cellular structure is to achieve a stable pore structure of the composite, which significantly affects its physical and mechanical properties.

When analyzing the issue, it becomes clear that obtaining the most efficient cellular concrete of low grades by average density is possible only by obtaining a stable porous structure, which contributes to obtaining a strong matrix of cellular concrete at constant density values [6].

The use of certain technological methods can improve the quality of the matrix of cellular concrete and increase its physical and mechanical characteristics. Such techniques include in particular: increasing the activity of the binder component, reducing the water to solids ratio, the use of chemical and mineral modifiers, mechanical activation of the binder, etc. [6].

An effective direction of light cellular concrete production is the use of hardening accelerants as well as quick-hardening binders, which include: gypsum, slag-silicate and others. The most promising binders are recognized slag-silicate binders, which are characterized by high strength (60-120 MPa) and rapid rate of strength gain at an early stage [7, 8].

In addition, the parameters of the cellular structure of the composite directly depend on the type of blowing agent and the quality of the foam obtained on its basis, which is characterized by dispersity, multiplicity and sustainability over time [9].

Improving the quality of the resulting foam at constant values of concentration of the blowing agent solution can be carried out with the addition of stabilizers, which include fine powders of mineral substances (fine grinded cement, microsilica, etc.), liquid glass, wood saponified resin (SDO), sulfuric acid iron, etc. [10]

One of the most important indicators influencing the quality of the obtained foam concrete is also the content of blowing agent in the composition of the composite. Thus, with a lack of blowing agent can be obtained composite with low construction-technical and physical–mechanical properties. On the contrary, an excess of blowing agent can lead to a slowing down of the hydration processes of the system and subsequently affect the physical and mechanical characteristics of cellular concrete [11, 12].

Considering the issue of obtaining effective constructional heat insulating material for the needs of the construction industry, it should be noted that according to the authors, a promising direction in this area is the development and production of highly efficient porous slag-silicate wood chip concrete with D600-D700 density mark on the basis of slag-silicate binder, coniferous wood chips and technical foam with optimal quality parameters.

In this regard, the relevant issue for the production of this type of concrete is to obtain a strong and uniform cellular structure, which is able to provide the required characteristics of the composite at constant values of density.

Positive results in the formation of the structure of porous wood chip concrete can be achieved, in particular, by improving the properties of technical foam by using stabilizer additives, namely bentonite clay, liquid glass and sodium humate.

On this basis, the purpose of this work is to select and study the properties of technical foam for obtaining porous wood chip concrete of high quality. In addition, the study of the cellular structure of the resulting composite and determination of the kinetics of hardening, the binder in the presence of surface active substances.

2 Materials and Methods

In this work we used blast furnace granulated slag containing 85% or more of the amorphous phase with a dispersion of 450–500 m^2/kg, a waste product of ferrous metallurgy, which was provided by Mechel PJSC, Moscow, Russian Federation.

A solution of sodium liquid glass from PROMSTECLOCENTRE LLC, Yekaterinburg, Russian Federation, with the following characteristics was used as a shutter as well as a stabilizing additive for technical foam: density is 1.31 g/cm^3; basicity modulus of liquid glass (Mo) is at the level of 2.

Bentonite clay with montmorillonite content not less than 85%. It was produced by Bentonit Khakassia LLC, Chernogorsk, Russian Federation, in accordance with TU 2164-005-49, 215, 611-2007. Bentonite clay was used as an additive stabilizer of technical foam.

Sodium humate with a humic acid content of at least 40% was also used as a foam stabilizer.

As a foaming agent for the preparation of technical foam was used foamer "PB Lux" production RHZ "NORDIX", Voskresensk, Russian Federation, with the following characteristics: density at 20 °C is at 1040–1100 kg/m^3; PH (pH product) is 8.0–10.5.

To study the kinetics of hydration of slag-silicate binder using sodium liquid glass as a shutter, were made two samples: a control (without the presence of blowing agent solution) and the study (with the addition of blowing agent solution with a concentration of 3,3% in an amount of 0,35% of the binder mass). The kinetics of binder hydration was investigated using isothermal calorimetry method on a TAM Air 3116–2 instrument, TA Instruments, with the following characteristics: measuring range: 5–90 °C; 0–10,000 kW; 0–1000 kJ; error: ±0,2 °C; ±20 μW.

Preparation of technical foam was made with a chopper "BOSH" with the following characteristics: motor power 500 W, bowl volume - 2L. Preparation of foam was made by consecutive addition of foam components in the chopper and further mixing for 60 s. Foam multiplicity was determined as the ratio of the resulting foam volume to the initial volume of the solution. Foam stability over time was determined by the volume of liquid released after exposure to a certain amount of it for 30 min. Determination of the characteristics of the technical foam was carried out according to the recommendations developed by the Research Institute of Reinforced Concrete of the Gosstroy of the USSR [13]. The compositions of the technical foam used in the experiments are presented in Table 1.

3 Results and Discussions

Evaluating the results of experiments on the data obtained, we can conclude that the best indicators of multiplicity and stability have foam samples with the addition of bentonite clay. Adding clay in an amount of 4% of the mass of the solution does not affect the multiplicity of the mixture and, at the same time, significantly reduces water separation to a level of 28,2%. Reduction of water separation can be explained by the fact that finely dispersed particles of bentonite clay prevent the flow of liquid from the film to the plateau boundary. In addition, due to rapid setting, they translate the film into a pseudo-solid state [14].

On the contrary, the addition of a mineral additive in excess of 4% leads to a decrease in the multiplicity of the mixture. This effect may occur due to chemisorption processes occurring between the active radicals of the surface active substance and the surface of the mineral particle. As a result of the process, the surface tension at the gas–liquid interface changes, which leads to the process of coalescence of air bubbles [14]. Dependence of foam multiplicity on the percentage of stabilizer additive is shown in Fig. 1. The dependence of water release from foam on the percentage of stabilizer additive is shown in Fig. 2.

Addition of liquid glass and sodium humate containing in its composition humic acid salts are designed to create in the system with foaming agent solution positively and negatively charged microparticles and aqua complexes adsorbed on the surface

Table 1 Compositions of technical foam

№ of the composition	Concentration of blowing agent solution, %	Name of stabilizer additive			Churning time, seconds	Foam characteristics	
		Quantity of liquid glass from the mass of the solution, %	Amount of bentonite clay from the mass of the solution, %	Quantity of sodium humate by mass of solution, %		Foam multiplicity	The amount of fluid released in 30 min, %
1	3,3	–	–	–	60	12,5	87,33
2	3,3	1,0	–	–		12,5	70,63
3	3,3	1,5	–	–		12,5	67,25
4	3,3	2,0	–	–		12,6	72,17
5	3,3	2,5	–	–		12,4	73,10
6	3,3	3,0	–	–		12,5	72,50
7	3,3	–	1,00	–		12,7	62,2
8	3,3	–	1,50	–		12,4	57,1
9	3,3	–	2,00	–		12,5	54,2
10	3,3	–	2,50	–		12,3	41,86
11	3,3	–	3,00	–		12,5	35,7
12	3,3	–	3,50	–		12,5	32,1
13	3,3	–	4,00	–		12,4	28,2
14	3,3	–	4,50	–		10,5	21,2
15	3,3	–	5,00	–		7,5	15,1
16	3,3	–	–	1,00		12,5	81,56
17	3,3	–	–	1,50		12,3	80,48
18	3,3	–	–	2,00		12,2	79,1
19	3,3	–	–	2,50		12,5	80,2
20	3,3	–	–	3,00		12,5	81,5

of thin films of foam bubbles and electrostatically interact with each other and the polar group of surface active substance structuring the foam. The stabilizing effect of these additives is less pronounced. This phenomenon may be related to the nature of interaction between the molecules of stabilizer additives and surfactant, which is an aqueous solution of anionic surfactants with functional additives.

At the second stage of the work we studied the effect of adding a concentrated solution of blowing agent on the heat release processes of slag-silicate binder with a sodium liquid glass as a binder. Heat release curves of the control sample and the sample with the addition of technical foam are shown in Fig. 3, 4 and 5.

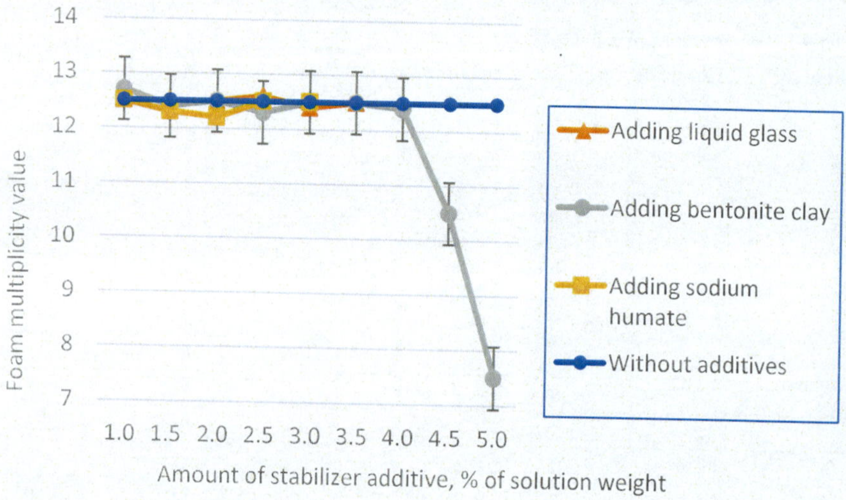

Fig. 1 Dependence of the foam multiplicity on the percentage of stabilizer additive

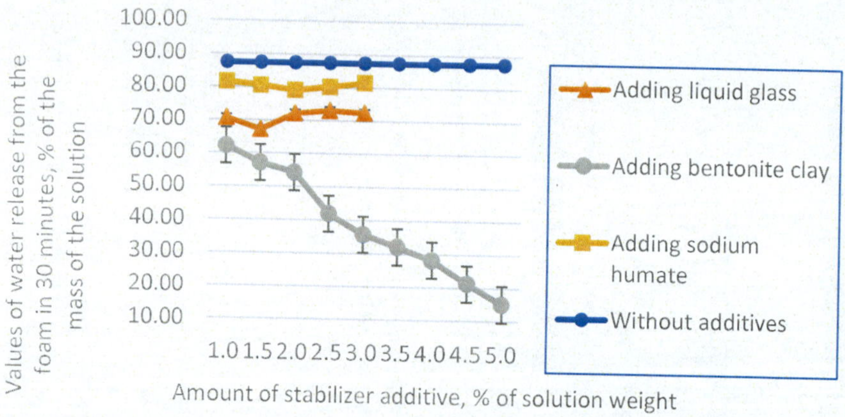

Fig. 2 Dependence of water release from the foam on the percentage of stabilizer additive

The measured values of heat release at characteristic points for the control sample and for the sample with blowing agent (per gram of mixture) are presented in Table 2.

The following conclusions can be drawn from the test results of the components for the manufacture of wood chip concrete:

For both samples studied, the fast reaction phase beginning immediately after the shutting of the mineral components by the liquid phase is characterized by intense heat release with a characteristic peak after about 9 min from the start of the reaction, followed by a decrease in the heat release intensity and transition to the induction period of the reaction.

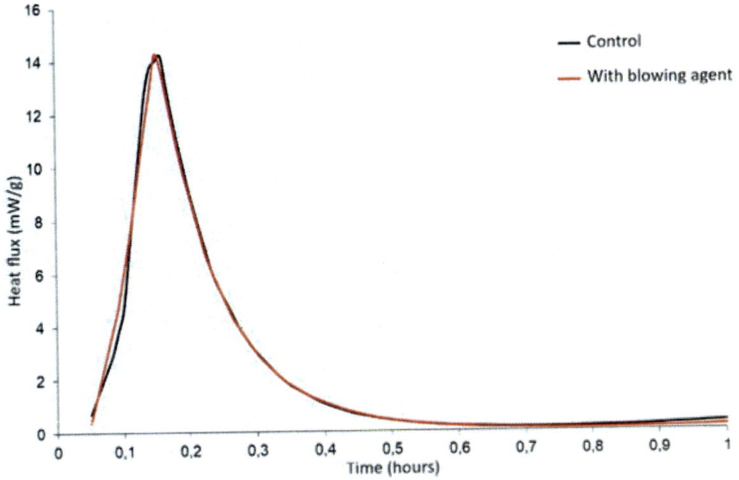

Fig. 3 Recorded heat fluxes for the studied samples in the interval 0–1 h. (in the range of 0–16 mW/g)

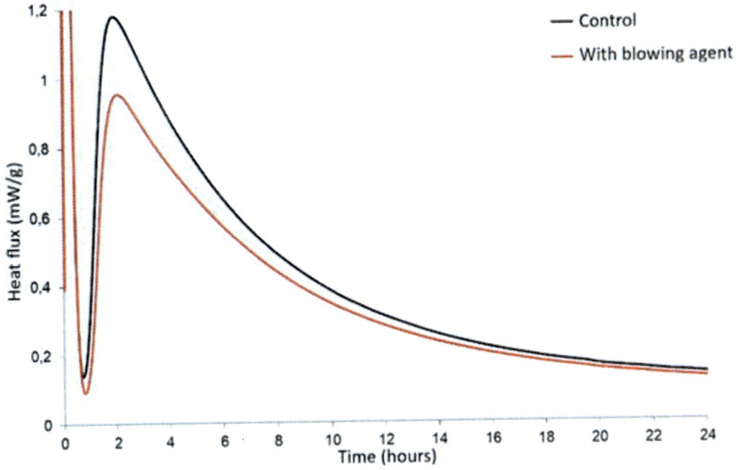

Fig. 4 Recorded heat fluxes for the studied samples in the interval 0–24 h. (in the range of 0–1.2 mW/g)

After completion of the induction period, the exothermic curves show two heat release peaks after 2 and 110 h (for the control sample) and after 2 and 128 h (for the sample with blowing agent) after the start of the reaction.

Summary thermal energy released during the reaction in the time interval from 3 min to 240 h from the time of mixing the components is 105,48 J/g (for the control sample) and 104,16 J/g (for the sample with blowing agent).

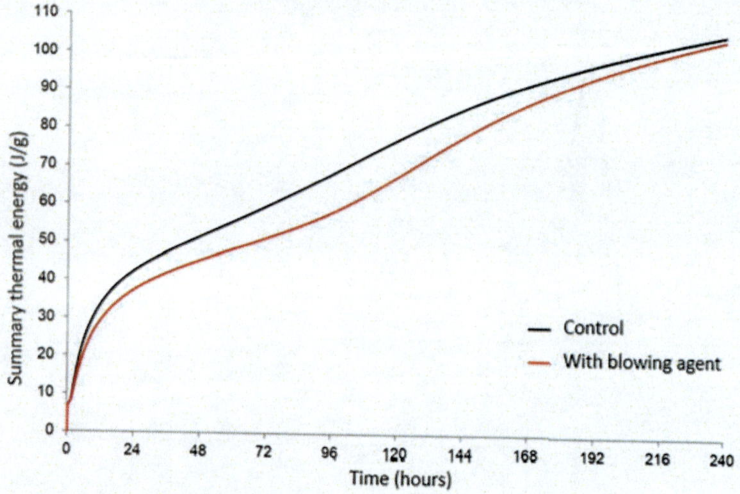

Fig. 5 Integral plots of the total heat release for the studied samples

Table 2 Measured values of heat release at characteristic points

Time, hour	Heat flux power Wt, W/g	Summary thermal energy for the control sample Qt, J/g	Summary heat energy for the sample with blowing agent Qt, J/g
1	0,3327	7,7704	7,618,482,622
2	1,1780	11,0901	9,993,978,226
4	0,8874	18,5598	16,21,118,842
8	0,4908	28,1333	24,5,804,393
12	0,3075	33,7095	29,61,657,973
24	0,1385	42,3029	37,50,324,934
72	0,0970	59,4760	51,23,573,038
96	0,1050	68,1828	58,09,636,849
120	0,1026	77,2728	67,66,824,337
144	0,0852	85,4628	78,26,760,791
168	0,0673	92,0525	87,16,628,375
192	0,0556	97,3501	94,15,034,219
216	0,0459	101,7440	99,67,621,657
240	0,0401	105,4823	104,1,579,945

After 240 h of reaction, the heat release continues at a level above 0,04 W/g of the mixture, indicating the continuation of structure formation processes.

The occurring decrease in thermal energy and slowing of hydration processes for the sample with the foamer relative to the control composition can be explained by the presence in the composition of surfactants, which due to their high adsorption capacity of molecules are able to slow down the processes of structure formation, and can also affect the morphology of the composite new formations [15].

4 Conclusions

1. As a result of the work, the composition of the foam with optimal quality parameters was selected. It was found that bentonite clay is an effective additive stabilizer for foam. When introduced in an amount of 4% of the mass of the solution the additive positively affects the stability of the foam, reducing the water separation to a value of 28, 2% without reducing the multiplicity of foam.
2. Analysis of hydration kinetics for the sample of binder with and without the addition of blowing agent was carried out. As a result of the tests it was found that for both samples the phase of rapid reactions, beginning immediately after the shutting of the mineral components of the liquid phase, is characterized by intense heat release with a characteristic peak after about 9 min from the start of the reaction with a subsequent decrease in the intensity of heat release and the transition to the induction period of the reaction. After 2 and 110 h, heat release peaks were observed for the control sample and after 2 and 128 h for the sample with blowing agent. During the entire observation period (from 3 min to 240 h), the total heat energy released for the control sample was at 105,48 J/g, while for the sample with blowing agent this value was at 104,16 J/g. It was found that the decrease in thermal energy and hydration kinetics for the sample with blowing agent relative to the control may be caused by the adsorption effect of surfactant molecules present in the blowing agent solution.

Acknowledgements The authors are grateful to the Research Institute of Building Materials and Technologies of the Moscow State University of Civil Engineering, for conducting research to determine the heat release by isothermal calorimetry. The authors would like to thank Mechel PJSC, Moscow, Russian Federation for the materials provided for the research. Experiments were carried out in the laboratory of the Department of Construction Materials Science of the Moscow State University of Civil Engineering.

References

1. Levy JP (1955) Lightweight concretes: preparation - properties – application. Gostroyizdat, Moscow, p 146
2. Jiang J, Lu Z, Niu Y, Li J, Zhang Y (2016) Study on the preparation and properties of high-porosity foamed concretes based on ordinary Portland cement. Mater Design 92:949–959
3. Nanazashvili IK (1984) Wood chip concrete - an effective building material. Stroyizdat, Moscow, p 121
4. Abramenko NI (1980) Porous cement wood chip concrete on wood aggregates. Research Institute of Concrete and Reinforced Concrete, Moscow, p 18

5. Shcherbakov AS (1966) Influence of properties of initial materials and technological factors on the strength of wood chip concrete. Forest Exploit Forest 32:17–20
6. Komar AG, Velichko EG, Belyakova JS (2001) On some aspects of the management of structure formation and properties of slag-silicate foam concrete. Constr Mater 7:12–15
7. Glukhovsky VD (ed) (1979) Alkaline and alkaline-alkaline-earth hydraulic binders and concretes. The Higher School, p 232
8. Bilek V, Hurta J, Done P, Zidek L (2016) Development of alkali-activated concrete for structures – mechanical properties and durability. Perspect Sci 7:190–194
9. Huang Z, Zhang T, Wen Z (2015) Proportioning and characterization of Portland cementbased ultra-lightweight foam concretes. Constr Build Mater 79:390–396
10. Laukaitis AA (2001) Prediction of some properties of low density cellular concrete. Build Mater 4:27–29
11. Shakhova LD, Balyasnikov VV (2002) Foamers for cellular concrete. SK Typography, Belgorod, p 147
12. Shakhova LD, Balyasnikov VV, Chernaya TI, et al (2000) Study of hydration processes of clinker minerals with additives of blowing agents of different nature. In: II international meeting on chemistry and technology of cement. Publishing house of Education «Information Center», St. Petersburg, pp 70–73
13. Recommendations for calculation and manufacture of products from porous wood chip concrete (1983) Research Institute of Concrete and Reinforced Concrete of the USSR, p 64
14. Abramzon AA, Zaichenko LP, Feingold SI (1888) Surface-active substances: synthesis, analysis, properties, application. Chem, p 200
15. Shakhova LD, Chernaya TN, Nesterova LL et al (2002) Study of the effect of the nature of the blowing agent on the hydration of SZA in the presence of gypsum by microscopic method. Proc NGA-SU 2(17):97–101

Author Index

P. Akimov et al. (eds.), *Proceedings of FORM 2022*, Lecture Notes in Civil Engineering
282, https://doi.org/10.1007/978-3-031-10853-2